The How and Why of
One Variable Calculus

The How and Why of
One Variable Calculus

Amol Sasane

London School of Economics, UK

Library of Congress Cataloging-in-Publication Data applied for.

ISBN: 9781119043386

A catalogue record for this book is available from the British Library.

Cover Image: Tuomas Kujansuu/iStockphoto

Set in 10/12pt, TimesLtStd by SPi Global, Chennai, India

1   2015

To my parents

Contents

Preface

Who is this book is for?

This book is meant as a textbook for an honours course in Calculus, and is aimed at first year students beginning studies at the university. The preparation assumed is high school level Mathematics. Any arguments not met before in high school (for example, geometric arguments à la Euclid) can be picked up along the way or simply skipped without any loss of continuity. This book may also be used as supplementary reading in a traditional methods-based Calculus course or as a textbook for a course meant to bridge the gap between Calculus and Real Analysis.

How should the student read the book?

Students reading the book should not feel obliged to study every proof at the first reading. It is more important to understand the theorems well, to see how they are used, and why they are interesting, than to spend all the time on proofs. So, while reading the book, one may wish, after reading the theorem statement, to first study the examples and solve a few relevant exercises, before returning to read the proof of that theorem.

The exercises are an integral part of studying this book. They are a combination of purely drill ones (meant for practising Calculus methods), and those meant to clarify the meanings of the definitions, theorems, and even to facilitate the goal of developing 'mathematical maturity'. The student should feel free to skip exercises that seem particularly challenging at the first instance, and return back to them now and again. Although detailed solutions are provided, the student should not be tempted to consult the given solution too soon. In the learning process leading to developing understanding, it is much better to think about the exercise (even if one does not find the answer oneself!), rather than look at the provided solution in order to understand how to solve it. In other words, it is the *struggle* to solve the exercise that turns out to be more important than the mere *knowledge* of the solution. After all, given a *new* problem, it will be the struggle that pays off, and not the knowledge of the solution of the (now irrelevant) *old* exercise! So the student should absolutely not feel discouraged if he or she doesn't manage to solve an exercise problem. Some of the exercises that are more abstract/technical/challenging as compared to the other exercises are indicated with an asterisk symbol ($*$).

Acknowledgements

I would like to thank Sara Maad Sasane (Lund University) for going through the entire manuscript, pointing out typos and mistakes, and offering insightful suggestions and comments. Thanks are also due to Lassi Paunonen (Tampere University) and Raymond Mortini (University of Lorraine-Metz) for many useful comments. A few pedagogical ideas in this book stem from some of the references listed at the end of this book. This applies also to the exercises. References are given in the section on notes at the end of the chapters, but no claim to originality is made in case there is a missing reference. The figures in this book have been created using xfig, Maple, and MATLAB. Finally, it is a pleasure to thank the editors and staff at Wiley, especially Debbie Jupe, Heather Kay, and Prachi Sinha Sahay. Thanks are also due to the project manager, Sangeetha Parthasarathy, for cheerfully and patiently overseeing the typesetting of the book.

Amol Sasane
London, 2014.

Introduction

What is Calculus?

Calculus is a branch of mathematics in which the focus is on two main things: given a real-valued function of a real variable, what is the rate of change of the function at a point (Differentiation), and what is the area under the graph of the function over an interval (Integration).

Differentiation and **Integration**

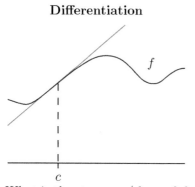

What is the steepness/slope of f at the point c?

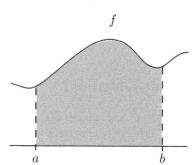

What is the area under the graph of f over an interval from a to b?

Differentiation	Integration
Differentiation is concerned with velocities, accelerations, curvatures, etc. These are rates of change of function values and are defined *locally*.	Integration is concerned with areas, volumes, average values, etc. These take into account the totality of function values, and are *not* defined locally.

We will see later on that the rate of change of f at c is defined by

$$f'(c) = \lim_{x \to c} \frac{f(x) - f(c)}{x - c},$$

and what matters is not what the function is doing *far away* from the point c, but rather the manner in which the function behaves in the *vicinity* of c. This is what we mean when we say that 'differentiation is a local concept'. On the other hand, we will learn that for nice functions, the area will be given by an expression that looks like

$$\int_a^b f(x)dx = \lim_{N \to \infty} \sum_{n=0}^N f\left(a + \frac{b-a}{N}n\right) \frac{b-a}{N},$$

and we see that in the above process, the values of the function over the entire interval from a to b *do matter*. In this sense integration is a 'non-local' or 'global' process.

Thus it seems that in Calculus, there are these two quite *different* topics of study. However there is a remarkable fact, known as that *Fundamental Theorem of Calculus*, which creates a bridge between these seemingly different worlds: it says, roughly speaking that the processes of differentiation and integration are inverses of each other:

$$\int_a^b f'(x)dx = f(b) - f(a) \quad \text{and} \quad \frac{d}{dx} \int_a^x f(\xi)d\xi = f(x).$$

This interaction between differentiation and integration provides a powerful body of under-standing and calculational technique, called 'Calculus'. Problems that would be otherwise computationally difficult can be solved mechanically using a few simple Calculus rules, and without the exertion of a great deal of penetrating thought.

Why study Calculus?

The reason why Calculus is a standard component of all scientific undergraduate education is because it is universally applicable in Physics, Engineering, Biology, Economics, and so on. Here are a few very simple examples:

(1) What is the escape velocity of a rocket on the surface of the Earth?

(2) If a hole of radius 1 cm is drilled along a diametrical axis in a solid sphere of radius 2 cm, then what is the volume of the body left over?

(3) If a strain of bacteria grows at a rate proportional to the amount present, and if the population doubles in an hour, then what is the population of bacteria at any time t?

(4) If the manufacturing cost of x lamps is given by $C(x) = 2700 - 100x$, and the revenue function is given by $R(x) = x - 0.03x^2$, then what is the number of lamps maximising the profit?

We will primarily be concerned with developing and understanding the tools of Calculus, but now and then in the exercises and examples chosen, we will consider a few toy models from various application areas to illustrate how the techniques of Calculus have universal applications.

What will we learn in this book?

This book is divided into six chapters, listed below.

(1) The real numbers.

(2) Sequences.

(3) Continuity.

(4) Differentiation.

(5) Integration.

(6) Series.

This covers the core component of a *single/one* variable Calculus course, where the basic object of study is a real-valued function of *one* real variable, and one studies the themes of differentiation and integration for such functions. On the other hand, in *multi/several* variable Calculus, the basic object of study is an $\mathbb{R}^m$-valued/vector valued function of several variables, and the themes of differentiation and integration for such functions. This book does not cover this latter subject.

We refrain from giving a brief gist of the contents of each of the chapters, since it won't make much sense to the novice at this stage, but instead we appeal to whatever previous exposure the student might have had in high school regarding these concepts. We will of course study each of these topics from scratch. We make one pertinent point though in the paragraph below.

A discussion of Calculus needs an ample supply of examples, which are typically through considering specific functions one meets in applications. The simplest among these illustrative functions are the algebraic functions, but it would be monotonous to just consider these. Much more interesting things happen with the so-called elementary transcendental functions such as the logarithm, exponential function, trigonometric functions, and so on. A rigorous definition of these unfortunately needs the very tools of Calculus that are being developed in this course. It would be a shame, however, if such rich examples centered around these functions have to wait till a rigorous treatment has been done. So we adopt a dual approach: we *will* choose to illustrate our definitions/theorems with these functions, and *not exclude* these functions from our preliminary discussion, hoping that the student has *some* exposure to the definitions (at whatever intuitive/rigour level) and properties of these transcendental functions. Later on, when the time is right (Chapter 5), we will give the precise mathematical definitions of these functions and prove the very properties that were accepted on faith in the initial parts of this book. This dual approach adopted by us has the advantage of not depriving the student of the nice illustrations of the results provided by these functions, and of preparing the student for the actual treatment of these functions later on. In any case, if the student meets a very unfamiliar property or manipulation involving these functions in the initial part of this book, it is safe to simply skip the relevant part and revisit it after Chapter 5 has been read.

How did Calculus arise?

Some preliminary ideas of Calculus are said to date back to as much as 2000 years ago when Archimedes determined areas using the Method of Exhaustion; see the following discussion and Figure 1.

The development of Differential and Integral Calculus is largely attributed to Newton (1642–1727) and Leibniz (1646–1716), and the foundations of the subject continued to be investigated into the 19th century, among others by Cauchy, Bolzano, Riemann, Weierstrass, Lebesgue, and so on.

We end this introduction with making a few remarks about the 'Method of Exhaustion', which besides treating this historical milestone in the development of Calculus, will also provide some motivation to begin our journey into Calculus with a study of the real numbers.

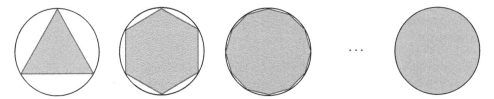

Figure 1.   Determination of the area of a circle using the Method of Exhaustion.

In Figure 1, it is clear that what we are doing is trying to obtain the area of a circle by inscribing polygons inside it, each time doubling the number of sides, hence 'exhausting' more and more of the circular area. The idea is then that if A is the area of the circle we seek, and a_n is the area of the polygon at the nth step, then for large n, a_n approximates A. As we have that $a_1 \leq a_2 \leq a_3 \leq \cdots$, and since a_n misses A by smaller and smaller amounts as n increases, we expect that A should the 'smallest' number exceeding the numbers $a_1, a_2, a_3, \cdots$. Does such a number always exist?

Obviously, one can question the validity of this heuristic approach to solving the problem. The objections are for example:

(1) We did not really define what we mean by the area enclosed.

(2) We are not sure about what properties of numbers we are allowed to use. For example, we seem to be needing the fact that 'if we have an increasing sequence of numbers, all of which are less than a certain number[1], then there is a smallest number which is bigger than each of the numbers $a_1, a_2, a_3, \cdots$'. Is this property true for rational numbers?

Such questions might seem frivolous to a scientist who is just interested in 'real world applications'. But such a sloppy attitude can lead to trouble. Indeed, some work done in the 16th to the 18th century relying on a mixture of deductive reasoning and intuition, involving vaguely defined terms, was later shown to be *incorrect*. To give the student a quick example of how things might easily go wrong, one might naively, but incorrectly, guess that the answer to question (2) above is yes. This prompts the question of whether there is a bigger set of numbers than the rational numbers for which the property happens to be true? The answer is yes, and this is the **real number system** $\mathbb{R}$.

Thus a thorough treatment of Calculus must start with a careful study of the number system in which the action of Calculus takes place, and this is the real number system $\mathbb{R}$, where our journey begins!

---

[1] imagine a square circumscribing the circle: then each of the numbers $a_1, a_2, a_3, \cdots$ are all less than the area of the square

Preliminary notation

$A := B$ or $B =: A$	A is defined to be B; A is defined by B
$\forall$	for all; for every
$\exists$	there exists
$\neg\, S$	negation of the statement S; it is not the case that S
$a \in A$	the element a belongs to the set A
$\emptyset$	the empty set containing no elements
$A \subset B$	A is a subset of B
$A \subsetneq B$	A is a subset of B, but is not equal to B
$A \backslash B$	the set of elements of A that do not belong to B
$A \cap B$	intersection of the sets A and B
$\bigcap_{i \in I} A_i$	intersection of the sets A_i, $i \in I$
$A \cup B$	union of the sets A and B
$\bigcup_{i \in I} A_i$	union of the sets A_i, $i \in I$
$A_1 \times \cdots \times A_n$	Cartesian product of the sets $A_1, \cdots, A_n$; $\{(a_1, \cdots, a_n) : a_1 \in A_1, \cdots, a_n \in A_n\}$

1

The real numbers

From the considerations in the Introduction, it is clear that in order to have a firm foundation of Calculus, one needs to study the real numbers carefully. We will do this in this chapter. The plan is as follows:

(1) An intuitive, visual picture of $\mathbb{R}$: the number line. We will begin our understanding of $\mathbb{R}$ intuitively as points on the 'number line'. This way, we will have a mental picture of $\mathbb{R}$, in order to begin stating the precise properties of the real numbers that we will need in the sequel. It is a legitimate issue to worry about the actual construction of the set of real numbers, and we will say something about this in Section 1.8.

(2) Properties of $\mathbb{R}$. Having a rough feeling for the real numbers as being points of the real line, we will proceed to state the precise properties of the real numbers we will need. So we will think of $\mathbb{R}$ as an undefined set for now, and just state rigorously what properties we need this set $\mathbb{R}$ to have. These desirable properties fall under three categories:

    (a) *the field axioms*, which tell us about what laws the arithmetic of the real numbers should follow,

    (b) *the order axiom*, telling us that comparison of real numbers is possible with an order $>$ and what properties this order relation has, and

    (c) *the Least Upper Bound Property of* $\mathbb{R}$, which tells us roughly that unlike the set of rational numbers, the real number line has 'no holes'. This last property is the most important one from the viewpoint of Calculus: it is the one which makes Calculus possible with real numbers. *If* rational numbers had this nice property, then we would not have bothered studying real numbers, and instead we would have just used rational numbers for doing Calculus.

The How and Why of One Variable Calculus, First Edition. Amol Sasane.
© 2015 John Wiley & Sons, Ltd. Published 2015 by John Wiley & Sons, Ltd.

(3) The construction of $\mathbb{R}$**.** Although we will think of real numbers intuitively as 'numbers that can be depicted on the number line', this is not acceptable as a rigorous mathematical definition. So one can ask:

> Is there really a set $\mathbb{R}$ that can be constructed which has the stipulated properties (2)(a), (b), and (c) (and which will be detailed further in Sections 1.2, 1.3, 1.4)?

The answer is yes, and we will make some remarks about this in Section 1.8.

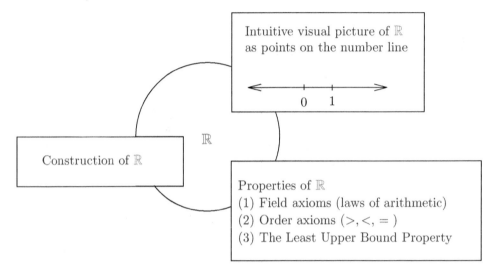

1.1   Intuitive picture of $\mathbb{R}$ as points on the number line

In elementary school, we learn about

the natural numbers $\mathbb{N} := \{1, 2, 3, \cdots\}$

the integers $\mathbb{Z} := \{\cdots, -3, -2, -1, 0, 1, 2, 3, \cdots\}$, and

the rational numbers $\mathbb{Q} := \left\{ \left[\dfrac{n}{d}\right] : n, d \in \mathbb{Z}, d \neq 0 \right\}$.

Incidentally, the rationale behind denoting the rational numbers by $\mathbb{Q}$ is that it reminds us of 'quotient', and $\mathbb{Z}$ for integers comes from the German word 'zählen' (meaning 'count'). In the above,

$$\left[\frac{n}{d}\right]$$

represents a whole family of 'equivalent fractions'; for example, $\dfrac{2}{4} = \dfrac{1}{2} = \dfrac{-3}{-6}$ etc.

We are accustomed to visualising these numbers on the 'number line'. What is the number line? It is any line in the plane, on which we have chosen a point O as the 'origin', representing the number 0, and chosen a unit length by marking off a point on the right of O, where the number 1 is placed. In this way, we get all the positive integers, $1, 2, 3, 4, \cdots$ by repeatedly marking off successively the unit length towards the right, and all the negative integers $-1, -2, -3, \cdots$ by repeatedly marking off successively the unit length towards the left.

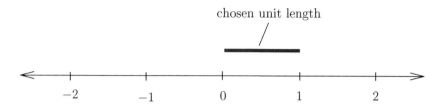

chosen unit length

Just like the integers can be depicted on the number line, we can also depict all rational numbers on it as follows. First of all, here is a procedure for dividing a unit length on the number line into d ($\in \mathbb{N}$) equal parts, allowing us to construct the rational number $1/d$ on the number line. See Figure 1.1.

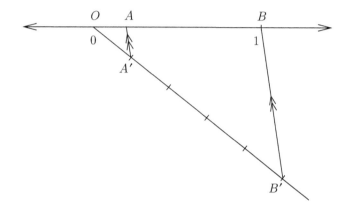

Figure 1.1 Construction of rational numbers: in the above picture, given the length 1 (that is, knowing the position of B), we can construct the length $1/5$, and so the point A corresponds to the rational number $1/5$.

The steps are as follows: Let the points O and B correspond to the numbers 0 and 1.

(1) Take any arbitrary length $\ell(OA')$ along a ray starting at O in any direction other than that of the number line itself.

(2) Let B' be a point on the ray such that $\ell(OB') = d \cdot \ell(OA')$.

(3) Draw AA' parallel to BB' to meet the number line at A.

Conclusion: From the similar triangles $\triangle OAA'$ and $\triangle OBB'$, we see that the length $\ell(OA) = 1/d$.

Having obtained $1/d$, we can now construct n/d on the number line for *any* $n \in \mathbb{Z}$, by repeating the length $1/d$ n times towards the right of 0 if $n > 0$, and towards the left $-n$ times from 0 if n is negative.

Hence, we can depict all the rational numbers on the number line. Does this exhaust the number line? That is, suppose that we start with all the points on the number line being coloured black, and suppose that at a later time, we colour all the rational ones by red: are there any black points left over? The answer is yes, and we demonstrate this below. We will show that there does 'exist', based on geometric reasoning, a point on the number line, whose square is 2, but we will also argue that this number, denoted by $\sqrt{2}$, is not a rational number.

First of all, the picture below shows that $\sqrt{2}$ exists as a point on the number line. Indeed, by looking at the right angled triangle $\triangle OBA$, Pythagoras's Theorem tells us that the length of the hypotenuse OA satisfies

$$(\ell(OA))^2 = (\ell(OB))^2 + (\ell(AB))^2 = 1^2 + 1^2 = 2,$$

and so $\ell(OA)$ is a number, denoted say by $\sqrt{2}$, whose square is 2. By taking O as the centre and radius $\ell(OA)$, we can draw a circle using a compass that intersects the number line at a point C, corresponding to the number $\sqrt{2}$. Is $\sqrt{2}$ a rational number? We show below that it isn't!

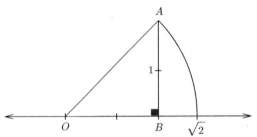

Exercise 1.1. Depict $-11/6$ and $\sqrt{3}$ on the number line.

Theorem 1.1 (An 'origami' proof of the irrationality of $\sqrt{2}$). *There is no rational number* $q \in \mathbb{Q}$ *such that* $q^2 = 2$.

Proof. Suppose that $\sqrt{2}$ is a rational number. Then some scaling of the triangle

by an integer will produce a similar triangle, all of whose sides are integers. Choose the smallest such triangle, say $\triangle ABC$, with integer lengths $\ell(BC) = \ell(AB) = n$, and $\ell(AC) = N$, $n, N \in \mathbb{N}$. Now do the following origami: fold along a line passing through A so that B lies on AC, giving rise to the point B' on AC. The 'crease' in the paper is actually the angle bisector AD of the angle $\angle BAC$.

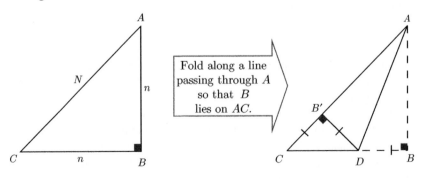

In $\triangle CB'D$, $\angle CB'D = 90°$, $\angle B'CD = 45°$. So $\triangle CB'D$ is an isosceles right triangle. We have $\ell(CB') = \ell(B'D) = \ell(AC) - \ell(AB') = N - n \in \mathbb{N}$, while

$$\ell(CD) = \ell(CB) - \ell(DB) = n - \ell(B'C) = n - (N - n) = 2n - N \in \mathbb{N}.$$

So $\triangle CB'D$ is similar to the triangle

has integer side lengths, and is smaller than $\triangle ABC$, contradicting the choice of $\triangle ABC$. So there is no rational number q such that $q^2 = 2$. □

A different proof is given in the exercise below.

Exercise 1.2. (∗) We offer a different proof of the irrationality of $\sqrt{2}$, and en route learn a technique to prove the irrationality of 'surds'.[1]

(1) Prove the Rational Zeros Theorem: Let $c_0, c_1, \cdots, c_d$ be $d \geq 1$ integers such that c_0 and c_d are not zero. Let $r = p/q$ where p, q are integers having no common factor and such that $q > 0$. Suppose that r is a zero of the polynomial $c_0 + c_1 x + \cdots + c_d x^d$. Then q divides c_d and p divides c_0.

(2) Show that $\sqrt{2}$ is irrational.

(3) Show that $\sqrt[3]{6}$ is irrational.

Thus, we have seen that the elements of $\mathbb{Q}$ can be depicted on the number line, and that not all the points on the number line belong to $\mathbb{Q}$. We think of $\mathbb{R}$ as *all* the points on the number line. As mentioned earlier, if we take out everything on the number line (the black points) except for the rational numbers $\mathbb{Q}$ (the red points), then there will be holes among the rational numbers (for example, there will be a missing black point where $\sqrt{2}$ lies on the number line). We can think of the real numbers as 'filling in' these holes between the rational numbers. We will say more about this when we make remarks about the construction of $\mathbb{R}$. Right now, we just have an intuitive picture of the set of real numbers as a bigger set than the rational numbers, and we think of the real numbers as points on the number line. Admittedly, this is certainly not a mathematical definition, and is extremely vague. In order to be precise, and to do Calculus rigorously, we just can't rely on this vague intuitive picture of the real numbers. So we now turn to the precise properties of the real numbers that we are allowed to use in

---

[1] Surds refer to irrational numbers that arise as the nth root of a natural number. The mathematician al-Khwarizmi (around 820 AD) called irrational numbers 'inaudible', which was later translated to the Latin *surdus* for 'mute'.

developing Calculus. While stating these properties, we will think of the set $\mathbb{R}$ as an (as yet) undefined set containing $\mathbb{Q}$ which will satisfy the properties of

(1) the field axioms (laws of arithmetic in $\mathbb{R}$),

(2) the order axioms (allowing us to compare real numbers with $>, <, =$), and

(3) the Least Upper Bound Property (making Calculus possible in $\mathbb{R}$),

stipulated below.

It is a pertinent question if one can construct (if there really exists) such a set $\mathbb{R}$ satisfying the above properties (1–3). The answer to this question is yes, but it is tedious. So in this first introductory course, we will not worry ourselves too much with it. It is a bit like the process of learning physics: typically one does not start with quantum mechanics and the structure of an atom, but with the familiar realm of classical mechanics. To consider another example, imagine how difficult it would be to learn a foreign language if one starts to painfully memorise systematically all the rules of grammar first; instead a much more fruitful method is to start practicing simple phrases, moving on to perhaps children's comic books, listening to pop music in that language, news, literature, and so on. Of course, along the way one picks up grammar and a formal study can be done at leisure later resulting in better comprehension. We will actually give some idea about the construction of the real numbers in Section 1.8. Right now, we just accept on faith that the construction of $\mathbb{R}$ possessing the properties we are about to learn can be done, and to have a concrete object in mind, we rely on our familiarity with the number line to think of the real numbers when we study the properties (1), (2), (3) listed above.

We also remark that property (3) (the Least Upper Bound Property) of $\mathbb{R}$ will turn out to be crucial for doing Calculus. The properties (1), (2) are also possessed by the rational number system $\mathbb{Q}$, but we will see that (3) fails for $\mathbb{Q}$.

1.2   The field axioms

The content of this section can be summarised in one sentence: $(\mathbb{R}, +, \cdot)$ forms a field. What does this mean? It is a compact way of saying the following. $\mathbb{R}$ is a set, equipped with two operations:

$$+ : \mathbb{R} \times \mathbb{R} \to \mathbb{R},$$

called *addition*, which sends a pair of real numbers (x, y) to their *sum* $x + y$, and the other operation is

$$\cdot : \mathbb{R} \times \mathbb{R} \to \mathbb{R},$$

called *multiplication*, which sends a pair of real numbers (x, y) to their *product* $x \cdot y$, and these two operations satisfy certain laws, called the 'field axioms'.[2] The field axioms for $\mathbb{R}$ are

---

[2] There are other number systems, for example, the rational numbers $\mathbb{Q}$ which also obey similar laws of arithmetic, and so $(\mathbb{Q}, +, \cdot)$ is also deemed to be a field. So the word 'field' is invented to describe the situation that one has a number system $\mathbb{F}$ with corresponding operations $+$ and $\cdot$ which obey the usual laws of arithmetic, rather than listing all of these laws.

listed below:

$$+ \begin{cases} \text{(F1) (Associativity)} & \text{For all } x, y, z \in \mathbb{R}, \ x + (y + z) = (x + y) + z. \\ \text{(F2) (Additive identity)} & \text{For all } x \in \mathbb{R}, \ x + 0 = x = 0 + x. \\ \text{(F3) (Inverses)} & \text{For all } x \in \mathbb{R}, \text{there exists} - x \in \mathbb{R} \\ & \text{such that } x + (-x) = 0 = -x + x. \\ \text{(F4) (Commutativity)} & \text{For all } x, y \in \mathbb{R}, \ x + y = y + x. \end{cases}$$

$$\cdot \begin{cases} \text{(F5) (Associativity)} & \text{For all } x, y, z \in \mathbb{R}, \ x \cdot (y \cdot z) = (x \cdot y) \cdot z. \\ \text{(F6) (Multiplicative identity)} & 1 \neq 0 \text{ and for all } x \in \mathbb{R}, \ x \cdot 1 = x = 1 \cdot x. \\ \text{(F7) (Inverses)} & \text{For all } x \in \mathbb{R} \backslash \{0\}, \text{ there exists } x^{-1} \in \mathbb{R} \\ & \text{such that } x \cdot x^{-1} = 1 = x^{-1} \cdot x. \\ \text{(F8) (Commutativity)} & \text{For all } x, y \in \mathbb{R}, \ x \cdot y = y \cdot x. \end{cases}$$

$$+, \cdot \begin{cases} \text{(F9) (Distributivity)} & \text{For all } x, y, z \in \mathbb{R}, \ x \cdot (y + z) = x \cdot y + x \cdot z. \end{cases}$$

With these axioms, it is possible to prove the usual arithmetic manipulations we are accustomed to. Here are a couple of examples.

Example 1.1. For every $a \in \mathbb{R}$, $a \cdot 0 = 0$.

Let $a \in \mathbb{R}$. Then we have $a \cdot 0 \overset{F2}{=} a \cdot (0 + 0) \overset{F9}{=} a \cdot 0 + a \cdot 0$. So with $x := a \cdot 0$, we have got $x + x = x$. Adding $-x$ on both sides (F3!), and using (F1) we obtain

$$0 = x + (-x) = (x + x) + (-x) \overset{F1}{=} x + (x + (-x)) \overset{F3}{=} x + 0 \overset{F2}{=} x = a \cdot 0,$$

completing the proof of the claim.   ◇

Example 1.2. If $a, b \in \mathbb{R}$, and $a \cdot b = 0$, then $a = 0$ or $b = 0$.

If $a = 0$, then we are done. Suppose that $a \neq 0$. By (F7), there exists a real number a^{-1} such that $a \cdot a^{-1} = a^{-1} \cdot a = 1$. Hence

$$b = 1 \cdot b = (a^{-1} \cdot a) \cdot b = a^{-1} \cdot (a \cdot b) = a^{-1} \cdot 0 = 0.$$

So if $a \neq 0$, then $b = 0$. Thus $\left(a, b \in \mathbb{R} \text{ such that } a \cdot b = 0 \right) \Rightarrow \left(a = 0 \text{ or } b = 0 \right)$.   ◇

Of course in this book, we will not do such careful justifications every time we need to manipulate real numbers. We have listed the above laws to once and for all stipulate the laws of arithmetic for real numbers that justify the usual calculational rules we are familiar with, so that we know the *source* of it all. For example, the student may wish to try his/her hand at producing a rigorous justification based on (F1) to (F9) of the following well known facts.

Exercise 1.3. (∗) Using the field axioms of $\mathbb{R}$, prove the following:

(1) Additive inverses are unique.

(2) For all $a \in \mathbb{R}$, $(-1) \cdot a = -a$.

(3) $(-1) \cdot (-1) = 1$.

1.3   Order axioms

We now turn to order axioms for the real numbers. This is the source of the inequality '>' that we are used to, enabling one to compare two real numbers. The relation $>$ between real numbers arises from a special subset $\mathbb{P}$ of the real numbers.

Order axiom. There exists a subset $\mathbb{P}$ of $\mathbb{R}$ such that

(O1) If $x, y \in \mathbb{P}$, then $x + y \in \mathbb{P}$ and $x \cdot y \in \mathbb{P}$.

(O2) For every $x \in \mathbb{R}$, *one and only one* of the following statements is true:

$\underline{1^\circ}$ $x = 0$.
$\underline{2^\circ}$ $x \in \mathbb{P}$.
$\underline{3^\circ}$ $-x \in \mathbb{P}$.

Definition 1.1 (Positive numbers). The elements of $\mathbb{P}$ are called *positive numbers*. For real numbers x, y, we say that

$x > y$ if $x - y \in \mathbb{P}$,
$x < y$ if $y - x \in \mathbb{P}$,
$x \geq y$ if $x = y$ or $x > y$,
$x \leq y$ if $x = y$ or $x < y$.

It is clear from (O2) that 0 is *not* a positive number. Also, from (O2) it follows that for real numbers x, y, *one and only one* of the following statements is true:

$\underline{1^\circ}$ $x = y$.

$\underline{2^\circ}$ $x > y$.

$\underline{3^\circ}$ $x < y$.

Why is this so? If $x \neq y$, then $x - y \neq 0$, and so by (O2), we have the mutually exclusive possibilities $x - y \in \mathbb{P}$ or $y - x = -(x - y) \in \mathbb{P}$ happening, that is, either $x > y$ or $x < y$.

Example 1.3. $1 > 0$.

We have three possible, mutually exclusive cases:

$\underline{1^\circ}$ $1 = 0$.

$\underline{2^\circ}$ $1 \in \mathbb{P}$.

$\underline{3^\circ}$ $-1 \in \mathbb{P}$.

As $1 \neq 0$, we know that 1° is not possible.

Suppose that 3° holds, that is, $-1 \in \mathbb{P}$. From Exercise 1.3(3), $(-1) \cdot (-1) = 1$. Using (O1), and the fact that $-1 \in \mathbb{P}$, it then follows that $1 = (-1) \cdot (-1) \in \mathbb{P}$. So if we assume that 3° holds, then we obtain that *both* 2° and 3° are true, which is impossible as it violates (O2).

Thus by (O2), the only remaining case, namely 2° must hold, that is, $1 \in \mathbb{P}$.   ◇

Exercise 1.4. (∗) Using the order axioms for $\mathbb{R}$, show the following:

(1) For all $a \in \mathbb{R}$, $a^2 \geq 0$.

(2) There is no real number x such that $x^2 + 1 = 0$.

Again, just like we can use the field axioms to justify arithmetic manipulations of real numbers, it is enough to know that if challenged, one can derive all the usual laws of manipulating inequalities among real numbers based on these order axioms, but we will not do this at every instance we meet an inequality.

From our intuitive picture of $\mathbb{R}$ as points on the number line, what is the set $\mathbb{P}$? $\mathbb{P}$ is simply the set of all points/real numbers to the right of the origin O.

Also, geometrically on the number line, the inequality $a < b$ between real numbers a, b means that b lies to the right of a on the number line.

1.4   The Least Upper Bound Property of $\mathbb{R}$

This property is crucial for proving the results of Calculus, and when studying the proofs of the key results (the Bolzano–Weierstrass Theorem, the Intermediate Value Theorem, the Extreme Value Theorem, and so on), we will gradually learn to appreciate the key role played by it.

Definition 1.2 (Upper bound of a set). Let S be a subset of $\mathbb{R}$. A real number u is said to be an *upper bound of S* if for all $x \in S$, $x \leq u$.

If we think of the set S as some blob on the number line, then u should be any point on the number line that lies to the right of the points of the blob.

Example 1.4.

(1) If $S = \{0, 1, 9, 7, 6, 1976\}$, then 1976 is an upper bound of S. In fact, any real number $u \geq 1976$ is an upper bound of S. So S has lots of upper bounds.

(2) Let $S := \{x \in \mathbb{R} : x < 1\}$. Then 1 is an upper bound of S. In fact, any real number $u \geq 1$ is an upper bound of S.

(3) If $S = \mathbb{R}$, then S has no upper bound. Why? Suppose that $u \in \mathbb{R}$ is an upper bound of $\mathbb{R}$. Consider $u + 1 \in S = \mathbb{R}$. Then

$$\underbrace{u + 1}_{\in S} \quad \leq \quad \underbrace{u}_{\text{upper bound of } S} \quad ,$$

and so $1 \leq 0$, a contradiction!

(4) Let $S = \emptyset$ (the empty set, containing no elements). *Every* $u \in \mathbb{R}$ is an upper bound. For if $u \in \mathbb{R}$ is not an upper bound of S, then there must exist an element $x \in S$ which prevents u from being an upper bound of S, that is,

$$\boxed{\text{it is not the case that } x \leq u}$$

But S has no elements at all, much less an element such that $\boxed{\cdots}$ holds.

(This is an example of a 'vacuous truth'. Consider the statement

$$\boxed{\text{Every man with 60000 legs is intelligent.}}$$

This is considered a true statement in Mathematics. The argument is: Can you show me a man with 60000 legs for which the claimed property (namely of being intelligent) is not true? No! Because there are no men with 60000 legs! By the same logic, even the statement

$$\boxed{\text{Every man with 60000 legs is not intelligent.}}$$

is true in Mathematics.)

Definition 1.3 (Set bounded above). If $S \subset \mathbb{R}$ and S has an upper bound (that is, the set of upper bounds of S is not empty), then S is said to be *bounded above*.

Example 1.5. The set $\mathbb{R}$ is not bounded above.

Each of the sets $\{0, 1, 9, 7, 6, 1976\}$, $\emptyset$, $\{x \in \mathbb{R} : x < 1\}$ is bounded above.

Similarly one can define the notions of a lower bound, and of a set being bounded below.

Definition 1.4 (Lower bound of a set; set bounded below). Let S be a subset of $\mathbb{R}$. A real number ℓ is said to be a *lower bound of S* if for all $x \in S$, $\ell \leq x$.

    If $S \subset \mathbb{R}$ and S has a lower bound (that is, the set of lower bounds of S is not empty), then S is said to be *bounded below*.

If we think of the set S as some blob on the number line, then ℓ should be any point on the number line that lies to the left of the points of the blob.

Example 1.6.

(1) If $S = \{0, 1, 9, 7, 6, 1976\}$, then 0 is a lower bound of S. In fact, any real number $\ell \leq 0$ serves as a lower bound of S. So S is bounded below.

(2) Let $S := \{x \in \mathbb{R} : x < 1\}$. Then S is not bounded below. Let us show this. Suppose that, on the contrary, S does have a lower bound, say $\ell \in \mathbb{R}$. Let $x \in S$. Then $\ell \leq x < 1$. We have

$$\ell - 1 < \ell \leq x < 1,$$

and so $\ell - 1 < 1$. Thus $\ell - 1 \in S$, and as ℓ is a lower bound of S, we must have $\ell \leq \ell - 1$, that is, $1 < 0$, a contradiction! So our original assumption that S is bounded below must be false. Thus S is not bounded below. (This claim was intuitively obvious too, since the set of points in S on the number line is the entire ray of points on the left of 1, leaving no room for points on $\mathbb{R}$ to be on the 'left of S'.)

(3) If $S = \mathbb{R}$, then S has no lower bound. Indeed, if $\ell \in \mathbb{R}$ is a lower bound of $\mathbb{R}$, then

$$\underbrace{\ell}_{\text{lower bound of } S} \leq \underbrace{\ell - 1}_{\in S},$$

and so $1 \leq 0$, a contradiction. Thus $\mathbb{R}$ is not bounded below.

(4) Let $S = \emptyset$ (the empty set, containing no elements). Every $\ell \in \mathbb{R}$ is a lower bound. If $\ell \in \mathbb{R}$ is not a lower bound of S, then there must exist an element $x \in S$ which prevents ℓ from being a lower bound of S, that is, it is not the case that $\ell \leq x$. But as S is empty, this is impossible. So S is bounded below. ◇

Definition 1.5 (Bounded set). Let $S \subset \mathbb{R}$. S is called *bounded* if S is bounded below and bounded above.

Example 1.7.

S	An upper bound	Bounded above?	A lower bound	Bounded below?	Bounded?
$\{0, 1, 9, 7, 6, 1976\}$	1976 Any $u \geq 1976$	Yes	0 Any $\ell \leq 0$	Yes	Yes
$\{x \in \mathbb{R} : x < 1\}$	1 Any $u \geq 1$	Yes	Does not exist	No	No
$\mathbb{R}$	Does not exist	No	Does not exist	No	No
$\emptyset$	Every $u \in \mathbb{R}$	Yes	Every $\ell \in \mathbb{R}$	Yes	Yes

We now introduce the notions of a least upper bound (also called supremum) and a greatest lower bound (also called infimum) of a subset S of $\mathbb{R}$.

Definition 1.6 (Supremum and infimum). Let S be a subset of $\mathbb{R}$.

(1) $u_* \in \mathbb{R}$ is called a *least upper bound of S* (or a *supremum of S*) if

    (a) u_* is an upper bound of S, and

    (b) if u is an upper bound of S, then $u_* \leq u$.

(2) $\ell_* \in \mathbb{R}$ is called a *greatest lower bound of S* (or an *infimum of S*) if

    (a) ℓ_* is a lower bound of S, and

    (b) if ℓ is a lower bound of S, then $\ell \leq \ell_*$.

Pictorially, the supremum is the leftmost point among the upper bounds, and the infimum is the rightmost point among the lower bounds of a set.

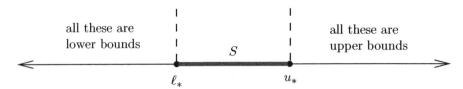

Example 1.8.

(1) If $S = \{0, 1, 9, 7, 6, 1976\}$, then $u_* = 1976$ is a least upper bound of S because

    (1) 1976 is an upper bound of S, and

    (2) if u is an upper bound of S, then $(S \ni) 1976 \leq u$, that is $u_* \leq u$.

Similarly, 0 is a greatest lower bound of S.

(2) Let $S = \{x \in \mathbb{R} : x < 1\}$. Then we claim that $u_* = 1$ is a least upper bound of S. Indeed we have:

    (a) 1 is an upper bound of S: If $x \in S$, then $x < 1 = u_*$.

    (b) Let u be an upper bound of S. We want to show that $u_* = 1 \leq u$. Suppose the contrary, that is, $1 > u$. Then there is a gap between u and 1.

(But then we see that this gap between u and 1 of course contains elements of S which are to *right* of the supposed upper bound u, and this should give the contradiction we seek.) To this end, let us consider the number $(1 + u)/2$. We have

$$\frac{1 + u}{2} < \frac{1 + 1}{2} = 1$$

and so $(1 + u)/2$ belongs to S. As u is an upper bound of S, we must have

$$\frac{1+u}{2} < u,$$

which upon rearranging gives $1 < u$, a contradiction.

S does not have a lower bound, and so certainly no greatest lower bound either (a greatest lower bound has to be first of all a lower bound!).

(3) $\mathbb{R}$ does not have a supremum, and no infimum either.

(4) $\emptyset$ has no supremum. (We intuitively expect this: every real number serves as an upper bound, but there is no smallest one among these!) Indeed, suppose on the contrary that $u_* \in \mathbb{R}$ is a supremum. Then $u_* - 1 \in \mathbb{R}$ is an upper bound of $\emptyset$ (since it is *some* real number, and we had seen that *all* real numbers are upper bounds of $\emptyset$). As u_* is a least upper bound, we must have $u_* \leq u_* - 1$, that is, $1 \leq 0$, a contradiction.

Similarly, $\emptyset$ has no infimum either. $\diamond$

A set may have many upper bounds and many lower bounds, but it is intuitively clear, based on our visual number line picture, that the supremum and infimum of a set, assuming they exist, must be unique. Here is a formal proof.

Theorem 1.2. *If a subset S of $\mathbb{R}$ has a supremum, then it is unique.*

Proof. Let u_*, u_*' be two supremums of S. Then as u_*' is, in particular, an upper bound, and since u_* is a least upper bound, we must have

$$u_* \leq u_*'. \tag{1.1}$$

Similarly, since u_* is, in particular, an upper bound, and since u_*' is a least upper bound, we must also have

$$u_*' \leq u_*. \tag{1.2}$$

From (1.1) and (1.2), it now follows that $u_* = u_*'$. $\square$

So when S has **a** supremum, then it is **the** supremum. Thus we can give it special notation (since we know what it means unambiguously):

$$\sup S.$$

Similarly, if a set S has an infimum, it is unique and is denoted by

$$\inf S.$$

Example 1.9. We have

$$\sup\{0, 1, 9, 7, 6, 1976\} = 1976,$$

$$\sup\{x \in \mathbb{R} : x < 1\} = 1,$$

$$\inf\{x \in \mathbb{R} : x \geq 1\} = 1.$$

To see the last equality, we note that 1 is certainly a lower bound of the set $S := \{x \in \mathbb{R} : x \geq 1\}$, and if ℓ is any lower bound, then as 1 is an element of the set S, we have $\ell \leq 1$.   ◇

Note that comparing the first two examples above, when $S := \{0, 1, 9, 7, 6, 1976\}$, we have

$$\sup S \in S,$$

while in the case of $S := \{x \in \mathbb{R} : x < 1\}$, we have

$$\sup S \notin S.$$

It will be convenient to keep track of when the supremum (or for that matter infimum) of a set *belongs to* the set. So we introduce the following definitions and corresponding notation.

Definition 1.7 (Maximum, minimum of a set).

(1) If $\sup S \in S$, then $\sup S$ is called a *maximum of S*, denoted by $\max S$.

(2) If $\inf S \in S$, then $\inf S$ is called a *minimum of S*, denoted by $\min S$.

Example 1.10.

S	Supremum	Maximum	Infimum	Minimum
$\{0, 1, 9, 7, 6, 1976\}$	1976	1976	0	0
$\{x \in \mathbb{R} : x < 1\}$	1	Does not exist	Does not exist	Does not exist
$\mathbb{R}$	Does not exist	Does not exist	Does not exist	Does not exist
$\emptyset$	Does not exist	Does not exist	Does not exist	Does not exist
$\{x \in \mathbb{R} : x \geq 1\}$	Does not exist	Does not exist	1	1

                                                                  ◇

Exercise 1.5. Provide the following information about the set S

An upper bound	A lower bound	Is S bounded?	$\sup S$	$\inf S$	$\max S$	$\min S$

where S is given by:

(1) $(0, 1] := \{x \in \mathbb{R} : 0 < x \leq 1\}$

(2) $[0, 1] := \{x \in \mathbb{R} : 0 \leq x \leq 1\}$

(3) $(0, 1) := \{x \in \mathbb{R} : 0 < x < 1\}$.

In the above Example 1.10, we note that if S is nonempty and bounded above, then its supremum exists. In fact, this is a fundamental property of the real numbers, called the *Least Upper Bound Property* of the real numbers, which we state below:

> If $S \subset \mathbb{R}$ is such that $S \neq \emptyset$ and S has an upper bound, then $\sup S$ exists.

Example 1.11.

(1) $S = \{0, 1, 9, 7, 6, 1976\}$ is a subset of $\mathbb{R}$, it is nonempty, and it has an upper bound. So the Least Upper Bound Property of $\mathbb{R}$ tells us that this set should have a least upper bound. This is indeed true, as we had seen earlier that S has 1976 as the supremum.

(2) $S = \{x \in \mathbb{R} : x < 1\}$ is a subset of $\mathbb{R}$, it is nonempty ($0 \in S$), and it has an upper bound (for example, 2). So the Least Upper Bound Property of $\mathbb{R}$ tells us that this set should have a least upper bound. This is indeed true, as we had seen earlier that 1 is the supremum of S.

(3) $S = \mathbb{R}$ is a subset of $\mathbb{R}$, it is nonempty, and it has no supremum. So what went wrong? Well, S is not bounded above.

(4) $S = \emptyset$ is a subset of $\mathbb{R}$ and it is bounded above. But it has no supremum. There is no contradiction to the Least Upper Bound Property because S is empty! ◇

Example 1.12. Let $S := \{x \in \mathbb{R} : x^2 \leq 2\}$. Clearly S is a subset of $\mathbb{R}$ and it is nonempty since $1 \in S$: $1^2 = 1 \leq 2$. Let us show that S is bounded above. In fact, 2 serves as an upper bound of S. Since if $x > 2$, then $x^2 > 4 > 2$. Thus if $x \in S$, then $x^2 \leq 2$, and so $x \leq 2$.

By the Least Upper Bound Property of $\mathbb{R}$, $u_* := \sup S$ exists in $\mathbb{R}$. Moreover, one can show that this u_* satisfies $u_*^2 = 2$ by showing that the cases $u_*^2 < 2$ and $u_*^2 > 2$ are both impossible.

First of all, $u_* \geq 1$ (as u_* is in particular an upper bound of S and $1 \in S$). Now define

$$r := u_* - \frac{u_*^2 - 2}{u_* + 2} = \frac{2(u_* + 1)}{u_* + 2} > 0. \tag{1.3}$$

Then, we have

$$r^2 - 2 = \frac{2(u_*^2 - 2)}{(u_* + 2)^2}. \tag{1.4}$$

$\underline{1°}$ Suppose $u_*^2 < 2$. Then (1.4) implies that $r^2 - 2 < 0$, and so $r \in S$. But from (1.3), $r > u_*$, contradicting the fact that u_* is an upper bound of S.

$\underline{2°}$ Suppose that $u_*^2 > 2$. If $r' > r$ (> 0), then $r'^2 = r' \cdot r' > r \cdot r' > r \cdot r = r^2$. From (1.4), $r^2 > 2$, and so from the above, we know that $r'^2 > 2$ as well. Hence $r' \notin S$. So we have shown that if $r' \in S$, then $r' \leq r$. This means that r is an upper bound of S. But this is impossible, since (1.3) shows that $r < u_*$, and u_* is the *least* upper bound of S.

So it must be the case that $u_*^2 = 2$. Note also that u_* is nonnegative (as $u_* \geq 1 \in S$). (We will denote this nonnegative $u_* \in \mathbb{R}$ satisfying $u_*^2 = 2$ by $\sqrt{2}$.) ◇

Example 1.13 ($\mathbb{Q}$ does not possess the Least Upper Bound Property). Consider the set $S := \{x \in \mathbb{Q} : x^2 \leq 2\}$. Clearly S is a subset of $\mathbb{Q}$ and it is nonempty since $1 \in S$: $1^2 = 1 \leq 2$. Let us show that S is bounded above. In fact, 2 serves as an upper bound of S. Since if $x > 2$, then $x^2 > 4 > 2$. Thus, if $x \in S$, then $x^2 \leq 2$, and so $x \leq 2$.

If $\mathbb{Q}$ has the Least Upper Bound Property, then the above nonempty subset of $\mathbb{Q}$ which is bounded above must possess a least upper bound $u_* := \sup S \in \mathbb{Q}$. Once again, just as in the previous example, we can show that this $u_* \in \mathbb{Q}$ must satisfy that $u_*^2 = 2$ (and we have given the details below). But we know that this is impossible as we had shown that there is no rational number whose square is 2.

First of all, $u_* \geq 1$ (as u_* is in particular an upper bound of S and $1 \in S$). Now define

$$r := u_* - \frac{u_*^2 - 2}{u_* + 2} = \frac{2(u_* + 1)}{u_* + 2} > 0. \tag{1.5}$$

Note also that as $u_* \in \mathbb{Q}$, the rightmost expression for r shows that $r \in \mathbb{Q}$ as well. Then, we have

$$r^2 - 2 = \frac{2(u_*^2 - 2)}{(u_* + 2)^2}. \tag{1.6}$$

$\underline{1°}$ Suppose $u_*^2 < 2$. Then (1.6) implies that $r^2 - 2 < 0$, and so $r \in S$. But from (1.5), $r > u_*$, contradicting the fact that u_* is an upper bound of S.

$\underline{2°}$ Suppose that $u_*^2 > 2$. If $r' > r \ (> 0)$, then $r'^2 = r' \cdot r' > r \cdot r' > r \cdot r = r^2$. From (1.6), $r^2 > 2$, and so from the above, we know that $r'^2 > 2$ as well. Hence $r' \notin S$. So we have shown that if $r' \in S$, then $r' \leq r$. This means that r is an upper bound of S. But this is impossible, since (1.5) shows that $r < u_*$, and u_* is the *least* upper bound of S.

So it must be the case that $u_*^2 = 2$. But as we mentioned earlier, this is impossible by Theorem 1.1. Hence $\mathbb{Q}$ does not possess the Least Upper Bound Property.   ◇

In order to get the useful results in Calculus (for example, the fact that for an increasing sequence of numbers bounded above, there must be a smallest number bigger than each of the terms of the sequence—a fact needed to calculate the area of a circle via the polygons inscribed within it as described in the Introduction), it turns out to be the case that the Least Upper Bound Property is indispensable. So it makes sense that when we set up the definitions and results in Calculus, we do not work with the rational number system $\mathbb{Q}$ (which regrettably does *not* possess the Least Upper Bound Property), but rather with the larger real number system $\mathbb{R}$, which does possess the Least Upper Bound Property.

Exercise 1.6. Let $a_1, a_2, a_3, \cdots$ be an infinite list (or sequence) of real numbers such that $a_n \leq a_{n+1}$ for all $n \in \mathbb{N}$, that is, the sequence is increasing. Also suppose that

$$S := \{a_n : n \in \mathbb{N}\}$$

is bounded above. Show that there is a smallest real number L that is bigger than each of the $a_n, n \in \mathbb{N}$.

Exercise 1.7.

(1) Let S be a nonempty subset of real numbers, which is bounded below. Let $-S$ denote the set of all real numbers $-x$, where x belongs to S. Prove that $\inf S$ exists and $\inf S = -\sup(-S)$.

(2) Conclude from here that $\mathbb{R}$ also has the 'Greatest Lower Bound Property':

> If S is a nonempty subset of $\mathbb{R}$ having a lower bound, then $\inf S$ exists.

Exercise 1.8. Let S be a nonempty subset of $\mathbb{R}$, which is bounded above, and let $\alpha > 0$. Show that $\alpha \cdot S := \{\alpha x : x \in X\}$ is also bounded above and that $\sup(\alpha \cdot S) = \alpha \cdot \sup S$. Similarly, if S is a nonempty subset of $\mathbb{R}$, which is bounded below and $\alpha > 0$, then show that $\alpha \cdot S$ is bounded below, and that $\inf(\alpha \cdot S) = \alpha \cdot \inf S$.

Exercise 1.9. Let A and B be nonempty subsets of $\mathbb{R}$ that are bounded above and such that $A \subset B$. Prove that $\sup A \leq \sup B$.

Exercise 1.10. For any nonempty bounded set S, prove that $\inf S \leq \sup S$, and that the equality holds if and only if S is a singleton set (that is, a set with cardinality 1).

Exercise 1.11. Let A and B be nonempty subsets of $\mathbb{R}$ that are bounded above. Prove that $\sup(A \cup B)$ exists and that $\sup(A \cup B) = \max\{\sup A, \sup B\}$.

Exercise 1.12. Determine whether the following statements are true or false.

(1) If u is an upper bound of S ($\subset \mathbb{R}$), and $u' < u$, then u' is not an upper bound of S.

(2) If u_* is the supremum of S ($\subset \mathbb{R}$), and $\epsilon > 0$, then $u_* - \epsilon$ is not an upper bound of S.

(3) Every subset of $\mathbb{R}$ has a maximum.

(4) Every subset of $\mathbb{R}$ has a supremum.

(5) Every bounded subset of $\mathbb{R}$ has a maximum.

(6) Every bounded subset of $\mathbb{R}$ has a supremum.

(7) Every bounded nonempty subset of $\mathbb{R}$ has a supremum.

(8) Every set that has a supremum is bounded above.

(9) For every set that has a maximum, the maximum belongs to the set.

(10) For every set that has a supremum, the supremum belongs to the set.

(11) For every set S that is bounded above, $|S|$ defined by $\{|x| : x \in S\}$ is bounded.

(12) For every set S that is bounded, $|S|$ defined by $\{|x| : x \in S\}$ is bounded.

(13) For every bounded set S, if $\inf S < x < \sup S$, then $x \in S$.

Exercise 1.13. Let A and B be nonempty subsets of $\mathbb{R}$ that are bounded above and define

$$A + B = \{x + y : x \in A \text{ and } y \in B\}.$$

Prove that $\sup(A + B)$ exists and that $\sup(A + B) = \sup A + \sup B$.

Exercise 1.14. Let S be a nonempty set of positive real numbers, and define

$$S^{-1} = \left\{ \frac{1}{x} : x \in S \right\}.$$

Show that S^{-1} is bounded above if and only if $\inf S > 0$. Furthermore, in case $\inf S > 0$, show that

$$\sup S^{-1} = \frac{1}{\inf S}.$$

We now prove the following theorem, which is called the *Archimedean property* of the real numbers.

Theorem 1.3 (Archimedean Property). *If $x, y \in \mathbb{R}$ and $x > 0$, then there exists an $n \in \mathbb{N}$ such that $y < nx$.*

If $y \leq 0$ to begin with, then the above is just the trivial statement that $n \cdot x > 0 \geq y$, which works with every $n \in \mathbb{N}$. So the interesting content of the theorem is when $y > 0$. Then the above is telling us, that no matter how small x is, if we keep 'tiling' the real line with multiples of the length x, then eventually we will surpass y. Here is a picture to bear in mind.

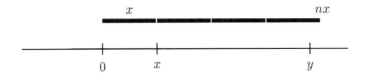

Proof. Suppose that it is not the case that

'there exists an $n \in \mathbb{N}$ such that $nx > y$'.

Then for *every* $n \in \mathbb{N}$, we must have $nx \leq y$. Let $S := \{nx : n \in \mathbb{N}\}$. Then S is a subset of $\mathbb{R}$, $S \neq \emptyset$ (indeed, $x = 1 \cdot x \in S$), and y is an upper bound of S. Thus, by the Least Upper Bound Property of $\mathbb{R}$, $u_* := \sup S$ exists. As $x > 0$, the number $u_* - x$ is smaller than the least upper bound u_* of S. Hence $u_* - x$ cannot be an upper bound of S, which means that there is an element $mx \in S$, for some $m \in \mathbb{N}$, which prevents $u_* - x$ from being an upper bound: $mx > u_* - x$. Rearranging, we obtain $u_* < mx + x = (m + 1)x \in S$, contradicting the fact that u_* is an upper bound of S. Thus, our original claim is false. In other words, there *does* exist an $n \in \mathbb{N}$ such that $nx > y$. $\quad\square$

Example 1.14. Let $S = \left\{ \frac{1}{n} : n \in \mathbb{N} \right\} = \left\{ 1, \frac{1}{2}, \frac{1}{3}, \cdots \right\}$. We claim that $\inf S = 0$.

Clearly 0 is a lower bound of S since all the elements of S are positive.

Suppose that ℓ is a lower bound of S. We want to show that $\ell \leq 0$. Suppose on the contrary that $\ell > 0$. Then by the Archimedean property (with the real numbers x and y taken as $x = 1 \ (> 0)$ and $y = 1/\ell$), there exists a $n \in \mathbb{N}$ such that

$$\frac{1}{\ell} = y < nx = n \cdot 1 = n,$$

and so

$$\frac{1}{n} < \ell,$$

contradicting the fact that ℓ is a lower bound of S. Thus, any lower bound of S must be less than or equal to 0. Hence 0 is the infimum of S.   ◇

Exercise 1.15. Provide the following information about the set S

An upper bound	A lower bound	Is S bounded?	sup S	inf S	max S	min S

where S is given by:

(1) $\left\{ \dfrac{1}{n} : n \in \mathbb{Z} \backslash \{0\} \right\}$

(2) $\left\{ \dfrac{n}{n+1} : n \in \mathbb{N} \right\}$

(3) $\left\{ (-1)^n \left(1 + \dfrac{1}{n} \right) : n \in \mathbb{N} \right\}$

Exercise 1.16. Let $S := \{ (xy - 1)^2 + x^2 : (x, y) \in \mathbb{R}^2 \}$.

(a) Show that S is bounded below.

(b) What is inf S? *Hint:* To justify your answer, consider $(x, y) = (1/n, n)$, $n \in \mathbb{N}$.

(c) Does min S exist?

Example 1.15 (The greatest integer part $\lfloor \cdot \rfloor$ of $x \in \mathbb{R}$). If we think of the real numbers as points of the line, then we see that along it, there are 'milestones' at each of the integers. So if we take any real number, then it lies between two milestones. We take $\lfloor x \rfloor$ to be the milestone immediately to the left of x—in other words, it is the 'greatest integer less than or equal to x'. So for example $\lfloor 3.1 \rfloor = 3$, $\lfloor 0 \rfloor = 0$, $\lfloor n \rfloor = n$ for all integers n, $\lfloor -3.1 \rfloor = -4$, etc.

Using the Archimedean Property, one can give a rigorous justification of the fact that every real number *has to* belong to an interval $[n, n+1)$ for some $n \in \mathbb{Z}$ (so that this $n = \lfloor x \rfloor$). By the Archimedean Property, there exists an $m_1 \in \mathbb{N}$ such that $m_1 \cdot 1 > x$. By the Archimedean

Property, there exists an $m_2 \in \mathbb{N}$ such that $m_2 \cdot 1 > -x$. So there are integers m_1, m_2 such that $-m_2 < x < m_1$. Among the finitely many integers $k \in \mathbb{Z}$ such that $-m_2 \le k \le m_1$, we take as $\lfloor x \rfloor$ the largest one such that it is also $\le x$.   ◊

Theorem 1.4 (Density of $\mathbb{Q}$ in $\mathbb{R}$). *If $a, b \in \mathbb{R}$, and $a < b$, then there exists a $r \in \mathbb{Q}$ such that $a < r < b$.*

This results says that '$\mathbb{Q}$ is dense in $\mathbb{R}$'. In everyday language, we may say for example, that 'These woods have a dense growth of birch trees', and the picture we then have in mind is that in any small area of the woods, we find a birch tree. A similar thing is conveyed by the above: no matter what 'patch' (described by the two numbers a and b) we take on the real line (thought of as the woods), we can find a rational number (analogous to birch trees) in that patch.

Proof. As $b - a > 0$ and since $1 \in \mathbb{R}$, by the Archimedean Property, there exists an $n \in \mathbb{N}$ such that $n(b - a) > 1$, that is, $na + 1 < nb$. Let $m := \lfloor na \rfloor + 1$. Then $\lfloor na \rfloor \le na < \lfloor na \rfloor + 1$, that is, $m - 1 \le na < m$. So

$$a < \frac{m}{n} \le \frac{na + 1}{n} < \frac{nb}{n} = b.$$

With $r := \dfrac{m}{n} \in \mathbb{Q}$, the proof of the theorem is complete.   □

Exercise 1.17 (Density of irrationals in $\mathbb{R}$). Show that if $a, b \in \mathbb{R}$ and $a < b$, then there exists an irrational number between a and b.

1.5   Rational powers of real numbers

Definition 1.8 (Integral powers of nonzero real numbers).

(1) Given $a \in \mathbb{R}$ and $n \in \mathbb{N}$, we define $a^n \in \mathbb{R}$ by $a^n := \underbrace{a \cdot a \cdots a}_{n \text{ times}}$.

(2) If $a \in \mathbb{R}$ and $a \ne 0$, then we define $a^0 := 1$.

(3) If $a \in \mathbb{R}$, $a \ne 0$ and $n \in \mathbb{N}$, then we define $a^{-n} := \left(\dfrac{1}{a}\right)^n$.

In this manner, all integral powers of nonzero real numbers is defined, and it can be checked that the following *laws of exponents* hold:

(E1) For all $a, b \in \mathbb{R}$ and all $n \in \mathbb{Z}$, with $a, b \ne 0$ if $n \le 0$, $(ab)^n = a^n b^n$.

(E2) For all $a \in \mathbb{R}$, all $m, n \in \mathbb{Z}$, with $a \ne 0$ if $m \le 0$ or $n \le 0$, $(a^m)^n = a^{mn}$ and $a^{m+n} = a^m a^n$.

(E3) For all $a, b \in \mathbb{R}$ with $0 \le a < b$ and $n \in \mathbb{N}$, $a^n < b^n$.

For example, (E3) can be shown like this: If $0 \le a < b$, then we have

$$a^2 = a \cdot a < a \cdot b < b \cdot b = b^2,$$
$$a^3 = a^2 \cdot a < b^2 \cdot a < b^2 \cdot b = b^3,$$
$$a^4 = a^3 \cdot a < b^3 \cdot a < b^3 \cdot b = b^4, \text{ and so on.}$$

One can also define fractional powers of positive real numbers. First we have the following:

Theorem 1.5 (Existence of nth roots). *For every $a \in \mathbb{R}$ with $a \ge 0$ and every $n \in \mathbb{N}$, there exists a unique $b \in \mathbb{R}$ such that $b \ge 0$ and $b^n = a$.*

This unique b is called the *nth root of a*, and is denoted by $a^{1/n}$ or $\sqrt[n]{a}$.

Proof. We will skip the details of the proof, which is similar to what we did to show that $\sqrt{2}$ exists in Example 1.12: the number b we seek is the supremum u_* of the set

$$S_a := \{x \in \mathbb{R} : x^n \le a\},$$

and u_* can be shown to exist by using the Least Upper Bound Property of $\mathbb{R}$.   □

Definition 1.9 (Fractional powers of positive real numbers). If $r \in \mathbb{Q}$ and

$$r = \frac{m}{n},$$

where $m, n \in \mathbb{Z}$ and $n > 0$, then for $a \in \mathbb{R}$ such that $a > 0$, we define

$$a^r := (a^m)^{1/n}.$$

It can be shown that if

$$r = \frac{m}{n} = \frac{p}{q},$$

with $p, q \in \mathbb{Z}$ and $q > 0$, then $(a^p)^{1/q} = (a^m)^{1/n}$, so that our notion of raising to rational powers is 'well-defined', that is, it does not depend on *which* particular integers m, n we take in the representation of the rational number r.

Later on, after having studied the logarithm function in Chapter 5, we will also extend the above definitions consistently to the case of real powers of positive real numbers.

1.6   Intervals

In Calculus, we will consider real-valued functions of a real variable, and develop results about these. It will turn out while doing so that we will keep meeting certain types of subsets of the real numbers (for example, subsets of this type will often be the 'domains' of our real-valued functions for which the results of Calculus hold). These special subsets of $\mathbb{R}$ are called 'intervals', and we give the definition below. Roughly speaking, these are the 'connected subsets' of the real line, namely subsets of $\mathbb{R}$ not having any 'holes/gaps'.

Definition 1.10 (Interval). An *interval* is a set consisting of all the real numbers between two given real numbers, or of all the real numbers on one side or the other of a given number. So an interval is a set of any of the following forms, where $a, b \in \mathbb{R}$:

$$(a, b) = \{x \in \mathbb{R} : a < x < b\}$$

$$[a, b] = \{x \in \mathbb{R} : a \leq x \leq b\}$$

$$(a, b] = \{x \in \mathbb{R} : a < x \leq b\}$$

$$[a, b) = \{x \in \mathbb{R} : a \leq x < b\}$$

$$(a, \infty) = \{x \in \mathbb{R} : a < x\}$$

$$[a, \infty) = \{x \in \mathbb{R} : a \leq x\}$$

$$(-\infty, b) = \{x \in \mathbb{R} : x < b\}$$

$$(-\infty, b] = \{x \in \mathbb{R} : x \leq b\}$$

$$(-\infty, \infty) = \mathbb{R}$$

In the above notation for intervals, a parenthesis '(' or ')' means that the respective endpoint is not included, and a square bracket '[' or ']' means that the endpoint is included. Thus $[0, 1)$ means the set of all real numbers x such that $0 \leq x < 1$. (Note that the use of the symbol ∞ in the notation for intervals is simply a matter of convenience and is not be taken as suggesting that there is a number ∞.)

Also, it will be convenient to give certain types of interval a special name.

Definition 1.11 (Open interval). An interval of the form (a, b), (a, ∞), $(-\infty, b)$, or $\mathbb{R}$ is called an *open interval*.

We note that if I is an open interval, then for every member $x \in I$, there exists a $\delta > 0$ such that $(x - \delta, x + \delta) \subset I$, that is, there is always some 'room' around x consisting only of elements of I.

Exercise 1.18. Show that if $a, b \in \mathbb{R}$, then the interval (a, b) has the following property:

for every $x \in (a, b)$, there exists a $\delta > 0$ such that $(x - \delta, x + \delta) \subset (a, b)$.

Show also that $[a, b]$ does not possess the above property.

Definition 1.12 (Compact interval). If $a, b \in \mathbb{R}$ and $a \leq b$, then we call $[a, b]$ a *compact interval*.

Note that $\mathbb{R} \backslash [a, b]$ is the union of two open intervals, namely $(-\infty, a)$ and (b, ∞) and that $[a, b]$ is a bounded set.

Exercise 1.19. If A_n, $n \in \mathbb{N}$, is a collection of sets, then $\bigcap_{n \in \mathbb{N}} A_n$ denotes their intersection:

$$\bigcap_{n \in \mathbb{N}} A_n = \{x : \forall n \in \mathbb{N}, \ x \in A_n\},$$

and $\bigcup_{n \in \mathbb{N}} A_n$ denotes their union: $\bigcup_{n \in \mathbb{N}} A_n = \{x : \exists n \in \mathbb{N} \text{ such that } x \in A_n\}$. Prove that

(1) $\emptyset = \bigcap_{n \in \mathbb{N}} \left(0, \frac{1}{n}\right)$.

(2) $\{0\} = \bigcap_{n \in \mathbb{N}} \left[0, \frac{1}{n}\right]$.

(3) $(0, 1) = \bigcup_{n \in \mathbb{N}} \left[\frac{1}{n+2}, 1 - \frac{1}{n+2}\right]$.

(4) $[0, 1] = \bigcap_{n \in \mathbb{N}} \left(-\frac{1}{n}, 1 + \frac{1}{n}\right)$.

1.7   Absolute value $|\cdot|$ and distance in $\mathbb{R}$

In Calculus, in order to talk about notions such as *rate of change, continuity, convergence, etc*, we will need a notion of 'closeness/distance' between real numbers. This is provided by the absolute value $|\cdot|$, and the distance between real numbers x and y is $|x - y|$. We give the definitions below.

Definition 1.13 (Absolute value and distance).

(1) The *absolute value* or *modulus* of a real number x is denoted by $|x|$, and it is defined as follows:
$$|x| = \begin{cases} x & \text{if } x \geq 0, \\ -x & \text{if } x < 0. \end{cases}$$

(2) The *distance* $d(x, y)$ between two real numbers x and y is the absolute value $|x - y|$ of their difference.

Thus, $|1| = 1$, $|0| = 0$, $|-1| = 1$, and the distance between the real numbers -1 and 1 is equal to $d(-1, 1) = |-1 - 1| = |-2| = 2$. The distance gives a notion of closeness of two points, which is crucial in the formalisation of the notions of analysis. We can now specify regions comprising points close to a certain point $c \in \mathbb{R}$ in terms of inequalities in absolute values, that is, by demanding that the distance of the points of the region, to the point c, is less than a certain positive number δ, say $\delta = 0.01$ or $\delta = 0.0000001$, and so on.

Theorem 1.6. *Let $c \in \mathbb{R}$ and $\delta > 0$. Then:*

$$\boxed{d(x, c) := |x - c| < \delta} \quad \Leftrightarrow \quad \boxed{c - \delta < x < c + \delta.}$$

Although the proof is trivial, it is worthwhile remembering Theorem 1.6, as such a manipulation will keep arising over and over again in our subsequent development of Calculus. See Figure 1.2.

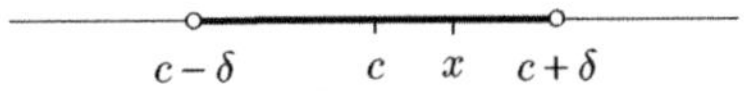

$$c-\delta \qquad c \quad x \quad c+\delta$$

Figure 1.2   The interval $I = (c - \delta, c + \delta) = \{x \in \mathbb{R} : |x - c| < \delta\}$ is the set of all points in $\mathbb{R}$ whose distance to the point c is strictly less than $\delta \, (> 0)$.

Proof.
($\Rightarrow$) Suppose that $|x - c| < \delta$. Then $x - c \le |x - c| < \delta$, and $-(x - c) \le |x - c| < \delta$. So $-\delta < x - c < \delta$, that is, $c - \delta < x < c + \delta$.

($\Leftarrow$) If $c - \delta < x < c + \delta$, then $x - c < \delta$ *and* $-(x - c) = c - x < \delta$. Thus $|x - c| < \delta$, because $|x - c|$ is either $x - c$ or $-(x - c)$, and in both cases the numbers are less than δ.   □

If we think of the real numbers as points on the number line, and we think about the integers as milestones, then it is clear that the distance between, say -1 and 3 should be 4 miles, and we observe that $4 = |-1 - 3|$. So taking $|x - y|$ as the distance between $x, y \in \mathbb{R}$ is a sensible thing to do, based on our visual picture of $\mathbb{R}$ as points on the number line (Figure 1.3).

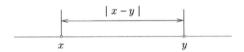

$$| x - y |$$
$$x \qquad\qquad y$$

Figure 1.3   Distance between real numbers.

Exercise 1.20. Show that a subset S of $\mathbb{R}$ is bounded if and only if there exists an $M \in \mathbb{R}$ such that for all $x \in S$, $|x| \le M$.

The following properties of the absolute value will be useful in the sequel.

Theorem 1.7. *If x, y are real numbers, then*

$$|x \cdot y| = |x| \cdot |y|, \tag{1.7}$$
$$|x + y| \le |x| + |y|. \tag{1.8}$$

(1.8) is called the *triangle inequality*.

Proof. We prove (1.7) by exhausting all possible cases:

$1°$ $x = 0$ or $y = 0$. Then $|x| = 0$ or $|y| = 0$, and so $|x| \, |y| = 0$. On the other hand, as $x = 0$ or $y = 0$, it follows that $xy = 0$ and so $|xy| = 0$.

$2°$ $x > 0$ and $y > 0$. Then $|x| = x$ and $|y| = y$, and so $|x| \, |y| = xy$. On the other hand, as $x > 0$ and $y > 0$, it follows that $xy > 0$ and so $|xy| = xy$.

$3°$ $x > 0$ and $y < 0$. Then $|x| = x$ and $|y| = -y$, and so $|x| \, |y| = x(-y) = -xy$. On the other hand, as $x > 0$ and $y < 0$, it follows that $xy < 0$ and so $|xy| = -xy$.

$\underline{4^\circ}$ $x < 0$ and $y > 0$. This follows from 3° above by interchanging x and y.

$\underline{5^\circ}$ $x < 0$ and $y < 0$. Then $|x| = -x$ and $|y| = -y$, and so $|x|\,|y| = (-x)(-y) = xy$. On the other hand, as $x < 0$ and $y < 0$, it follows that $xy > 0$ and so $|xy| = xy$.

This proves (1.7).

Next we prove (1.8). First observe that from the definition of $|\cdot|$, it follows that for any real $x \in \mathbb{R}$, $|x| \geq x$: indeed if $x \geq 0$, then $|x| = x$, while if $x < 0$, then $-x > 0$, and so we have that $|x| = -x > 0 > x$.

From (1.7), we also have $|-x| = |-1 \cdot x| = |-1||x| = 1|x| = |x|$, for all $x \in \mathbb{R}$, and so it follows that $|x| = |-x| \geq -x$ for all $x \in \mathbb{R}$.

We have the following cases:

$\underline{1^\circ}$ $x + y \geq 0$. Then we have that $|x + y| = x + y$. Since $|x| \geq x$ and $|y| \geq y$, we obtain $|x| + |y| \geq x + y = |x + y|$.

$\underline{2^\circ}$ $x + y < 0$. Then $|x + y| = -(x + y)$. Since $|x| \geq -x$ and $|y| \geq -y$, it follows that $|x| + |y| \geq -x + (-y) = -(x + y) = |x + y|$.

This proves (1.8). $\square$

Using these, it is easy to check that the 'metric/distance function' defined by

$$d(x, y) = |x - y|, \quad x, y \in \mathbb{R},$$

satisfies the following properties:

(D1) (Positive definiteness) For all $x, y \in \mathbb{R}$, $d(x, y) \geq 0$. If $d(x, y) = 0$ then $x = y$.

(D2) (Symmetry) For all $x, y \in \mathbb{R}$, $d(x, y) = d(y, x)$.

(D3) (Triangle inequality) For all $x, y, z \in \mathbb{R}$, $d(x, z) \leq d(x, y) + d(y, z)$.

The reason (D3) is called the triangle inequality is that, for triangles in Euclidean geometry of the plane, we know that the sum of the lengths of two sides of a triangle is at least as much as the length of the third side: so for the points X, Y, Z in a plane forming the three vertices of a triangle: we know that $\ell(XZ) \leq \ell(XY) + \ell(YZ)$; see Figure 1.4. (D3) reminds us of this triangle inequality, and hence the name.

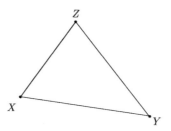

Figure 1.4 How the triangle inequality gets its name.

Exercise 1.21. Prove that if x, y are real numbers, then $||x| - |y|| \leq |x - y|$.

Exercise 1.22 (When does equality hold in the triangle inequality?).

(1) Show the generalised triangle inequality: if $n \in \mathbb{N}$ and $a_1, \cdots, a_n$ are real numbers, then $|a_1 + \cdots + a_n| \leq |a_1| + \cdots + |a_n|$.

(2) $(*)$ We say that the numbers $a_1, \cdots, a_n$ *have the same sign* if either of the following two cases is true:

    $1°$ $a_1 \geq 0, \cdots, a_n \geq 0$.
    $2°$ $a_1 \leq 0, \cdots, a_n \leq 0$.

In other words, the numbers have the same sign if on the number line either they all lie on the right of 0 including 0, or they all lie on the left of 0 including 0. Show that equality holds in the generalised triangle inequality if and only if the numbers have the same sign.
Hint: Consider the $n = 2$ case first.

Exercise 1.23. For $a, b \in \mathbb{R}$, show that $\max\{a, b\} = \frac{a+b+|a-b|}{2}$ and $\min\{a, b\} = \frac{a+b-|a-b|}{2}$.

1.8 $(*)$ Remark on the construction of $\mathbb{R}$

Natural numbers

Although we get familiar with the numbers $0, 1, 2, 3, \cdots$ from an early age, we don't learn its abstract construction in elementary school. Such an abstract construction can be given using set theory. One associates

0 with the empty set $\emptyset$,

1 with $\{\emptyset\}$,

2 with $\{\emptyset, \{\emptyset\}\}$,

3 with $\{\emptyset, \{\emptyset, \{\emptyset\}\}\}$,

and so on.

In this manner, we obtain $0, 1, 2, 3, \cdots$, in other words, the set $\mathbb{N} \cup \{0\}$, and one can also define addition via a successor function and establish the usual arithmetic laws of addition (commutativity, associativity etc.).

Integers

We can introduce the integers as pairs (m, n), where $m, n \in \mathbb{N} \cup \{0\}$, where (m, n) and (a, b) are considered to be defining the same integer if

$$m + b = n + a.$$

Then $n \in \mathbb{N} \cup \{0\}$ can be identified with $(n, 0) \in \mathbb{Z}$ and $(0, n) \in \mathbb{Z}$ is thought of as the non-positive integer $-n$, $n \in \mathbb{N} \cup \{0\}$. So -1 is $(0, 1) = (2, 3) = (1975, 1976)$, and so on.

Rational numbers

The rational numbers $\mathbb{Q}$ can be defined using pairs of integers, where the second integer is not zero, and (m, n), (a, b) are considered identical if $mb = na$.

Real numbers

What about the construction of the real number system $\mathbb{R}$?

In this book, we treat the real number system $\mathbb{R}$ as a given. But one might wonder if we can take the existence of real numbers on faith alone. It turns out that a mathematical proof of its existence can be given.

There are several ways of doing this. One is by a method called 'completion of $\mathbb{Q}$', where one considers 'Cauchy sequences' in $\mathbb{Q}$, and defines $\mathbb{R}$ to be 'equivalence classes of Cauchy sequences under a certain equivalence relation'. We refer the interested student to [**S2**, Problem 1, p. 588] or [**R**, Exercises 24, 25, p. 82] for details about this.

Another way, which is more intuitive, is via '(Dedekind) Cuts', where we identify each real number by means of two sets A and B associated with it: A is the set of rationals less than the real number we are defining, and B is set of rational numbers at least as big as the real number we are trying to identify. In other words, if we view the rational numbers lying on the number line, and think of the sets A and B (described above) corresponding to a real number, then this real number is the place along this rational number line where it can be cut, with A lying on the left side of this cut, and B lying on the right side of this cut. See Figure 1.5. More precisely, a *cut* (A, B) in $\mathbb{Q}$ is a pair of subsets A, B of $\mathbb{Q}$ such that $A \bigcup B = \mathbb{Q}$, $A \neq \emptyset$, $B \neq \emptyset$, $A \bigcap B = \emptyset$, if $a \in A$ and $b \in B$ then $a < b$, and A contains no largest element. $\mathbb{R}$ is then taken as the set of all cuts (A, B). Here are two examples of cuts:

$$(A, B) = \Big(\{r \in \mathbb{Q} : r < 0\},\ \{r \in \mathbb{Q} : r \geq 0\} \Big) \quad \text{(giving the real number '0')}$$

$$(A, B) = \Big(\{r \in \mathbb{Q} : r \leq 0 \text{ or } r^2 < 2\},\ \{r \in \mathbb{Q} : r > 0 \text{ and } r^2 \geq 2\} \Big) \quad (\text{`}\sqrt{2}\text{'}).$$

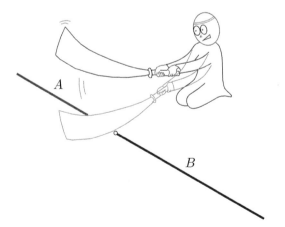

Figure 1.5   Dedekind cut.

It turns out that $\mathbb{R}$ is a field containing $\mathbb{Q}$, and it can be shown to possess the Least Upper Bound Property. The interested reader is referred to the Appendix to Chapter 1 in the classic textbook by Walter Rudin [**R**].

1.9   Functions

The concept of a 'function' is fundamental in Mathematics and in particular in Calculus. So in this section, we will quickly review:

(1) the definition of a function, and standard terminology associated with functions, such as the domain/codomain/range of a function, injective/one-to-one functions, surjective/onto functions, bijective functions/one-to-one correspondences, graph of a function;

(2) Cartesian geometry (which will allow us to visualise functions $f : D\ (\subset \mathbb{R}) \to \mathbb{R}$, by looking at their graphs);

(3) some examples.

Informal view of functions

Let X, Y be sets. A function $f : X \to Y$ is a rule that sends each $x \in X$ to one and only one corresponding point $f(x) \in Y$.

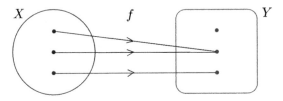

Some terminology:

(1) The set X is called the *domain of* f.

(2) The set Y is called the *codomain of* f.

(3) The set $\{y \in Y :$ there exists an $x \in X$ such that $y = f(x)\}$ is called the *range of* f. Note that the range of f is a subset of the codomain.

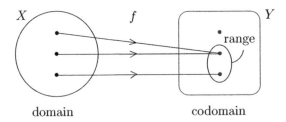

We disqualify rules that assign multiple points of Y to a point of X, and also those that miss out assigning points of Y to some points of X, from being legitimate functions.

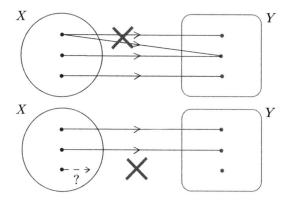

Example 1.16. Let

$$X := \{\text{all students in the classroom}\},$$

$$Y := \mathbb{R},$$

and $f : X \rightarrow Y$ be given by $f(x) = $ height in centimeters of student x, $x \in X$. Then f is a function. Indeed, each person has a unique height: one person can't have two heights. Note that there can exist of course two persons having the same height. The domain of this function is the set X of all students in the classroom, and the codomain is the set of *all* real numbers. On the other hand, it is clear that the range of f is a much smaller subset of $\mathbb{R}$: it is the finite set consisting of the heights of the students in the classroom.   ◇

Formal definition of a function

Let X, Y be sets. A *function* $f : X \rightarrow Y$ is a subset R of

$$X \times Y := \{(x, y) : x \in X, \ y \in Y\}$$

with the following two properties:

(1) For every $x \in X$, there exists a $y \in Y$ such that $(x, y) \in R$.

(2) If (x_1, y_1) and (x_2, y_2) belong to R, and if $x_1 = x_2$, then $y_1 = y_2$.

In plain English, the first requirement above, says that each x in X is sent by f to *some* element of Y (so that no elements of X are 'left out' by the function f), and the second requirement says each element of X is sent to only *one* corresponding element of Y (that is, it is not the case that some element of X is sent to more than one element of Y).

Functions are sometimes also called *maps* or *mappings*. We say for a function $f : X \rightarrow Y$ that 'f maps X to Y', and if $x \in X$, then we also say 'f maps x to $f(x)$', written

$$x \mapsto f(x) \ \text{ or } \ x \xmapsto{f} f(x).$$

In (one variable) Calculus, usually $X, Y \subset \mathbb{R}$.

Exercise 1.24. Let $f, g : \mathbb{R} \to \mathbb{R}$ be given by

$$f(x) = 1 + x^2,$$
$$g(x) = 1 - x^2,$$

$x \in \mathbb{R}$. Compute the following:

(1) $f(3) + g(3)$.

(2) $f(3) - 3 \cdot g(3)$.

(3) $f(3) \cdot g(3)$.

(4) $(f(3))/(g(3))$.

(5) $f(g(3))$.

(6) For $a \in \mathbb{R}$, $f(a) + g(-a)$.

(7) For $t \in \mathbb{R}$, $f(t) \cdot g(-t)$.

Classification of functions

We will now learn about three important classes of functions:

(1) injective or one-to-one functions,

(2) surjective or onto functions, and

(3) bijective functions or one-to-one correspondences.

Let $f : X \to Y$. Then f is called

Injective/ One-to-one	Surjective/Onto	Bijective/ One-to-one correspondence
if for every x_1 and x_2 in X such that $x_1 \neq x_2$, $f(x_1) \neq f(x_2)$.	if for each $y \in Y$ there exists an $x \in X$ such that $f(x) = y$.	if it is both injective and surjective.
'Distinct points have distinct images'.	'Codomain=Range'	

See Figure 1.6.

Example 1.17. The height function $f : X \to \mathbb{R}$ we considered earlier in Example 1.16 will not be one-to-one whenever the set X contains two students having the same height. On the other hand, the height function will be injective if all the students have distinct heights.

Also, the height function is clearly not onto. For example, there is no student whose height is $-399 \in Y = \mathbb{R}$!

As the height function is not onto, it cannot be bijective either.   ◇

injective
but not onto

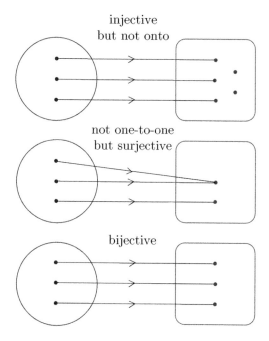

not one-to-one
but surjective

bijective

Figure 1.6   Injective/one-to-one functions, surjective/onto functions, bijective functions/one-to-one correspondences.

Example 1.18.

(1) Consider function $f_1 : \mathbb{R} \to [0, \infty)$ given by $f_1(x) = x^2$, $x \in \mathbb{R}$.

Then f_1 is not one-to-one (for example, because $f_1(-1) = 1 = f_1(1)$).
But f_1 is onto, since for every $y \in [0, \infty)$, $x := \sqrt{y} \in \mathbb{R}$ and $f_1(x) = f_1(\sqrt{y}) = (\sqrt{y})^2 = y$.

(2) Consider the function $f_2 : [0, \infty) \to \mathbb{R}$ given by $f_2(x) = \sqrt{x}$, $x \in [0, \infty)$.

Then f_2 is injective, because if $f_2(x_1) = f_2(x_2)$ for some $x_1, x_2 \geq 0$, then $\sqrt{x_1} = \sqrt{x_2}$, and so $x_1 = (\sqrt{x_1})^2 = (\sqrt{x_2})^2 = x_2$.
But f_2 is not surjective, since f_2 never assumes negative values.

(3) The function $f_3 : \mathbb{R} \to \mathbb{R}$ given by $f_3(x) = 2x$, $x \in \mathbb{R}$ is a bijection.

Indeed, f_3 is injective (since if $f_3(x_1) = f_3(x_2)$ for some $x_1, x_2 \in \mathbb{R}$, then $2x_1 = 2x_2$, and so $x_1 = x_2$), and f_3 is surjective (if $y \in \mathbb{R}$, then $f_3(y/2) = 2 \cdot (y/2) = y$).   ◇

Exercise 1.25. Show that the map $f : \mathbb{R} \to \mathbb{R}$ given by $f(x) = x|x|$, $x \in \mathbb{R}$, is a one-to-one correspondence.

Graph of a function. Review of Cartesian geometry

Definition 1.14 (Graph of a function). Let $f : X \to Y$ be a function. Then the *graph of f* is the set $\{(x, f(x)) : x \in X\}$.

The graph of f is a subset of $X \times Y$. When X and Y are both subsets of $\mathbb{R}$, then we can visualise the function f by 'plotting' its graph in the Cartesian plane $\mathbb{R}^2$. Let us recall how this is done.

First of all the word 'Cartesian' comes from the name of the mathematician **Decartes** (16th century AD), who described points in the plane with two real numbers, we recall this below.

We first draw two mutually perpendicular lines in the plane, intersecting at a point O called the *origin*. The horizontal line is called the *x-axis*, and the vertical line is called the *y-axis*.

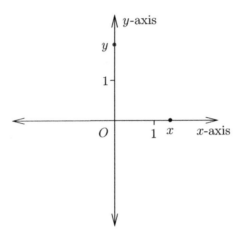

We choose unit lengths along the x-axis and y-axis, we label the number 1 on the x-axis to the right of the origin, and we label the number 1 on the y-axis above the origin. Thus, any $x \in \mathbb{R}$ is determined on the x-axis, and any $y \in \mathbb{R}$ is determined on the y-axis.

Any point $P = (x, y) \in \mathbb{R} \times \mathbb{R} =: \mathbb{R}^2$ can be depicted in the Cartesian plane by taking it to be the intersection point of the vertical line ℓ_x passing through the point x on the x-axis, and of the horizontal line ℓ_y passing through the point y on the y-axis.

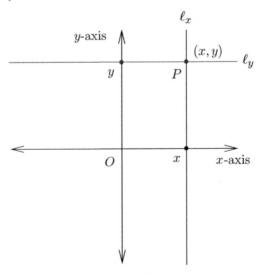

The number x is called the *x-coordinate* of $P = (x, y)$, and the number y is called the *y-coordinate* of $P = (x, y)$.

Exercise 1.26. Suppose that $f : \mathbb{R} \to \mathbb{R}$ has a graph as shown in Figure 1.7. Sketch the graphs of the functions $g_1, g_2, g_3, g_4, g_5, g_6 : \mathbb{R} \to \mathbb{R}$, defined for $x \in \mathbb{R}$ by

$$g_1(x) := f(x+1),$$
$$g_2(x) := f(x-1),$$
$$g_3(x) := f(2x),$$
$$g_4(x) := f(x/2),$$
$$g_5(x) := f(-x),$$
$$g_6(x) := -f(x).$$

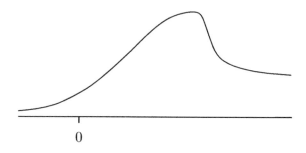

0

Figure 1.7   The graph of f.

Here are a few examples of functions and their graphs.

Example 1.19.

(1) (Sequence) Let $f_1 : \mathbb{N} \to \mathbb{R}$ be given by $f_1(n) = \dfrac{1}{n}$, $n \in \mathbb{N}$.

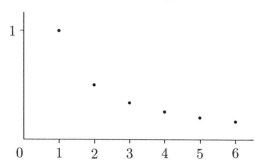

(2) (Constant function) Let $f_2 : \mathbb{R} \to \mathbb{R}$ be given by $f_2(x) = c$, $x \in \mathbb{R}$. (Here $c \in \mathbb{R}$ is fixed, say $c = 1$.)

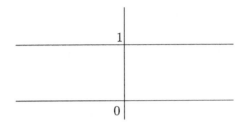

(3) (Identity function) Let $f_3 : \mathbb{R} \to \mathbb{R}$ be given by $f_3(x) = x$, $x \in \mathbb{R}$. See the picture on the left in the following figure. More generally, a *linear function* $L : \mathbb{R} \to \mathbb{R}$ is a function of the form

$$L(x) = mx + c, \quad x \in \mathbb{R},$$

for some constants m and c. The number m is called the *slope of L*, and c is referred to as the *y-axis intercept*. Note that $L(0) = c$, so that the graph of L does intersect the y-axis at the point $(0, c)$ in the Cartesian plane. Also, we note that for all distinct real numbers x_2, x_1,

$$\frac{L(x_2) - L(x_1)}{x_2 - x_1} = \frac{mx_2 + c - (mx_1 + c)}{x_2 - x_1} = m.$$

See the picture on the right below.

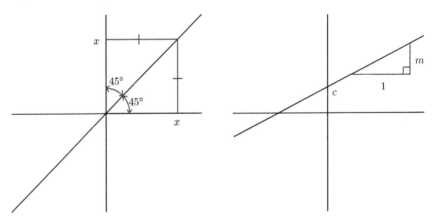

(4) (Absolute value/modulus function) Consider the absolute value/modulus function $|\cdot|$ from $\mathbb{R}$ to $\mathbb{R}$, $x \mapsto |x|$, $x \in \mathbb{R}$. The graph has a 'corner' at $x = 0$:

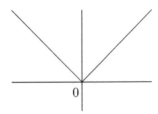

(5) (Integer and fractional part) Consider the greatest integer part function $\lfloor \cdot \rfloor : \mathbb{R} \to \mathbb{R}$, given by $x \mapsto \lfloor x \rfloor$, $x \in \mathbb{R}$. There are 'jumps' or 'discontinuities' at the integer points. Similarly, one can define the *fractional part* $\{ \cdot \} : \mathbb{R} \to \mathbb{R}$ by

$$\{x\} := x - \lfloor x \rfloor, \quad x \in \mathbb{R}.$$

For example, we have $\{-3.05\} = -3.05 - (-4) = 0.95$, $\{-3\} = -3 - (-3) = 0$, and $\{3.05\} = 3.05 - 3 = 0.05$.

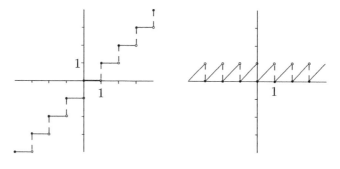

◇

Plotting with `Maple`

Using a computer package like `Maple` (or `Mathematica`), it is possible to plot the graphs of functions for specified intervals. The basic syntax of the plot command is

> `plot(f,range);`

There are many options to the plot command, and the best way to get familiar with this is to experiment with it, and to use the 'help' option. For example,

> `plot(x^2, x = -9..9);`

displays the graph of $x \mapsto x^2$ for $x \in [-9,9]$; see the picture below. In the above command, we indicated the range of x by writing 'x=-9..9'. Maple automatically chooses a scale on the vertical axis.

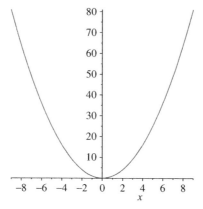

Polynomial functions

The simplest functions in Calculus are the constant function $x \mapsto 1$, the identity function given by $x \mapsto x$, and pointwise products of this, namely the *power functions* $x \mapsto x^n$, where $n \in \mathbb{N}$ is fixed. Linear combinations of these are called the *polynomials*, that is, a polynomial $p : \mathbb{R} \to \mathbb{R}$ is a function

$$p(x) = c_0 \cdot 1 + c_1 \cdot x + c_2 \cdot x^2 + c_3 \cdot x^3 + \cdots + c_d \cdot x^d, \quad x \in \mathbb{R},$$

where the *coefficients* are the (fixed) numbers $c_0, c_1, c_2, c_3, \cdots, c_d \in \mathbb{R}$, and $c_d \neq 0$. $d \in \{0, 1, 2, 3, \cdots\}$ is then called the *degree of p*. For example,

$$x^6 - 3 \cdot x^4 + 2 \cdot x^2 - \frac{1}{3},$$

is a polynomial of degree 6. If all the coefficients are zeros, then we say that p is the *zero polynomial*, and its degree is taken to be 0.

Exercise 1.27. Use Maple (or an equivalent computer program) to plot the graph of the polynomial p, where

$$p(x) = x^6 - 3 \cdot x^4 + 2 \cdot x^2 - \frac{1}{3},$$

for $x \in \left(-\frac{3}{2}, \frac{3}{2}\right)$. Can you explain the symmetry in the resulting picture?

Rational functions

A function $r : D \ (\subset \mathbb{R}) \to \mathbb{R}$ of the form

$$r(x) = \frac{n(x)}{d(x)}, \quad x \in D,$$

where n, d are fixed polynomials and $D := \mathbb{R} \setminus \{\zeta \in \mathbb{R} : d(\zeta) = 0\}$, is called a *rational function*. The polynomial n is called the *numerator polynomial of r*, and d is called the *denominator polynomial of r*. For example,

$$x \mapsto \frac{1}{1 + x^2} : \mathbb{R} \to \mathbb{R} \quad \text{and} \quad x \mapsto \frac{2 - x}{1 - x^2} : \mathbb{R} \setminus \{-1, 1\} \to \mathbb{R}$$

are rational functions. The graphs are displayed below.

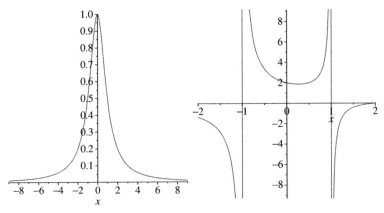

Example 1.20 (The circle). Consider all $(x, y) \in \mathbb{R}^2$ such that $x^2 + y^2 = r^2$ for some fixed $r > 0$. Then for $x \in [-r, r]$, $y = \sqrt{r^2 - x^2}$ or $y = -\sqrt{r^2 - x^2}$. (Recall that if $a \geq 0$, then $\sqrt{a}$ denotes the unique positive square root of a. For example, $\sqrt{9} = 3$.) So we can view the circle as made up of the graphs of *two* functions:

$$f_+ : [-r, r] \to \mathbb{R} \text{ given by } f_+(x) = \sqrt{r^2 - x^2}, \ x \in [-r, r], \text{ and}$$
$$f_- : [-r, r] \to \mathbb{R} \text{ given by } f_-(x) = -\sqrt{r^2 - x^2}, \ x \in [-r, r]. \qquad \diamondsuit$$

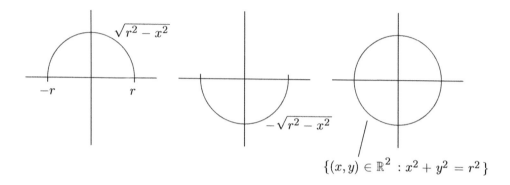

$$\{(x, y) \in \mathbb{R}^2 : x^2 + y^2 = r^2\}$$

Polynomials and rational functions, and related functions such as the square root function, and their combinations, are loosely called *algebraic functions*. Later on in Chapter 5, we will learn about some other functions that often arise in applications, such as the logarithm, the exponential, and trigonometric functions, and these are examples of 'non-algebraic' functions, or 'transcendental functions'.

Inverse functions

If one has a bijective function f from X to Y, then we can imagine a picture where points from X are taken to points in Y by f. But now if we start from any point y in Y, since the function is surjective, there has to be a point x in X which is sent to $y \in Y$, and moreover, since f is injective, we know that this x is unique. So we can 'reverse the arrow' that takes x to y under f. In this way, we get a new rule/function that takes elements from Y to elements in X, by just reversing all the old arrows of the bijective f (taking elements of X to those in Y). This map is called the 'inverse function of f', denoted by f^{-1}. We summarise this below.

If $f : X \to Y$ is a bijection, then the *inverse function*

$$f^{-1} : Y \to X$$

is defined as follows. Given $y \in Y$, there is an $x \in X$ such that $f(x) = y$ (since f is surjective), and moreover, this x is *unique* (since f is injective). So for $y \in Y$, we define $f^{-1}(y) := x$, where x is the *unique* element in X such that $f(x) = y$. It is easy to see that

$$f(f^{-1}(y)) = y \text{ for all } y \in Y, \text{ and}$$
$$f^{-1}(f(x)) = x \text{ for all } x \in X.$$

Example 1.21. Let $f : [0, 1] \to [1, 3]$ be given by $f(x) = 2x + 1, 0 \le x \le 1$. It can be checked that f is bijective. Then the inverse of f is $f^{-1} : [1, 3] \to [0, 1]$, given by

$$f^{-1}(y) = \frac{y - 1}{2}, \quad 1 \le y \le 3.$$

◇

The graphs of f and f^{-1} are displayed below.

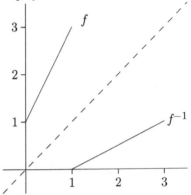

In the picture above, we notice that the graph of f^{-1} is just the reflection of the graph of f in about the line $y = x$ in the plane. This is no coincidence. The sequence of pictures in Figure 1.8 gives a key step towards explaining this: we look at the two points (a, a), (b, b), and note that

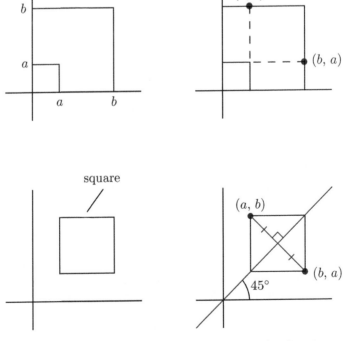

Figure 1.8   The point (b, a) is the reflection of the point (a, b) in the $y = x$ line.

that the line joining (a, b) to (b, a) is the diagonal of a square, the other diagonal of the square being the line $y = x$, and so the point (a, b) is obtained by reflecting the point (b, a) about the line $y = x$. Bearing this fact in mind, we finally note that

$$(y, x) \in \text{ graph of } f^{-1}$$

$$\Updownarrow$$

$$x = f^{-1}(y)$$

$$\Updownarrow$$

$$f(x) = y$$

$$\Updownarrow$$

$$(x, y) \in \text{ graph of } f.$$

And this completes the explanation of the fact that the graph of f^{-1} is just the reflection of the graph of f about the line $y = x$ in the plane.

Example 1.22 (The nth root function $\sqrt[n]{\cdot}$) Let $n \in \mathbb{N}$ be fixed. Let the function $f : [0, \infty) \to [0, \infty)$ be given by $f(x) = x^n$, $x \geq 0$. Then f is one-to-one because if $0 \leq a < b$, then the law of exponents (E3) on page 20 gives $a^n < b^n$. It is also onto by Theorem 1.5. Thus f is bijective, and its inverse is the nth root function $f^{-1} = \sqrt[n]{\cdot} : [0, \infty) \to [0, \infty)$ given by $f^{-1}(x) = \sqrt[n]{x}$, $x \geq 0$. Taking $n = 2$, the graphs of $f := x^2$ and its inverse $f^{-1} = \sqrt{\cdot}$ are shown in Figure 1.9. $\diamondsuit$

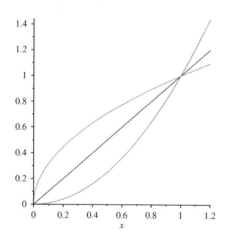

Figure 1.9   The graph of $f := x^2$ and its inverse $f^{-1} = \sqrt{\cdot}$.

Inverse of an injective map. Even if a function f fails to be bijective, but is only injective (and not necessarily surjective), we can define an inverse function from its *range* (and not its codomain) to the domain of f. We explain this now.

Let I be an interval, and $f : I \to \mathbb{R}$ be injective/one-to-one. Let us denote by $f(I)$ the range of f, that is,

$$f(I) := \{f(x) : x \in I\}.$$

We define the *inverse function* $f^{-1} : f(I) \rightarrow I$ as follows.

$$f^{-1}(y) := x, \text{ where } x \text{ is the unique element in } I \text{ such that } f(x) = y.$$

Again, since (y, x) belongs to graph of f^{-1} if and only if $f(x) = y$, that is, if and only if (x, y) belongs to graph of f, the graph of f^{-1} is obtained from the graph of f by reflection in the $y = x$ line.

1.10   (∗) Cardinality

This section is independent of the rest of the subject matter of Calculus, and if the reader so desires, it may be skipped.

For finite sets, we can compare sizes by just counting the number of elements, and this is referred to as the *cardinality of the set*: for example, the set $\{A, B, C, \cdots, Z\}$ of alphabet letters in the English language has cardinality 26, while the cardinality of $\{0, 1, 2, 3, 4, 5, 6, 7, 8, 9\}$ is 10. Note that finite sets of the same cardinality can be put in a one-to-one correspondence, that is, we can define a bijection between the two sets. Sets that do not have finite cardinality are called *infinite sets*. One can then ask the natural question: can any two infinite sets always also be put in a one-to-one correspondence? For example, we know that the set $\mathbb{N}$ is infinite, and now suppose that we have another infinite set S. Then can we always establish a bijection between the elements of $\mathbb{N}$ and those of S? In other words can we 'list' the elements of S, as the first element of S, the second element of S, and so on? The answer, perhaps surprisingly, is no! For example, such a bijection fails to exist if we take $S = \mathbb{R}$, and this is the content of Theorem 1.11 below. But first, the above discussion motivates the following definition.

Definition 1.15 (Countable set). An infinite set S is said to be *countable* if there is a bijective map from $\mathbb{N}$ onto S.

Example 1.23. Clearly if we consider the identity map $n \mapsto n : \mathbb{N} \rightarrow \mathbb{N}$, then we see that $\mathbb{N}$ is countable.

A non-trivial example is that also the set $\mathbb{Z}$ of integers is countable. This is best seen by means of a picture, as shown in Figure 1.10.

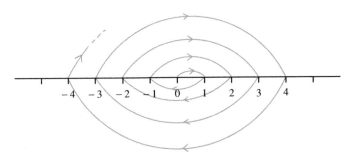

Figure 1.10   Countability of $\mathbb{Z}$.

Clearly the resulting map from $\mathbb{N}$ to $\mathbb{Z}$ is injective (since each integer is crossed by the spiral path *only once* ever—having crossed an integer, the subsequent distance of the path to the origin *increases*), and surjective (since every integer will be crossed by the spiral path *some*time). ◇

Let us show that the set $\mathbb{Q}$ of rational numbers is countable. To this end, we will need the following two auxiliary results, which are interesting in their own right.

Lemma 1.8. *Every infinite subset of a countable set is countable.*

Proof. First let us show that any infinite subset S of $\mathbb{N}$ is countable.

Let $a_1 := \min S$. If $a_1, \cdots, a_k$ have been constructed, then define

$$a_{k+1} := \min(S \setminus \{a_1, \cdots, a_k\}).$$

Then $a_1 < a_2 < a_3 < \cdots$. Define $\varphi : \mathbb{N} \to S$ by $\varphi(n) = a_n$, $n \in \mathbb{N}$. Then φ is injective (because if $n < m$, then $\varphi(n) < \varphi(m)$). Also, φ is surjective. Indeed, for each element $m \in S$, there are only finitely many natural numbers, and much less elements of S, which are smaller than m. If the number of such elements of S that are smaller than m is n_m, then it is clear that $\varphi(n_m + 1) = m$.

Now let S be countable and let T be an infinite subset of S. Let $\varphi : S \to \mathbb{N}$ be a bijection. There is a bijection from T to the range of[3] $\varphi|_T$. But the range of $\varphi|_T$ is a subset of $\mathbb{N}$, and so it is countable. Hence T is countable too. $\qquad \square$

Lemma 1.9. *If A, B are countable, then $A \times B$ is also countable.*

Proof. Since A and B are countable, we can list their elements:

$$A = \{a_1, a_2, a_3, \cdots\},$$
$$B = \{b_1, b_2, b_3, \cdots\}.$$

Arrange the elements of $A \times B$ in an array and list them by following the path as shown below.

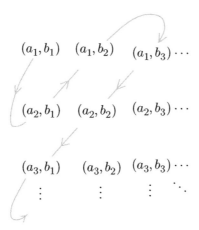

The resulting map from $\mathbb{N}$ to $A \times B$ is clearly surjective (since every element (a_n, b_m) is hit by the zigzag path *sometime*), and it is also injective (since the zigzag path never hits a point after having crossed it because it moves on to a parallel antidiagonal below). $\qquad \square$

---

[3] Here $\varphi|_T$ denotes the restriction of φ to T. In general, if $f : X \to Y$ is a function and S is a subset of X, then the *restriction of f to S* is the function $f|_S : S \to Y$ given by $f|_S(x) = f(x)$ for all $x \in S$.

We are now ready to show the countability of the rationals.

Theorem 1.10. $\mathbb{Q}$ *is countable.*

Proof. Each $q \in \mathbb{Q}$ can be written uniquely as $q = \frac{n}{d}$, where $n, d \in \mathbb{Z}$, $d > 0$ and the greatest common divisor $\gcd(n, d)$ of n, d is equal that is 1 (that is, n and d have no common factor besides 1, or n, d are coprime/relatively prime). Hence we can consider $\mathbb{Q}$ as a subset of $\mathbb{Z} \times \mathbb{Z}$. But $\mathbb{Z}$ is countable, and so by the previous part, $\mathbb{Z} \times \mathbb{Z}$ is countable. Consequently, $\mathbb{Q}$ is countable. □

Theorem 1.11. $\mathbb{R}$ *is uncountable.*

Proof. Using Lemma 1.8, we see that it is enough to show that $[0, 1]$ ($\subset \mathbb{R}$) is uncountable. Suppose, on the contrary, that $[0, 1]$ is countable. Let $x_1, x_2, x_3, \cdots$ be an enumeration of $[0, 1]$. For each $n \in \mathbb{N}$, construct a subinterval $[a_n, b_n]$ of $[0, 1]$, that does not contain x_n, inductively as follows:

Initially, $a_0 := 0$, $b_0 := 1$.
Suppose that for $k \geq 0$, a_k, b_k have been chosen. Choose a_{k+1}, b_{k+1} like this:

If $x_{k+1} \leq a_k$ or $x_{k+1} \geq b_k$, then

$$a_{k+1} := a_k + \frac{b_k - a_k}{3},$$

$$b_{k+1} := a_k + 2 \cdot \frac{b_k - a_k}{3}.$$

If $a_k < x_{k+1} < b_k$, then

$$a_{k+1} := x_{k+1} + \frac{b_k - x_{k+1}}{3},$$

$$b_{k+1} := x_{k+1} + 2 \cdot \frac{b_k - x_{k+1}}{3}.$$

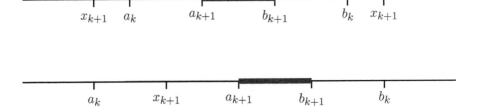

Then for all $n \in \mathbb{N}$, $[a_n, b_n] \neq \emptyset$ and $x_n \notin [a_n, b_n]$.

Moreover, $0 < a_1 < a_2 < a_3 < \cdots < a_n < \cdots < b_n < b_{n-1} < \cdots < b_2 < b_1 < 1$.

Let $a := \sup_{n \in \mathbb{N}} a_n$ and $b := \inf_{n \in \mathbb{N}} b_n$.

Then $a \leq b$, and so $[a, b] \neq \emptyset$. Also, for all $n \in \mathbb{N}$, $[a, b] \subset [a_n, b_n]$ and $x_n \notin [a_n, b_n]$. So for all $n \in \mathbb{N}$, $x_n \notin [a, b]$. Thus the points in $[a, b]$ ($\subset [0, 1]$) are missing from the enumeration, a contradiction! $\square$

Notes

The discussion in Example 1.4.(4) is based on [**J**, Page 10]. The picture in Figure 1.5 is inspired by [**P**, Figure 1.3, page 12]. Exercise 1.24 is based on [**A**, Exercise 1.5.2].

2

Sequences

The notion of a sequence occurs in ordinary conversation. For example, when one says 'an unfortunate sequence of events', we imagine a *first* event, followed by a *second* event, followed by a *third* one, and so on.

Similarly, a sequence of real numbers is an infinite list

$$a_1, a_2, a_3, \cdots$$

of real numbers, where

a_1 is the first number/member/term of the sequence,

a_2 is the second term of the sequence,

a_3 is the third term of the sequence, and so on.

For example,

$$1, \frac{1}{2}, \frac{1}{3}, \cdots$$

is a sequence of real numbers, where 1 is the first term, $1/2$ is the second term, and in general, the nth term is $1/n$, $n \in \mathbb{N}$.

If in the sequence

$$a_1, a_2, a_3, \cdots,$$

we think of a_1 as $f(1)$, a_2 as $f(2)$, a_3 as $f(3)$, and so on, then it becomes clear that a sequence is a special type of function, namely one with domain $\mathbb{N}$ and codomain $\mathbb{R}$.

Definition 2.1 (Sequence). A *sequence* is a function $f : \mathbb{N} \to \mathbb{R}$.

Only the notation is somewhat unusual. Instead of writing $f(n)$ for the value of f at a natural number n, we write a_n. The entire sequence is then referred to with the notation

$$(a_n)_{n \in \mathbb{N}}.$$

The How and Why of One Variable Calculus, First Edition. Amol Sasane.
© 2015 John Wiley & Sons, Ltd. Published 2015 by John Wiley & Sons, Ltd.

The nth term a_n of a sequence may be defined explicitly by a formula involving n, as in the example given above:

$$a_n = \frac{1}{n}, \quad n \in \mathbb{N}.$$

It might also sometimes be defined recursively. For example,

$$a_1 = 1, \quad a_{n+1} = \frac{n}{n+1} a_n \text{ for } n \in \mathbb{N}.$$

(Write down the first few terms of this sequence.)

Example 2.1. Here are a couple of examples of sequences, and we have also displayed the first few terms.

$(1)_{n \in \mathbb{N}}$ $1, 1, 1, \cdots$

$\left(\dfrac{1}{n}\right)_{n \in \mathbb{N}}$ $1, \dfrac{1}{2}, \dfrac{1}{3}, \cdots$

$((-1)^n)_{n \in \mathbb{N}}$

$-1, 1, -1, -1, 1, -1, \cdots$

$(n)_{n \in \mathbb{N}}$ $1, 2, 3, \cdots$

$\left(1 + \dfrac{1}{2} + \dfrac{1}{3} + \cdots + \dfrac{1}{n}\right)_{n \in \mathbb{N}}$

$1, 1 + \dfrac{1}{2}, 1 + \dfrac{1}{2} + \dfrac{1}{3}, \cdots$

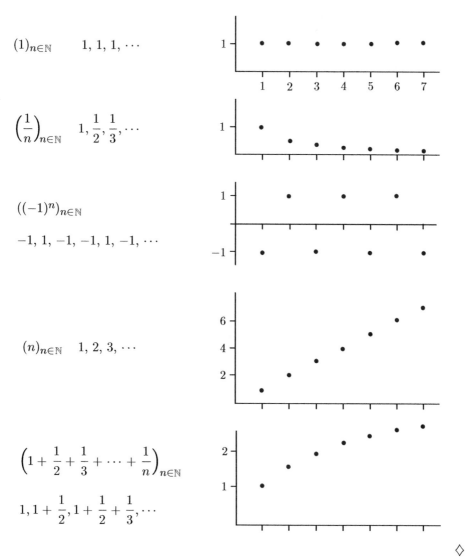

◇

What do we want to know about sequences? In Calculus, we want to know 'the limiting behavior' of the sequence, that is, what a_n behaves like for large n, and in particular, whether a_n gets closer and closer to some number L (called the *limit* of the sequence at hand).

What is the motivation for studying the limiting behavior of sequences? For example, the terms of the sequence might be the sum of the areas of the rectangles in the picture on the left below, or they might be the slopes of the chords in the picture on the right, and we might be interested in the limiting behavior because we want to calculate the area under the graph (left picture) or the instantaneous rate of change of function at the point c (right picture). Thus we want to know what happens when n increases to the sequence $(a_n)_{n\in\mathbb{N}}$ where

(Left picture) $a_n = \displaystyle\sum_{k=1}^{n-1} m_k \cdot \frac{k}{n}$, here $m_k :=$ height of kth shaded rectangle,

(Right picture) $a_n = \dfrac{f(c + \frac{1}{n}) - f(c)}{\frac{1}{n}}$.

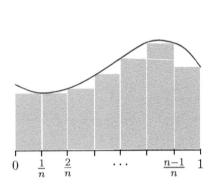 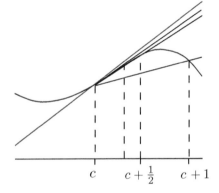

2.1   Limit of a convergent sequence

We want to give a precise definition for

'the sequence $(a_n)_{n\in\mathbb{N}}$ is convergent with limit L' or '$\displaystyle\lim_{n\to\infty} a_n = L$'.

Intuitively, by the above, we mean that there is a number L such that the terms of the sequence are getting 'closer and closer' or are 'settling down' to L for larger and larger values of n. If there is no such finite number L to which the terms of the sequence get arbitrarily close, then the sequence is said to diverge.

For example, the sequence $\left(\dfrac{1}{n}\right)_{n\in\mathbb{N}}$ seems to be convergent with limit 0, that is,

$$\lim_{n\to\infty} \frac{1}{n} = 0.$$

This is consistent with the idea of convergence that we have in mind: a sequence $(a_n)_{n \in \mathbb{N}}$ converges to some real number L, if the terms a_n get 'closer and closer' to L as n 'increases without bound'.

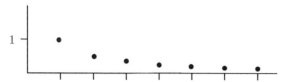

The problem with such a characterisation is its imprecision. Exactly what does it mean when we say that the terms of a sequence get 'closer and closer' or 'as close as we like' or 'arbitrarily close' to some number L? Even if we accept this ambiguity, how would we use the definition to prove theorems that involve sequences?

The terms of the sequence $\left(1 + \dfrac{1}{n} \right)_{n \in \mathbb{N}}$ are

$$2, \frac{3}{2}, \frac{4}{3}, \frac{5}{4} \cdots,$$

and the first few are plotted below.

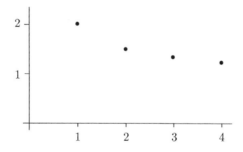

The terms of this sequence get 'closer and closer' to 0 (indeed the distance to 0 keeps decreasing), but

$$\lim_{n \to \infty} \left(1 + \frac{1}{n} \right) \neq 0,$$

rather

$$\lim_{n \to \infty} \left(1 + \frac{1}{n} \right) = 1.$$

One might say 'but clearly the terms do not get *arbitrarily* close to 0, but they *do* get arbitrarily close to 1!'

Moreover, we would also like to say that a sequence is convergent with limit L even if the *adjacent* terms of the sequence do not always *reduce* their distance to L, but it is nevertheless true that the distance to the limit can be made arbitrarily small provided we go far enough in the sequence: an example is the sequence

$$\left(\frac{n \bmod 5}{n} \right)_{n \in \mathbb{N}}.$$

Here, $n \bmod 5$ denotes the remainder obtained when n is divided by 5. The graph of the sequence is shown below.

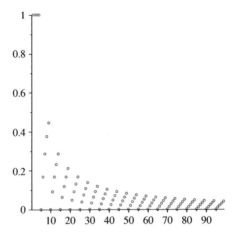

We notice that the limit of this sequence turns out to be 0, despite the fact that any two successive terms may not always reduce the distance to 0. However, given any small distance $\epsilon > 0$, there is some index N beyond which all the terms of the sequence *do* lie within a distance of ϵ from 0. In other words, the sequence *is* settling down to the value 0.

Based on the above examples, we would like to say that a sequence is deemed to be convergent with limit L if

'No matter what distance ϵ is specified, there is an index N beyond which all the terms $a_{N+1}, a_{N+2}, a_{N+3}, \cdots$ all have a distance smaller than ϵ to L'.

In other words,

$\forall \epsilon > 0$	$\exists N \in \mathbb{N}$	such that $\forall n > N$,	$\lvert a_n - L \rvert < \epsilon$
for every	there is	such that all terms	have distance to L
specified distance ϵ	an index	beyond that index	less than ϵ

(In the above, we have used the symbol '$\forall$', which is read 'for every'. Also the symbol '$\exists$' means 'there exists a/an'.)

With these introductory remarks, we now have the following concrete, precise mathematical definition for the convergence/divergence of a sequence.

Definition 2.2 (Convergent/Divergent sequence; limit). A sequence $(a_n)_{n \in \mathbb{N}}$ is said to be *convergent with limit L* ($\in \mathbb{R}$) if for every $\epsilon > 0$, there exists[1] an $N \in \mathbb{N}$ such that for all $n \in \mathbb{N}$ with $n > N$, $\lvert a_n - L \rvert < \epsilon$. Then we write

$$\lim_{n \to \infty} a_n = L.$$

If there is no $L \in \mathbb{R}$ such that $\lim_{n \to \infty} a_n = L$, then $(a_n)_{n \in \mathbb{N}}$ is called *divergent*.

---

[1] depending on ϵ

The picture below gives the geometric meaning of the definition of a sequence being convergent with limit L.

There exists an L

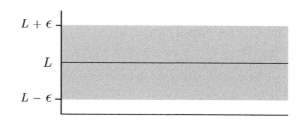

such that no matter
what $\epsilon > 0$ we pick
and consider a shaded strip
of width ϵ around
the horizontal line passing
through L,

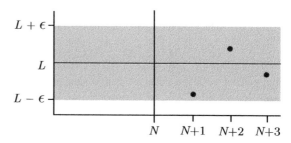

there exists an index N
such that all terms with indices
$n > N$ lie in that strip.

Had we chosen a smaller ϵ,
then perhaps a larger N'
would work.

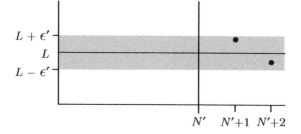

Let us consider some simple examples in order to illustrate the definition.

Example 2.2. $(1)_{n \in \mathbb{N}}$ is convergent with limit 1. We want to check if the following holds:

$$\forall \epsilon > 0,\ \exists N \in \mathbb{N} \text{ such that } \forall n > N,\ |a_n - L| < \epsilon. \tag{2.1}$$

Well, given $\epsilon > 0$, we have that $|a_n - L| = |1 - 1| = |0| = 0 < \epsilon$ *always*, that is for *all* $n \in \mathbb{N}$! So any $N \in \mathbb{N}$ works. Pictorially, no matter what the width of the shaded region is, *all* the terms of the sequence lie in that shaded strip. So for example, $N = 1$ works.

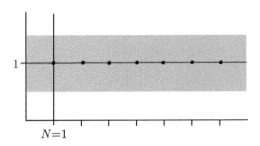

$N=1$

Here is a rigorous proof of '$\lim\limits_{n\to\infty} a_n = 1$':

Let $\epsilon > 0$.
Let N be any natural number, say $N = 1$.
Then for all $n > N = 1$, we have $|a_n - L| = |1 - 1| = |0| = 0 < \epsilon$.

So we have checked that the statement in (2.1) holds.   ◇

Example 2.3. $\left(\dfrac{1}{n}\right)_{n\in\mathbb{N}}$ is a convergent sequence with limit 0.

Before one proceeds to give rigorous proof, we often need to do some rough work. Recall that in order to check the claim, we need to verify

$$\forall \epsilon > 0, \ \exists N \in \mathbb{N} \ \text{such that} \ \forall n > N, \ |a_n - L| < \epsilon. \tag{2.2}$$

Thus given $\epsilon > 0$, the task is to find a special index N such that the inequality $|a_n - L| < \epsilon$ is satisfied for all $n > N$. So in order to find this N, we will work backwards, by first starting with the inequality $|a_n - L| < \epsilon$, and making an educated guess about what N is likely to work. Then we will proceed to give a formal proof.

(*Rough work:* Let $\epsilon > 0$. We want an N such that for all $n > N$, $|a_n - L| < \epsilon$, that is,

$$\left|\frac{1}{n} - 0\right| = \frac{1}{n} < \epsilon,$$

that is, $n > 1/\epsilon$. So we guess that we can take any $N \in \mathbb{N}$ such that $N > 1/\epsilon$, because then for $n > N, n > N > 1/\epsilon$, and we may retrace the steps above.)

Rigorous argument:

Let $\epsilon > 0$.
Let $N \in \mathbb{N}$ be such that $N > 1/\epsilon$.
(We use the Archimedean Property here with $y := 1/\epsilon$, $x := 1$: by Theorem 1.3, there exists an $N \in \mathbb{N}$ such that $Nx > y$, that is, $N > 1/\epsilon$.)
Then for all $n \in \mathbb{N}$ with $n > N$, we have

$$|a_n - L| = \left|\frac{1}{n} - 0\right| = \frac{1}{n} < \frac{1}{N} < \epsilon.$$

So $\lim\limits_{n\to\infty} \dfrac{1}{n} = 0$.   ◇

Example 2.4. $\left(1 + \dfrac{1}{n}\right)_{n \in \mathbb{N}}$ is a convergent sequence with limit 1.

(*Rough work:* $|a_n - L| = \left|1 + \dfrac{1}{n} - 1\right| = \left|\dfrac{1}{n}\right| = \dfrac{1}{n} < \epsilon$ for $n > N > \dfrac{1}{\epsilon}$.)

Rigorous argument:

Let $\epsilon > 0$.

Let $N \in \mathbb{N}$ be such that $N > 1/\epsilon$.

Then for all $n \in \mathbb{N}$ with $n > N$, we have

$$|a_n - L| = \left|1 + \dfrac{1}{n} - 1\right| = \left|\dfrac{1}{n}\right| = \dfrac{1}{n} < \dfrac{1}{N} < \epsilon.$$

So $\displaystyle \lim_{n \to \infty} \left(1 + \dfrac{1}{n}\right) = 1$.

We note that it is **not** the case that $\displaystyle \lim_{n \to \infty} \left(1 + \dfrac{1}{n}\right) = 0$. For, if on the contrary,

$$\lim_{n \to \infty} \left(1 + \dfrac{1}{n}\right) = 0,$$

then the following statement holds:

$$\forall \epsilon > 0, \ \exists N \in \mathbb{N} \text{ such that } \forall n > N, \ |a_n - 0| = \left|1 + \dfrac{1}{n} - 0\right| = 1 + \dfrac{1}{n} < \epsilon.$$

But if we take $\epsilon = 1 > 0$, then the above gives the existence of an $N \in \mathbb{N}$ such that

$$\forall n > N, \ 1 + \dfrac{1}{n} < \epsilon = 1.$$

If we take $n = N + 1$, then this last inequality gives the contradiction that

$$\dfrac{1}{N+1} < 0.$$

(We will soon learn in Theorem 2.1 that in fact if a sequence is convergent with a certain limit L, then it cannot converge to any *other* limit L'. So in light of this result, the last paragraph above is superfluous: indeed, since we proved that

$$\lim_{n \to \infty} \left(1 + \dfrac{1}{n}\right) = 1,$$

we immediately know that for any $L' \neq 1$, it cannot be the case that

$$\lim_{n \to \infty} \left(1 + \dfrac{1}{n}\right) = L',$$

and in particular, with $L' := 0 \neq 1$, we surely know that $\displaystyle \lim_{n \to \infty} \left(1 + \dfrac{1}{n}\right) \neq 0$.)   ◇

Here is an example of a divergent sequence.

Example 2.5. $((-1)^n)_{n \in \mathbb{N}}$ is divergent.

We will prove this by contradiction. Let $((-1)^n)_{n \in \mathbb{N}}$ be convergent with limit L. Then,

$$\forall \epsilon > 0, \ \exists N \in \mathbb{N} \text{ such that } \forall n > N, \ |a_n - L| = |(-1)^n - L| < \epsilon.$$

Take $\epsilon = 1/2$. (This choice is motivated by hindsight—we want to arrive at a contradiction, and it will turn out that this choice of ϵ delivers the contradiction. In order to make this transparent, let us keep working with a general ϵ in our argument below, and at a crucial last step, we will see the rationale behind our choice of $\epsilon = 1/2$!)

Then there exists an $N \in \mathbb{N}$ such that for all $n > N$, $|(-1)^n - L| < \epsilon$. But if we take any *even* $n > N$ (for example, $2N, 4N, 6N, 8N, \cdots$), then we obtain

$$|(-1)^n - L| = |1 - L| < \epsilon. \tag{2.3}$$

(This inequality says that the distance of L to 1 is less than ϵ.) On the other hand, if we take any *odd* $n > N$ (for example, $2N + 1, 4N + 1, 6N + 1, 8N + 1, \cdots$), then

$$|(-1)^n - L| = |-1 - L| < \epsilon. \tag{2.4}$$

(This inequality says that the distance of L to -1 is less than ϵ.)

So pictorially, our L is supposed to lie in an interval about 1 with width 2ϵ, and in an interval about -1 with width 2ϵ. But such intervals will not overlap if $\epsilon = 1/2$ (in fact any positive $\epsilon \leq 1$ will do the job!), and this will give us the contradiction.

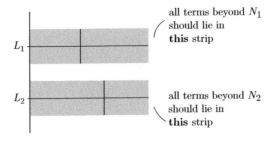

Indeed, we have, using (2.3) and (2.4) that

$$2 = |-1 - L + L - 1| \leq |-1 - L| + |L - 1| < \epsilon + \epsilon = 2\epsilon = 2 \cdot \frac{1}{2} = 1,$$

a contradiction. Consequently, the sequence $((-1)^n)_{n \in \mathbb{N}}$ is divergent.   ◇

The notation

$$\lim_{n \to \infty} a_n$$

suggests that the limit of a convergent sequence is unique. Indeed this is the case, and we prove this below.

Theorem 2.1. *A convergent sequence has a unique limit.*

Proof. Let $(a_n)_{n \in \mathbb{N}}$ be a convergent sequence with limits L_1 and L_2, with $L_1 \neq L_2$.

Let

$$\epsilon := \frac{|L_1 - L_2|}{3} > 0,$$

where the positivity of the ϵ defined above follows from the fact that $L_1 \neq L_2$. Since L_1 is a limit of the sequence $(a_n)_{n\in\mathbb{N}}$, $\exists N_1 \in \mathbb{N}$ such that

$$\text{for all } n > N_1, \ |a_n - L_1| < \epsilon.$$

Since L_2 is a limit of the sequence $(a_n)_{n\in\mathbb{N}}$, $\exists N_2 \in \mathbb{N}$ such that

$$\text{for all } n > N_2, \ |a_n - L_2| < \epsilon.$$

Consequently for $n > N_1 + N_2$, we have $n > N_1$ and $n > N_2$, and so

$$|L_1 - L_2| = |L_1 - a_n + a_n - L_2| \leq |L_1 - a_n| + |a_n - L_2| < \epsilon + \epsilon = 2\epsilon = \frac{2}{3}|L_1 - L_2|.$$

So we arrive at the contradiction that $1 < \frac{2}{3}$. Hence our original assumption was incorrect, and so a convergent sequence must have a unique limit. $\qquad\square$

Checking whether a sequence is convergent or not by using the definition is cumbersome. In the rest of the chapter, we will learn ways of deducing the convergence without having to do this hard work. Instead, we will establish results that allow us to deduce the convergence based on certain properties possessed by the sequence. One example of such a result is:

$$\text{Bounded and monotone sequences are convergent.}$$

So in the next section, among other things, we will study what is meant by a bounded sequence, a monotone sequence, and also see a proof of the result stated above.

Exercise 2.1. $(*)$

(1) Can the limit of a convergent sequence be one of the terms of the sequence?

(2) If none of the terms of a convergent sequence equal its limit, then prove that the terms of the sequence cannot consist of a finite number of distinct values.

(3) Prove that the sequence $((-1)^n)_{n\in\mathbb{N}}$ is divergent using the above.

Exercise 2.2. In each of the cases listed below, give an example of a divergent sequence $(a_n)_{n\in\mathbb{N}}$ that satisfies the given conditions. Suppose that $L = 1$.

(1) For all $\epsilon > 0$, there exists an N such that for infinitely many $n > N$, $|a_n - L| < \epsilon$.

(2) There exists an $\epsilon > 0$ and an $N \in \mathbb{N}$ such that for all $n > N$, $|a_n - L| < \epsilon$.

Exercise 2.3. Let S be a nonempty subset of $\mathbb{R}$ such that S is bounded above. Show that there exists a sequence $(a_n)_{n\in\mathbb{N}}$ contained in S (that is, $a_n \in S$ for all $n \in \mathbb{N}$) and which is convergent with limit equal to $\sup S$.

Exercise 2.4. Let $(a_n)_{n\in\mathbb{N}}$ be a sequence such that for all $n \in \mathbb{N}$, $a_n \geq 0$. Prove that if $(a_n)_{n\in\mathbb{N}}$ is convergent with limit L, then $L \geq 0$.

Exercise 2.5. Which of the following statements mean the same as 'it is not the case that the sequence $(a_n)_{n\in\mathbb{N}}$ is convergent to L'?

☐ (A) $\forall \epsilon > 0, \exists N \in \mathbb{N}$ such that $\forall n \in \mathbb{N}$ such that $n > N$, $|a_n - L| \geq \epsilon$.

☐ (B) $\forall \epsilon > 0, \exists N \in \mathbb{N}$ such that $\forall n \in \mathbb{N}$ such that $n \leq N$, $|a_n - L| \geq \epsilon$.

☐ (C) $\exists \epsilon > 0$, such that $\forall N \in \mathbb{N}$, $\exists n \in \mathbb{N}$ such that $n > N$ but $|a_n - L| \geq \epsilon$.

☐ (D) $\exists \epsilon > 0, \exists N \in \mathbb{N}$, such that $\forall n \in \mathbb{N}$ such that $n > N$, $|a_n - L| \geq \epsilon$.

2.2   Bounded and monotone sequences

Bounded sequences

Definition 2.3 (Bounded sequence). A sequence $(a_n)_{n\in\mathbb{N}}$ is said to be *bounded* if there exists an $M > 0$ such that

$$\text{for all } n \in \mathbb{N}, \quad |a_n| \leq M. \tag{2.5}$$

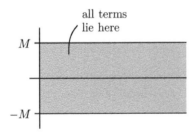

Note that a sequence is bounded if and only if the set $S = \{a_n : n \in \mathbb{N}\}$ is bounded. (See Exercise 1.20 on page 24.)

Example 2.6.

(1) $(1)_{n\in\mathbb{N}}$ is bounded, since $|1| = 1 \leq 1$ for all $n \in \mathbb{N}$.

(2) $\left(\dfrac{1}{n}\right)_{n\in\mathbb{N}}$ is bounded, since $\left|\dfrac{1}{n}\right| = \dfrac{1}{n} \leq 1$ for all $n \in \mathbb{N}$.

(3) $((-1)^n)_{n\in\mathbb{N}}$ is bounded, since $|(-1)^n| = 1 \leq 1$ for all $n \in \mathbb{N}$.

(4) $(n)_{n\in\mathbb{N}}$ is not bounded.

(If there exists an $M > 0$ such that for all $n \in \mathbb{N}$, $|a_n| = |n| = n \leq M$, then this contradicts the Archimedean Property: we know there exists an $N \in \mathbb{N}$ such that with $x := 1, N = N \cdot x > M =: y$.)

(5) The sequence $(a_n)_{n \in \mathbb{N}}$ is bounded, where

$$a_n = \frac{1}{1^1} + \frac{1}{2^2} + \frac{1}{3^3} + \cdots + \frac{1}{n^n}, \quad n \in \mathbb{N}.$$

Indeed this can be seen as follows:

$$|a_n| = \left| \frac{1}{1^1} + \frac{1}{2^2} + \frac{1}{3^3} + \cdots + \frac{1}{n^n} \right| = \frac{1}{1^1} + \frac{1}{2^2} + \frac{1}{3^3} + \cdots + \frac{1}{n^n}$$

$$\leq \frac{1}{1^1} + \frac{1}{2^2} + \frac{1}{2^3} + \cdots + \frac{1}{2^n}$$

$$= \frac{1}{1^1} + \frac{1}{2}\left(1 - \frac{1}{2}\right) + \frac{1}{2^2}\left(1 - \frac{1}{2}\right) + \cdots + \frac{1}{2^{n-1}}\left(1 - \frac{1}{2}\right)$$

$$= 1 + \frac{1}{2} - \frac{1}{2^2} + \frac{1}{2^2} - \frac{1}{2^3} + \cdots + \frac{1}{2^{n-1}} - \frac{1}{2^n}$$

$$= 1 + \frac{1}{2} - \frac{1}{2^n} < \frac{3}{2}.$$

Thus all the (positive) terms are bounded above by $\frac{3}{2}$, and so the sequence is bounded. $\diamondsuit$

The sequences $(1)_{n \in \mathbb{N}}$, $(1/n)_{n \in \mathbb{N}}$ are convergent, and we have shown above that they are also bounded. This is not a coincidence, and in the next theorem we show that the set of all convergent sequences is contained in the set of all bounded sequences.

Theorem 2.2. *If a sequence is convergent, then it is bounded.*

Proof. Let $(a_n)_{n \in \mathbb{N}}$ be a convergent sequence with limit L. Let $\epsilon := 1 > 0$. Then there exists an $N \in \mathbb{N}$ such that for all $n > N$, $|a_n - L| < \epsilon = 1$. Consequently, for all $n > N$, we have that $|a_n| = |a_n - L + L| \leq |a_n - L| + |L| < 1 + |L|$. So all the terms with index beyond N lie in the shaded strip below.

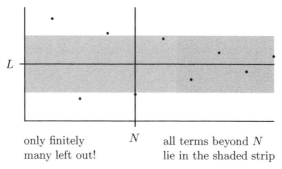

only finitely N all terms beyond N
many left out!   lie in the shaded strip

But only finitely many are left out, and surely for $n = 1, \cdots, N$,

$$|a_n| \leq \max\{|a_1|, \cdots, |a_N|\}.$$

So if we set $M := \max\{|a_1|, \ldots, |a_N|, 1 + |L|\}$, then for all $n \in \mathbb{N}$

$$|a_n| \leq M,$$

and so $(a_n)_{n \in \mathbb{N}}$ is bounded. $\qquad\square$

Thus:

$$\text{convergent} \;\Rightarrow\; \text{bounded}.$$

But the reverse implication is not true, since for example, $((-1)^n)_{n\in\mathbb{N}}$ is bounded, but not convergent. So:

$$\text{convergent} \;\not\Leftarrow\; \text{bounded}.$$

But we will see soon enough that if we add the property of being 'monotone' to boundedness, then we do get convergence:

$$\text{bounded \textbf{and} 'monotone'} \;\Rightarrow\; \text{convergent}.$$

We will now study what we mean by a monotone sequence before proving this last implication.

Exercise 2.6.

(1) Let $(b_n)_{n\in\mathbb{N}}$ be a bounded sequence. Prove that $(b_n/n)_{n\in\mathbb{N}}$ is convergent with limit 0.

(2) Is the sequence $((\sin n)/n)_{n\in\mathbb{N}}$ convergent?

Exercise 2.7. $(*)$ If $(a_n)_{n\in\mathbb{N}}$ is a convergent sequence with limit L, then prove that the sequence $(s_n)_{n\in\mathbb{N}}$, where

$$s_n = \frac{a_1 + \cdots + a_n}{n}, \quad n \in \mathbb{N},$$

is also convergent with limit L. Give an example of a sequence $(a_n)_{n\in\mathbb{N}}$ such that $(s_n)_{n\in\mathbb{N}}$ is convergent but $(a_n)_{n\in\mathbb{N}}$ is divergent.

Monotone sequences

Definition 2.4 (Increasing, decreasing, and monotone sequences).

(1) A sequence $(a_n)_{n\in\mathbb{N}}$ is said to be *increasing* if for all $n \in \mathbb{N}$, $a_n \leq a_{n+1}$, that is, if $a_1 \leq a_2 \leq a_3 \leq \cdots$.

(2) A sequence $(a_n)_{n\in\mathbb{N}}$ is said to be *decreasing* if for all $n \in \mathbb{N}$, $a_n \geq a_{n+1}$, that is, if $a_1 \geq a_2 \geq a_3 \geq \cdots$.

(3) A sequence is said to be *monotone* if it is increasing or decreasing.

Here are some examples.

Example 2.7.

Sequence	Is it increasing?	Is it decreasing?	Is it monotone?
$(n)_{n\in\mathbb{N}}$	Yes	No	Yes
$\left(\dfrac{1}{1^1} + \dfrac{1}{2^2} + \dfrac{1}{3^3} + \cdots + \dfrac{1}{n^n}\right)_{n\in\mathbb{N}}$	Yes	No	Yes
$(1)_{n\in\mathbb{N}}$	Yes	Yes	Yes
$((-1)^n)_{n\in\mathbb{N}}$	No	No	No
$\left(\dfrac{1}{n}\right)_{n\in\mathbb{N}}$	No	Yes	Yes

The following theorem can be useful in showing that sequences converge when one does not know the limit beforehand. This is the central result of this section on bounded and monotone sequences.

Theorem 2.3. *If a sequence is monotone and bounded, then it is convergent.*

Proof.
$1°$ We will first consider the case of *increasing* sequences that are bounded. Let $(a_n)_{n\in\mathbb{N}}$ be an increasing and bounded sequence. We want to show that $(a_n)_{n\in\mathbb{N}}$ is convergent. But with what limit?

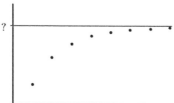

The picture above suggests that the limit should be the smallest number bigger than each of the terms of this sequence, and if we recall Exercise 1.6, we know that this is the supremum of the set $\{a_n : n \in \mathbb{N}\}$. Since $(a_n)_{n\in\mathbb{N}}$ is bounded, it follows that the set $S := \{a_n : n \in \mathbb{N}\}$ has an upper bound and so $\sup S$ exists. We show that in fact $(a_n)_{n\in\mathbb{N}}$ converges to $\sup S$. Let $\epsilon > 0$. Since $\sup S - \epsilon < \sup S$, it follows that $\sup S - \epsilon$ is *not* an upper bound for S, and so $\exists a_N \in S$ such that $\sup S - \epsilon < a_N$, that is $\sup S - a_N < \epsilon$. Since $(a_n)_{n\in\mathbb{N}}$ is an increasing sequence, for $n > N$, we have $a_N \le a_n$. Since $\sup S$ is an upper bound for S, $a_n \le \sup S$ and so $|a_n - \sup S| = \sup S - a_n$, Thus for $n > N$, $|a_n - \sup S| = \sup S - a_n \le \sup S - a_N < \epsilon$.

$2°$ If $(a_n)_{n\in\mathbb{N}}$ is a *decreasing* and bounded sequence, then clearly $(-a_n)_{n\in\mathbb{N}}$ is an increasing sequence. Furthermore if $(a_n)_{n\in\mathbb{N}}$ is bounded, then $(-a_n)_{n\in\mathbb{N}}$ is bounded as well $(|-a_n| = |a_n| \le M)$. Hence by the case considered above, it follows that $(-a_n)_{n\in\mathbb{N}}$ is a convergent sequence with limit

$$\sup\{-a_n : n \in \mathbb{N}\} = -\inf\{a_n : n \in \mathbb{N}\} = -\inf S,$$

where $S = \{a_n : n \in \mathbb{N}\}$ (see Exercise 1.7 on page 17). So given $\epsilon > 0$, there exists an $N \in \mathbb{N}$ such that for all $n > N, |-a_n - (-\inf S)| < \epsilon$, that is, $|a_n - \inf S| < \epsilon$. Thus $(a_n)_{n\in\mathbb{N}}$ is convergent with limit $\inf S$.   □

Exercise 2.8. Fill in the blanks in the following proof of the fact that *every bounded decreasing sequence of real numbers converges.*

Let $(a_n)_{n\in\mathbb{N}}$ be a bounded decreasing sequence of real numbers. Let ℓ_* be the _____ lower bound of $\{a_n : n \in \mathbb{N}\}$. The existence of ℓ_* is guaranteed by the _____ of the set of real numbers. We show that ℓ^* is the _____ of $(a_n)_{n\in\mathbb{N}}$. Taking $\epsilon > 0$, we must show that there exists a positive integer N such that _____ for all $n > N$. Since $\ell_* + \epsilon > \ell_*$, $\ell_* + \epsilon$ is not _____ of $\{a_n : n \in \mathbb{N}\}$. Therefore there exists N with _____ $\le a_N <$ _____. Since $(a_n)_{n\in\mathbb{N}}$ is _____, we have for all $n \ge N$ that $\ell_* - \epsilon < \ell_* \le$ _____ $\le a_N < \ell_* + \epsilon$, and so $|a_n - \ell_*| < \epsilon$.   □

Note that the result in Theorem 2.3 gives a sufficient condition for convergence: namely by knowing the *properties* of monotonicity and boundedness (which can be checked by just looking at the terms a_n of the sequence), we can deduce convergence. We do not need to make a guess about what the limit of the sequence is, and we do not need to check the cumbersome Definition 2.2. Here is an example of the use of this result.

Example 2.8. We had seen earlier that the sequence $(a_n)_{n\in\mathbb{N}}$ given by

$$a_n = \frac{1}{1^1} + \frac{1}{2^2} + \frac{1}{3^3} + \cdots + \frac{1}{n^n}, \quad n \in \mathbb{N}$$

is monotone (indeed, it is increasing since

$$a_{n+1} - a_n = \frac{1}{(n+1)^{n+1}} > 0$$

for all $n \in \mathbb{N}$) and bounded (see Example 2.6.(5) on page 55). Thus it follows from Theorem 2.3 that this sequence is convergent. (Although it is known that this sequence is convergent to some limit $L \in \mathbb{R}$, which is the supremum of the terms of the sequence,

$$L = \sup_{n\in\mathbb{N}} \left(\frac{1}{1^1} + \frac{1}{2^2} + \frac{1}{3^3} + \cdots + \frac{1}{n^n} \right),$$

it is so far not even known if the limit[2] L is rational or irrational, and this is still an open problem in mathematics!) $\diamond$

We remark that although [boundedness and monotonicity] is a sufficient condition for convergence, it is *not necessary*, as illustrated in the following example.

Example 2.9 (Convergence $\not\Rightarrow$ (Monotone and bounded)). Consider the sequence $(a_n)_{n\in\mathbb{N}}$ given by $a_n := \frac{(-1)^n}{n}$, $n \in \mathbb{N}$. Then this sequence is convergent with limit 0: given any $\epsilon > 0$, with $N \in \mathbb{N}$ such that $N > 1/\epsilon$, we have for all $n > N$ that $|a_n - 0| = |\frac{(-1)^n}{n} - 0| = \frac{1}{n} < \frac{1}{N} < \epsilon$. Although the sequence is bounded (all convergent sequences are!), it is not monotone: $a_1 = -1 < a_2 = \frac{1}{2} > a_3 = -\frac{1}{3}$. So the sequence is neither increasing (see the second inequality above), nor decreasing (see the first inequality above). $\diamond$

The following table gives a summary of the valid implications, and gives counterexamples for implications that are not true. See also the Venn diagram after the table.

Question	Answer	Reason/Counterexample
Is every convergent sequence bounded?	Yes	Theorem 2.2
Is every bounded sequence convergent?	No	$((-1)^n)_{n\in\mathbb{N}}$ is bounded, but not convergent.
Is every convergent sequence monotone?	No	$(\frac{(-1)^n}{n})_{n\in\mathbb{N}}$ is convergent, but not monotone.
Is every monotone sequence convergent?	No	$(n)_{n\in\mathbb{N}}$ is not convergent.
Is every bounded **and** monotone sequence convergent?	Yes	Theorem 2.3

---

[2] Also associated with this sequence is the interesting identity $\sum_{n=1}^{\infty} \frac{1}{n^n} = \int_0^1 \frac{1}{x^x} dx$; see Exercise 5.61.

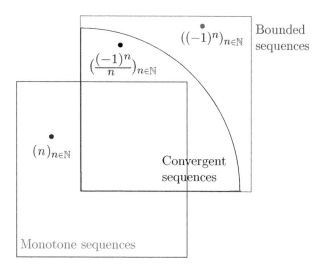

Exercise 2.9. Let $(a_n)_{n\in\mathbb{N}}$ be defined by $a_1 = 1$ and $a_n = \dfrac{2n+1}{3n}a_{n-1}, n \geq 2.$

(1) Show that $(a_n)_{n\in\mathbb{N}}$ is bounded.

(2) Show that $(a_n)_{n\in\mathbb{N}}$ is decreasing.

(3) Conclude that $(a_n)_{n\in\mathbb{N}}$ is convergent.

Exercise 2.10. Given a bounded sequence $(a_n)_{n\in\mathbb{N}}$, define

$$\ell_k = \inf\{a_n : n \geq k\} \text{ and } u_k = \sup\{a_n : n \geq k\}, \quad k \in \mathbb{N}.$$

Show that the sequences $(\ell_n)_{n\in\mathbb{N}}, (u_n)_{n\in\mathbb{N}}$ are bounded and monotone, and conclude that they are convergent. Their respective limits are called the *limit superior* and *limit inferior*, respectively, and are denoted by $\liminf\limits_{n\to\infty} a_n$ and $\limsup\limits_{n\to\infty} a_n$.

2.3 Algebra of limits

In this section, we will learn that if we 'algebraically' combine the terms of convergent sequences, then the new sequence that is obtained, is again convergent, and the limit of this sequence is the same algebraic combination of the limits. In this manner, we can sometimes prove the convergence of complicated sequences by breaking them down and writing them as an algebraic combination of simple sequences. Thus, we conveniently apply arithmetic rules to compute the limits of sequences if the terms are the sum, product, quotient of terms of simpler sequences with a known limit. For instance, using the formal definition of a limit, one can show that the sequence $(a_n)_{n\in\mathbb{N}}$ defined by

$$a_n = \frac{4n^2 + 9}{3n^2 + 7n + 11}$$

converges to $\dfrac{4}{3}$. However, it is simpler to observe that

$$a_n = \frac{n^2\left(4 + \dfrac{9}{n^2}\right)}{n^2\left(3 + \dfrac{7}{n} + \dfrac{11}{n^2}\right)} = \frac{4 + \dfrac{9}{n^2}}{3 + \dfrac{7}{n} + \dfrac{11}{n^2}},$$

and by a repeated application of Theorem 2.4 given below, we obtain

$$\lim_{n\to\infty} a_n = \frac{\lim\limits_{n\to\infty}\left(4 + \dfrac{9}{n^2}\right)}{\lim\limits_{n\to\infty}\left(3 + \dfrac{7}{n} + \dfrac{11}{n^2}\right)} = \frac{\lim\limits_{n\to\infty} 4 + \lim\limits_{n\to\infty}\dfrac{9}{n^2}}{\lim\limits_{n\to\infty} 3 + \lim\limits_{n\to\infty}\dfrac{7}{n} + \lim\limits_{n\to\infty}\dfrac{11}{n^2}} = \frac{4 + 0}{3 + 0 + 0} = \frac{4}{3}.$$

Theorem 2.4. *If* $(a_n)_{n\in\mathbb{N}}$ *and* $(b_n)_{n\in\mathbb{N}}$ *are convergent sequences, then the following hold:*

(1) *For all* $\alpha \in \mathbb{R}$, $(\alpha a_n)_{n\in\mathbb{N}}$ *is a convergent sequence and* $\lim\limits_{n\to\infty} \alpha a_n = \alpha \lim\limits_{n\to\infty} a_n$.

(2) $(|a_n|)_{n\in\mathbb{N}}$ *is a convergent sequence and* $\lim\limits_{n\to\infty} |a_n| = \left|\lim\limits_{n\to\infty} a_n\right|$.

(3) $(a_n + b_n)_{n\in\mathbb{N}}$ *is convergent and* $\lim\limits_{n\to\infty}(a_n + b_n) = \lim\limits_{n\to\infty} a_n + \lim\limits_{n\to\infty} b_n$.

(4) $(a_n b_n)_{n\in\mathbb{N}}$ *is a convergent sequence and* $\lim\limits_{n\to\infty} a_n b_n = \left(\lim\limits_{n\to\infty} a_n\right)\left(\lim\limits_{n\to\infty} b_n\right)$.

(5) *For all* $k \in \mathbb{N}$, $(a_n^k)_{n\in\mathbb{N}}$ *is a convergent sequence and* $\lim\limits_{n\to\infty} a_n^k = \left(\lim\limits_{n\to\infty} a_n\right)^k$.

(6) *If for all* $n \in \mathbb{N}$, $b_n \neq 0$ *and* $\lim\limits_{n\to\infty} b_n \neq 0$, *then* $\left(\dfrac{1}{b_n}\right)_{n\in\mathbb{N}}$ *is convergent, and*

$$\lim_{n\to\infty} \frac{1}{b_n} = \frac{1}{\lim\limits_{n\to\infty} b_n}.$$

Proof. Let $(a_n)_{n\in\mathbb{N}}$ and $(b_n)_{n\in\mathbb{N}}$ converge to L_a and L_b, respectively.

(1) If $\alpha = 0$, then $\alpha a_n = 0$ for all $n \in \mathbb{N}$ and clearly $(0)_{n\in\mathbb{N}}$ is a convergent sequence with limit 0. Thus

$$\lim_{n\to\infty} \alpha a_n = 0 = 0 L_a = \alpha \lim_{n\to\infty} a_n.$$

If $\alpha \neq 0$, then given $\epsilon > 0$, let $N \in \mathbb{N}$ be such that for all $n > N$,

$$|a_n - L_a| < \frac{\epsilon}{|\alpha|},$$

that is

$$|\alpha a_n - \alpha L_a| = |\alpha|\,|a_n - L_a| < |\alpha|\frac{\epsilon}{|\alpha|} = \epsilon.$$

Hence $(\alpha a_n)_{n\in\mathbb{N}}$ is convergent with limit αL_a, that is,

$$\lim_{n\to\infty} \alpha a_n = \alpha L_a = \alpha \lim_{n\to\infty} a_n.$$

(2) Given $\epsilon > 0$, let $N \in \mathbb{N}$ be such that for all $n > N$, $|a_n - L_a| < \epsilon$. Then we have for all $n > N$: $||a_n| - |L_a|| \leq |a_n - L_a| < \epsilon$. Hence $(|a_n|)_{n \in \mathbb{N}}$ is convergent with limit $|L_a|$, that is,

$$\lim_{n \to \infty} |a_n| = |L_a| = |\lim_{n \to \infty} a_n|.$$

(3) Given $\epsilon > 0$, let $N_a \in \mathbb{N}$ be such that for all $n > N_a$,

$$|a_n - L_a| < \frac{\epsilon}{2}.$$

Let $N_b \in \mathbb{N}$ be such that for all $n > N_b$,

$$|b_n - L_b| < \frac{\epsilon}{2}.$$

Then for all $n > N := \max\{N_a, N_b\}$, we have

$$|a_n + b_n - (L_a + L_b)| = |a_n - L_a + b_n - L_b| \leq |a_n - L_a| + |b_n - L_b| < \frac{\epsilon}{2} + \frac{\epsilon}{2} = \epsilon.$$

Hence $(a_n + b_n)_{n \in \mathbb{N}}$ is convergent with limit $L_a + L_b$, that is,

$$\lim_{n \to \infty} (a_n + b_n) = L_a + L_b = \lim_{n \to \infty} a_n + \lim_{n \to \infty} b_n.$$

(4) Note that

$$|a_n b_n - L_a L_b| = |a_n b_n - L_a b_n + L_a b_n - L_a L_b| \leq |a_n b_n - L_a b_n| + |L_a b_n - L_a L_b|$$

$$= |a_n - L_a| |b_n| + |L_a| |b_n - L_b|. \tag{2.6}$$

Given $\epsilon > 0$, we need to find a N such that for all $n > N$,

$$|a_n b_n - L_a L_b| < \epsilon.$$

This can be achieved by finding an N such that each of the summands in (2.6) is less than $\epsilon/2$ for $n > N$. This can be done as follows.

Step 1. Since $(b_n)_{n \in \mathbb{N}}$ is convergent, by Theorem 2.2 it follows that it is bounded: $\exists M > 0$ such that for all $n \in \mathbb{N}$, $|b_n| \leq M$. Let $N_a \in \mathbb{N}$ be such that for all $n > N_a$,

$$|a_n - L_a| < \frac{\epsilon}{2M}.$$

Step 2. Let $N_b \in \mathbb{N}$ be such that for all $n > N_b$,

$$|b_n - L_b| < \frac{\epsilon}{2(|L_a| + 1)}.$$

(We add $+1$ in the denominator to take care of the case when $L_a = 0$.) Thus we have that for $n > N := \max\{N_a, N_b\}$,

$$|a_n b_n - L_a L_b| \leq |a_n - L_a| |b_n| + |L_a| |b_n - L_b| < \frac{\epsilon}{2M} M + |L_a| \frac{\epsilon}{2(|L_a| + 1)}$$

$$< \frac{\epsilon}{2} + \frac{\epsilon}{2} = \epsilon.$$

So $(a_n b_n)_{n \in \mathbb{N}}$ is a convergent sequence with limit $L_a L_b$, that is,

$$\lim_{n \to \infty} a_n b_n = L_a L_b = \left(\lim_{n \to \infty} a_n \right) \left(\lim_{n \to \infty} b_n \right).$$

(5) This can be shown by using induction on k and part 4 above. It is trivially true with $k = 1$. Suppose that it holds for some k: then $(a_n^k)_{n \in \mathbb{N}}$ is convergent and

$$\lim_{n \to \infty} a_n^k = \left(\lim_{n \to \infty} a_n \right)^k.$$

Hence by part 4 above applied to the sequences $(a_n)_{n \in \mathbb{N}}$ and $(a_n^k)_{n \in \mathbb{N}}$, we obtain that the sequence $(a_n \cdot a_n^k)_{n \in \mathbb{N}}$ is convergent and

$$\lim_{n \to \infty} a_n a_n^k = \left(\lim_{n \to \infty} a_n \right) \left(\lim_{n \to \infty} a_n^k \right) = \left(\lim_{n \to \infty} a_n \right) \left(\lim_{n \to \infty} a_n \right)^k = \left(\lim_{n \to \infty} a_n \right)^{k+1}.$$

Thus $(a_n^{k+1})_{n \in \mathbb{N}}$ is convergent and $\lim_{n \to \infty} a_n^{k+1} = \left(\lim_{n \to \infty} a_n \right)^{k+1}.$

(6) Let $N_1 \in \mathbb{N}$ be such that for all $n > N_1$, $|b_n - L_b| < \dfrac{|L_b|}{2}$. Thus for all $n > N_1$,

$$|L_b| - |b_n| \le ||L_b| - |b_n|| \le |b_n - L_b| < \frac{|L_b|}{2},$$

and so $|b_n| \ge \dfrac{|L_b|}{2}$. Let $\epsilon > 0$, and let $N_2 \in \mathbb{N}$ be such that for all $n > N_2$,

$$|b_n - L_b| < \frac{\epsilon |L_b|^2}{2}.$$

Hence for $n > N := \max\{N_1, N_2\}$, we have

$$\left| \frac{1}{b_n} - \frac{1}{L_b} \right| = \frac{|b_n - L_b|}{|b_n| \, |L_b|} < \frac{\epsilon |L_b|^2}{2} \frac{2}{|L_b| \, |L_b|} = \epsilon.$$

So $\left(\dfrac{1}{b_n} \right)_{n \in \mathbb{N}}$ is convergent and $\lim_{n \to \infty} \dfrac{1}{b_n} = \dfrac{1}{L_b} = \dfrac{1}{\lim_{n \to \infty} b_n}.$ □

Example 2.10. Consider the sequence $(a_n)_{n \in \mathbb{N}}$, where

$$a_n := \frac{1}{n^3} + \frac{2^2}{n^3} + \frac{3^2}{n^3} + \cdots + \frac{n^2}{n^3}, \quad n \in \mathbb{N}.$$

A student observes that

$$\lim_{n \to \infty} \frac{1}{n^3} = 0, \quad \lim_{n \to \infty} \frac{2^2}{n^3} = 0, \quad \lim_{n \to \infty} \frac{3^2}{n^3} = 0, \quad \cdots \quad , \quad \lim_{n \to \infty} \frac{n^2}{n^3} = 0,$$

and hastily concludes that

'by the Algebra of Limits, $\lim_{n \to \infty} a_n = 0 + 0 + 0 + \cdots + 0 = 0$'.

Where does the error in this argument lie?

Note that by Theorem 2.4.(3), we do have that the termwise sum of a *finite fixed* number of sequences is convergent with the limit of the sum being the sum of the limits. In other words, if

$$a_{n,1} \xrightarrow{n\to\infty} L_1,$$

$$a_{n,2} \xrightarrow{n\to\infty} L_2,$$

$$a_{n,3} \xrightarrow{n\to\infty} L_3,$$

$$\cdots$$

$$a_{n,k} \xrightarrow{n\to\infty} L_k,$$

then we do have that $a_{n,1} + a_{n,2} + a_{n,3} + \cdots + a_{n,k} \xrightarrow{n\to\infty} L_1 + L_2 + L_3 + \cdots + L_k.$

However, in the application above, the number of sequences was not fixed. In fact, knowing the following formula for the sum of squares (which can easily be shown by induction)

$$1^2 + 2^2 + 3^2 + \cdots + n^2 = \frac{n(n+1)(2n+1)}{6}, \quad n \in \mathbb{N},$$

we have

$$
\begin{aligned}
a_n &= \frac{1}{n^3} + \frac{2^2}{n^3} + \frac{3^2}{n^3} + \cdots + \frac{n^2}{n^3} = \frac{1^2 + 2^2 + 3^2 + \cdots + n^2}{n^3} \\
&= \frac{n(n+1)(2n+1)/6}{n^3} = \frac{1}{6}\left(1 + \frac{1}{n}\right)\left(2 + \frac{1}{n}\right),
\end{aligned}
$$

and so by the Algebra of Limits,

$$\lim_{n\to\infty} a_n = \lim_{n\to\infty} \frac{1}{6}\left(1 + \frac{1}{n}\right)\left(2 + \frac{1}{n}\right) = \frac{1}{6}\cdot(1+0)\cdot(2+0) = \frac{1}{3}. \qquad \diamondsuit$$

Exercise 2.11. Is the following manipulation justified based on Theorem 2.4?

$$\lim_{n\to\infty}\left(1 + \frac{1}{n}\right)^n = \left(\lim_{n\to\infty}\left(1 + \frac{1}{n}\right)\right)^n = \left(1 + \lim_{n\to\infty}\frac{1}{n}\right)^n = (1+0)^n = 1^n = 1.$$

Exercise 2.12. Suppose that the sequence $(a_n)_{n\in\mathbb{N}}$ is convergent, and assume that the sequence $(b_n)_{n\in\mathbb{N}}$ is bounded. Prove that the sequence $(c_n)_{n\in\mathbb{N}}$ defined by

$$c_n = \frac{a_n b_n + 5n}{a_n^2 + n}, \quad n \in \mathbb{N},$$

is convergent, and find its limit.

Exercise 2.13. Let $(a_n)_{n \in \mathbb{N}}$ be a convergent sequence with limit L and suppose that $a_n \geq 0$ for all $n \in \mathbb{N}$. Prove that the sequence $(\sqrt{a_n})_{n \in \mathbb{N}}$ is also convergent, with limit $\sqrt{L}$.

Hint: First show that $L \geq 0$. Let $\epsilon > 0$. If $L = 0$, then choose $N \in \mathbb{N}$ large enough so that for $n > N$, $|a_n - L| = a_n < \epsilon^2$. If $L > 0$, then choose $N \in \mathbb{N}$ large enough so that for $n > N$, $|\sqrt{a_n} - \sqrt{L}||\sqrt{a_n} + \sqrt{L}| = |a_n - L| < \epsilon\sqrt{L}$.

Exercise 2.14. Show that $(\sqrt{n^2 + n} - n)_{n \in \mathbb{N}}$ is a convergent sequence and find its limit.
Hint: 'Rationalise the numerator' by using $\sqrt{n^2 + n} + n$.

Exercise 2.15.

(1) Prove that if $(a_n)_{n \in \mathbb{N}}$ and $(b_n)_{n \in \mathbb{N}}$ are convergent sequences such that for all $n \in \mathbb{N}$, $a_n \leq b_n$, then

$$\lim_{n \to \infty} a_n \leq \lim_{n \to \infty} b_n.$$

Hint: Use Exercise 2.4 on page 54.

(2) With the same notation as in Exercise 2.10, show that for a bounded sequence $(a_n)_{n \in \mathbb{N}}$,

$$\liminf_{n \to \infty} a_n \leq \limsup_{n \to \infty} a_n.$$

Give an example of a bounded sequence to show that there can be a strict inequality here.

2.4   Sandwich theorem

Another useful theorem for proving that sequences are convergent and in determining their limits is the so-called Sandwich Theorem. Roughly speaking, it says that if a sequence is 'sandwiched' between two convergent sequences with the *same* limit, then the sandwiched sequence is also convergent with the same limit.

Theorem 2.5 (Sandwich theorem). *Let $(a_n)_{n \in \mathbb{N}}$, $(b_n)_{n \in \mathbb{N}}$ be convergent sequences with the same limit, that is,*

$$\lim_{n \to \infty} a_n = \lim_{n \to \infty} b_n.$$

If $(c_n)_{n\in\mathbb{N}}$ is a third sequence such that

$$\text{for all } n \in \mathbb{N}, \ a_n \le c_n \le b_n,$$

then $(c_n)_{n\in\mathbb{N}}$ is also convergent with the same limit, that is,

$$\lim_{n\to\infty} a_n = \lim_{n\to\infty} c_n = \lim_{n\to\infty} b_n.$$

Proof. Let L denote the common limit of $(a_n)_{n\in\mathbb{N}}$ and $(b_n)_{n\in\mathbb{N}}$:

$$\lim_{n\to\infty} a_n = L = \lim_{n\to\infty} b_n.$$

Given $\epsilon > 0$, let $N_a \in \mathbb{N}$ be such that for all $n > N_a$, $|a_n - L| < \epsilon$. Hence for $n > N_a$,

$$L - a_n \le |L - a_n| = |a_n - L| < \epsilon,$$

and so $L - a_n < \epsilon$, that is,

$$L - \epsilon < a_n.$$

Let $N_b \in \mathbb{N}$ be such that for all $n > N_b$, $|b_n - L| < \epsilon$. So for $n > N_b$, $b_n - L < \epsilon$, that is,

$$b_n < L + \epsilon.$$

Thus for $n > N := \max\{N_a, N_b\}$, we have

$$L - \epsilon < a_n \le c_n \le b_n < L + \epsilon,$$

and so $L - \epsilon < c_n < L + \epsilon$. Consequently, $c_n - L < \epsilon$ and $-(c_n - L) < \epsilon$, and so

$$|c_n - L| < \epsilon.$$

This proves that $(c_n)_{n\in\mathbb{N}}$ is convergent with limit L. $\square$

Example 2.11 (The geometric progression).

The aim of this example is to show that if $|r| < 1$, then $\lim_{n\to\infty} r^n = 0$.

First let us consider the case when $r \in (0,1)$. Then $h := \dfrac{1}{r} - 1 > 0$. For $n \in \mathbb{N}$,

$$\frac{1}{r^n} = \underbrace{(1+h)^n \ge 1 + nh}_{(*)} \ge nh. \tag{2.7}$$

One can show the inequality $(*)$ using induction as follows. Clearly when $n = 1$,

$$(1+h)^1 = 1 + h = 1 + 1 \cdot h.$$

If $(1+h)^n \ge 1 + nh$ for some n, then

$$(1+h)^{n+1} = (1+h)^n(1+h) \ge (1+nh)(1+h) = 1 + (n+1)h + nh^2 \ge 1 + (n+1)h,$$

and so the inequality is true for all n.

Hence we obtain $0 \leq r^n \leq \dfrac{1}{nh}$ for all $n \in \mathbb{N}$. Since

$$\lim_{n \to \infty} 0 = 0 = \lim_{n \to \infty} \frac{1}{nh},$$

it follows by the Sandwich Theorem that $\lim\limits_{n \to \infty} r^n = 0$ too.

When $r = 0$, $r^n = 0$ for all $n \in \mathbb{N}$, and so clearly $\lim\limits_{n \to \infty} r^n = 0$.

Now suppose that $|r| < 1$. Then $|r| \in [0, 1)$, and so by the above,

$$\lim_{n \to \infty} |r|^n = 0.$$

By the Algebra of Limits, $\lim\limits_{n \to \infty} -|r|^n = 0$ as well. Since

$$-|r|^n \leq r^n \leq |r|^n \quad \text{for all } n \in \mathbb{N},$$

it follows again by the Sandwich Theorem that $\lim\limits_{n \to \infty} r^n = 0$.

As a consequence of the above, we can show that if $r \in (-1, 1)$, then the 'sequence of partial sums' $(1 + r + r^2 + \cdots + r^n)_{n \in \mathbb{N}}$ converges because

$$1 + r + r^2 + \cdots + r^n = \frac{(1 - r)(1 + r + r^2 + \cdots + r^n)}{1 - r}$$

$$= \frac{1 + r + \cdots + r^n - (r + r^2 + \cdots + r^{n+1})}{1 - r}$$

$$= \frac{1 - r^{n+1}}{1 - r} = \frac{1 - r \cdot r^n}{1 - r},$$

and so $\lim\limits_{n \to \infty} (1 + r + r^2 + \cdots + r^n) = \lim\limits_{n \to \infty} \dfrac{1 - r \cdot r^n}{1 - r} = \dfrac{1 - r \cdot 0}{1 - r} = \dfrac{1}{1 - r}.$ ◇

Example 2.12. $\lim\limits_{n \to \infty} a^{1/n} = 1$ for $a > 1$.

For concreteness, let us take $a = 2$, but the proof is the same, mutatis mutandis[3], for any $a > 1$. As $2 > 1$, we have $2^{1/n} > 1$ for all $n \in \mathbb{N}$. So we can write $2^{1/n} = 1 + h$, where $h := 2^{1/n} - 1 > 0$. Thus

$$2 = (1 + h)^n = 1 + nh + \underbrace{\binom{n}{2} h^2 + \cdots + h^n}_{>0} > 1 + nh,$$

(where the inequality above can also be shown as the justification of $(*)$ in (2.7)), and so $1 > nh$. This gives

$$\frac{1}{n} > h = 2^{1/n} - 1 > 0 \quad \text{for all } n \in \mathbb{N},$$

and by the Sandwich Theorem, $\lim\limits_{n \to \infty} (2^{1/n} - 1) = 0$, that is, $\lim\limits_{n \to \infty} 2^{1/n} = 1$.   ◇

---

[3] Latin phrase meaning 'changing only those things that need to be changed'

Example 2.13. For any $a, b \in \mathbb{R}$, $\lim\limits_{n \to \infty} (|a|^n + |b|^n)^{\frac{1}{n}} = \max\{|a|, |b|\}$.

Let $M := \max\{|a|, |b|\}$. Then $|a| \leq M$ gives $|a|^n \leq M^n$, and similarly $|b|^n \leq M^n$. Thus $|a|^n + |b|^n \leq 2M^n$, and so

$$(|a|^n + |b|^n)^{1/n} \leq 2^{1/n}M.$$

Also, $|a|^n + |b|^n \geq M^n$ gives $(|a|^n + |b|^n)^{1/n} \geq M$. So we have

$$M \leq (|a|^n + |b|^n)^{1/n} \leq 2^{1/n}M \text{ for all } n \in \mathbb{N}.$$

Since $\lim\limits_{n \to \infty} 2^{1/n} = 1$, we have

$$\lim_{n \to \infty} M = M = \lim_{n \to \infty} (2^{1/n}M),$$

and so it follows from the Sandwich Theorem that

$$\lim_{n \to \infty} (|a|^n + |b|^n)^{1/n} = M = \max\{|a|, |b|\}.$$

In particular, with $a = 27$ and $b = 2014$, we have that

$$\lim_{n \to \infty} (27^n + 2014^n)^{\frac{1}{n}} = 2014,$$

that is, the sequence $2041, 2014.180975, 2014.001618, 2014.000016, \cdots$ is convergent with limit 2014. ◇

Exercise 2.16. Prove that the sequence $\left(\dfrac{n!}{n^n}\right)_{n \in \mathbb{N}}$ is convergent and that $\lim\limits_{n \to \infty} \dfrac{n!}{n^n} = 0$.

Hint: Observe that $0 \leq \dfrac{n!}{n^n} = \dfrac{1}{n} \cdot \dfrac{2}{n} \cdot \cdots \cdot \dfrac{n}{n} \leq \dfrac{1}{n} \cdot 1 \cdots \cdots 1 \leq \dfrac{1}{n}$.

Exercise 2.17. Prove that for all $k \in \mathbb{N}$, $\lim\limits_{n \to \infty} \dfrac{1^k + 2^k + 3^k + \cdots + n^k}{n^{k+2}} = 0$.

Exercise 2.18 ($\lim\limits_{n \to \infty} n^{\frac{1}{n}} = 1$).

(1) Using induction, prove that if $x \geq -1$ and $n \in \mathbb{N}$, then $(1+x)^n \geq 1 + nx$.

(2) Show that for all $n \in \mathbb{N}$, $1 \leq n^{\frac{1}{n}} < (1 + \sqrt{n})^{\frac{2}{n}} \leq (1 + \frac{1}{\sqrt{n}})^2$.
    Hint: Take $x = \frac{1}{\sqrt{n}}$ in the inequality above.

(3) Prove that $(n^{\frac{1}{n}})_{n \in \mathbb{N}}$ is convergent and find its limit.

Exercise 2.19. Let $(a_n)_{n \in \mathbb{N}}$ be a sequence contained in the interval (a, b) (that is, for all $n \in \mathbb{N}$, $a < a_n < b$). If $(a_n)_{n \in \mathbb{N}}$ is convergent with limit L, then prove that $L \in [a, b]$.

Hint: Use Exercise 2.4 on page 54.

Give an example to show that L need not belong to (a, b).

Exercise 2.20. Let $(a_n)_{n\in\mathbb{N}}$ be a convergent sequence, and let $(b_n)_{n\in\mathbb{N}}$ satisfy $|b_n - a_n| < \frac{1}{n}$ for all $n \in \mathbb{N}$. Show that $(b_n)_{n\in\mathbb{N}}$ is also convergent. What is its limit?

Hint: Observe that $-\frac{1}{n} + a_n < b_n < a_n + \frac{1}{n}$ for all $n \in \mathbb{N}$.

Exercise 2.21. (∗) See Exercises 2.10 and 2.15. Prove that a bounded sequence $(a_n)_{n\in\mathbb{N}}$ is convergent if and only if

$$\liminf_{n\to\infty} a_n = \limsup_{n\to\infty} a_n.$$

Moreover, then $\lim_{n\to\infty} a_n = \liminf_{n\to\infty} a_n = \limsup_{n\to\infty} a_n.$

2.5 Subsequences

In this section, we prove an important result in Calculus, called the Bolzano–Weierstrass Theorem, which says that

Every bounded sequence has a convergent 'subsequence'.

We begin this section by defining what we mean by a subsequence of a sequence.

Definition 2.5 (Subsequence of a sequence). Let $(a_n)_{n\in\mathbb{N}}$ be a sequence, and suppose that $n_1 < n_2 < n_3 < \cdots$ is a strictly increasing sequence of natural numbers. Then $(a_{n_k})_{k\in\mathbb{N}}$ is called a *subsequence of* $(a_n)_{n\in\mathbb{N}}$. Thus the terms of the subsequence are $a_{n_1}, a_{n_2}, a_{n_3}, \cdots$.

Example 2.14. For example, the sequence $(a_{n^2})_{n\in\mathbb{N}} = \left(\frac{1}{n^2}\right)_{n\in\mathbb{N}}$

$$1, \frac{1}{4}, \frac{1}{9}, \frac{1}{16}, \frac{1}{25}, \cdots$$

is a subsequence of the sequence $(a_n)_{n\in\mathbb{N}} = \left(\frac{1}{n}\right)_{n\in\mathbb{N}}$. However, the sequence

$$\frac{1}{9}, \frac{1}{4}, \frac{1}{16}, \frac{1}{25}, \cdots$$

is *not* a subsequence of $\left(\frac{1}{n^2}\right)_{n\in\mathbb{N}}$, since terms of subsequence are not in the same order as the original sequence:

$$a_3 = \frac{1}{9}, \quad a_2 = \frac{1}{4},$$

and $3 > 2!$ But

$$\frac{1}{9}, \frac{1}{4}, \frac{1}{25}, \cdots$$

is a subsequence of

$$1, \frac{1}{4}, \frac{1}{9}, \frac{1}{4}, \frac{1}{25}, \cdots.$$

The sequences

$$((-1)^{2n})_{n\in\mathbb{N}} \quad \text{(that is, the constant sequence } 1, 1, 1, \cdots \text{) and}$$

$$((-1)^{2n-1})_{n\in\mathbb{N}} \quad \text{(that is, the constant sequence } -1, -1, -1, \cdots \text{)}$$

are both subsequences of $((-1)^n)_{n\in\mathbb{N}}$. Here are some more examples:

$n_1 < n_2 < n_3 < \cdots$	Subsequence of $(a_n)_{n\in\mathbb{N}}$		Subsequence of $(\frac{1}{n})_{n\in\mathbb{N}}$
$1 < 2 < 3 < \cdots$	$(a_n)_{n\in\mathbb{N}}$	$a_1, a_2, a_3, \cdots$	$1, \frac{1}{2}, \frac{1}{3}, \cdots$
$2 < 3 < 4 < \cdots$	$(a_{n+1})_{n\in\mathbb{N}}$	$a_2, a_3, a_4, \cdots$	$\frac{1}{2}, \frac{1}{3}, \frac{1}{4}, \cdots$
$2 < 4 < 6 < 8 < \cdots$	$(a_{2n})_{n\in\mathbb{N}}$	$a_2, a_4, a_6, a_8, \cdots$	$\frac{1}{2}, \frac{1}{4}, \frac{1}{6}, \frac{1}{8}, \cdots$
$2 < 4 < 8 < 16 < \cdots$	$(a_{2^n})_{n\in\mathbb{N}}$	$a_2, a_4, a_8, a_{16}, \cdots$	$\frac{1}{2}, \frac{1}{4}, \frac{1}{8}, \frac{1}{16}, \cdots$
$1 < 4 < 27 < 64 < \cdots$	$(a_{n^n})_{n\in\mathbb{N}}$	$a_1, a_4, a_{27}, a_{64}, \cdots$	$1, \frac{1}{4}, \frac{1}{27}, \frac{1}{64}, \cdots$
$2 < 3 < 5 < 7 < \cdots$	$(a_{p_n})_{n\in\mathbb{N}}$ (p_n denotes the nth prime)	$a_2, a_3, a_5, a_7, \cdots$	$\frac{1}{2}, \frac{1}{3}, \frac{1}{5}, \frac{1}{7}, \cdots$
$1 < 2 < 6 < 24 < \cdots$	$(a_{n!})_{n\in\mathbb{N}}$	$a_1, a_2, a_6, a_{24}, \cdots$	$1, \frac{1}{2}, \frac{1}{6}, \frac{1}{24}, \cdots$

$\Diamond$

Exercise 2.22. Is $\left(\dfrac{1}{n^4}\right)$ a subsequence of $\left(\dfrac{1}{n^2}\right)_{n\in\mathbb{N}}$? Is $\left(\dfrac{1}{n^3}\right)$ a subsequence of $\left(\dfrac{1}{n^2}\right)_{n\in\mathbb{N}}$?

Exercise 2.23. (∗) Beginning with 2 and 7, the sequence $2, 7, 1, 4, 7, 4, 2, 8, 2, 8, \ldots$ is constructed by multiplying successive pairs of its terms and adjoining the result as the next one or two members of the sequence depending on whether the product is a one- or two-digit number. Thus we start with 2 and 7, giving the product 14, and so the next two terms are $1, 4$. Proceeding in this manner, we get subsequent terms as follows:

$$\underline{2, 7}$$
$$2, 7, 1, 4$$
$$2, \underline{7, 1}, 4$$
$$2, 7, 1, 4, 7$$
$$2, 7, \underline{1, 4}, 7$$
$$2, 7, 1, 4, 7, 4$$
$$2, 7, 1, \underline{4, 7}, 4$$
$$2, 7, 1, 4, 7, 4, 2, 8$$
$$2, 7, 1, 4, \underline{7, 4}, 2, 8$$
$$2, 7, 1, 4, 7, 4, 2, 8, 2, 8$$
$$\cdots$$

Prove that this sequence has the constant subsequence $6, 6, 6, \ldots$.

Hint: Show that 6 appears an infinite number of times as follows. Since the terms $2, 8, 2, 8$ are adjacent, they give rise to the adjacent terms $1, 6, 1, 6$ at some point, which in turn give rise to the adjacent terms $6, 6, 6$ eventually, and so on. Proceeding in this way, find out if you get a loop containing the term 6.

Theorem 2.6. *A subsequence of a convergent sequence is convergent with the same limit.*

Thus if $(a_n)_{n\in\mathbb{N}}$ is a convergent sequence with limit L, then any subsequence of $(a_n)_{n\in\mathbb{N}}$ is also convergent with the limit L.

Proof. Let $(a_{n_k})_{k\in\mathbb{N}}$ be a subsequence of a convergent sequence $(a_n)_{n\in\mathbb{N}}$ with limit L. Given $\epsilon > 0$, let $N \in \mathbb{N}$ be such that for all $n > N$, $|a_n - L| < \epsilon$. Since the sequence $n_1 < n_2 < n_3 < \cdots$ of natural numbers is increasing, it follows that there exists a $K \in \mathbb{N}$ such that $n_K > N$. Then for all $k > K$, $n_k > n_K > N$. Hence for $k > K$, $|a_{n_k} - L| < \epsilon$, and so $(a_{n_k})_{k\in\mathbb{N}}$ is convergent with limit L. $\qquad\qquad\qquad\qquad\qquad\qquad\qquad\square$

Example 2.15. From Example 2.14 and the fact that $\lim\limits_{n\to\infty} \dfrac{1}{n} = 0$, it follows that

$$\lim_{n\to\infty} \frac{1}{n+1} = \lim_{n\to\infty} \frac{1}{2n} = \lim_{n\to\infty} \frac{1}{n^2} = \lim_{n\to\infty} \frac{1}{2^n} = \lim_{n\to\infty} \frac{1}{n^n} = \lim_{n\to\infty} \frac{1}{p_n} = \lim_{n\to\infty} \frac{1}{n!} = 0.$$

In the above p_n denotes the nth prime number.

Let us give a proof of the fact that $((-1)^n)_{n\in\mathbb{N}}$ is divergent based on Theorem 2.6. Suppose on the contrary, that $((-1)^n)_{n\in\mathbb{N}}$ is convergent with limit L. Then the terms with odd indices give the subsequence $-1, -1, -1, \cdots$, which is convergent with limit -1, and so (by uniqueness of limits!) $L = -1$. On the other hand, the terms with even indices give the subsequence $1, 1, 1, \cdots$, which is convergent with limit 1, and so $L = 1$. So we have arrived at the contradiction that $-1 = L = 1$. Hence $((-1)^n)_{n\in\mathbb{N}}$ is divergent. $\qquad\qquad\diamond$

Example 2.16 ('The harmonic series diverges'.). Consider $(s_n)_{n\in\mathbb{N}}$, where

$$s_n := 1 + \frac{1}{2} + \frac{1}{3} + \cdots + \frac{1}{n}, \quad n \in \mathbb{N}.$$

Suppose that $(s_n)_{n\in\mathbb{N}}$ is convergent with limit L. Then its subsequence $(s_{2n})_{n\in\mathbb{N}}$ would also be convergent with limit L, and so by the Algebra of Limits, the sequence $(s_{2n} - s_n)_{n\in\mathbb{N}}$ must converge to $L - L = 0$. But

$$s_{2n} - s_n = 1 + \frac{1}{2} + \frac{1}{3} + \cdots + \frac{1}{n} + \frac{1}{n+1} + \cdots + \frac{1}{2n} - \left(1 + \frac{1}{2} + \frac{1}{3} + \cdots + \frac{1}{n}\right)$$

$$= \frac{1}{n+1} + \cdots + \frac{1}{2n} > \underbrace{\frac{1}{2n} + \cdots + \frac{1}{2n}}_{n\ \text{times}} = n \cdot \frac{1}{2n} = \frac{1}{2}.$$

Hence $|(s_{2n} - s_n) - 0| = s_{2n} - s_n > \dfrac{1}{2}$, showing that it is *not* the case that

$$\lim_{n\to\infty} (s_{2n} - s_n) = 0,$$

a contradiction. So $\left(1 + \dfrac{1}{2} + \dfrac{1}{3} + \cdots + \dfrac{1}{n}\right)_{n\in\mathbb{N}}$ diverges. $\qquad\qquad\diamond$

Exercise 2.24. Recall the convergent sequence $(a_n)_{n\in\mathbb{N}}$ from Exercise 2.9 on page 59:

$$a_1 = 1 \text{ and } a_n = \frac{2n+1}{3n}a_{n-1} \text{ for } n \geq 2.$$

What is its limit?

Exercise 2.25. Determine if the following statements are true or false.

(1) Every subsequence of a convergent real sequence is convergent.

(2) Every subsequence of a divergent real sequence is divergent.

(3) Every subsequence of a bounded real sequence is bounded.

(4) Every subsequence of an unbounded real sequence is unbounded.

(5) Every subsequence of a monotone real sequence is monotone.

(6) Every subsequence of a nonmonotone real sequence is nonmonotone.

(7) If every subsequence of a real sequence converges, the sequence itself converges.

(8) If $(a_{2n})_{n\in\mathbb{N}}$ and $(a_{2n+1})_{n\in\mathbb{N}}$ both converge, then $(a_n)_{n\in\mathbb{N}}$ converges.

(9) If $(a_{2n})_{n\in\mathbb{N}}$ and $(a_{2n+1})_{n\in\mathbb{N}}$ both converge to the same limit, then $(a_n)_{n\in\mathbb{N}}$ converges.

Exercise 2.26. $(*)$ Show that if $(a_n)_{n\in\mathbb{N}}$ is a sequence that does not converge to L, then there exists an $\epsilon > 0$ and there exists a subsequence $(a_{n_k})_{k\in\mathbb{N}}$ of $(a_n)_{n\in\mathbb{N}}$ such that for all $k \in \mathbb{N}$, $|a_{n_k} - L| \geq \epsilon$.

Exercise 2.27. Consider the sequence $(a_n)_{n\in\mathbb{N}}$ given by

$$a_1 = \sqrt{2} \quad \text{and}$$
$$a_{n+1} = \sqrt{2 + a_n} \quad \text{for all } n \in \mathbb{N}.$$

Thus the first few terms of the sequence are $\sqrt{2}, \sqrt{2+\sqrt{2}}, \sqrt{2+\sqrt{2+\sqrt{2}}}, \cdots$.

(a) Show that for all $n \in \mathbb{N}$, $a_n \leq 2$. *Hint:* Use induction on n.

(b) Show that $(a_n)_{n\in\mathbb{N}}$ is increasing. *Hint:* Consider $a_{n+1}^2 - a_n^2$.

(c) Is $(a_n)_{n\in\mathbb{N}}$ convergent? If so, find its limit.

Theorem 2.7. *Every sequence has a monotone subsequence.*

Before giving the formal proof, we give an illustration of the idea behind this proof[4]. If $(a_n)_{n\in\mathbb{N}}$ is the given sequence, then imagine that there is an infinite chain of hotels along a line, where the nth hotel has height a_n, and at the horizon, there is a sea. A hotel is said to have the

---

[4] This illustrative analogy stems from [**B**]. The proof seems to go back to [**N3**]. See also [**N4**].

seaview property if it is higher than all hotels following it (so that from the roof of the hotel, one can view the sea). Now there are only two possibilities, as illustrated below.

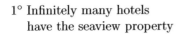

1° Infinitely many hotels
     have the seaview property

2° Finitely many hotels
     have the seaview property

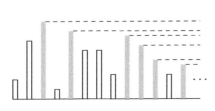
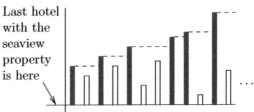

Last hotel
with the
seaview
property
is here

1° There are infinitely many hotels with the seaview property.	2° There are finitely many hotels with the seaview property.
Then by taking successively the heights of the hotels with the seaview property we get a *decreasing* subsequence.	Then after the last hotel with the seaview property, one can start with any hotel and then always find one that is at least as high, which is taken as the next hotel, and then finding yet another that is at least as high as that one, and so on. The heights of these hotels form an *increasing* subsequence.

Proof. Let $(a_n)_{n\in\mathbb{N}}$ be a sequence, and let

$$S = \{m \in \mathbb{N}: \text{ for all } n > m, \ a_n < a_m\}.$$

(This is the collection of indices of hotels with the seaview property.)
Then we have the following two cases.

1° S is infinite.
    Arrange the elements of S in increasing order: $n_1 < n_2 < n_3 < \cdots$. Then $(a_{n_k})_{k\in\mathbb{N}}$ is a *decreasing* subsequence of $(a_n)_{n\in\mathbb{N}}$.

2° S is finite.
    If S empty, then define $n_1 = 1$, and otherwise let $n_1 = \max S + 1$. Define inductively $n_{k+1} = \min\{m \in \mathbb{N} : m > n_k \text{ and } a_m \geq a_{n_k}\}$. ($n_{k+1}$ is the index of the first hotel blocking the view from the top of the n_kth hotel.) The minimum exists as $\{m \in \mathbb{N} : m > n_k \text{ and } a_m \geq a_{n_k}\}$ is a nonempty subset of $\mathbb{N}$. Indeed otherwise if it were empty, then $n_k \in S$, and this is not possible if S was empty, and also impossible if S was not empty, since $n_k > \max S$.) Then $(a_{n_k})_{k\in\mathbb{N}}$ is an *increasing* subsequence of $(a_n)_{n\in\mathbb{N}}$.

Thus every sequence $(a_n)_{n\in\mathbb{N}}$ has a monotone subsequence.   □

An important consequence of the above theorem is the following result.

Theorem 2.8 (Bolzano–Weierstrass Theorem). *Every bounded sequence has a convergent subsequence.*

Proof. Let $(a_n)_{n\in\mathbb{N}}$ be a bounded sequence. Then there exists an $M > 0$ such that for all $n \in \mathbb{N}$, $|a_n| \le M$. From Theorem 2.7 above, it follows that the sequence $(a_n)_{n\in\mathbb{N}}$ has a monotone subsequence, say $(a_{n_k})_{k\in\mathbb{N}}$. Then clearly for all $k \in \mathbb{N}$, $|a_{n_k}| \le M$, and so the sequence $(a_{n_k})_{k\in\mathbb{N}}$ is also bounded. Since $(a_{n_k})_{k\in\mathbb{N}}$ is monotone and bounded, it follows from Theorem 2.3 that it is convergent. $\qquad\qquad\qquad\qquad\qquad\qquad\qquad\qquad\qquad\qquad\qquad\qquad\qquad\quad$ □

Example 2.17. ('Compactness' of $[a, b]$.) Consider any sequence $(a_n)_{n\in\mathbb{N}}$ in $[a, b]$, that is,

$$\text{for all } n \in \mathbb{N}, \ a_n \in [a, b], \ \text{ or equivalently, } a \le a_n \le b.$$

Then $(a_n)_{n\in\mathbb{N}}$ is bounded, and so it has a convergent subsequence, say $(a_{n_k})_{k\in\mathbb{N}}$. Then for all $k \in \mathbb{N}$, $a \le a_{n_k} \le b$. Consequently,

$$a \le \lim_{k\to\infty} a_{n_k} \le b$$

as well (using Exercise 2.15). Thus we obtain the following conclusion:

Every sequence in $[a, b]$ has a convergent subsequence,
and the limit of this subsequence belongs to $[a, b]$. $\qquad\qquad\qquad$ ◇

Example 2.18. Consider the sequence $(a_n)_{n\in\mathbb{N}}$ of fractional parts of integral multiples of $\sqrt{2}$, defined by $a_n := \{n\sqrt{2}\} := n\sqrt{2} - \lfloor n\sqrt{2}\rfloor$, for $n \in \mathbb{N}$. The terms of the sequence $(a_n)_{n\in\mathbb{N}}$ are as follows:

$$\sqrt{2} = 1.414213 \ldots \qquad a_1 = 0.414213 \ldots$$
$$2\sqrt{2} = 2.828427 \ldots \qquad a_2 = 0.828427 \ldots$$
$$3\sqrt{2} = 4.242640 \ldots \qquad a_3 = 0.242640 \ldots$$
$$4\sqrt{2} = 5.656854 \ldots \qquad a_4 = 0.656854 \ldots$$
$$5\sqrt{2} = 7.071067 \ldots \qquad a_5 = 0.071067 \ldots$$
$$\ldots$$

The sequence $(a_n)_{n\in\mathbb{N}}$ is bounded: indeed, $0 \le a_n < 1$. So by the Bolzano–Weierstrass Theorem it has a convergent subsequence[5]. $\qquad\qquad\qquad\qquad\qquad\qquad\qquad\qquad$ ◇

Exercise 2.28. Does the sequence $(\sin n)_{n\in\mathbb{N}}$ have a convergent subsequence? What about the sequence $(n)_{n\in\mathbb{N}}$?

---

[5] In fact, it can be shown that these fractional parts a_n are 'dense' in $(0, 1)$. Thus given any number $L \in (0, 1)$, there exists a subsequence of the sequence $(a_n)_{n\in\mathbb{N}}$ above that converges to L.

Exercise 2.29. ($*$) Consider the bounded divergent sequence $((-1)^n)_{n\in\mathbb{N}}$. Note that there exist two subsequences $(-1, -1, -1, \ldots$ and $1, 1, 1, \ldots)$ that have distinct limits $(-1 \neq 1)$. In this exercise, we show that this is a general phenomenon. Show that if $(a_n)_{n\in\mathbb{N}}$ is bounded and divergent, then it has two subsequences that converge to distinct limits.

Hint: Use the Bolzano–Weierstrass theorem twice, and also Exercise 2.26.

2.6 Cauchy sequences and completeness of $\mathbb{R}$

Another manifestation of the Least Upper bound Property of $\mathbb{R}$ is the 'completeness of $\mathbb{R}$', which says that Cauchy sequences in $\mathbb{R}$ are convergent. Let us begin with the notion of a Cauchy sequence.

Definition 2.6 (Cauchy sequence). A sequence $(a_n)_{n\in\mathbb{N}}$ of real numbers is said to be a *Cauchy sequence* if for every $\epsilon > 0$, there exists a $N \in \mathbb{N}$ such that whenever $m, n > N$, $|a_n - a_m| < \epsilon$.

Roughly speaking, we can make the terms of the sequence arbitrarily close to each other provided we go far enough in the sequence.

Example 2.19. The sequence $\left(\dfrac{1}{n}\right)_{n\in\mathbb{N}}$ is Cauchy. Indeed, we have

$$\left|\frac{1}{n} - \frac{1}{m}\right| \leq \frac{1}{n} + \frac{1}{m} < \frac{1}{N} + \frac{1}{N} = \frac{2}{N} \quad \text{whenever } n, m > N.$$

Thus given $\epsilon > 0$, we can choose $N \in \mathbb{N}$ larger than $\dfrac{2}{\epsilon}$, so that we then have

$$\left|\frac{1}{n} - \frac{1}{m}\right| < \frac{2}{N} < \epsilon$$

for all $n, m > N$. Consequently, $\left(\dfrac{1}{n}\right)_{n\in\mathbb{N}}$ is Cauchy.   ◇

Example 2.20. $(n)_{n\in\mathbb{N}}$ is *not* Cauchy.

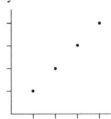

Indeed, if $n \neq m$, then $|n - m| \geq 1$. So for a positive $\epsilon < 1$, for example $\epsilon = 1/2$, there does not exist an $N \in \mathbb{N}$ such that for all $n, m > N$, we have that

$$|a_n - a_m| = |n - m| < \epsilon = 1/2.$$ ◇

Exercise 2.30. Show that if $(a_n)_{n\in\mathbb{N}}$ is a Cauchy sequence, then $(a_{n+1} - a_n)_{n\in\mathbb{N}}$ converges to 0.

Example 2.21 (Cauchyness is *not* the same as *consecutive* terms getting closer). This example shows that for a sequence $(a_n)_{n\in\mathbb{N}}$ to be Cauchy, it is not enough that $(a_{n+1} - a_n)_{n\in\mathbb{N}}$ converges to 0.

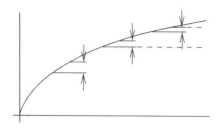

We will show that $(\sqrt{n})_{n\in\mathbb{N}}$ is not Cauchy. Suppose on the contrary that it is. Let us take $\epsilon := 1/2 > 0$, and let $N \in \mathbb{N}$ be such that for all $n, m > N$, $|\sqrt{n} - \sqrt{m}| < \epsilon = 1/2$. Take $n = 4N^2$ and $m = N^2$. Then

$$|a_n - a_m| = |\sqrt{4N^2} - \sqrt{N^2}| = |2N - N| = N \geq 1,$$

a contradiction. On the other hand, the consecutive terms of $(\sqrt{n})_{n\in\mathbb{N}}$ do get arbitrarily close:

$$a_{n+1} - a_n = \sqrt{n+1} - \sqrt{n} = \frac{1}{\sqrt{n+1} + \sqrt{n}} \xrightarrow{n\to\infty} 0. \qquad\qquad \diamondsuit$$

The next result says that the property of being a Cauchy sequence (henceforth referred to as 'Cauchyness') is a *necessary* condition for convergence.

Lemma 2.9. *Every convergent sequence is Cauchy.*

Proof. Let $(a_n)_{n\in\mathbb{N}}$ be a sequence of real numbers that converges to L. Let $\epsilon > 0$. (We want to find N, which guarantees for $n, m > N$ that $|a_n - a_m| < \epsilon$. But we *do* know that the terms a_n, a_m can both be made close to L if n, m are large enough. So we introduce L artificially: $|a_n - a_m| = |a_n - L + L - a_m|$ and use the triangle inequality to complete the argument. The details are given below.)

Then there exists an $N \in \mathbb{N}$ such that for $n > N$, we have $|a_n - L| < \dfrac{\epsilon}{2}$.

Thus for $n, m > N$, we have

$$|a_n - a_m| = |a_n - L + L - a_m| \leq |a_n - L| + |a_m - L| < \frac{\epsilon}{2} + \frac{\epsilon}{2} = \epsilon.$$

So the sequence $(a_n)_{n\in\mathbb{N}}$ is a Cauchy sequence. $\qquad\qquad\square$

We have so far seen that in $\mathbb{R}$,

$$\{\text{ convergent sequences }\} \subset \{\text{ Cauchy sequences }\}.$$

This raises the tempting question of whether the reverse inclusion is true too:

$$\{\text{ convergent sequences }\} \overset{?}{\supset} \{\text{ Cauchy sequences }\}.$$

Now we will prove the remarkable fact that in $\mathbb{R}$, Cauchyness turns out to be also a *sufficient* condition for the sequence to be convergent. In other words, in $\mathbb{R}$, every Cauchy sequence is convergent. This is a very useful fact since, in order to prove that a sequence is convergent using the definition, we would need to guess what the limit is. In contrast, checking whether or not a sequence is Cauchy needs only knowledge of the terms of the sequence, and no guesswork regarding the limit is needed. So this is a powerful technique for proving existence results (for example, in the Theory of Differential Equations).

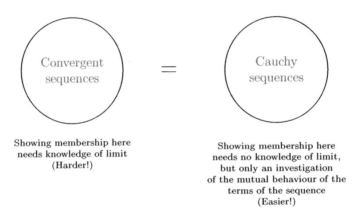

Theorem 2.10. *Every Cauchy sequence in $\mathbb{R}$ is convergent.*

Proof. There are three main steps. First, we show that every Cauchy sequence is bounded. Then we use the Bolzano–Weierstrass theorem to conclude that it must have a convergent subsequence. Finally we show that a Cauchy sequence having a convergent subsequence must itself be convergent. Suppose that $(a_n)_{n \in \mathbb{N}}$ is a Cauchy sequence.

Step 1. Choose a positive ϵ, say $\epsilon = 1$. Then there exists an $N \in \mathbb{N}$ such that for all n, $m > N$, $|a_n - a_m| < \epsilon = 1$. In particular, with $m = N + 1 > N$, and $n > N$, $|a_n - a_{N+1}| < 1$. Hence by the triangle inequality, for all $n > N$,

$$|a_n| = |a_n - a_{N+1} + a_{N+1}| \leq |a_n - a_{N+1}| + |a_{N+1}| < 1 + |a_{N+1}|.$$

On the other hand, for $n \leq N$, $|a_n| \leq \max\{|a_1|, \ldots, |a_N|, |a_{N+1}| + 1\} =: M$. Consequently, $|a_n| \leq M$ $(n \in \mathbb{N})$, that is, the sequence $(a_n)_{n \in \mathbb{N}}$ is bounded.

Step 2. By the Bolzano–Weierstrass Theorem, $(a_n)_{n \in \mathbb{N}}$ has a convergent subsequence $(a_{n_k})_{k \in \mathbb{N}}$ that is convergent, to L, say.

Step 3. Finally, we show that $(a_n)_{n \in \mathbb{N}}$ is also convergent with limit L. Let $\epsilon > 0$. Then there exists an $N \in \mathbb{N}$ such that for all $n, m > N$,

$$|a_n - a_m| < \frac{\epsilon}{2}. \tag{2.8}$$

Also, since $(a_{n_k})_{k \in \mathbb{N}}$ converges to L, we can find an $n_K > N$ such that $|a_{n_K} - L| < \dfrac{\epsilon}{2}$.

Taking $m = n_K$ in (2.8), we have for all $n > N$ that

$$|a_n - L| = |a_n - a_{n_K} + a_{n_K} - L| \leq |a_n - a_{n_K}| + |a_{n_K} - L| < \frac{\epsilon}{2} + \frac{\epsilon}{2} = \epsilon.$$

Thus $(a_n)_{n \in \mathbb{N}}$ is also convergent with limit L, and this completes the proof. □

Owing to the property that

$$\{\text{ Cauchy sequences in } \mathbb{R} \,\} = \{\text{ Convergent sequences in } \mathbb{R} \,\},$$

we say that $\mathbb{R}$ *is complete*. However, $\mathbb{Q}$ is not complete, since

$$\{\text{ Cauchy sequences in } \mathbb{Q} \,\} \overset{\supset}{\not\subset} \{\text{ Convergent sequences in } \mathbb{Q} \,\},$$

and we show this in the following example.

Example 2.22 ($\mathbb{Q}$ is not complete). Consider the sequence $(a_n)_{n \in \mathbb{N}}$ in $\mathbb{Q}$ defined by $a_1 = 3/2$, and for $n > 1$, recursively by

$$a_{n+1} = \frac{4 + 3a_n}{3 + 2a_n}.$$

Then it can be shown that $(a_n)_{n \in \mathbb{N}}$ is bounded below by $\sqrt{2}$ by induction, and that $(a_n)_{n \in \mathbb{N}}$ is monotone decreasing.

(A) $a_n \geq \sqrt{2}$ for all n.

If $n = 1$, then $a_1 = \frac{3}{2} \geq \sqrt{2}$ (as $\frac{9}{4} \geq 2$). If $a_n \geq \sqrt{2}$ for some n, then

$$a_{n+1}^2 - 2 = \frac{(4 + 3a_n)^2}{(3 + 2a_n)^2} - 2$$

$$= \frac{16 + 24a_n + 9a_n^2 - 18 - 24a_n - 8a_n^2}{(3 + 2a_n)^2} = \frac{a_n^2 - 2}{(3 + 2a_n)^2} \geq 0.$$

So this gives, since $a_{n+1} \geq 0$, that $a_{n+1} \geq \sqrt{2}$, and the claim follows by induction.

(B) $a_n \geq a_{n+1}$ for all n. We have

$$a_n - a_{n+1} = a_n - \frac{4 + 3a_n}{3 + 2a_n} = \frac{3a_n + 2a_n^2 - 4 - 3a_n}{3 + 2a_n} = \frac{2(a_n^2 - 2)}{3 + 2a_n} \geq 0,$$

where the last inequality follows from part (A).

So this sequence is convergent in $\mathbb{R}$. Hence it is also Cauchy in $\mathbb{R}$. But as each term a_n is a rational number for all $n \in \mathbb{N}$, it follows that $(a_n)_{n \in \mathbb{N}}$ is also Cauchy in $\mathbb{Q}$. However, we now show that $(a_n)_{n \in \mathbb{N}}$ is not convergent in $\mathbb{Q}$. Suppose, on the contrary, that $(a_n)_{n \in \mathbb{N}}$ converges to $L \in \mathbb{Q}$. Then from the recurrence relation, we obtain using the Algebra of Limits that

$$L = \frac{4 + 3L}{3 + 2L},$$

and so $L^2 = 2$. As L must be positive (the sequence is bounded below by $\sqrt{2}$), it follows that $L = \sqrt{2}$. But this is a contradiction, since we know that there is no rational number whose square is 2.

(Alternately, consider the real number c with the decimal expansion

$$c = 0.101001000100001\cdots.$$

This number c is irrational because it has a nonterminating and nonrepeating decimal expansion. We will prove this later when we treat decimal expansions in the chapter on series; see Exercise 6.8 on page 304 and the Appendix on page 335. If we consider the sequence of rational numbers obtained by truncation, namely

$$0.1$$

$$0.101$$

$$0.101001$$

$$0.1010010001$$

$$0.101001000100001$$

$$\cdots$$

then this sequence converges with limit c.)   ◇

Exercise 2.31. Which of the following statements is/are true?

☐ (A) A Cauchy sequence in $\mathbb{R}$ is always convergent in $\mathbb{R}$.

☐ (B) A convergent sequence in $\mathbb{R}$ is always Cauchy in $\mathbb{R}$.

☐ (C) A Cauchy sequence in $\mathbb{R}$ is always bounded.

☐ (D) A monotone and bounded sequence in $\mathbb{R}$ is always Cauchy in $\mathbb{R}$.

Exercise 2.32. Which of the following is always true for a real sequence $(a_n)_{n\in\mathbb{N}}$?

(1) If the sequence $(a_n^2)_{n\in\mathbb{N}}$ is Cauchy, then $(a_n)_{n\in\mathbb{N}}$ is Cauchy.

(2) If the sequence $(a_n)_{n\in\mathbb{N}}$ is Cauchy, then $(a_n^2)_{n\in\mathbb{N}}$ is Cauchy.

2.7 (∗) Pointwise versus uniform convergence

Let $I \subset \mathbb{R}$ be an interval, and let

$$f_n : I \to \mathbb{R} \quad (n \in \mathbb{N}),$$

$$f : I \to \mathbb{R}$$

be functions. Thus we have a sequence $n \mapsto f_n$ of *functions*, the first term is the function f_1, the second term is the function f_2, the third term is the function f_3, and so on. Then there are two natural notions of convergence of the sequence $(f_n)_{n\in\mathbb{N}}$ of *functions* to the *function f*:

(1) Pointwise convergence and

(2) uniform convergence.

If we fix an $x \in I$, then $f_n(x)$ is a number for each $n \in \mathbb{N}$, and so $(f_n(x))_{n\in\mathbb{N}}$ is a sequence of real *numbers*. If this sequence $(f_n(x))_{n\in\mathbb{N}}$ of numbers converges to $f(x)$, for each $x \in I$, then we say that the convergence is 'pointwise'.

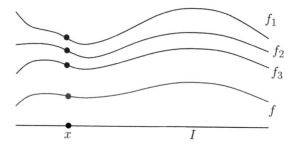

Definition 2.7 (Pointwise convergence). Let I be any set and $f, f_n : I \to \mathbb{R}$ $(n \in \mathbb{N})$ be functions. The sequence $(f_n)_{n\in\mathbb{N}}$ is said to *converge pointwise to f* if

$$\forall x \in I, \ \forall \epsilon > 0, \ \exists N \in \mathbb{N} \text{ such that } \forall n > N, \ |f_n(x) - f(x)| < \epsilon.$$

Example 2.23. Let $I := \mathbb{R}$, and for $x \in I = \mathbb{R}$, let

$$f_n(x) := \frac{x}{n} \quad (n \in \mathbb{N}),$$

$$f(x) := 0.$$

Figure 2.1 shows the graphs of the functions.

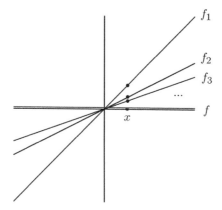

Figure 2.1 The sequence of functions $(f_n)_{n\in\mathbb{N}}$ converges pointwise to f.

It is clear that if we fix any $x \in \mathbb{R}$, then

$$\lim_{n \to \infty} f_n(x) = \lim_{n \to \infty} \frac{x}{n} = x \lim_{n \to \infty} \frac{1}{n} = x \cdot 0 = 0 = f(x).$$

So $(f_n)_{n \in \mathbb{N}}$ converges pointwise to 0. Let us have a closer look at this. Fix an $x \in \mathbb{R}$. Let $\epsilon > 0$ be given. Take $N \in \mathbb{N}$ such that $N > \frac{|x|+1}{\epsilon}$ (again this is obtained by working backwards!). Then for $n > N$,

$$|f_n(x) - f(x)| = \left| \frac{x}{n} - 0 \right| = \frac{|x|}{n} \le \frac{|x|}{N} < \frac{|x|\epsilon}{|x|+1} < \epsilon.$$

Note that the N we required to guarantee that

$$n > N \quad \Rightarrow \quad |f_n(x) - f(x)| < \epsilon$$

depends on the x fixed at the outset. (An $N < \frac{|x|}{\epsilon}$ won't do here!) In Calculus, it is convenient to distinguish the case when this

'dependence of N on which x we take'

is **absent**. We call this 'uniform' convergence.

Definition 2.8 (Uniform convergence). Let I be any set and $f, f_n : I \to \mathbb{R}$ ($n \in \mathbb{N}$) be functions. The sequence $(f_n)_{n \in \mathbb{N}}$ is said to *converge uniformly to f* if

$$\forall \epsilon > 0, \ \exists N \in \mathbb{N} \text{ such that } \forall n > N, \ \forall x \in I, \ |f_n(x) - f(x)| < \epsilon.$$

So the *same N* works for all x! Pictorially, this means the following. Consider the graph of the limit function f. Given any $\epsilon > 0$, we can translate the graph of f upwards by ϵ, and downwards by ϵ, in order to obtain the strip between the dotted lines shown in below. This strip precisely represents the region where the graph of a function f_n lies if it is to satisfy

$$\forall x \in I, \ |f_n(x) - f(x)| < \epsilon.$$

(Why?) If the convergence is uniform, then given any $\epsilon > 0$, there should exist an N such that the graphs of $f_{N+1}, f_{N+2}, f_{N+3}, \cdots$ all lie in this strip.

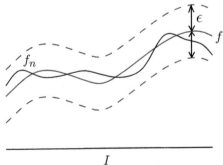

Example 2.24. Consider again the sequence of functions from Example 2.23. From the picture below, it is visibly clear that $(f_n)_{n \in \mathbb{N}}$ does not converge uniformly to f. Indeed, whatever width of strip we look at around the graph of f, and no matter which n we take, it is not the case that the graph of f_n lies entirely inside the strip—some portion of the graph of f_n always 'sticks out'.

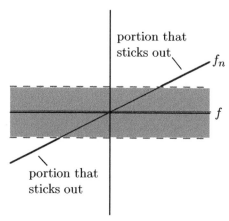

portion that
sticks out

f_n

f

portion that
sticks out

Here is a rigorous proof. Suppose that $(f_n)_{n \in \mathbb{N}}$ converges uniformly to f. Let $\epsilon = 1 > 0$. Then there exists an $N \in \mathbb{N}$ such that for all $x \in \mathbb{R}$, and all $n > N$, $|f_n(x) - f(x)| < 1$. Take $x = 2N + 2$. Then the above gives us that for all $n > N$,

$$\left| \frac{2N + 2}{n} - 0 \right| < 1.$$

In particular, for $n = N + 1$,

$$\frac{2N + 2}{N + 1} = \frac{2(N + 1)}{N + 1} = 2 < 1,$$

a contradiction! $\diamondsuit$

Clearly, if $(f_n)_{n \in \mathbb{N}}$ converges to f uniformly, then it converges pointwise to f too. (Indeed, if for every $\epsilon > 0$ there exists an $N \in \mathbb{N}$ such that for all $n > N$ and for *all* $x \in I$, $|f_n(x) - f(x)| < \epsilon$, and we take *any particular fixed* $x_* \in I$, then also, we have that for every $\epsilon > 0$ there exists an $N \in \mathbb{N}$ such that for all $n > N$, $|f_n(x_*) - f(x_*)| < \epsilon$: in other words, for this $x_* \in I$,

$$\lim_{n \to \infty} f_n(x_*) = f(x_*).$$

But the choice of $x_* \in I$ was arbitrary. So

$$\forall x \in I, \quad \lim_{n \to \infty} f_n(x) = f(x).$$

Hence $(f_n)_{n \in \mathbb{N}}$ converges pointwise to f. So here is an 'algorithm' to check uniform convergence:

(1) First find for each $x \in I$, $\lim\limits_{n \to \infty} f_n(x)$ and call the limit $f(x)$.

(2) Find a 'uniform bound' on $|f_n(x) - f(x)|$, namely

$$\sup_{x \in I} |f_n(x) - f(x)| =: M_n.$$

This delivers to us the sequence $(M_n)_{n \in \mathbb{N}}$ of numbers, and to check uniform convergence, it suffices to check that

$$\lim_{n \to \infty} M_n = 0.$$

We will justify this last claim in Exercise 2.33.

Example 2.25. Let $I := \mathbb{R}$, and for $x \in I = \mathbb{R}$, let

$$f_n(x) = \frac{\sin(nx)}{n}, \quad n \in \mathbb{N},$$

$$f(x) = 0.$$

Clearly for each $x \in \mathbb{R}$ we have

$$-\frac{1}{n} \le \frac{\sin(nx)}{n} \le \frac{1}{n}, \quad n \in \mathbb{N},$$

and so by the Sandwich Theorem, $\lim_{n \to \infty} f_n(x) = 0$.

Thus if $f : \mathbb{R} \to \mathbb{R}$ is the constant function equal to 0 everywhere, then $(f_n)_{n \in \mathbb{N}}$ converges pointwise to f.

Is the convergence uniform? We suspect that the answer is 'Yes', based on the Figure 2.2: if we look at a strip of an arbitrarily small width around the graph of the zero function f, it is clear that eventually the graphs of f_n do lie in this strip.

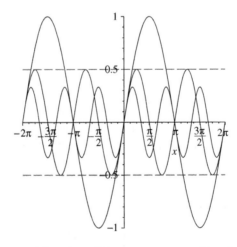

Figure 2.2   The sequence of functions $(f_n)_{n \in \mathbb{N}}$ converges uniformly to f: with $\epsilon = 1/2$, we see that the graphs of $f_3, f_4, \cdots$ all lie in the strip of width ϵ about the graph of the zero function f.

In fact, $M_n = \sup_{x \in \mathbb{R}} |f_n(x) - f(x)| = \sup_{x \in \mathbb{R}} \dfrac{|\sin(nx)|}{n} = \dfrac{1}{n}$, and as

$$\lim_{n \to \infty} M_n = \lim_{n \to \infty} \frac{1}{n} = 0,$$

it follows that $(f_n)_{n \in \mathbb{N}}$ converges uniformly to f.   ◇

Exercise 2.33. (∗)

(1) Suppose that I is an interval and $f_n : I \to \mathbb{R}$ ($n \in \mathbb{N}$) is a sequence that is pointwise convergent to $f : I \to \mathbb{R}$. Let the numbers $a_n := \sup\{|f_n(x) - f(x)| : x \in I\}$ ($n \in \mathbb{N}$) all exist. Prove that $(f_n)_{n \in \mathbb{N}}$ converges uniformly to f if and only if $\lim_{n \to \infty} a_n = 0$.

(2) Let $f_n : (0, \infty) \to \mathbb{R}$ be given by $f_n(x) = xe^{-nx}$ for $x \in (0, \infty)$ and $n \in \mathbb{N}$. Show that the sequence $(f_n)_{n \in \mathbb{N}}$ converges uniformly on $(0, \infty)$.

Exercise 2.34. Let $f_n : [0, 1] \to \mathbb{R}$ be defined by $f_n(x) = \dfrac{x}{1 + nx}$ for $x \in [0, 1]$. Does $(f_n)_{n \in \mathbb{N}}$ converge uniformly on $[0, 1]$?

Exercise 2.35. For $n \in \mathbb{N}$, let $f_n : (0, 1) \to \mathbb{R}$ be defined by $f_n(x) = x^n$, $x \in (0, 1)$.

(1) Does the sequence $(f_n)_{n \in \mathbb{N}}$ converge pointwise to some function?

(2) Is the convergence uniform?

(3) Sketch the graphs of the first few terms of the sequence, and explain visually your answer to part (2) above.

Pointwise versus uniform convergence. In order to better highlight the difference between pointwise and uniform convergence, let us consider the following two statements, where $I \subset \mathbb{R}$ is an interval, $f_n : I \to \mathbb{R}$ ($n \in \mathbb{N}$) and $f : I \to \mathbb{R}$ are functions.

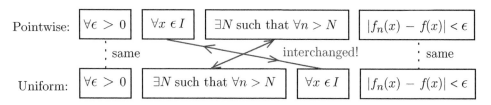

Can you spot the difference? What has changed is the order of

$$\boxed{\forall x \in X} \text{ and } \boxed{\exists N \in \mathbb{N} \text{ such that } \forall n > N.}$$

Order of the phrases 'for every' and 'there exists' (called quantifiers) matters in mathematical statements. This seemingly small alteration of interchanging the order of quantifiers makes a world of difference. Indeed, even in everyday language, the two statements:

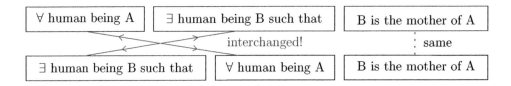

mean totally different things! In the latter, there is a person who is the mother to all human beings, a statement that is obviously false. The former statement is true, since it asserts for every person A we take, there exists (depending on *which* person A we have chosen) another person B who is the mother of A.

This is the same sort of a difference between the uniform convergence requirement, namely:

$$\forall \epsilon > 0, \ \exists N \in \mathbb{N} \text{ such that } \forall n > N, \ \forall x \in I, \ |f_n(x) - f(x)| < \epsilon.$$

and the pointwise convergence requirement, namely

$$\forall \epsilon > 0, \ \forall x \in I, \ \exists N \in \mathbb{N} \text{ such that } \forall n > N, \ |f_n(x) - f(x)| < \epsilon.$$

In the former, the same N works for all $x \in I$, while in the latter, the N might depend on the x in question.

Why bother with uniform convergence?

Uniform convergence often implies that the limit function inherits the 'nice' properties possessed by the terms of the sequence. This is not guaranteed to happen if one has mere *pointwise* convergence. For instance, we will see later on that if a sequence $(f_n)_{n \in \mathbb{N}}$ of continuous functions f_n ($n \in \mathbb{N}$) converges uniformly to a function f, then f is also continuous; see Proposition 3.9. Morally, the reason nice things can happen with uniform convergence is that we can exchange two 'limiting processes', which is *not* always allowed when one just has pointwise convergence. The following exercises demonstrate the precariousness of exchanging limiting processes arbitrarily.

Exercise 2.36. (∗) Let $f_n : \mathbb{R} \to \mathbb{R}$ be given by $f_n(x) = 1 - \dfrac{1}{(1 + x^2)^n}$ for $x \in \mathbb{R}, n \in \mathbb{N}$.

Show that the sequence $(f_n)_{n \in \mathbb{N}}$ of continuous functions converges pointwise to the function

$$f(x) = \begin{cases} 1 & \text{if } x \neq 0, \\ 0 & \text{if } x = 0, \end{cases}$$

which is discontinuous at 0.

Exercise 2.37 (∗)

(1) For $n \in \mathbb{N}$ and $m \in \mathbb{N}$, set $a_{m,n} = \dfrac{m}{m+n}$.

Show that for each fixed n, $\lim\limits_{m \to \infty} a_{m,n} = 1$, while for each fixed m, $\lim\limits_{n \to \infty} a_{m,n} = 0$.

Is $\lim\limits_{m \to \infty} \lim\limits_{n \to \infty} a_{m,n} = \lim\limits_{n \to \infty} \lim\limits_{m \to \infty} a_{m,n}$?

(2) Let $f_n : \mathbb{R} \to \mathbb{R}$ be defined by $f_n(x) = \dfrac{\sin(nx)}{\sqrt{n}}$, $x \in \mathbb{R}$, $n \in \mathbb{N}$.

Show that $(f_n)_{n \in \mathbb{N}}$ converges pointwise to the zero function f. However, show that $(f_n')_{n \in \mathbb{N}}$ does not converge pointwise to (the zero function) f'.

(3) Let $f_n : [0, 1] \to \mathbb{R}$ $(n \in \mathbb{N})$ be defined by $f_n(x) = nx(1 - x^2)^n$ $(x \in [0, 1])$. Show that $(f_n)_{n \in \mathbb{N}}$ converges pointwise to the zero function f. However, show that

$$\lim_{n \to \infty} \int_0^1 f_n(x)dx = \frac{1}{2} \neq 0 = \int_0^1 \lim_{n \to \infty} f_n(x)dx.$$

Remark 2.1. Besides the preservation of continuity under uniform convergence, one also has the following results associated with uniform convergence, which we will not establish in this book (but instead we refer the interested student to Rudin's book [**R**] for details).

Proposition 2.11. *If $f_n : [a, b] \to \mathbb{R}$ $(n \in \mathbb{N})$ is a sequence of Riemann-integrable functions on $[a, b]$ which converges uniformly to $f : [a, b] \to \mathbb{R}$, then f is also Riemann-integrable on $[a, b]$, and moreover*

$$\int_a^b f(x)dx = \lim_{n \to \infty} \int_a^b f_n(x)dx.$$

Proposition 2.12. *Let $f_n : (a, b) \to \mathbb{R}$ $(n \in \mathbb{N})$ be a sequence of differentiable functions on (a, b), such that there exists a point $c \in (a, b)$ for which $(f_n(c))_{n \in \mathbb{N}}$ converges. If the sequence $(f_n')_{n \in \mathbb{N}}$ converges uniformly to g on (a, b), then $(f_n)_{n \in \mathbb{N}}$ converges uniformly to a differentiable function f on (a, b), and moreover,*

$$f'(x) = g(x) \ \text{ for all } x \in (a, b).$$

Notes

Exercise 2.23 is from [**L**, 1.1.6]. The illustration of the proof of Theorem 2.7 is based on [**B**].

3

Continuity

Let I be an interval in $\mathbb{R}$. So I is a set of the form (a, b) or $[a, b]$ or $(-\infty, b)$, etc. Among all possible functions $f : I \rightarrow \mathbb{R}$, there is a 'nice' class of functions, namely ones that are *continuous on I*.

What's so nice about continuous functions? Continuous functions have properties that make them easy to work with in Calculus. For example, we will see that continuous functions possess two important properties, given by the Intermediate Value Theorem, and the Extreme Value Theorem. We will learn the statements and proofs of these in the course of this chapter. Functions that aren't continuous may fail to possess these properties.

Many bizarre functions make appearances in Calculus, and in order to avoid falling into pitfalls with simplistic thinking, we need definitions and assumptions of theorems to be stated carefully and clearly.

3.1 Definition of continuity

In everyday speech, a 'continuous' process is one that proceeds without gaps of interruptions or sudden changes.

What does it mean for a function $f : \mathbb{R} \rightarrow \mathbb{R}$ to be continuous? Roughly, f is said to be continuous on I if f has 'no breaks' at any point of I. If a break does occur in f, then this break will occur at some point of I. So we realise that in order to define continuity, we need to define what is meant by the notion of 'f being continuous *at a point* $c \in I$'.

Thus (based on this visual view of continuity), we first try to give the formal definition of the continuity of a function *at a point* below. Next, if a function is continuous at *each* point, then it will be called continuous. If a function has a break at a point c, then even if points x are close to c, the points $f(x)$ do not get close to $f(c)$. See Figure 3.1.

So 'no break in f at c' should mean that $f(x)$ stays close to $f(c)$ whenever x is close to c. This motivates the following definition of continuity, which guarantees that if a function is continuous at a point c, then we can make $f(x)$ as close as we like to $f(c)$, by choosing x sufficiently close to c. See Figure 3.2.

The How and Why of One Variable Calculus, First Edition. Amol Sasane.
© 2015 John Wiley & Sons, Ltd. Published 2015 by John Wiley & Sons, Ltd.

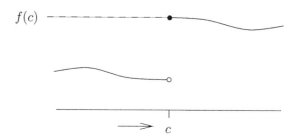

$f(c)$

Figure 3.1   A function with a break at c. If x lies to the left of c, then $f(x)$ is not close to $f(c)$, no matter how close x comes to c.

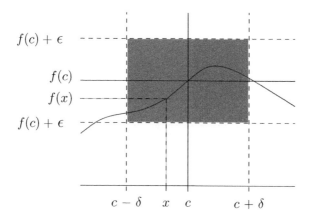

$f(c) + \epsilon$

$f(c)$

$f(x)$

$f(c) + \epsilon$

$c - \delta \quad x \quad c \qquad c + \delta$

Figure 3.2   The definition of continuity of a function at point c. If the function is continuous at c, then given any $\epsilon > 0$ (which determines a strip around the line $y = f(c)$ of width 2ϵ), there exists a $\delta > 0$ (which determines an interval $(c - \delta, c + \delta)$ of width 2δ around the point c) such that whenever x lies in this interval (so that x satisfies $c - \delta < x < c + \delta$, that is, $|x - c| < \delta$), then $f(x)$ satisfies $f(c) - \epsilon < f(x) < f(c) + \epsilon$, that is, $|f(x) - f(c)| < \epsilon$.

Definition 3.1 (Continuity at a point; Continuous function).
Let I be an interval in $\mathbb{R}$, $c \in I$ and $f : I \to \mathbb{R}$.

The function f is *continuous at c* if for every $\epsilon > 0$, there exists a $\delta > 0$ such that for all $x \in I$ satisfying $|x - c| < \delta$, $|f(x) - f(c)| < \epsilon$.

The function f is *continuous (on I)* if for every $x \in I$, f is continuous at x.

Remark 3.1.
(1) **Continuity is a 'local' concept**. That is, we can decide the continuity of f on an interval by looking at each point of the domain f and checking if f is continuous at that point, and moreover, what matters for continuity of f at a point, roughly speaking, is what the function is doing 'locally' in arbitrarily small neighbourhoods of the point, that is, 'near the point', and what happens away from the point is irrelevant.

(2) **History of the notion of continuity**. In the early development of Calculus, there was no rigorous definition of continuity offered. Only in the 18th century mathematicians started

examining this notion, in connection with Fourier's work on the theory of heat, where discontinuous functions arose naturally in various kinds of physical problems. A satisfactory mathematical definition of continuity was first formulated by Cauchy in 1821.

Example 3.1 (The constant function).

$f : \mathbb{R} \to \mathbb{R}$ given by $f(x) = 1$ for all $x \in \mathbb{R}$ is continuous.

Let $c \in \mathbb{R} = (-\infty, \infty)$. Let $\epsilon > 0$. For $x \in \mathbb{R}$, we have $|f(x) - f(c)| = |1 - 1| = 0 < \epsilon$. So any $\delta > 0$ will do! For example, take $\delta = 1$. Then if $x \in \mathbb{R}$ and $|x - c| < \delta = 1$, we have:

$$|f(x) - f(c)| = |1 - 1| = |0| = 0 < \epsilon.$$

So f is continuous at c. Since the choice of $c \in \mathbb{R}$ was arbitrary, it follows that f is continuous on $\mathbb{R}$. See the picture below.   ◇

Example 3.2 (The identity function).

$f : \mathbb{R} \to \mathbb{R}$ given by $f(x) = x$ for all $x \in \mathbb{R}$ is continuous.

Let $c \in \mathbb{R}$. Let $\epsilon > 0$.

(*Rough work*: $|f(x) - f(c)| = |x - c| < \delta \leq \epsilon$, if for example, $\delta := \epsilon$.)

Let $\delta = \epsilon$. Then if $x \in \mathbb{R}$ and $|x - c| < \delta$, we have:

$$|f(x) - f(c)| = |x - c| < \delta = \epsilon.$$

So f is continuous at c. Since the choice of $c \in \mathbb{R}$ was arbitrary, it follows that f is continuous on $\mathbb{R}$. See the following picture.   ◇

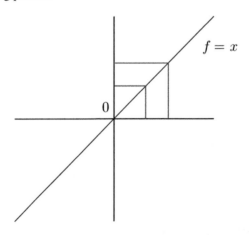

Example 3.3 (The Heaviside[1] function).

Let $Y : \mathbb{R} \to \mathbb{R}$ be given by

$$Y(x) = \begin{cases} 1 & \text{if } x > 0, \\ 0 & \text{if } x \leq 0. \end{cases}$$

From the graph of Y displayed below, we see clearly that there is a 'break' or 'jump' at $x = 0$, and so we guess that Y is not continuous at 0. Let us show this using the definition of continuity at a point.

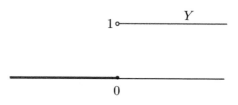

Suppose that Y is continuous at 0. Let $\epsilon = \frac{1}{2} > 0$. Suppose that there exists a $\delta > 0$ such that whenever $|x - 0| < \delta$, we have $|Y(x) - Y(0)| = |Y(x) - 0| < \epsilon = \frac{1}{2}$. Take $x = \frac{\delta}{2}$. Then $|x - 0| = |\frac{\delta}{2} - 0| = \frac{\delta}{2} < \delta$, and so we must have

$$|Y(x) - Y(0)| = |Y(\frac{\delta}{2}) - 0| = |1 - 0| = 1 < \epsilon = \frac{1}{2},$$

a contradiction. So Y is not continuous at 0.   ◇

Example 3.4 (The reciprocal function).

$h : (0, \infty) \to \mathbb{R}$ given by $h(x) = \frac{1}{x}$ for all $x \in (0, \infty)$ is continuous (on $(0, \infty)$). See Figure 3.3.

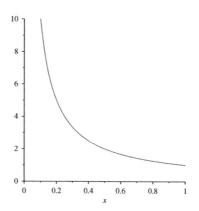

Figure 3.3 $x \mapsto \frac{1}{x} : (0, \infty) \to \mathbb{R}$ is continuous on $(0, \infty)$.

---

[1] Named after the mathematical physicist Oliver Heaviside (1850–1925)

Let $c \in (0, \infty)$. Let $\epsilon > 0$. (*Rough work*: We want a $\delta > 0$ such that whenever $|x - c| < \delta$, $|h(x) - h(c)| < \epsilon$. We have

$$|h(x) - h(c)| = \left| \frac{1}{x} - \frac{1}{c} \right| = \frac{|x - c|}{|x||c|}.$$

We know that if x is close to c, then the numerator $|x - c|$ can be made small. But what about the denominator $|x||c|$. Well, $|c| > 0$ is just a constant, and so it is harmless really. What about $|x|$? If it gets small, then it has the effect of making $|h(x) - h(c)|$ big, something that we want to avoid. But we note that when x is close to c, $|x|$ will be close to $|c|$, and so $|x|$ can be bounded below by some positive constant. Indeed, by the triangle inequality,

$$|c| - |x| \le ||c| - |x|| \le |c - x| = |x - c|,$$

and so if we choose the $\delta \le |c|/2$, then for x satisfying $|x - c| < \delta$ we will obtain from the above that $|x| \ge |c| - |c - x| \ge |c| - \delta \ge |c| - |c|/2 = |c|/2$. So for such x,

$$|h(x) - h(c)| = \frac{|x - c|}{|x||c|} < \frac{\delta}{(|c|/2) \cdot |c|},$$

and the last quantity can be made smaller than ϵ by further ensuring that the δ also satisfies that $\delta < \epsilon \frac{|c|^2}{2}$. Hence $\delta := \min\{\frac{|c|}{2}, \epsilon \frac{|c|^2}{2}\}$ should do the job! We remark that this is just one choice among many other equally good δs which will also work. End of *Rough Work*.)

Set $\delta = \min\left\{ \dfrac{c}{2}, \dfrac{\epsilon c^2}{2} \right\}$ (> 0). Then if $x \in (0, \infty)$ and $|x - c| < \delta$, we have

$$|c| - |x| \le |x - c| < \delta \le \frac{|c|}{2}$$

and so $\dfrac{|c|}{2} < |x|$, that is, $\dfrac{1}{|x|} < \dfrac{2}{|c|}$. Thus if $x \in (0, \infty)$ and $|x - c| < \delta$, then

$$\left| \frac{1}{x} - \frac{1}{c} \right| = \frac{|c - x|}{|x| \, |c|} = \frac{|x - c|}{|x| \, |c|} < \delta \cdot \frac{2}{|c|} \cdot \frac{1}{|c|} = \frac{2\delta}{c^2} \le \epsilon.$$

So f is continuous at c. Since the choice of $c \in (0, \infty)$ was arbitrary, it follows that f is continuous on $(0, \infty)$.   ◇

Exercise 3.1. Let the function $f : \mathbb{R} \to \mathbb{R}$ be given by $f(x) = x^2$.

(1) Prove that f is continuous at 0.

(2) (∗) Suppose that c is a nonzero real number. Prove that f is continuous at c.

In Exercise 3.7, we will give a slick proof of the fact that f is continuous on $\mathbb{R}$.

Exercise 3.2. Let $f : \mathbb{R} \to \mathbb{R}$ be such that $f(x + y) = f(x) + f(y)$ for all $x, y \in \mathbb{R}$.

(1) Let f be continuous at some real number c. Prove that f is continuous on $\mathbb{R}$.

Hint: Since f is continuous at c, given $\epsilon > 0$, $\exists \delta > 0$ such that for all $x \in \mathbb{R}$ satisfying $|x - c| < \delta$, $|f(x) - f(c)| < \epsilon$. Show that given any other point $c' \in \mathbb{R}$, the function f is continuous at c' by showing that the same δ works (for this ϵ).

(2) Give an example of such a continuous, additive function.

Exercise 3.3. Suppose that $f : \mathbb{R} \to \mathbb{R}$ and there exists an $M > 0$ such that for all $x \in \mathbb{R}$, $|f(x)| \leq M|x|$. Prove that f is continuous at 0. *Hint:* Find $f(0)$.

Exercise 3.4. Let $f : \mathbb{R} \to \mathbb{R}$ be defined by

$$f(x) = \begin{cases} 0 & \text{if } x \text{ is rational,} \\ 1 & \text{if } x \text{ is irrational.} \end{cases}$$

Prove that for every $c \in \mathbb{R}$, f is not continuous at c.

Hint: Use the fact that there are irrational numbers arbitrarily close to any rational number and rational numbers arbitrarily close to any irrational number.

Exercise 3.5. Let $f : (a, b) \to \mathbb{R}$ be a continuous function. Prove that if for some $c \in (a, b)$, $f(c) > 0$, then there exists a $\delta > 0$ such that for all $x \in (c - \delta, c + \delta), f(x) > 0$.

Exercise 3.6. Show that in the definition of continuity of a function at a point, we may replace the symbol $<$ with $\leq$, that is, the following statements are equivalent for $f : I \to \mathbb{R}$, and c belonging to the interval I:

(1) $\forall \epsilon > 0$, $\exists \delta > 0$ such that whenever $x \in I$ satisfies $|x - c| < \delta$, $|f(x) - f(c)| < \epsilon$.

(2) $\forall \epsilon > 0$, $\exists \delta > 0$ such that whenever $x \in I$ satisfies $|x - c| < \delta$, $|f(x) - f(c)| \leq \epsilon$.

(3) $\forall \epsilon > 0$, $\exists \delta > 0$ such that whenever $x \in I$ satisfies $|x - c| \leq \delta$, $|f(x) - f(c)| \leq \epsilon$.

(4) $\forall \epsilon > 0$, $\exists \delta > 0$ such that whenever $x \in I$ satisfies $|x - c| \leq \delta$, $|f(x) - f(c)| < \epsilon$.

3.2   Continuous functions preserve convergence

In Example 3.4, we had to work hard in order to prove the continuity of the reciprocal function. We will now learn about a result that will make life considerably simpler. Roughly speaking, this results says that a function is continuous at a point if and only if it preserves convergence of sequences with limit c.

Theorem 3.1. *Let I be an interval in $\mathbb{R}$, $c \in I$ and $f : I \to \mathbb{R}$. Then*

$$\boxed{f \text{ is continuous at } c}$$

if and only if

$$\boxed{\begin{array}{l} \textit{for every convergent sequence } (x_n)_{n \in \mathbb{N}} \textit{ contained in } I \textit{ with limit } c, \\ (f(x_n))_{n \in \mathbb{N}} \textit{ is convergent with limit } f(c). \end{array}} \tag{3.1}$$

Proof.

Only if: Suppose that f is continuous at $c \in I$.

Let $(x_n)_{n \in \mathbb{N}}$ be a convergent sequence contained in I with limit c.

Since f is continuous at $c \in I$, given $\epsilon > 0$, there exists a $\delta > 0$ such that for all $x \in I$ satisfying $|x - c| < \delta$, $|f(x) - f(c)| < \epsilon$.

As $(x_n)_{n \in \mathbb{N}}$ is convergent with limit c, there exists an $N \in \mathbb{N}$ such that for all $n > N$, $|x_n - c| < \delta$.

Consequently for $n > N$, $|f(x_n) - f(c)| < \epsilon$. So $(f(x_n))_{n \in \mathbb{N}}$ is convergent with limit $f(c)$. This completes the proof of the 'Only if' part.

If: Now suppose that (3.1) holds. Then we need to show that f is continuous at c and we prove this by contradiction. Assume that f is not continuous at c, that is,

$$\neg[\forall \epsilon > 0 \ \exists \delta > 0 \ \text{such that} \ \forall x \in I \ \text{such that} \ |x - c| < \delta, \ |f(x) - f(c)| < \epsilon]$$

that is, $\exists \epsilon > 0$ such that $\forall \delta > 0$ $\exists x \in I$ such that $|x - c| < \delta$ but $|f(x) - f(c)| \geq \epsilon$. Hence if $\delta = \frac{1}{n}$, then we can find an $x_n \in I$ such that we have $|x_n - c| < \delta = \frac{1}{n}$, but $|f(x_n) - f(c)| \geq \epsilon$.

Claim 1: The sequence $(x_n)_{n \in \mathbb{N}}$ is contained in I and is convergent with limit c.

We have for all $n \in \mathbb{N}$ that $|x_n - c| < 1/n$, that is, $c - \dfrac{1}{n} < x_n < c + \dfrac{1}{n}$.

As $\lim\limits_{n \to \infty} c - \dfrac{1}{n} = c = \lim\limits_{n \to \infty} c + \dfrac{1}{n}$, the Sandwich Theorem gives $\lim\limits_{n \to \infty} x_n = c$ too.

Claim 2: The sequence $(f(x_n))_{n \in \mathbb{N}}$ does not converge to $f(c)$.

Indeed, for all $n \in \mathbb{N}$, we have $|f(x_n) - f(c)| \geq \epsilon$. Thus for instance, $\frac{\epsilon}{2} > 0$, but it is not possible to find a large enough $N \in \mathbb{N}$ such that for all $n > N$, we have $|f(x_n) - f(c)| < \frac{\epsilon}{2}$ (for if this were possible, then we would arrive at the contradiction $\epsilon \leq |f(x_n) - f(c)| < \frac{\epsilon}{2}$).

Claims 1 and **2** show that (3.1) does not hold, a contradiction. Hence f is continuous at c. $\qquad\qquad\square$

Let us revisit some of our examples from the previous section in light of this result.

Example 3.5 (The reciprocal function). Recall h from Example 3.4. Let $c \in (0, \infty)$ and $(x_n)_{n \in \mathbb{N}}$ be any convergent sequence in $(0, \infty)$ with limit c. Then by the Algebra of Limits, $(h(x_n)) = (1/x_n)$ is convergent with limit $1/c = h(c)$. By Theorem 3.1, it follows that h is continuous at c. As the choice of $c \in (0, \infty)$ was arbitrary, h is continuous on $(0, \infty)$. Done! $\qquad\qquad\diamondsuit$

Example 3.6 (The Heaviside function). Recall Y from Example 3.3. Consider the convergent sequence $(1/n)_{n \in \mathbb{N}}$ with limit 0. Then $(Y(1/n))_{n \in \mathbb{N}} = (1)_{n \in \mathbb{N}}$ is convergent with limit $1 \neq 0 = Y(0)$.

But if Y *was* continuous at 0, then by Theorem 3.1, $(Y(1/n))_{n \in \mathbb{N}}$ should have been convergent with limit $Y(0) = 0$. Thus we conclude that Y is not continuous at 0. $\qquad\diamondsuit$

Exercise 3.7. Recall Exercise 3.1: $f : \mathbb{R} \to \mathbb{R}$ is given by $f(x) = x^2$ for $x \in \mathbb{R}$. Using the characterisation of continuity provided in Theorem 3.1, prove that f is continuous on $\mathbb{R}$.

Exercise 3.8. Let $c \in \mathbb{R}$, $\delta > 0$ and $f : (c - \delta, c] \to \mathbb{R}$ be continuous and strictly increasing on $(c - \delta, c)$. Show that f is strictly increasing on $(c - \delta, c]$.

Exercise 3.9. Prove that if $f : \mathbb{R} \to \mathbb{R}$ is continuous and $f(x) = 0$ if x is rational, then $f(x) = 0$ for all $x \in \mathbb{R}$. Revisit Exercise 3.4.

Hint: Given any real number c, there exists a sequence of rational numbers $(r_n)_{n \in \mathbb{N}}$ that converges to c.

Exercise 3.10. Let $f : \mathbb{R} \to \mathbb{R}$ be a function that preserves divergent sequences, that is, for every divergent sequence $(x_n)_{n \in \mathbb{N}}$, $(f(x_n))_{n \in \mathbb{N}}$ is divergent as well. Prove that f is one-to-one.

Hint: Let x_1, x_2 be distinct real numbers, and consider the sequence $x_1, x_2, x_1, x_2, \ldots$.

Exercise 3.11. Let I be an interval, $c \in I$, and $f : I \to \mathbb{R}$. Show that the following are equivalent:

(1) f is continuous at c.

(2) For every sequence $(x_n)_{n \in \mathbb{N}}$ contained in I such that $(x_n)_{n \in \mathbb{N}}$ converges to c, the sequence $(f(x_n))_{n \in \mathbb{N}}$ converges.

Exercise 3.12. Consider the function $f : \mathbb{R} \to \mathbb{R}$ defined by

$$ f(x) = \begin{cases} x & \text{if } x \text{ is rational,} \\ -x & \text{if } x \text{ is irrational.} \end{cases} $$

Prove that f is continuous only at 0.

Hint: For every rational number, there is a sequence of irrational numbers that converges to it, and for every irrational number, there is a sequence of rational numbers that converges to it.

Exercise 3.13. (*) Every nonzero rational number x can be uniquely written as $x = n/d$, where n, d denote integers without any common divisors and $d > 0$. When $x = 0$, we take $d = 1$ and $n = 0$. Consider the function $f : \mathbb{R} \to \mathbb{R}$ defined by

$$ f(x) = \begin{cases} 0 & \text{if } x \text{ is irrational,} \\ \dfrac{1}{d} & \text{if } x \left(= \dfrac{n}{d} \right) \text{ is rational.} \end{cases} $$

Prove that f is discontinuous at every rational number, and continuous at every irrational number.

Hint: For an irrational number x, given any $\epsilon > 0$, and any interval $(N, N + 1)$ containing x, show that there are just finitely many rational numbers r in $(N, N + 1)$ for which $f(r) \geq \epsilon$. Use this to show the continuity at irrationals.

Exercise 3.14. (*) Let $f : \mathbb{R} \to \mathbb{R}$ be a continuous function such that for all $x, y \in \mathbb{R}$, $f(x + y) = f(x) + f(y)$.

Show that there exists a real number a such that for all $x \in \mathbb{R}$, $f(x) = ax$.

Hint: Show first that for natural numbers n, $f(n) = nf(1)$. Extend this to integers n, and then to rational numbers n/d. Finally use the density of $\mathbb{Q}$ in $\mathbb{R}$ to prove the claim.

Exercise 3.15. Determine all continuous functions $f : \mathbb{R} \to \mathbb{R}$ such that for all $x \in \mathbb{R}$,

$$f(x) + f(2x) = 0.$$

Hint: Show that $f(x) = -f\left(\dfrac{x}{2}\right) = f\left(\dfrac{x}{4}\right) = -f\left(\dfrac{x}{8}\right) = \cdots.$

Exercise 3.16. Give an example of

(1) an interval I,

(2) a continuous function $f : I \to \mathbb{R}$, and

(3) a Cauchy sequence $(x_n)_{n \in \mathbb{N}}$

for which $(f(x_n))_{n \in \mathbb{N}}$ is not a Cauchy sequence in $\mathbb{R}$. (Later on, in Exercise 3.43, we will show that 'uniformly continuous' functions do preserve Cauchyness, even if continuous functions may not.)

Exercise 3.17. Determine if the following statements are always true for two continuous functions $f, g : \mathbb{R} \to \mathbb{R}$.

(1) If $f\left(\dfrac{1}{2n+7}\right) = g\left(\dfrac{n}{n^2+1}\right)$ for all $n \in \mathbb{N}$, then $f(0) = g(0)$.

(2) If $f(n) = g(n^2)$ for all n, and $\lim\limits_{n \to \infty} g(n) = L$, then $\lim\limits_{n \to \infty} f(n)$ also exists, and is equal to L.

Using Theorem 3.1, we obtain the following useful result that says that the pointwise sum, product, etc. of continuous functions is continuous. But before we state this result, we introduce some convenient notation.

Let I be an interval in $\mathbb{R}$. Given functions $f : I \to \mathbb{R}$ and $g : I \to \mathbb{R}$, we define:

(1) If $\alpha \in \mathbb{R}$, then we define the function $\alpha f : I \to \mathbb{R}$ by

$$(\alpha f)(x) = \alpha f(x), \quad x \in I.$$

(2) We define the *absolute value of* f, $|f| : I \to \mathbb{R}$ by

$$|f|(x) = |f(x)|, \quad x \in I.$$

(3) The *sum of* f *and* g, $f + g : I \to \mathbb{R}$ is defined by

$$(f + g)(x) = f(x) + g(x), \quad x \in I.$$

(4) The *product of* f *and* g, $fg : I \to \mathbb{R}$ is defined by

$$(fg)(x) = f(x)g(x), \quad x \in I.$$

(5) If $k \in \mathbb{N}$, then we define the *kth power of* f, $f^k : I \to \mathbb{R}$ by

$$f^k(x) = (f(x))^k, \quad x \in I.$$

(6) If for all $x \in I$, $g(x) \neq 0$, then we define $\dfrac{1}{g} : I \rightarrow \mathbb{R}$ by

$$\left(\frac{1}{g}\right)(x) = \frac{1}{g(x)}, \quad x \in I.$$

Theorem 3.2. *Let I be an interval in $\mathbb{R}$ and let $c \in I$. Suppose that $f : I \rightarrow \mathbb{R}$ and $g : I \rightarrow \mathbb{R}$ are continuous at c. Then:*

(1) *For all $\alpha \in \mathbb{R}$, αf is continuous at c.*

(2) *$|f|$ is continuous at c.*

(3) *$f + g$ is continuous at c.*

(4) *fg is continuous at c.*

(5) *For all $k \in \mathbb{N}$, f^k is continuous at c.*

(6) *If for all $x \in I$, $g(x) \neq 0$, then $\frac{1}{g}$ is continuous at c.*

Proof. Suppose that $(x_n)_{n \in \mathbb{N}}$ is a convergent sequence contained in I, with limit c. Since f and g are continuous at c, from Theorem 3.1, it follows that $(f(x_n))_{n \in \mathbb{N}}$ and $(g(x_n))_{n \in \mathbb{N}}$ are convergent with limits $f(c)$ and $g(c)$, respectively. Hence from Theorem 2.4, it follows that:

(1) $(\alpha \cdot f(x_n))_{n \in \mathbb{N}}$ is convergent with limit $\alpha \cdot f(c)$, that is, $((\alpha f)(x_n))_{n \in \mathbb{N}}$ is convergent with limit $(\alpha f)(c)$. So from Theorem 3.1, it follows that αf is continuous at c.

(2) $(|f(x_n)|)_{n \in \mathbb{N}}$ is convergent with limit $|f(c)|$, that is, $(|f|(x_n))_{n \in \mathbb{N}}$ is convergent with limit $|f|(c)$. So from Theorem 3.1, it follows that $|f|$ is continuous at c.

(3) $(f(x_n) + g(x_n))_{n \in \mathbb{N}}$ is convergent to $f(c) + g(c)$, that is, $((f + g)(x_n))_{n \in \mathbb{N}}$ is convergent with limit $(f + g)(c)$. So from Theorem 3.1, it follows that $f + g$ is continuous at c.

(4) $(f(x_n)g(x_n))_{n \in \mathbb{N}}$ is convergent with limit $f(c)g(c)$, that is, $((fg)(x_n))_{n \in \mathbb{N}}$ is convergent with limit $(fg)(c)$. So from Theorem 3.1, it follows that fg is continuous at c.

(5) $((f(x_n))^k)_{n \in \mathbb{N}}$ is convergent with limit $(f(c))^k$, that is, $(f^k(x_n))_{n \in \mathbb{N}}$ is convergent with limit $f^k(c)$. So from Theorem 3.1, it follows that f^k is continuous at c.

(6) $(\frac{1}{g(x_n)})_{n \in \mathbb{N}}$ is convergent with limit $\frac{1}{g(c)}$ (since for all $x \in I$, $g(x) \neq 0$, in particular $g(x_n) \neq 0$ and $g(c) \neq 0$), that is, $((\frac{1}{g})(x_n))_{n \in \mathbb{N}}$ is convergent with limit $(\frac{1}{g})(c)$. So from Theorem 3.1, it follows that $\frac{1}{g}$ is continuous at c. $\qquad \square$

Example 3.7 (Polynomials are continuous). As $f : \mathbb{R} \rightarrow \mathbb{R}$ given by $f(x) = x$ for $x \in \mathbb{R}$ is continuous (see Example 3.2 on page 88), it follows that for all $k \in \mathbb{N}$, x^k is continuous. Thus given arbitrary scalars $c_0, c_1, \cdots, c_d$ in $\mathbb{R}$, it follows that the functions

$c_0 \cdot 1, c_1 \cdot x, \cdots, c_d \cdot x^d$ are continuous. Hence the *polynomial function* $p : \mathbb{R} \to \mathbb{R}$ defined by $p(x) = c_0 + c_1 x + \cdots + c_d x^d$, $x \in \mathbb{R}$, is continuous. ◇

Example 3.8 (The reciprocal function). Let us revisit the function h we had considered earlier in Example 3.4. As $x \overset{g}{\mapsto} x : (0, \infty) \to \mathbb{R}$ is continuous, and since $g(x) = x \neq 0$ for all $x \in (0, \infty)$, it follows that $h = \frac{1}{g} : (0, \infty) \to \mathbb{R}$, given by

$$h(x) = \frac{1}{x}, \quad x > 0,$$

is continuous too. ◇

Exercise 3.18. Show that the rational function $f : \mathbb{R} \to \mathbb{R}$ defined by

$$f(x) = \frac{x^2}{1 + x^2}, \quad x \in \mathbb{R},$$

is continuous on $\mathbb{R}$.

The composition of continuous functions is continuous. Let D_f, D_g be intervals in $\mathbb{R}$, and $f : D_f \to \mathbb{R}$, $g : D_g \to \mathbb{R}$ be two functions such that

$$f(D_f) := \{f(x) : x \in D_f\} \subset D_g,$$

that is, the range of f is contained in the domain of g. Then the *composition of g with f*, denoted by $g \circ f$, is the function $g \circ f : D_f \to \mathbb{R}$ defined by

$$(g \circ f)(x) = g(f(x)), \quad x \in D_f.$$

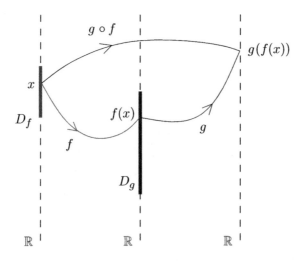

Theorem 3.3. *Let D_f, D_g be intervals in $\mathbb{R}$, and $f : D_f \to \mathbb{R}$, $g : D_g \to \mathbb{R}$ be two functions such that*

(1) $f(D_f) := \{f(x) : x \in D_f\} \subset D_g$,

(2) *f is continuous at c, and*

(3) *g is continuous at $f(c)$ $(\in D_g)$.*

Then their composition $g \circ f : D_f \to \mathbb{R}$ is continuous at c.

Proof. Let $(x_n)_{n \in \mathbb{N}}$ be any sequence in D_f with limit c. As f is continuous at c, $(f(x_n))_{n \in \mathbb{N}}$ converges to $f(c)$. But for all $n \in \mathbb{N}$, $f(x_n) \in f(D_f) \subset D_g$, and $f(c) \in f(D_f) \subset D_g$. As g is continuous at $f(c)$, $(g(f(x_n)))_{n \in \mathbb{N}}$ converges to $g(f(c))$, that is, $((g \circ f)(x_n))_{n \in \mathbb{N}}$ converges to $(g \circ f)(c)$. Hence $g \circ f$ is continuous at c. $\qquad\qquad\square$

Example 3.9. We know that the polynomial function $x \xmapsto{p} 1 + x^2 : \mathbb{R} \to \mathbb{R}$ is continuous, and that the reciprocal function $x \xmapsto{h} 1/x : (0, \infty) \to \mathbb{R}$ is continuous. Also, the range of p, $p(\mathbb{R}) = \{1 + x^2 : x \in \mathbb{R}\} \subset (0, \infty) = $ domain of h. So their composition, namely the rational function

$$x \xmapsto{hop} \frac{1}{1 + x^2} : \mathbb{R} \to \mathbb{R}$$

is continuous too.

More generally, any rational function $r : D_r\ (\subset \mathbb{R}) \to \mathbb{R}$,

$$r(x) = \frac{n(x)}{d(x)}, \quad x \in D_r,$$

where n, d are fixed polynomials and $D_r := \mathbb{R} \backslash \{\zeta \in \mathbb{R} : d(\zeta) = 0\}$, is continuous on its domain D_r. $\qquad\qquad\diamond$

Exercise 3.19. Let $f : \mathbb{R} \to \mathbb{R}$ be defined by $f(x) = |x + 1| - |x|$ for $x \in \mathbb{R}$. Find $\lim_{x \to -2} (f \circ f)(x)$.

Exercise 3.20. Determine if the following statements are always true for $f, g : \mathbb{R} \to \mathbb{R}$ and $a \in \mathbb{R}$.

(1) If $g \circ f$ is continuous at a, then f is continuous at a and g is continuous at $f(a)$.

(2) If $g \circ f$ is continuous at a, then f is continuous at a or g is continuous at $f(a)$.

(3) If $g \circ f$ isn't continuous at a, then f isn't continuous at a and g isn't continuous at $f(a)$.

(4) If $g \circ f$ isn't continuous at a, then f isn't continuous at a or g isn't continuous at $f(a)$.

Exercise 3.21. Show that the function $f : \mathbb{R} \to \mathbb{R}$ given by

$$f(x) = \begin{cases} x \sin \frac{1}{x} & \text{if } x \neq 0, \\ 0 & \text{if } x = 0, \end{cases}$$

is continuous. Use Maple/Mathematica or some other equivalent program to plot the graph of f.

Exercise 3.22. Let I be an interval, and let the functions $f, g : I \to \mathbb{R}$ be continuous on I. Define $\max\{f, g\} : I \to \mathbb{R}$ by $(\max\{f, g\})(x) = \max\{f(x), g(x)\}$, $x \in I$. Is $\max\{f, g\}$ continuous on I? *Hint:* Exercise 1.23.

In the next two sections, we will learn two fundamental results concerning continuous functions $f : [a, b] \to \mathbb{R}$ on a compact interval $[a, b]$, namely:

(1) The Intermediate Value Theorem, saying that f assumes all the values between $f(a)$ and $f(b)$ (the *'intermediate* values'). Geometrically, this means the following. Consider the graph of f in the Cartesian plane. If we choose any number y lying between $f(a)$ and $f(b)$ and draw a horizontal line through the point y on the y-axis, then this horizontal line must meet the graph of f at some point. This is 'clear' since f, being continuous, should have a graph having 'no breaks'.

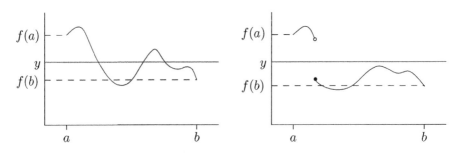

Figure 3.4   The Intermediate Value Theorem: the picture on the right shows that the condition of continuity can't be dropped.

(2) The Extreme Value Theorem, saying that f has a maximiser and a minimiser on $[a, b]$ (that is, f assumes the *extreme* values of the range $f([a, b])$). Geometrically, this means that if we consider the graph of f, then there must be a point in $c \in [a, b]$, where the graph y coordinate is highest, and a point $d \in [a, b]$ where the graph y coordinate is lowest.

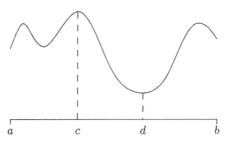

Although these two properties might seem 'obvious' when interpreted geometrically, they require proofs. We will see that the Least Upper Bound Property of $\mathbb{R}$ will be used crucially in the proofs. We will begin with the Intermediate Value Property.

3.3 Intermediate Value Theorem

Roughly speaking, the Intermediate Value Theorem says that a continuous function on a compact interval cannot 'hop over' intermediate values. For instance, if the height of a mountain is 1976 meters above sea level, then given any number between 0 and 1976, say 399, there must exist a point on the mountain that is exactly 399 meters above sea level. The picture shown in Figure 3.4 shows that the continuity of the function is an essential requirement.

Theorem 3.4 (Intermediate Value Theorem). *If $f : [a, b] \to \mathbb{R}$ is continuous and $y \in \mathbb{R}$ lies between $f(a)$ and $f(b)$, (that is, $f(a) \leq y \leq f(b)$ or $f(b) \leq y \leq f(a)$), then there exists a $c \in [a, b]$ such that $f(c) = y$.*

Proof. Consider first the case that $f(a) \leq y \leq f(b)$, and define

$$S_y = \{x \in [a, b] : f(x) \leq y\}.$$

(Pictorially, this set can be visualised like this: imagine again the horizontal line through y, and look at the portion of the graph of f that lies below y. S_y is the shadow on the x axis of this portion with a light source very high up above.)

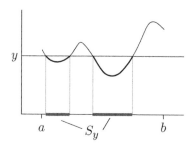

Clearly S_y is a subset of $\mathbb{R}$, it is nonempty (since $a \in S_y$) and S_y is bounded above (by b). So by the Least Upper Bound Property of $\mathbb{R}$, $c := \sup S_y$ exists. As b is an upper bound of S_y, and c is the *least* upper bound of S_y, clearly $c \leq b$. As $a \in S_y$, we also know that $a \leq c$. Summarising, we have $c \in [a, b]$. We now claim that this c does the job.

Claim: $f(c) = y$.

We will show that $f(c) \leq y$ and $f(c) \geq y$, and this will prove the claim.

That $f(c) \leq y$: Note that for every $n \in \mathbb{N}$, $c - \frac{1}{n}$ is not an upper bound of S_y. So there must be an $x_n \in S_y$ such that $x_n > c - \frac{1}{n}$. Hence we have for all n that

$$c - \frac{1}{n} < x_n \leq c.$$

By the Sandwich Theorem, $\lim\limits_{n\to\infty} x_n = c$, and since f is continuous, also

$$\lim_{n\to\infty} f(x_n) = f(c).$$

As $f(x_n) \leq y$, $n \in \mathbb{N}$ (since $x_n \in S_y$), we have $f(c) = \lim\limits_{n\to\infty} f(x_n) \leq y$.

That $f(c) \geq y$: If $c = b$, then we are done, since $y \leq f(b) = f(c)$. So we suppose that $c < b$. Define for $n \in \mathbb{N}$

$$x_n := c + \frac{b-c}{n} \quad \left(\leq c + \frac{b-c}{1} = b \right).$$

Then $x_n \in [a, b]$, and $(x_n)_{n\in\mathbb{N}}$ is convergent with limit c. As f is continuous, $(f(x_n))_{n\in\mathbb{N}}$ converges to $f(c)$. But $x_n > c$ for each $n \in \mathbb{N}$, and so $x_n \notin S_y$ for each n. Hence for all n, $f(x_n) > y$. Thus $f(c) \geq y$.

Consequently $f(c) = y$, proving the claim. Thus the proof of the theorem is complete when $f(a) \leq y \leq f(b)$.

Now suppose that $f(b) \leq y \leq f(a)$. Then $(-f)(a) \leq -y \leq (-f)(b)$. By the continuity of f, $-f$ is continuous too. So applying the previous result (with $-f$ instead of f, and $-y$ instead of y), it follows that there is a $c \in [a, b]$ such that $(-f)(c) = -y$, that is, $-f(c) = -y$, and so $f(c) = y$. This completes the proof. $\qquad\square$

Example 3.10. Consider the polynomial $p : \mathbb{R} \to \mathbb{R}$ given by

$$p(x) = x^{2014} + x^{1976} - \frac{1}{399}, \quad x \in \mathbb{R}.$$

Then p is continuous, and

$$p(0) = 0 + 0 - \frac{1}{399} = -\frac{1}{399} < 0,$$

$$p(1) = 1 + 1 - \frac{1}{399} > 0.$$

As $p(0) \leq y := 0 \leq p(1)$, and since $p : [0, 1] \to \mathbb{R}$ is continuous, it follows by the Intermediate Value Theorem, that there exists a $c \in [0, 1]$ such that $p(c) = 0$. In other words, p has a real root in $[0, 1]$.

More generally, one can show that *any* odd degree polynomial p with real coefficients must have at least one real root. The reason is that for large positive values of x, $p(x)$ will have the same sign as the leading coefficient c_d, while for large[2] negative values of x, $p(x)$

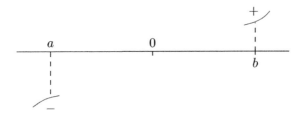

---

[2] That is, $x < 0$ and $|x|$ large

will have the opposite sign as that of c_d (since d is *odd*). Consequently, p must vanish somewhere in between these two extremes of large positive and negative xs.   ◇

Example 3.11. At any given instant of time, there exists a pair of diametrically opposite points on the equator of the earth, which have the same temperature.

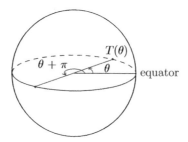

Let $T(\theta)$ denote the surface temperature at the point on the equator with longitude θ. Then $\theta \mapsto T(\theta)$ is continuous on the interval $[0, 2\pi]$ (with longitude measured in radians[3]). Note that $T(0) = T(2\pi)$. Let $S : [0, \pi] \to \mathbb{R}$ be given by

$$S(\theta) = T(\theta) - T(\theta + \pi), \quad \theta \in [0, \pi].$$

Then S is continuous, and

$$S(\pi) = T(\pi) - T(2\pi) = T(\pi) - T(0) = -(T(0) - T(\pi)) = -S(0).$$

So 0 lies between $S(\pi)$ and $S(0) = -S(\pi)$. By the Intermediate Value Theorem, there exists a $\theta_* \in [0, \pi]$ such that $S(\theta_*) = 0$, that is, $T(\theta_*) = T(\theta_* + \pi)$.   ◇

Exercise 3.23. Suppose that $f : [0, 1] \to \mathbb{R}$ is a continuous function such that for all $x \in [0, 1]$, $0 \le f(x) \le 1$. Prove that there exists at least one $c \in [0, 1]$ such that $f(c) = c$.
Hint: Consider $g(x) = f(x) - x$, and use the Intermediate Value Theorem.

Exercise 3.24. Let $f : [0, 1] \to \mathbb{R}$ be continuous. Show that there exists a $c \in [0, 1]$ such that $f(c) - f(1) = (f(0) - f(1))c$. *Hint:* Consider $f(x) - f(1) - (f(0) - f(1))x$.

Exercise 3.25. Consider a flat pancake of arbitrary shape. Show that there is a straight line cut that divides the pancake into two parts having equal areas. Can the direction of the straight line cut be chosen arbitrarily?

---

[3] We will define the radian angle measure later on in Chapter 5; if this is unfamiliar, one may just think of T as a function on the interval $[0, 360]$, with the angle θ measured in degrees.

Exercise 3.26. True or false? There is real number x such that $x^{399} + \dfrac{1976}{1 + x^2(\cos x)^2} = 28$.

Exercise 3.27. At 8:00 a.m. on Saturday, a hiker begins walking up the side of a mountain to his weekend campsite. On Sunday morning at 8:00 a.m., he walks back down the mountain along the same trail. It takes him one hour to walk up, but only half an hour to walk down. At some point on his way down, he realises that he was at the same spot at exactly the same time on Saturday. Prove that he is right.

Hint: Let $u(t)$ and $d(t)$ be the position functions for the walks up and down, and apply the Intermediate Value Theorem to $f(t) = u(t) - d(t)$.

Exercise 3.28. Show that the polynomial function $p(x) = 2x^3 - 5x^2 - 10x + 5$ has a real root in the interval $[-1, 2]$.

Exercise 3.29. Let $f : [a, b] \to \mathbb{R}$ be continuous and such that for all $x \in [a, b]$, $f(x) \neq 0$. Show that f assumes only positive values or f assumes only negative values.

Exercise 3.30. Let $f : \mathbb{R} \to \mathbb{R}$ be continuous. If $S := \{f(x) : x \in \mathbb{R}\}$ is neither bounded above nor bounded below, prove that $S = \mathbb{R}$.

Hint: If $y \in \mathbb{R}$, then since S is neither bounded above nor bounded below, there exist x_0, $x_1 \in \mathbb{R}$ such that $f(x_0) < y < f(x_1)$.

Exercise 3.31. (*) Show that given any continuous function $f : \mathbb{R} \to \mathbb{R}$, there exists an $x_0 \in [0, 1]$ and an $m \in \mathbb{Z} \backslash \{0\}$ such that $f(x_0) = mx_0$. In other words, the graph of f intersects some nonhorizontal line $y = mx$ at some point x_0 in $[0, 1]$.

Hint: If $f(0) = 0$, take $x_0 = 0$ and any $m \in \mathbb{Z} \backslash \{0\}$. If $f(0) > 0$, then choose $N \in \mathbb{N}$ satisfying $N > f(1)$, and apply the Intermediate Value Theorem to the continuous function $g(x) = f(x) - Nx$ on the interval $[0, 1]$. If $f(0) < 0$, then first choose a $N \in \mathbb{N}$ such that $N > -f(1)$, and consider the function $g(x) = f(x) + Nx$, and proceed in a similar manner.

Exercise 3.32. (*) Prove that there does not exist a continuous function $f : \mathbb{R} \to \mathbb{R}$ which assumes rational values at irrational numbers, and irrational values at rational numbers, that is, $f(\mathbb{Q}) \subset \mathbb{R} \backslash \mathbb{Q}$ and $f(\mathbb{R} \backslash \mathbb{Q}) \subset \mathbb{Q}$.

Hint: Note that for each $m \in \mathbb{Z} \backslash \{0\}$, there is no $x_0 \in \mathbb{R}$ such that $f(x_0) = mx_0$.

Inverse functions and continuity

Theorem 3.5. *Suppose that* $f : [a, b] \to \mathbb{R}$ *is strictly increasing and continuous. Then* $f([a, b]) = [f(a), f(b)]$, *and* $f^{-1} : [f(a), f(b)] \to \mathbb{R}$ *is strictly increasing and continuous.*

Remark 3.2.

(1) As f is strictly increasing, it is injective, and so the inverse function f^{-1} is defined on the range $f([a,b])$. The above result says in particular that this range $f([a,b])$ is the interval $[f(a),f(b)]$, and so f^{-1} has domain $[f(a),f(b)]$.

(2) Analogous to the above result for strictly *increasing* functions, the following version of the result for strictly *decreasing* functions is also true.

If $f : [a,b] \to \mathbb{R}$ is strictly decreasing and continuous, then $f([a,b]) = [f(b),f(a)]$, and $f^{-1} : [f(b),f(a)] \to \mathbb{R}$ is strictly decreasing and continuous.

Proof. $f([a,b]) = [f(a),f(b)]$: Since $a < b$, we have $f(a) < f(b)$ as f is strictly increasing. Let $y \in [f(a),f(b)]$. By the Intermediate Value Theorem, there exists a $c \in [a,b]$ such that $f(c) = y$. Hence $[f(a),f(b)] \subset f([a,b])$. Also, if $x \in [a,b]$, then $f(x) \in [f(a),f(b)]$ as f is increasing. So $f([a,b]) \subset [f(a),f(b)]$.

f^{-1} is strictly increasing: If $f(a) \le y_1 < y_2 \le f(b)$, and $y_1 = f(x_1)$, $y_2 = f(x_2)$ for some $x_1, x_2 \in [a,b]$, then $x_1 < x_2$. (Otherwise, $x_1 \ge x_2$ implies $f(x_1) \ge f(x_2)$, a contradiction.) But $x_1 = f^{-1}(y_1)$ and $x_2 = f^{-1}(y_2)$. So f^{-1} is strictly increasing.

f^{-1} is continuous: Let $y_* = f(x_*) \in [f(a),f(b)]$, and let $\epsilon > 0$.

Q. What $\delta > 0$ guarantees that whenever $|y - y_*| < \delta$, we have $|f^{-1}(y) - x_*| < \epsilon$?
A. We read this off from the picture shown below:

$$\delta = \min\{f(x_* + \epsilon) - f(x_*),\ f(x_*) - f(x_* - \epsilon)\} > 0.$$

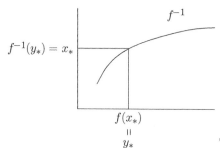
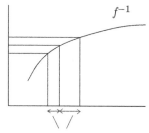

Take δ to be the smaller among these two!

With this δ,

$$y_* + \delta = f(x_*) + \delta \le f(x_* + \epsilon)\ \text{and}$$
$$y_* - \delta = f(x_*) - \delta \ge f(x_* - \epsilon).$$

So if $y_* - \delta < y < y_* + \delta$, then since f^{-1} is strictly increasing,

$$x_* - \epsilon = f^{-1}(f(x_* - \epsilon)) \le f^{-1}(y_* - \delta)$$
$$< f^{-1}(y)$$
$$< f^{-1}(y_* + \delta) \le f^{-1}(f(x_* + \epsilon)) = x_* + \epsilon,$$

that is, $|f^{-1}(y) - x_*| < \epsilon$. □

Example 3.12. The nth root function $\sqrt[n]{\cdot} : [0, \infty) \to [0, \infty)$, which is the inverse of the strictly increasing, continuous function $\cdot^n : [0, \infty) \to [0, \infty)$, is strictly increasing and continuous too.

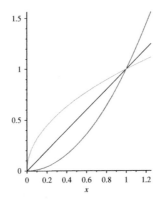

Thus it also follows for example that the composition of the two continuous functions, given by $x \mapsto 1 + x^2 : \mathbb{R} \to \mathbb{R}$ and $x \mapsto \sqrt{x} : [0, \infty) \to \mathbb{R}$, namely $x \mapsto \sqrt{1 + x^2} : \mathbb{R} \to \mathbb{R}$, is continuous. $\diamond$

Exercise 3.33. Let $f : [0, \infty) \to \mathbb{R}$ be the function defined by

$$f(x) = \frac{1}{1 + x^2}, \quad x \in [0, \infty).$$

Show that f is strictly decreasing and that $f([0, \infty)) = (0, 1]$. Find an expression for the inverse function $f^{-1} : (0, 1] \to [0, \infty)$ and explain why f^{-1} is continuous on $(0, 1]$. Sketch the graphs of f and f^{-1}.

Continuous functions preserve connectedness

Intervals are 'special' subsets of $\mathbb{R}$ in the sense that they are 'connected' or that 'they have no holes in them'. We make this precise below.

Definition 3.2 (Interval property). A subset $I \subset \mathbb{R}$ is said to have the *interval property* if for all $x, y \in I$ with $x < y$, $[x, y] \subset I$.

The following picture shows that 'I having no holes' means exactly that 'I has the interval property'.

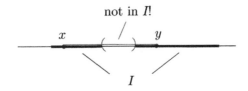

It is clear that intervals possess the interval property. We now show that there are no others!

Theorem 3.6. *If $I \subset \mathbb{R}$ has the interval property, then I is an interval.*

Proof. If $I = \emptyset$, then $I = (a, a)$ for any $a \in \mathbb{R}$. Suppose that $I \neq \emptyset$. Then we have the following cases:

I is bounded				I is bounded below but not above		I is bounded above but not below		I is neither bounded above nor below
$a := \inf I$ $b := \sup I$				$a := \inf I$		$b := \sup I$		
$a \in I$ $b \in I$	$a \in I$ $b \notin I$	$a \notin I$ $b \in I$	$a \notin I$ $b \notin I$	$a \in I$	$a \notin I$	$b \in I$	$b \notin I$	
$I = [a, b]$	$I = [a, b)$	$I = (a, b]$	$I = (a, b)$	$I = [a, \infty)$	$I = (a, \infty)$	$I = (-\infty, b]$	$I = (-\infty, b)$	$I = (-\infty, \infty)$

Each of the above claims can be checked. For example, let us consider the case when I is bounded and $a, b \notin I$. Then we need to show that $I \subset (a, b)$ and that $(a, b) \subset I$.

1° Let $x \in I$. Then $a = \inf I \leq x \leq \sup I = b$. But $a \notin I$ and $b \notin I$, while $x \in I$. So we can't have equalities in $a \leq x \leq b$. Hence $a < x < b$, that is, $x \in (a, b)$.

2° Let $x \in (a, b)$. Then $a < x < b$. Since $a := \inf I$ and $b := \sup I$, it follows that there exist $x_*, y_* \in I$ such that $a \leq x_* < x < y_* \leq b$. As I has the interval property, $[x_*, y_*] \subset I$, and in particular x (which belongs to $[x_*, y_*]$) belongs to I. □

In light of this characterisation of intervals, we have the following consequence of the Intermediate Value Theorem.

Corollary 3.7. *Let $I \subset \mathbb{R}$ be an interval and $f : I \to \mathbb{R}$ be continuous. Then $f(I)$ is an interval.*

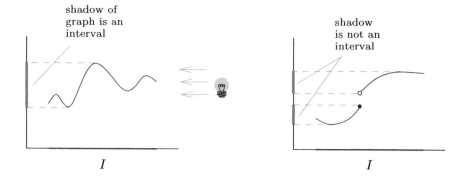

shadow of graph is an interval

I

shadow is not an interval

I

Proof. Let $y_1, y_2 \in f(I)$ be such that $y_1 < y_2$. Then there exist $a, b \in I$ such that $y_1 = f(a)$, $y_2 = f(b)$. Let $y \in [y_1, y_2]$. By the Intermediate Value Theorem, there exists a point c, belonging to the compact interval with endpoints a and b, such that $f(c) = y$. So $y \in f(I)$. As the choice of $y \in [y_1, y_2]$ was arbitrary, it follows that $[y_1, y_2] \subset f(I)$. Hence $f(I)$ has the interval property, and consequently, $f(I)$ is an interval. □

The above corollary can be summarised by saying that 'continuous functions preserve connectedness'.

Exercise 3.34. In each of the following cases, give an example of a continuous function $f : S \to \mathbb{R}$ such that $f(S) = T$, or explain why such an f can't exist.

(1) $S = (0, 1)$, $T = (0, 1]$.

(2) $S = (0, 1)$, $T = \{0, 1\}$.

3.4   Extreme Value Theorem

Theorem 3.8 (Extreme Value Theorem). *If $f : [a, b] \to \mathbb{R}$ is continuous, then*

(1) $S := \{f(x) : x \in [a, b]\} =: f([a, b]) = $ *range of f is bounded.*

(2) $\sup S$ *and* $\inf S$ *exist.*

(3) $\sup S$ *and* $\inf S$ *are attained, that is, there exist $c, d \in [a, b]$ such that* $f(c) = \sup S = \max S$ *and $f(d) = \inf S = \min S$.*

Thus, in the above conclusion, we have $f(c) \geq f(x)$ for all $x \in [a, b]$ (so that c is a *maximiser* of f), and we have $f(d) \leq f(x)$ for all $x \in [a, b]$ (so that d is a *minimiser* of f).

Note that continuity of f says something locally about f at each point of its domain. However, the conclusion says something globally about f. This miracle happens because $[a, b]$ is 'compact'. We will later see examples that show that maximisers/minimisers may fail to exist if either $[a, b]$ is not compact or if f is not continuous. First, let us prove the Extreme Value Theorem.

Proof.

(1) We first show that f is bounded, that is, $S := \{f(x) : x \in [a, b]\}$ is bounded. Suppose that S is not bounded. Let $n \in \mathbb{N}$. Then this n is not an upper bound of S. So there exists some $x_n \in [a, b]$ such that $|f(x_n)| > n$. In this way, we get a sequence $(x_n)_{n \in \mathbb{N}}$. Since $a \leq x_n \leq b$ for all $n \in \mathbb{N}$, $(x_n)_{n \in \mathbb{N}}$ is bounded. By the Bolzano–Weierstrass Theorem (Theorem 2.8), it follows that it has a convergent subsequence, say $(x_{n_k})_{k \in \mathbb{N}}$, that converges to some limit L. Since for all k, we have $a \leq x_{n_k} \leq b$, it follows that $a \leq L \leq b$, that is, $L \in [a, b]$. As f is continuous in particular at L, $(f(x_{n_k}))_{k \in \mathbb{N}}$ is convergent and in particular bounded. So there must exist an $M > 0$ such that for all $k \in \mathbb{N}$, $(n_k <) |f(x_{n_k})| \leq M$, a contradiction. Thus S is bounded.

(2) S is not empty (since $f(a) \in S$!). S is bounded. So by the Least Upper Bound Property of $\mathbb{R}$, $\sup S$ exists, and by the Greatest Lower Bound Property of $\mathbb{R}$, $\inf S$ exists too.

(3) We claim that there exists a $c \in [a, b]$ such that $f(x) = M := \sup S$. Let $n \in \mathbb{N}$. Then $M - \frac{1}{n}$ is not an upper bound of S. So there exists a $y_n \in S$ such that $M - \frac{1}{n} < y_n \leq M$. As this y_n belongs to the range S of f, $y_n = f(x_n)$ for some $x_n \in [a, b]$. By Bolzano–Weierstrass Theorem, there is a subsequence, say $(x_{n_k})_{k \in \mathbb{N}}$, of $(x_n)_{n \in \mathbb{N}}$ which converges with limit, say c. As $a \leq x_{n_k} \leq b$ for all k, it follows that $c \in [a, b]$. We have $M - \frac{1}{n_k} < f(x_{n_k}) \leq M$ for all k. So by the Sandwich Theorem, we conclude that $(f(x_{n_k}))_{k \in \mathbb{N}}$ is convergent with limit M. But since f is continuous at c, and since $(x_{n_k})_{k \in \mathbb{N}}$ converges to c, we must have $(f(x_{n_k}))_{k \in \mathbb{N}}$ is convergent with limit $f(c)$. By the uniqueness of limits, $f(c) = M = \sup S$. But $f(c) \in S$. So $\max S$ exists.

Finally, consider $-f : [a, b] \to \mathbb{R}$. As f is continuous, $-f$ is continuous too. By the above, there exists a $d \in [a, b]$ such that

$$(-f)(d) = \sup\{(-f)(x) : x \in [a, b]\}$$
$$= \sup\{-f(x) : x \in [a, b]\} = \sup(-S) = -\inf S,$$

and so $f(d) = \inf S$. Since $f(d) \in S$, it follows that $\min S$ exists. □

Example 3.13. There is no continuous function $f : [0, 1] \to \mathbb{R}$ onto $\mathbb{R}$. Indeed, by the Extreme Value Theorem, there exist m, M such that for all $x \in [0, 1]$, $m \leq f(x) \leq M$, that is, the range $f([0, 1])$ of f is a bounded set, and so it can't equal the unbounded set $\mathbb{R}$.   ◇

Example 3.14.

(1) Let $f_1 : (0, 1) \to \mathbb{R}$ be defined by $f_1(x) := \frac{1}{x}$ for $x \in (0, 1)$.
Then f_1 is continuous, but $(0, 1)$ is not a compact interval. We have

$$f_1((0, 1)) = \{1/x : x \in (0, 1)\} = \{y : y > 1\} = (1, \infty),$$

and so $\sup f_1((0, 1)) = \sup(1, \infty)$ does not exist. Also,

$$\inf f_1((0, 1)) = \inf(1, \infty) = 1,$$

but it is not attained, that is, there does not exist a $d \in (0, 1)$ such that $f(d) = 1$. (Indeed, for all $d \in (0, 1)$, $f(d) = 1/d > 1$.)

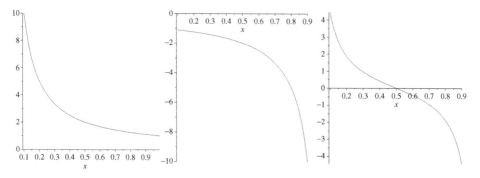

$$\text{Graphs of } x \mapsto \frac{1}{x}, \frac{1}{x - 1}, \frac{x - \frac{1}{2}}{x(x - 1)} : (0, 1) \to \mathbb{R}.$$

(2) Let $f_2 : (0, 1) \to \mathbb{R}$ given by $f_2(x) = \dfrac{1}{x-1}$, $x \in (0, 1)$.

Then it can be shown that $f_2((0, 1)) = (-\infty, -1)$, and so $\sup f_2((0, 1)) = -1$, but it is not attained, and $\inf f_2((0, 1))$ does not exist.

(3) Similarly, if we consider $f_3 : (0, 1) \to \mathbb{R}$ given by

$$f_3(x) = \frac{x - \frac{1}{2}}{x(x-1)}, \quad x \in (0, 1),$$

then it can be shown that $f_3((0, 1)) = \mathbb{R}$, and so neither $\sup f_3((0, 1))$ nor $\inf f_3((0, 1))$ exist.

(4) Let $f_4 : (0, 1) \to \mathbb{R}$ be given by $f_4(x) = x$, $x \in (0, 1)$. Then f_4 is continuous, but $(0, 1)$ is not compact, and

$$f_4((0, 1)) = \{f(x) : 0 < x < 1\} = \{x : 0 < x < 1\} = (0, 1).$$

$f_4((0, 1))$ is bounded, $\sup(0, 1) = 1$, but there is no $c \in (0, 1)$ such that $f_4(c) = 1$, and $\inf(0, 1) = 0$, but there is no $d \in (0, 1)$ such that $f_4(d) = 0$.

(5) Let $f_5 : [0, 1] \to \mathbb{R}$ be given by

$$f_5(x) = \begin{cases} 2x & \text{if } 0 \le x < \frac{1}{2}, \\ 0 & \text{if } x = \frac{1}{2}, \\ 2 - 2x & \text{if } \frac{1}{2} < x \le 1. \end{cases}$$

Then $[0, 1]$ is compact, but f_5 is *not* continuous. We have $f_5([0, 1]) = [0, 1)$, and there is no $c \in [0, 1]$ such that $f_5(c) = \sup f_5([0, 1]) = 1$.

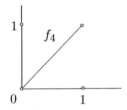
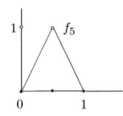
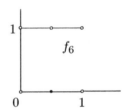

(6) (Continuity or compactness is not necessary for existence of maximisers and minimisers.) Let $f_6 : (0, 1) \to \mathbb{R}$ be given by

$$f_6(x) = \begin{cases} 1 & \text{if } 0 < x < \frac{1}{2}, \\ 0 & \text{if } x = \frac{1}{2}, \\ 1 & \text{if } \frac{1}{2} < x < 1. \end{cases}$$

Then $(0, 1)$ is not compact, and f is not continuous. But $f_6([0, 1]) = \{0, 1\}$, and there do exist maximisers and a minimiser:

$$f(1/2) = 0 = \inf f((0, 1)), \quad \text{and} \quad f(3/4) = 1 = \sup f((0, 1)).$$

We summarise the above examples in the following table.

Function $f : I \to \mathbb{R}$	I compact?	f continuous?	$\sup f(I)$ exists?	$\inf f(I)$ exists?	$\sup f(I)$ attained?	$\inf f(I)$ attained?
f_1	No	Yes	No	Yes	-	No
f_2	No	Yes	Yes	No	No	-
f_3	No	Yes	No	No	-	-
f_4	No	Yes	Yes	Yes	No	No
f_5	Yes	No	Yes	Yes	No	Yes
f_6	No	No	Yes	Yes	Yes	Yes

The utility of the Extreme Value Theorem in Optimisation

The Extreme Value Theorem (and its multivariable generalisation saying that a real-valued continuous function on a 'compact set' $K \subset \mathbb{R}^n$ has a maximiser and a minimiser) is useful in Optimisation Theory. In Optimisation Theory, one often meets *necessary* conditions for maximisers, that is, results of the following form:

> If x_* is a maximiser of $f : S \to \mathbb{R}$,
> then x_* satisfies $\boxed{* * *}$.

(Where $\boxed{* * *}$ are certain mathematical conditions, such as the Lagrange multiplier equations.) Now such a result has limited use, since even if we find all x_* which satisfy $\boxed{* * *}$, we can't conclude that there is one among these is actually a maximiser. But if we had an existence result (for example, the Extreme Value Theorem: suppose we know that S is compact and that f is continuous), then we know that a maximiser exists, and so we know that it must be among the (few) x_* in S that satisfy $\boxed{* * *}$. For example, we will later on learn that:

> If x_* is a maximiser of $f : (a,b) \to \mathbb{R}$, then $f'(x_*) = 0$.

As an example, consider $f : [0, \pi/2] \to \mathbb{R}$, given by $f(x) = \cos x + \frac{x}{2}$, $x \in [0, \pi/2]$. Then $f(0) = 1$, $f(\pi/2) = \frac{\pi}{4} < 1$, and $f(\pi/3) = \frac{1}{2} + \frac{\pi}{6} > \frac{1}{2} + \frac{3}{6} = 1 = f(0)$. By the Extreme Value Theorem, f has a maximiser $x_* \in [0, \pi/2]$. But the above calculation shows that $x_* \neq 0$ and $x_* \neq \pi/2$. Thus $x_* \in (0, \pi/2)$. Hence $f'(x_*) = 0$, that is, $-\sin x_* + \frac{1}{2} = 0$, and so $x_* = \pi/6$.

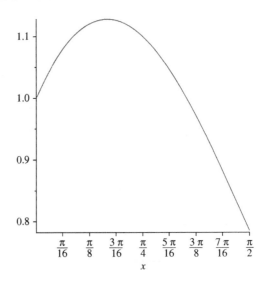

Exercise 3.35. A function $f : \mathbb{R} \to \mathbb{R}$ is called *periodic* if there exists a $T > 0$ such that for all $x \in \mathbb{R}, f(x+T) = f(x)$. If $f : \mathbb{R} \to \mathbb{R}$ is continuous and periodic, then prove that f is bounded, that is, the set $S = \{f(x) : x \in \mathbb{R}\}$ is bounded.

Exercise 3.36. Let $f : [a, b] \to \mathbb{R}$ be continuous on $[a, b]$, and define f_* as follows:

$$f_*(x) = \begin{cases} f(a) & \text{if } x = a, \\ \max\{f(y) : y \in [a, x]\} & \text{if } x \in (a, b]. \end{cases}$$

(1) Show that f_* is a well-defined function.

(2) If $f : [0, 1] \to \mathbb{R}$ is given by $f(x) = x - x^2$, then find f_*.

Exercise 3.37. True or false? If $f : [a, b] \to \mathbb{R}$ is continuous and $f(x) > 0$ for all $x \in [a, b]$, then f is in fact 'bounded away from 0', that is, there exists a $\delta > 0$ such that $f(x) \geq \delta$ for all $x \in [a, b]$.

Exercise 3.38. Let $f : [0, 3] \to [3, 9]$ be a continuous function such that $f(0) = 3$ and $f(3) = 6$. Which of the following statements is/are always true?

☐ (A) There exists a unique $c \in [0, 3]$ such that $f(c) = 4$.

☐ (B) The range of f contains the interval $[3, 6]$.

☐ (C) $f([0, 3]) = [3, 6]$.

☐ (D) There cannot exist a $c \in [0, 3]$ such that $f(c) = 9$.

Exercise 3.39. Let $f : [a, b] \to \mathbb{R}$ be continuous. Show that for any $c_1, \cdots, c_n \in [a, b]$, there is a $c \in [a, b]$ such that

$$f(c) = \frac{f(c_1) + \cdots + f(c_n)}{n}.$$

3.5   Uniform convergence and continuity

Recall that when we learnt about uniform convergence of a sequence $(f_n)_{n\in\mathbb{N}}$ of functions $f_n : I \to \mathbb{R}$, $n \in \mathbb{N}$, to $f : I \to \mathbb{R}$, we said that the limit f inherits some properties possessed by the terms of the sequence. Here is an instance of this, concerning the property of continuity.

Proposition 3.9. *Let I be an interval, and $f_n : I \to \mathbb{R}$, $n \in \mathbb{N}$, be continuous functions such that the sequence $(f_n)_{n\in\mathbb{N}}$ converges uniformly to $f : I \to \mathbb{R}$. Then f is continuous.*

Proof. Let $c \in I$ and $\epsilon > 0$. Let $N \in \mathbb{N}$ be such that that for all $n > N$, and all $x \in I$, we have $|f_n(x) - f(x)| < \epsilon/3$. In particular, for all $x \in I$, $|f_{N+1}(x) - f(x)| < \epsilon/3$. As f_{N+1} is continuous, we can find a $\delta > 0$ such that for all $x \in I$ satisfying $|x - c| < \delta$, $|f_{N+1}(x) - f_{N+1}(c)| < \epsilon/3$. So for all $x \in X$ satisfying $|x - c| < \delta$, we have using the triangle inequality that

$$|f(x) - f(c)| = |f(x) - f_{N+1}(x) + f_{N+1}(x) - f_{N+1}(c) + f_{N+1}(c) - f(c)|$$
$$\leq |f(x) - f_{N+1}(x)| + |f_{N+1}(x) - f_{N+1}(c)| + |f_{N+1}(c) - f(c)|$$
$$< \epsilon/3 + \epsilon/3 + \epsilon/3 = \epsilon.$$

Hence f is continuous at c. Since the choice of $c \in I$ was arbitrary, it follows that f is continuous on I. $\qquad\square$

3.6   Uniform continuity

Roughly speaking, we use the adjective 'uniform' in Calculus whenever 'the same thing works everywhere'. We have already seen one instance of this when we discussed *uniform* convergence of a sequence of functions. Now we will learn about uniform continuity.

Recall that a function $f : I \to \mathbb{R}$ is said to be continuous at a point c of an interval I if for every $\epsilon > 0$, there exists a $\delta > 0$ such that whenever $x \in I$ satisfied $|x - c| < \delta$, we have that $|f(x) - f(c)| < \epsilon$. And f is called continuous on I if for every $c \in I$, f is continuous at c, that is:

$\forall \epsilon > 0, \forall c \in I, \exists \delta > 0$ such that if $x \in I$ satisfies $|x - c| < \delta$, then $|f(x) - f(c)| < \epsilon$.

In the above long statement, we note that the choice of δ might depend on which $c \in I$ we consider. For a 'uniformly' continuous function on I, it doesn't! That is, given an $\epsilon > 0$, the same δ (depending only on ϵ) works everywhere in I, irrespective of which $c \in I$ we have considered. The precise definition is given below.

Definition 3.3 (Uniformly continuous function). Let $I \subset \mathbb{R}$ be an interval. $f : I \to \mathbb{R}$ is said to be *uniformly continuous* if for every $\epsilon > 0$, there exists a $\delta > 0$ such that for all $x, y \in I$ satisfying $|x - y| < \delta$, there holds that $|f(x) - f(y)| < \epsilon$.

Note that in the definition we are introducing the notion of uniform continuity of a function *on a set* (I), and not *at a point*.

Example 3.15. $x \mapsto x : \mathbb{R} \to \mathbb{R}$ is uniformly continuous. Let $\epsilon > 0$. Set $\delta = \epsilon > 0$. Then for all $x, y \in \mathbb{R}$ such that $|x - y| < \delta$, we have $|f(x) - f(y)| = |x - y| < \delta = \epsilon$. ◇

The name makes sense because:

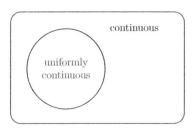

Theorem 3.10. *If $f : I \to \mathbb{R}$ is uniformly continuous on I, then f is continuous on I.*

Proof. Let $c \in I$. Let $\epsilon > 0$. Let $\delta > 0$ be such that whenever $x, y \in I$ and $|x - y| < \delta$, $|f(x) - f(y)| < \epsilon$. In particular, whenever $x \in I$ and $|x - c| < \delta$, $|f(x) - f(c)| < \epsilon$. This shows that f is continuous at c. But $c \in I$ was arbitrary. Hence f is continuous on I. □

Here is an example of a function that *is not* uniformly continuous, but is continuous.

Example 3.16 (Uniform Continuity and Continuity are distinct notions). Let $f : \mathbb{R} \to \mathbb{R}$ be given by $f(x) = x^2, x \in \mathbb{R}$. Then f is continuous on $\mathbb{R}$. But let us show that it is not uniformly continuous on $\mathbb{R}$.

Suppose, on the contrary, that f is uniformly continuous on $\mathbb{R}$. Let $\epsilon := 1 > 0$. Then there exists a $\delta > 0$ such that whenever $x, y \in \mathbb{R}$ satisfy $|x - y| < \delta$, we have $|f(x) - f(y)| = |x^2 - y^2| < \epsilon = 1$. Let $N \in \mathbb{N}$ be such that $N > 1/\delta$. Then with $x := n + \frac{1}{n}$ and $y := n$, for $n > N$, we have $|x - y| = \frac{1}{n} < \frac{1}{N} < \delta$, and so

$$ 1 > \left| \left(n + \frac{1}{n} \right)^2 - n^2 \right| = \frac{1}{n} \cdot \left(2n + \frac{1}{n} \right) = 2 + \frac{1}{n^2} > 2, $$

a contradiction. This is also clear in an intuitive manner pictorially:

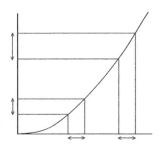

◇

Exercise 3.40. Let $f : (0, 1) \to \mathbb{R}$ be defined by $f(x) = \frac{1}{x}, 0 < x < 1$. Then f is continuous on $(0, 1)$. Show that f is not uniformly continuous on $(0, 1)$.

Hint: Consider $x = \frac{1}{n}$ and $y = \frac{1}{2n}$ for large natural numbers n.

In the above example, we have seen that there are continuous functions that are not uniformly continuous. But the following result tells us that if we are working with a *compact* interval, then mere continuity is enough to conclude (the stronger property) of uniform continuity.

Proposition 3.11. *If $f : [a, b] \to \mathbb{R}$ is continuous, then f is uniformly continuous.*

Proof. We argue by contradiction. So let us suppose that f is *not* uniformly continuous. Then

$$\neg \left(\forall \epsilon > 0 \ \exists \delta > 0 \quad \text{such that if } x, y \in [a, b] \text{ and } |x - y| < \delta, \text{ then } |f(x) - f(y)| < \epsilon \right)$$

that is,

$$\exists \epsilon > 0 \text{ such that } \forall \delta > 0, \exists x, y \in [a, b] \text{ such that } |x - y| < \delta, \text{ but } |f(x) - f(y)| \geq \epsilon.$$

In particular, if we take $\delta = 1/n$, then there exist $x_n, y_n \in [a, b]$ such that

$$|x_n - y_n| < 1/n \text{ but } |f(x_n) - f(y_n)| \geq \epsilon,$$

for all $n \in \mathbb{N}$. By using the compactness of $[a, b]$, the sequence $(x_n)_{n \in \mathbb{N}}$ has a convergent subsequence, say $(x_{n_k})_{k \in \mathbb{N}}$, converging to, say $x \in [a, b]$. Also from the inequality $|x_n - y_n| < \frac{1}{n}$, we obtain $x_{n_k} - \frac{1}{n_k} < y_{n_k} < x_{n_k} + \frac{1}{n_k}$ for all k, and so by the Sandwich Theorem, $(y_{n_k})_{k \in \mathbb{N}}$ is also convergent to x. Also, by the continuity of f, we have $(f(x_{n_k}))_{k \in \mathbb{N}}$ and $(f(y_{n_k}))_{k \in \mathbb{N}}$ both converge to $f(x)$. Hence $|f(x_{n_k}) - f(y_{n_k})|$ converges to $|f(x) - f(x)| = 0$. But on the other hand, from $|f(x_n) - f(y_n)| \geq \epsilon$ for all n, we obtain $0 = |f(x) - f(x)| \geq \epsilon > 0$, a contradiction. $\square$

Example 3.17. $f : [0, \infty) \to \mathbb{R}$, given by $f(x) = \sqrt{x}$ for $x \geq 0$, is uniformly continuous. Note that for $x, y > 0$,

$$|f(x) - f(y)| = |\sqrt{x} - \sqrt{y}| = \frac{|x - y|}{\sqrt{x} + \sqrt{y}}.$$

For $|x - y|$ small, the numerator can be made small too, but for x, y close to zero, the right hand side can perhaps be large. So a direct approach with the definition will be messy. But let us now see how by splitting the domain $[0, \infty)$ into two parts, namely the compact interval $[0, 1]$, and $[1, \infty)$, and using the previous result over the compact interval, we can complete the proof. (Note that for $x, y > 1$, the right hand side can be estimated above easily by $|x - y|/2$, and in that region $[1, \infty)$ we can easily use the definition to show uniform continuity.) Let $\epsilon > 0$.

(a) f is uniformly continuous on $[0, 1]$, since it is continuous on the compact interval $[0, 1]$. Thus given $\epsilon > 0$, there exists a $\delta_1 > 0$ such that if $x, y \in [0, 1]$ satisfy $|x - y| < \delta_1$, $|f(x) - f(y)| = |\sqrt{x} - \sqrt{y}| < \frac{\epsilon}{2}$.

(b) Let us show that f is uniformly continuous on $[1, \infty)$. Given $\epsilon > 0$, set $\delta_2 := \epsilon$. Then if $x, y \geq 1$, and $|x - y| < \delta_2$, we have

$$|f(x) - f(y)| = |\sqrt{x} - \sqrt{y}| = \frac{|x - y|}{\sqrt{x} + \sqrt{y}} < \frac{\delta_2}{1 + 1} = \frac{\delta_2}{2} = \frac{\epsilon}{2}.$$

Let $\delta := \min\{\delta_1, \delta_2\}$ (> 0).

If $0 \leq x, y \leq 1$ and $|x - y| < \delta$, then by (a), $|f(x) - f(y)| = |\sqrt{x} - \sqrt{y}| < \dfrac{\epsilon}{2} < \epsilon$.

If $x, y \geq 1$ and $|x - y| < \delta$, then by (b), $|f(x) - f(y)| = |\sqrt{x} - \sqrt{y}| < \dfrac{\epsilon}{2} < \epsilon$.

Now let $x \leq 1 \leq y$ and $|x - y| < \delta$. Then by (a) and (b),

$$|f(x) - f(y)| = |\sqrt{x} - \sqrt{y}| = |\sqrt{x} - \sqrt{1} + \sqrt{1} - \sqrt{y}|$$
$$\leq |\sqrt{x} - \sqrt{1}| + |\sqrt{1} - \sqrt{y}| < \frac{\epsilon}{2} + \frac{\epsilon}{2} = \epsilon,$$

where the rightmost inequality follows since $|x - 1| < \delta$ and since $|y - 1| < \delta$.

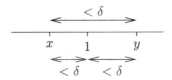

$\diamondsuit$

Exercise 3.41. Prove that the function $f : \mathbb{R} \to \mathbb{R}$ defined by $f(x) = |x|$ $(x \in \mathbb{R})$ is uniformly continuous.

Exercise 3.42. A function $f : \mathbb{R} \to \mathbb{R}$ is said to be *(globally) Lipschitz continuous* if there exists a number $L > 0$ such that for all $x, y \in \mathbb{R}$, $|f(x) - f(y)| \leq L|x - y|$. Prove that every Lipschitz continuous function is uniformly continuous. Is the converse true?

Exercise 3.43. Let I be an interval and let $f : I \to \mathbb{R}$ be uniformly continuous. Show that if $(x_n)_{n \in \mathbb{N}}$ is a Cauchy sequence, then $(f(x_n))_{n \in \mathbb{N}}$ is a Cauchy sequence. Compare this with Exercise 3.16.

Exercise 3.44. Let $f, g : I \to \mathbb{R}$ be uniformly continuous functions on the interval I.

(1) Show that $f + g$ is also uniformly continuous on I.

(2) Is fg also always uniformly continuous on I?

(3) In addition to the assumed uniform continuity of f, g, if f, g are also bounded, then show that fg is uniformly continuous on I.

Exercise 3.45. Let $I \subset \mathbb{R}$ be an interval and $f : I \to \mathbb{R}$ be continuous on I. Which of the following statements is/are true?

☐ (A) f is uniformly continuous on I.

☐ (B) $\forall x \in I \; \forall \epsilon > 0, \; \exists \delta > 0$ such that $\forall y \in I$ satisfying $|y - x| < \delta$, $|f(y) - f(x)| < \epsilon$.

☐ (C) For every $x \in I, f$ is continuous at x.

☐ (D) If $(x_n)_{n \in \mathbb{N}}$ is a sequence in I converging to $x \in I$, $(f(x_n))_{n \in \mathbb{N}}$ converges to $f(x)$.

3.7   Limits

We now develop some convenient terminology to describe the behaviour of functions near a finite value of the real variable or as the real variable 'tends to $+\infty$ or $-\infty$'. The meaning of these hitherto undefined notions will become clearer as we plough through this section and the numerous examples.

Definition 3.4 (Limit of a function). Let $c \in (a,b)$ and $f : (a,b)\backslash\{c\} \to \mathbb{R}$. We say that f *has a limit* $L \ (\in \mathbb{R})$ *at* c, written

$$\lim_{x \to c} f(x) = L$$

if for every $\epsilon > 0$, there exists a $\delta > 0$ such that whenever $x \in (a,b)$ satisfies $0 < |x - c| < \delta$, we have $|f(x) - L| < \epsilon$.

This will be a handy concept when we study differentiation. But first let us see a simple example.

Example 3.18. Let $f : \mathbb{R}\backslash\{1\} \to \mathbb{R}$ be given by

$$f(x) = \frac{x^2 - 1}{x - 1}, \quad x \neq 1.$$

We claim that $\lim_{x \to 0} f(x) = 2$.

Indeed, if $\epsilon > 0$, then taking $\delta := \epsilon > 0$, we have for $x \in \mathbb{R}$ satisfying $0 < |x - 1| < \delta$ that

$$|f(x) - 2| = \left| \frac{x^2 - 1}{x - 1} - 2 \right| \overset{(x \neq 1)}{=} \left| \frac{(x-1)(x+1)}{(x-1)} - 2 \right| = |(x+1) - 2| = |x - 1| < \delta = \epsilon.$$

$\diamondsuit$

The following result links continuity of a function at a point with the limiting behaviour of the function at that point.

Theorem 3.12. *Let $f : (a,b) \to \mathbb{R}$ and $c \in (a,b)$. Then f is continuous at c if and only if $\lim_{x \to c} f(x) = f(c)$.*

This is expected, since for a continuous function, $f(x)$ stays close to $f(c)$ when x is close to c.

Proof.
The proof is entirely elementary; it is just a translation of the relevant definitions.

('Only if' part:) Suppose that f is continuous at c. Let $\epsilon > 0$. Then there exists a $\delta > 0$ such that whenever $x \in (a,b)$ satisfies $|x - c| < \delta$, we have $|f(x) - f(c)| < \epsilon$. In particular, if $x \in (a,b)$ is such that $0 < |x - c| < \delta$, then too we have $|f(x) - f(c)| < \epsilon$. So, $\lim_{x \to c} f(x) = f(c)$.

('If' part:) Now suppose that

$$\lim_{x \to c} f(x) = f(c).$$

Let $\epsilon > 0$. Then there exists a $\delta > 0$ such that for all $x \in (a,b)$ such that $0 < |x - c| < \delta$, we have $|f(x) - f(c)| < \epsilon$. But if $|x - c| = 0$, that is, if $x = c$, then we have that $|f(x) - f(c)| = |f(c) - f(c)| = |0| = 0 < \epsilon$. So for $x \in (a,b)$ with $|x - c| < \delta$, we have $|f(x) - f(c)| < \epsilon$. Consequently f is continuous at c. $\square$

Exercise 3.46.

(1) Prove or disprove: There exists a continuous function $f : \mathbb{R} \to \mathbb{R}$ such that

$$f(x) = \frac{x^3 - 3x - 2}{x - 2}, \quad x \neq 2.$$

(2) Prove or disprove: There exists a continuous function $f : \mathbb{R} \to \mathbb{R}$ such that

$$f(x) = \begin{cases} x & \text{if } x < 2, \\ 2x & \text{if } x > 2. \end{cases}$$

We can also recast the definition of the limit of a function in terms of sequences.

Theorem 3.13. *Let $c \in (a, b)$, $f : (a, b) \backslash \{c\} \to \mathbb{R}$, and $L \in \mathbb{R}$. Then the following are equivalent:*

(1) $\lim_{x \to c} f(x) = L$.

(2) *For every $(x_n)_{n \in \mathbb{N}}$ in $(a, b) \backslash \{c\}$ with $\lim_{n \to \infty} x_n = c$, we have $\lim_{n \to \infty} f(x_n) = L$.*

Proof. (2) $\Rightarrow$ (1): Suppose that

$$\neg \left(\lim_{x \to c} f(x) = L \right),$$

that is,

$$\neg \left(\forall \epsilon > 0 \; \exists \delta > 0 \text{ such that if } x \in (a, b) \text{ and } 0 < |x - c| < \delta, \text{ then } |f(x) - L| < \epsilon \right).$$

Thus there exists an $\epsilon > 0$ such that for every $\delta > 0$, there exists an $x \in (a, b)$ (depending on δ) such that $0 < |x - c| < \delta$, but $|f(x) - L| \geq \epsilon$. Taking $\delta = \frac{1}{n}$, $n \in \mathbb{N}$, we thus find a sequence $(x_n)_{n \in \mathbb{N}}$ such that

$$\lim_{n \to \infty} x_n = c,$$

such that for all $n \in \mathbb{N}$, $x_n \neq c$, and $|f(x_n) - L| \geq \epsilon$. But this last condition implies that

$$\neg \left(\lim_{n \to \infty} f(x_n) = L \right).$$

(1) $\Rightarrow$ (2): Now suppose that

$$\lim_{x \to c} f(x) = L.$$

Let $(x_n)_{n \in \mathbb{N}}$ be a sequence contained in $(a, b) \backslash \{c\}$ such that

$$\lim_{n \to \infty} x_n = c.$$

We want to show that

$$\lim_{n \to \infty} f(x_n) = L.$$

Let $\epsilon > 0$. Then there exists a $\delta > 0$ such that whenever $x \in (a, b)$ satisfies $0 < |x - c| < \delta$, we have $|f(x) - L| < \epsilon$. Also, there exists an $N \in \mathbb{N}$ such that whenever $n > N$, $0 < |x_n - c| < \delta$. Consequently, for $n > N$, $|f(x_n) - L| < \epsilon$. Hence $\lim_{n \to \infty} f(x_n) = L$. $\qquad \square$

Since convergent sequences have a unique limit, we obtain the following.

Corollary 3.14. *Let* $c \in (a, b)$ *and* $f : (a, b) \setminus \{c\} \to \mathbb{R}$. *If* f *has a limit at* c, *then the limit is unique.*

Moreover, using the algebra of limits for real sequences, it follows that the same sort of results carry over to limits of functions.

Theorem 3.15 (Algebra of Limits). *Let* $c \in (a, b)$, *and* $f, g : (a, b) \setminus \{c\} \to \mathbb{R}$ *be such that* $\lim_{x \to c} f(x)$ *and* $\lim_{x \to c} g(x)$ *exist. Then:*

(1) $\lim_{x \to c} (f + g)(x)$ *exists and* $\lim_{x \to c} (f + g)(x) = \lim_{x \to c} f(x) + \lim_{x \to c} g(x)$.

(2) $\lim_{x \to c} (fg)(x)$ *exists and* $\lim_{x \to c} (fg)(x) = \left(\lim_{x \to c} f(x) \right) \left(\lim_{x \to c} g(x) \right)$.

(3) *If* $\lim_{x \to c} g(x) \neq 0$, *then* $\lim_{x \to c} \left(\dfrac{1}{g} \right)(x)$ *exists, and* $\lim_{x \to c} \left(\dfrac{1}{g} \right)(x) = \dfrac{1}{\lim_{x \to c} g(x)}$.

Proof. As an example, let us prove (2). Let $(x_n)_{n \in \mathbb{N}}$ be any sequence in $(a, b) \setminus \{c\}$ such that

$$\lim_{n \to \infty} x_n = c.$$

Then

$$\lim_{n \to \infty} f(x_n) = \lim_{x \to c} f(x) \text{ and } \lim_{n \to \infty} g(x_n) = \lim_{x \to c} g(x).$$

Thus $(f(x_n) + g(x_n))_{n \in \mathbb{N}}$ is convergent, with limit

$$\lim_{x \to c} f(x) + \lim_{x \to c} g(x),$$

that is, $((f + g)(x_n))_{n \in \mathbb{N}}$ is convergent with limit

$$\lim_{x \to c} f(x) + \lim_{x \to c} g(x).$$

So $\lim_{x \to c} (f + g)(x)$ exists, and is equal to $\lim_{x \to c} f(x) + \lim_{x \to c} g(x)$. □

In the rest of this section, we study some more notions of limits of functions, which will be convenient to use in the rest of this book.

Definition 3.5 (Limit of a function). Let $c \in (a, b)$ and $f : (a, b) \setminus \{c\} \to \mathbb{R}$. Then we have the following definitions:

(1) $\boxed{\lim_{x \to c} f(x) = \infty}$ if

$\forall M > 0, \ \exists \delta > 0$ such that if $x \in (a, b)$ and $0 < |x - c| < \delta$, then $f(x) > M$.

(2) $\boxed{\lim_{x \to c} f(x) = -\infty}$ if

$\forall M > 0, \ \exists \delta > 0$ such that if $x \in (a, b)$ and $0 < |x - c| < \delta$, then $f(x) < -M$.

Example 3.19.

(1) $\lim\limits_{x \to 0} \dfrac{1}{|x|} = \infty$.

Let $M > 0$. Let $\delta = \dfrac{1}{M} > 0$. If $0 < |x - 0| < \delta$, then $f(x) = \dfrac{1}{|x|} > \dfrac{1}{\delta} = \dfrac{1}{1/M} = M$.

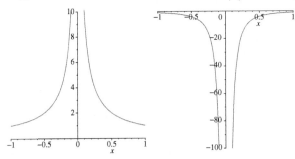

The graphs of $\dfrac{1}{|x|}$ and $-\dfrac{1}{x^2}$.

(2) $\lim\limits_{x \to 1} -\dfrac{1}{x^2} = -\infty$.

Indeed, if $M > 0$, then with $\delta := 1/\sqrt{M} > 0$, we have that whenever $0 < |x - 0| < \delta$, $0 < x^2 < \delta^2 = 1/M$, and so $f(x) = -1/x^2 < -M$. $\diamondsuit$

Now we will define *one-sided*, but *finite* limits.

Definition 3.6 (Limit of a function). Let $c \in (a, b)$ and $f : (a, b) \to \mathbb{R}$. Then we have the following definitions:

(1) $\boxed{\lim\limits_{x \searrow a} f(x) = L \text{ or } \lim\limits_{x \to a+} f(x) = L}$ if

   $\forall \epsilon > 0, \ \exists \delta > 0$ such that if $x \in (a, b)$ and $0 < x - a < \delta$, then $|f(x) - L| < \epsilon$.

(2) $\boxed{\lim\limits_{x \nearrow b} f(x) = L \text{ or } \lim\limits_{x \to b-} f(x) = L}$ if

   $\forall \epsilon > 0, \ \exists \delta > 0$ such that if $x \in (a, b)$ and $0 < b - x < \delta$, then $|f(x) - L| < \epsilon$.

Example 3.20.

(1) $\lim\limits_{x \to 0+} \lfloor x \rfloor = 0$.

   Let $\epsilon > 0$. Set $\delta = 1$. If $0 < x - 0 < 1$, then

$$|f(x) - 0| = |\lfloor x \rfloor - 0| = |0 - 0| = 0 < \epsilon.$$

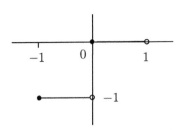

(2) $\lim\limits_{x\to 0-} \lfloor x \rfloor = -1$.

Let $\epsilon > 0$. Set $\delta = 1$. If $0 < 0 - x < 1$, that is, if $-1 < x < 0$, then

$$|f(x) - (-1)| = |\lfloor x \rfloor + 1| = |-1 + 1| = 0 < \epsilon. \qquad \diamondsuit$$

Similarly, we can define one-sided, and *infinite* limits.

Definition 3.7 (Limit of a function). Let $c \in (a, b)$ and $f : (a, b) \to \mathbb{R}$. Then we have the following definitions:

(1) $\boxed{\lim\limits_{x\searrow a} f(x) = \infty \text{ or } \lim\limits_{x\to a+} f(x) = \infty}$ if

$\forall M > 0,\ \exists \delta > 0$ such that if $x \in (a, b)$ and $0 < x - a < \delta$, then $f(x) > M$

(2) $\boxed{\lim\limits_{x\searrow a} f(x) = -\infty \text{ or } \lim\limits_{x\to a+} f(x) = -\infty}$ if

$\forall M > 0,\ \exists \delta > 0$ such that if $x \in (a, b)$ and $0 < x - a < \delta$, then $f(x) < -M$.

(3) $\boxed{\lim\limits_{x\nearrow b} f(x) = \infty \text{ or } \lim\limits_{x\to b-} f(x) = \infty}$ if

$\forall M > 0,\ \exists \delta > 0$ such that if $x \in (a, b)$ and $0 < b - x < \delta$, then $f(x) > M$.

(4) $\boxed{\lim\limits_{x\nearrow b} f(x) = -\infty \text{ or } \lim\limits_{x\to b-} f(x) = -\infty}$ if

$\forall M > 0,\ \exists \delta > 0$ such that if $x \in (a, b)$ and $0 < b - x < \delta$, then $f(x) < -M$.

Example 3.21.

(1) $\lim\limits_{x\to 0+} \dfrac{1}{x} = \infty$.

Let $M > 0$. Set $\delta = 1/M$. If $0 < x - 0 < \delta$, then $\dfrac{1}{x} > \dfrac{1}{\delta} = \dfrac{1}{1/M} = M$.

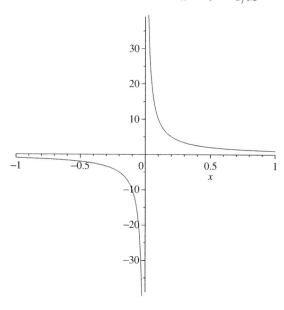

(2) $\lim\limits_{x\to 0-} \dfrac{1}{x} = -\infty$.

Let $M > 0$. Set $\delta = 1/M$. If $0 < 0 - x < \delta$, that is, $-1/x > 1/\delta$, then

$$f(x) = \frac{1}{x} < -\frac{1}{\delta} = -\frac{1}{1/M} = -M. \qquad \diamondsuit$$

Finally, let us define the notions of limiting behaviours of functions as 'the real variable x tends to ∞'.

Definition 3.8 (Limit of a function). Let $f : (a, \infty) \to \mathbb{R}$. Then we have the following definitions.

(1) $\boxed{\lim\limits_{x\to\infty} f(x) = L}$ if $\forall \epsilon > 0$, $\exists R > a$ such that if $x > R$, then $|f(x) - L| < \epsilon$.

(2) $\boxed{\lim\limits_{x\to\infty} f(x) = \infty}$ if $\forall M > 0$, $\exists R > a$ such that if $x > R$, then $f(x) > M$.

(3) $\boxed{\lim\limits_{x\to\infty} f(x) = -\infty}$ if $\forall M > 0$, $\exists R > a$ such that if $x > R$, then $f(x) < -M$.

Example 3.22.

(1) $\lim\limits_{x\to\infty} \dfrac{1}{1+x^2} = 0$.

Let $\epsilon > 0$. (Let us do some 'rough work'. We have

$$|f(x) - 0| = \frac{1}{1+x^2} < \frac{1}{1+R^2}$$

if $x > R$, and the leftmost quantity can be made smaller than ϵ if

$$R > \sqrt{\frac{1}{\epsilon} - 1}.$$

This works if $\epsilon \leq 1$. But we also need to cover the case that $\epsilon > 1$. Well, in that case, we would like an R so that for all $x > R, f(x) < \epsilon$. But $f(x)$ is *always* ≤ 1, and so *any* choice of R would do when $\epsilon > 1$.) So we set

$$R = \begin{cases} 1 & \text{if } \epsilon > 1, \\ \sqrt{\dfrac{1}{\epsilon} - 1} & \text{if } (0 <)\epsilon \leq 1. \end{cases}$$

If $\epsilon \geq 1$, we have for all $x > R$ that

$$|f(x) - 0| = \frac{1}{1+x^2} < 1 \leq \epsilon;$$

and on the other hand, if $\epsilon < 1$, then we have for $x > R$ that

$$|f(x) - 0| = \frac{1}{1+x^2} < \frac{1}{1+R^2} = \frac{1}{1 + \dfrac{1}{\epsilon} - 1} = \epsilon.$$

(2) $\lim\limits_{x\to\infty} x = +\infty$.

Let $M > 0$. Set $R = M$. Then for all $x > R$, we have $f(x) = x > R = M$. $\qquad \diamondsuit$

Similarly, we can also define the notions of limiting behaviours of functions as 'the real variable x tends to $-\infty$'.

Definition 3.9 (Limit of a function). Let $f : (-\infty, b) \to \mathbb{R}$. Then we have the following definitions.

(1) $\boxed{\lim_{x \to -\infty} f(x) = L}$ if $\forall \epsilon > 0$, $\exists R < b$ such that if $x < R$, then $|f(x) - L| < \epsilon$.

(2) $\boxed{\lim_{x \to -\infty} f(x) = \infty}$ if $\forall M > 0$, $\exists R < b$ such that if $x < R$, then $f(x) > M$.

(3) $\boxed{\lim_{x \to -\infty} f(x) = -\infty}$ if $\forall M > 0$, $\exists R < b$ such that if $x < R$, then $f(x) < -M$.

Example 3.23. $\lim_{x \to -\infty} x = -\infty$.

Let $M > 0$. Set $R = -M$. Then for all $x < R$, we have $f(x) = x < R = -M$. ◇

Summarising, we have displayed our list of definitions in the table on page 122.

Exercise 3.47. Let $f(a, c) \cup (c, b) \to \mathbb{R}$ be such that

$$\lim_{x \to c+} f(x) \quad \text{and} \quad \lim_{x \to c-} f(x)$$

exist. Show that $\lim_{x \to c} f(x)$ exists if and only if $\lim_{x \to c+} f(x) = \lim_{x \to c-} f(x)$.

Exercise 3.48. In each of the following cases, calculate the limit if it exists.

(1) $\lim_{x \to 0} \dfrac{|x|}{x + 1}$.

(2) $\lim_{x \to 1} (\lfloor x \rfloor - x)$.

(3) $\lim_{x \to 0} x \lfloor x \rfloor$.

(4) $\lim_{x \to 0} \sin \dfrac{1}{x}$.

Exercise 3.49. Let the polynomials A, B of degrees $\alpha, \beta \in \mathbb{N}$ be given by

$$A(x) = a_0 + a_1 x + \cdots + a_\alpha x^\alpha,$$
$$B(x) = b_0 + b_1 x + \cdots + b_\beta x^\beta,$$

where a_α and b_β are nonzero. Show that

$$\lim_{x \to \infty} \frac{A(x)}{B(x)} = \begin{cases} 0 & \text{if } \alpha < \beta, \\[2mm] \dfrac{a_\alpha}{b_\beta} & \text{if } \alpha = \beta, \\[2mm] +\infty & \text{if } \alpha > \beta \text{ and } \dfrac{a_\alpha}{b_\beta} > 0, \\[2mm] -\infty & \text{if } \alpha > \beta \text{ and } \dfrac{a_\alpha}{b_\beta} < 0. \end{cases}$$

$$c \in (a,b), f : (a,b) \setminus \{c\} \to \mathbb{R}, L \in \mathbb{R}$$

$\displaystyle\lim_{x \to c} f(x) = L$	$\forall \epsilon > 0 \ \exists \delta > 0$ such that if $x \in (a,b)$ and $0 < \|x - c\| < \delta$ then $\|f(x) - L\| < \epsilon$
$\displaystyle\lim_{x \to c} f(x) = \infty$	$\forall M > 0 \ \exists \delta > 0$ such that if $x \in (a,b)$ and $0 < \|x - c\| < \delta$ then $f(x) > M$
$\displaystyle\lim_{x \to c} f(x) = -\infty$	$\forall M > 0 \ \exists \delta > 0$ such that if $x \in (a,b)$ and $0 < \|x - c\| < \delta$ then $f(x) < -M$
$\displaystyle\lim_{x \to a+} f(x) = L$	$\forall \epsilon > 0 \ \exists \delta > 0$ such that if $x \in (a,b)$ and $0 < x - a < \delta$ then $\|f(x) - L\| < \epsilon$
$\displaystyle\lim_{x \to a+} f(x) = \infty$	$\forall M > 0 \ \exists \delta > 0$ such that if $x \in (a,b)$ and $0 < x - a < \delta$ then $f(x) > M$
$\displaystyle\lim_{x \to a+} f(x) = -\infty$	$\forall M > 0 \ \exists \delta > 0$ such that if $x \in (a,b)$ and $0 < x - a < \delta$ then $f(x) < -M$
$\displaystyle\lim_{x \to b-} f(x) = L$	$\forall \epsilon > 0 \ \exists \delta > 0$ such that if $x \in (a,b)$ and $0 < b - x < \delta$ then $\|f(x) - L\| < \epsilon$
$\displaystyle\lim_{x \to b-} f(x) = \infty$	$\forall M > 0 \ \exists \delta > 0$ such that if $x \in (a,b)$ and $0 < b - x < \delta$ then $f(x) > M$
$\displaystyle\lim_{x \to b-} f(x) = -\infty$	$\forall M > 0 \ \exists \delta > 0$ such that if $x \in (a,b)$ and $0 < b - x < \delta$ then $f(x) < -M$

$$f : (a, \infty) \to \mathbb{R}, L \in \mathbb{R}$$

$\displaystyle\lim_{x \to \infty} f(x) = L$	$\forall \epsilon > 0 \ \exists R > a$ such that if $x > R$ then $\|f(x) - L\| < \epsilon$
$\displaystyle\lim_{x \to \infty} f(x) = \infty$	$\forall M > 0 \ \exists R > a$ such that if $x > R$ then $f(x) > M$
$\displaystyle\lim_{x \to \infty} f(x) = -\infty$	$\forall M > 0 \ \exists R > a$ such that if $x > R$ then $f(x) < -M$

$$f : (-\infty, b) \to \mathbb{R}, L \in \mathbb{R}$$

$\displaystyle\lim_{x \to -\infty} f(x) = L$	$\forall \epsilon > 0 \ \exists R < b$ such that if $x < R$ then $\|f(x) - L\| < \epsilon$
$\displaystyle\lim_{x \to -\infty} f(x) = \infty$	$\forall M > 0 \ \exists R < b$ such that if $x < R$ then $f(x) > M$
$\displaystyle\lim_{x \to -\infty} f(x) = -\infty$	$\forall M > 0 \ \exists R < b$ such that if $x < R$ then $f(x) < -M$

Exercise 3.50. Let $a \in \mathbb{R}$. Which of the following statements is/are true?

□ (A) $\lim\limits_{x \to a} \dfrac{x}{x+a} = \dfrac{1}{2}$ if $a \neq 0$.

□ (B) $\lim\limits_{x \to a} \dfrac{x}{x+a} = 1$ if $a = 0$.

□ (C) $\lim\limits_{x \to \infty} \dfrac{x}{x+a} = 1$.

□ (D) $\lim\limits_{x \to -\infty} \dfrac{x}{x+a} = -1$.

Exercise 3.51 (Partial Fraction Expansion). Sometimes one would like to decompose a rational function into simpler rational functions; the need arises, for example when one wants to take 'inverse Laplace transforms' or in order to find the integral of a rational function. For example, the rational function

$$\frac{1}{(x+1)(x+2)(x+3)}$$

can be decomposed into the sum of elementary rational functions having the form $\dfrac{A}{(s-\alpha)^k}$:

$$\frac{1}{(x+1)(x+2)(x+3)} = \frac{1/2}{x+1} + \frac{-1}{x+2} + \frac{1/2}{x+3}.$$

In general, if the rational function $r = N/D$ is *strictly proper*, (that is, the degree of the denominator D is strictly bigger than that of the numerator), then such a decomposition is always possible:

$$R(x) = \frac{N(x)}{D(x)} = \frac{N(x)}{C(x-\alpha_1)^{m_1} \cdots (x-\alpha_K)^{m_K}}$$

$$= \sum_{k=1}^{K} \left(\frac{A_{k,1}}{x-\alpha_k} + \frac{A_{k,2}}{(x-\alpha_k)^2} + \cdots + \frac{A_{k,m_k}}{(x-\alpha_k)^{m_k}} \right). \tag{3.2}$$

The decomposition in (3.2) is called a *partial fraction expansion of R*. (If the rational function is not strictly proper, we can first divide the numerator N by the denominator D in order to get a remainder N': $N = Q \cdot D + N'$, and then write

$$R = \frac{N}{D} = \frac{Q \cdot D + N'}{D} = Q + \frac{N'}{D},$$

and carry out the partial fraction expansion on the (now strictly proper) rational function $R' := \frac{N'}{D}$.) How do we find the coefficients $A_{k,\ell}$? One way is to get a whole bunch of linear equations in the unknowns $A_{k,\ell}$ by 'taking a common denominator' on the right hand side, and 'comparing coefficients' on both sides of various powers of x. But this can be cumbersome. The aim of this exercise is to show a somewhat slicker method, based on taking limits.

(1) Consider first the case when the denominator polynomial has *distinct* real roots, so that

$$R(x) = \frac{N(x)}{D(x)} = \frac{N(x)}{C(x - \alpha_1) \cdots (x - \alpha_d)} = \frac{A_1}{x - \alpha_1} + \cdots + \frac{A_d}{x - \alpha_d}.$$

If we multiply throughout by $x - \alpha_1$, where $x \neq \alpha_1$, then

$$(x - \alpha_1)R(x) = A_1 + \frac{A_2(x - \alpha_1)}{x - \alpha_2} + \cdots + \frac{A_d(x - \alpha_1)}{x - \alpha_d},$$

and passing the limit on both sides as $x \to \alpha_1$ gives immediately that

$$A_1 = \lim_{x \to \alpha_1} (x - \alpha_1)R(x).$$

Similarly all the other coefficients can also be found out. Use this method to find the partial fraction expansion of

$$\frac{2x + 1}{x^2 - 2x - 3}.$$

(2) If there are repeated roots, then the procedure is similar. It is enough to consider each summand in the partial fraction expansion. If S_k is given by

$$S_k(x) = \frac{A_{k,1}}{x - \alpha_k} + \frac{A_{k,2}}{(x - \alpha_k)^2} + \cdots + \frac{A_{k,m_k}}{(x - \alpha_k)^{m_k}},$$

then we start from the highest power term coefficient first. Thus

$$A_{k,m_k} = \lim_{x \to \alpha_k} (x - \alpha_k)^{m_k} S_k(x).$$

Then we simply take this last term from the right hand side over to the left hand side, and repeat the procedure:

$$\widetilde{S}_k(x) := S_k(x) - \frac{A_{k,m_k}}{(x - \alpha_k)^{m_k}} = \frac{A_{k,1}}{x - \alpha_k} + \frac{A_{k,2}}{(x - \alpha_k)^2} + \cdots + \frac{A_{k,m_k-1}}{(x - \alpha_k)^{m_k-1}},$$

so that

$$A_{k,m_k-1} = \lim_{x \to \alpha_k} (x - \alpha_k)^{m_k-1} \widetilde{S}_k(x).$$

Use this method to find the partial fraction expansion of

$$\frac{x^2 + 3x + 9}{(x + 1)(x - 2)^2}.$$

Notes

Exercise 3.27 is based on [L, 6.2.5]. Exercise 3.34 is based on [A2, Exercise 4.28].

4

Differentiation

Given a function $f : (a,b) \to \mathbb{R}$, and a point $c \in (a,b)$, consider the *difference quotient* for $x \in (a,b)$, $x \neq c$:

$$\frac{f(x) - f(c)}{x - c}.$$

Geometrically, this number represents the slope of the chord passing through the points $(c,f(c))$ and $(x,f(x))$ on the graph of f.

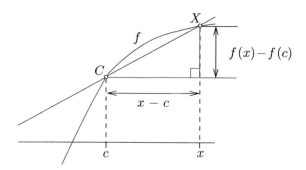

The number

$$\frac{f(x) - f(c)}{x - c}$$

is a quotient of differences (hence the name difference quotient), which is the slope of the chord XC, where $X \equiv (x,f(x))$ and $C \equiv (c,f(c))$.

What happens when x approaches c? We expect that if the above difference quotient has a limit L, this L is the instantaneous rate of change of f at c, and geometrically is the slope of the 'tangent line' to the graph of f at the point $(c,f(c))$.

The How and Why of One Variable Calculus, First Edition. Amol Sasane.
© 2015 John Wiley & Sons, Ltd. Published 2015 by John Wiley & Sons, Ltd.

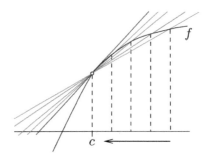

Notice also that if we 'zoom into' the graph of f around $(c, f(c))$, then the graph of f seems to coincide with the tangent line, and intuitively, the tangent line seems to provide a 'straight line or linear approximation' to f for x near c.

Definition 4.1 (Derivative of a function that is differentiable at a point). Let $f : (a, b) \to \mathbb{R}$ and $c \in (a, b)$. If there exists an $L \in \mathbb{R}$ such that

$$\lim_{x \to c} \frac{f(x) - f(c)}{x - c} = L,$$

that is, for all $\epsilon > 0$, there exists a $\delta > 0$ such that whenever $x \in (a, b)$ satisfies $0 < |x - c| < \delta$, we have

$$\left| \frac{f(x) - f(c)}{x - c} - L \right| < \epsilon,$$

then we say that f is *differentiable at the point* c, with *derivative* $f'(c) := L$ at c. Sometimes one writes

$$\frac{df}{dx}(c)$$

instead of $f'(c)$. (Note that the limit L, if it exists, is unique.)

If f is differentiable at each point in (a, b), then we say that f is *differentiable* $(on(a, b))$. We then denote the map

$$x \mapsto f'(x) : (a, b) \to \mathbb{R}$$

by f' or by $\dfrac{df}{dx}$, and call this *function* f' the *derivative of* f $(on(a, b))$.

Exercise 4.1. Use the definition to find $f'(x)$, where $f(x) := \sqrt{x^2 + 1}, x \in \mathbb{R}$.

Exercise 4.2. (∗) Let $f : (0, \infty) \to \mathbb{R}$ be a function and let $c > 0$. Show that the following are equivalent:

(1) f is differentiable at c.

(2) $\displaystyle\lim_{k \to 1} \frac{f(kc) - f(c)}{k - 1}$ exists.

Moreover, show that if (1) or (2) holds, then $f'(c) = \dfrac{1}{c} \displaystyle\lim_{k \to 1} \frac{f(kc) - f(c)}{k - 1}$.

Exercise 4.3. Let $f : (-a, a) \to \mathbb{R}$ be differentiable, and let f be an even function, that is, for all $x \in (-a, a), f(-x) = f(x)$. Show that f' is an odd function, that is, $f'(-x) = -f'(x)$ for all $x \in (-a, a)$. What is $f'(0)$?

Interpretation as the instantaneous speed. Suppose that $f : \mathbb{R} \to \mathbb{R}$ is the function describing the position $f(t)$ at time t of a particle moving along the real line.

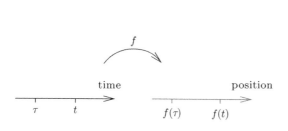
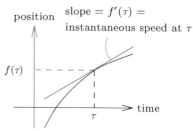

Then the difference quotient

$$\frac{f(t) - f(\tau)}{t - \tau}$$

is the average speed of the particle over the time interval $[\tau, t]$. If we take ts closer and closer to τ, so that we are looking at shorter and shorter time intervals $[\tau, t]$, then the average speeds approach the *instantaneous* speed at τ, and so the instantaneous speed at τ is

$$\lim_{t \to \tau} \frac{f(t) - f(\tau)}{t - \tau} = f'(\tau).$$

Example 4.1 (Uniform motion with constant speed v). Let $f : \mathbb{R} \to \mathbb{R}$ be given by $f(t) = vt + x_0, t \in \mathbb{R}$. Here v (which we will see is the constant instantaneous speed) is a fixed real number, and so is x_0 (which is the 'initial' position at time $t = 0$). Let $\tau \in \mathbb{R}$.

Questions: Is f differentiable at τ? If so, what is $f'(\tau)$?

For $t \neq \tau$, we have

$$\frac{f(t) - f(\tau)}{t - \tau} = \frac{vt + \cancel{x_0} - v\tau - \cancel{x_0}}{t - \tau} = \frac{v(t - \tau)}{t - \tau} = v.$$

Let $\epsilon > 0$. Take any $\delta > 0$. Then whenever $0 < |t - \tau| < \delta$, we have

$$\left| \frac{f(t) - f(\tau)}{t - \tau} - v \right| = |v - v| = 0 < \epsilon.$$

So f is differentiable at τ, and the derivative of f at τ is the (number)

$$f'(\tau) = v = \frac{df}{dt}(\tau).$$

As $\tau \in \mathbb{R}$ was arbitrary, the derivative (function) f' of f

$$\frac{df}{dt} \quad \text{or} \quad f' : \mathbb{R} \to \mathbb{R}$$

is given by $f'(t) = v$ for all $t \in \mathbb{R}$. In other words, f' is the constant function, taking value v everywhere on $\mathbb{R}$. $\diamondsuit$

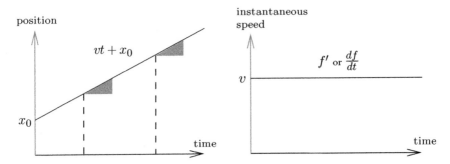

Example 4.2 (Instantaneous speed of a body falling freely under gravity). Let $f : \mathbb{R} \to \mathbb{R}$ be given by

$$f(t) = \frac{1}{2}gt^2, \quad t \in \mathbb{R}.$$

$g > 0$ is a fixed real number, and we will see that this is the constant 'acceleration' due to gravity. Let $\tau \in \mathbb{R}$.

Questions: Is f differentiable at τ? If so, what is $f'(\tau)$?

For $t \neq \tau$, we have

$$\frac{f(t) - f(\tau)}{t - \tau} = \frac{gt^2 - g\tau^2}{2 \cdot (t - \tau)} = \frac{g \cdot (t - \tau) \cdot (t + \tau)}{2 \cdot (t - \tau)} = \frac{g}{2} \cdot (t + \tau).$$

When t is close to τ, then $t + \tau \approx 2\tau$. So we *guess* that $f'(\tau) = g\tau$. (We have not proved this last equality yet; we are still at the 'rough work' stage and we will make this rigorous below.)

For $t \neq \tau$, we have

$$\left| \frac{f(t) - f(\tau)}{t - \tau} - g\tau \right| = \left| \frac{g}{2} \cdot (t + \tau) - g\tau \right| = \frac{g}{2} \cdot |t - \tau|.$$

Let $\epsilon > 0$. Set $\delta := \frac{2\epsilon}{g} > 0$. Then whenever $0 < |t - \tau| < \delta$, we have

$$\left| \frac{f(t) - f(\tau)}{t - \tau} - g\tau \right| = \frac{g}{2} \cdot |t - \tau| < \frac{g}{2} \cdot \delta = \frac{g}{2} \cdot \frac{2\epsilon}{g} = \epsilon.$$

So f is differentiable at τ, and the derivative of f at τ is the (number)

$$f'(\tau) = g\tau = \frac{df}{dt}(\tau).$$

As $\tau \in \mathbb{R}$ was arbitrary, the derivative (function) f' of f

$$\frac{df}{dt} \quad \text{or} \quad f' : \mathbb{R} \to \mathbb{R}$$

is given by $f'(t) = gt$ for all $t \in \mathbb{R}$. Thus, the instantaneous speed changes linearly with time. The rate of change of instantaneous speed is called the acceleration. Hence the acceleration is

$$(f')'(\tau) = g, \quad \tau \in \mathbb{R}.$$

(If you like, this follows from the previous example, with $v := g$ and $x_0 := 0$.) So the acceleration (due to gravity) is constant, and is equal to g. ◇

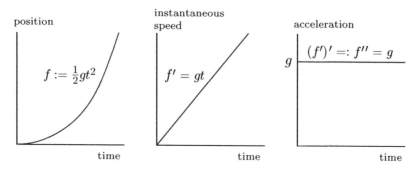

position

instantaneous speed

acceleration

$f := \frac{1}{2}gt^2$

$f' = gt$

$(f')' =: f'' = g$

time

time

time

Tangent line as a linear approximation. If $f : (a, b) \to \mathbb{R}$ is differentiable at $c \in (a, b)$, then given an $\epsilon > 0$, we have for all x sufficiently close to c, but different from c,

$$\left| \frac{f(x) - f(c)}{x - c} - f'(c) \right| < \epsilon,$$

and so $|f(x) - f(c) - f'(c) \cdot (x - c)| < \epsilon|x - c|$. So if ϵ were a number that one considers tiny, we have for $x \approx c$ that $|f(x) - f(c) - f'(c)(x - c)| \approx 0$, and so

$$f(x) \approx f(c) + \underbrace{f'(c)(x - c)}_{\substack{\textit{linear} \text{ in the} \\ \text{increment } x - c}}$$

This explains the picture below, where the graph of f near the point c looks like a straight line, with slope $f'(c)$, passing through the point $(c, f(c))$.

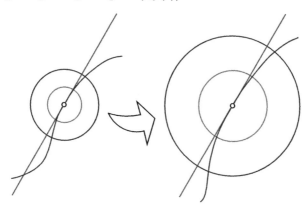

Exercise 4.4. Using a linear approximation, determine approximately the relative speed to which a particle at rest must be accelerated in order to increase its mass by 1%. You may use the fact that the mass m_v at relative speed v is related to the mass m_0 at rest by

$$m_v = \frac{m_0}{\sqrt{1 - v^2/c^2}},$$

where c is the speed of light.

Exercise 4.5 (Differentiation is a local process). Let $f, g : (a, b) \rightarrow \mathbb{R}$ be two functions that are differentiable at $c \in (a, b)$, and such that they coincide in a small neighbourhood of c, that is, there is a $\delta > 0$ (no matter how minuscule) so that for all $x \in (c - \delta, c + \delta), f(x) = g(x)$. Then show that $f'(c) = g'(c)$.

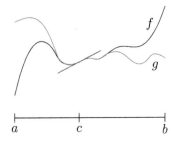

(This should be intuitively clear: the derivative at c is geometrically the slope of the tangent line to the graph of the function, and if f, g coincide in a neighborhood of c, they have the same graph around the point c, and hence the same tangent line—so same slope too!)

Continuity versus differentiability.

Theorem 4.1. *If* $f : (a, b) \rightarrow \mathbb{R}$ *is differentiable at* $c \in (a, b)$, *then* f *is continuous at* c.

So

$$\boxed{\text{differentiability of } f \text{ at } c} \Rightarrow \boxed{\text{continuity of } f \text{ at } c.}$$

Hence if a function is not continuous at a point, then it can't be differentiable there.

Proof. Let $\epsilon > 0$. Then there exists a $\delta_1 > 0$ such that for all $x \in (a, b)$ such that $0 < |x - c| < \delta_1$, we have

$$\left| \frac{f(x) - f(c)}{x - c} - f'(c) \right| < \epsilon,$$

that is, $|f(x) - f(c) - f'(c) \cdot (x - c)| < \epsilon|x - c|$, and so, using the triangle inequality, we obtain

$$|f(x) - f(c)| < \epsilon|x - c| + |f'(c)||x - c| = (\epsilon + |f'(c)|)|x - c|.$$

Of course if x is close to c, we can make the right hand side as small as we like, and in particular, less than ϵ. But how close precisely is enough? Clearly, $|x - c|$ ought to be less than $\epsilon/(\epsilon + |f'(c)|)$. This was all rough work, explaining that the following δ, was not pulled out of a hat, but is actually quite sensible.

Define

$$\delta := \min \left\{ \delta_1, \frac{\epsilon}{\epsilon + |f'(c)|} \right\}.$$

Then for all $x \in (a, b)$ such that $0 < |x - c| < \delta$, we have

$$|f(x) - f(c)| < (\epsilon + |f'(c)|)|x - c| \le (\epsilon + |f'(c)|)\frac{\epsilon}{\epsilon + |f'(c)|} = \epsilon.$$

The first inequality holds since $\delta < \delta_1$, while the latter as $|x - c| < \delta \le \epsilon/(\epsilon + |f'(c)|)$. (If $x = c$, then $|f(x) - f(c)| = |f(c) - f(c)| = |0| = 0 < \epsilon$ is trivially true.) Hence we have shown that for every $\epsilon > 0$, there exists a $\delta > 0$ such that whenever $x \in (a, b)$ satisfies $|x - c| < \delta$, there holds $|f(x) - f(c)| < \epsilon$, that is, f is continuous at c.   □

However:

$$\boxed{\text{continuity of } f \text{ at } c} \not\Rightarrow \boxed{\text{differentiability of } f \text{ at } c.}$$

that is, the converse of the theorem is not true, and the following example demonstrates this.

Example 4.3. The function $f : \mathbb{R} \to \mathbb{R}$ defined by $f(x) = |x|$ $(x \in \mathbb{R})$ is (uniformly) continuous since

$$\left| f(x) - f(y) \right| = \left| |x| - |y| \right| \le |x - y|.$$

But let us now show that f is not differentiable at 0. If it were, then with $\epsilon := \frac{1}{2} > 0$, there exists a $\delta > 0$ such that whenever $0 < |x| < \delta$, we have

$$\left| \frac{|x|}{x} - f'(0) \right| < \epsilon = \frac{1}{2}.$$

In particular, if we take $x = \delta/2$, then we obtain

$$|1 - f'(0)| < \frac{1}{2}. \tag{4.1}$$

On the other hand, if we take $x = -\delta/2$, we also get

$$|-1 - f'(0)| < \frac{1}{2}. \tag{4.2}$$

But from (4.1), (4.2), and the triangle inequality, we now obtain

$$2 = |-1 - 1| = |-1 - f'(0) + f'(0) - 1| \le |-1 - f'(0)| + |f'(0) - 1| < \frac{1}{2} + \frac{1}{2} = 1,$$

a contradiction. (The lack of differentiability of $|\cdot|$ at 0 is visually obvious, since one can't draw a tangent to the graph at the 'corner' at $(0, 0)$.)   ◇

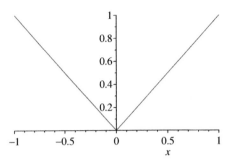

Remark 4.1. (∗) In connection with Theorem 4.1, we might wonder how badly behaved continuous functions might be with respect to the notion of differentiability. It turns out that there are functions that are *continuous everywhere*, but *differentiable nowhere*! One construction is that of the so-called *blancmange function* obtained by taking the basic *sawtooth* function f_1,

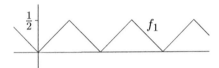

and constructing $f_2, f_3, \ldots$ by setting $f_n(x) = \left(\dfrac{1}{2}\right)^{n-1} f_1(2^{n-1}x)$, and adding these:

$$b(x) = \sum_{n=1}^{\infty} f_n(x), \quad x \in \mathbb{R}.$$

Then it can be shown that b is continuous on $\mathbb{R}$, but not differentiable at any $x \in \mathbb{R}$. (Intuitively, the graph of the function has no breaks, but no matter which point one takes, the function is so jagged that it fails to be differentiable at that point, and one can't draw a tangent line to the graph of the function at any point.)

Exercise 4.6. Let $f : \mathbb{R} \to \mathbb{R}$ be defined by

$$f(x) = \begin{cases} x^2 & \text{if } x \in \mathbb{Q}, \\ 0 & \text{if } x \in \mathbb{R} \backslash \mathbb{Q}. \end{cases}$$

Show that f is differentiable at 0. What can you say about the differentiability of f at nonzero real numbers?

Exercise 4.7. If $f : (a, b) \to \mathbb{R}$ is differentiable at $c \in (a, b)$, then show that

$$\lim_{h \searrow 0} \frac{f(c + h) - f(c - h)}{2h}$$

exists and equals $f'(c)$. Is the converse true?

Exercise 4.8 (Differentiable but not continuously differentiable example). Consider the function $f : \mathbb{R} \rightarrow \mathbb{R}$ defined by $f(0) = 0$ and for $x \neq 0$,

$$f(x) = x^2 \sin \frac{1}{x}.$$

Prove that f is differentiable, but f' is not continuous at 0.

Exercise 4.9. Let $f : (a, b) \rightarrow \mathbb{R}$ and let $c \in (a, b)$. Which of the following statements is/are true?

□ (A) If f is continuous at c, then f is differentiable at c.

□ (B) If f is differentiable at c, then f is continuous at c.

□ (C) If f is continuous at c, then f^2 is continuous at c.

□ (D) If f^2 is differentiable at c, then f is differentiable at c. (*Hint:* Consider $|x|$.)

The algebra of derivatives. We have already seen the 'Algebra of Limits' and the 'Algebra of Continuous Functions', and by now we know that when we say 'Algebra of $\cdots$', we mean that the relevant concept is well-behaved/respects the algebraic combinations such as taking pointwise sums, products, and so on.

One can show the following result about rules for differentiating the sum and product of differentiable functions.

Proposition 4.2 (Algebra of Derivatives). *Let $f, g : (a, b) \rightarrow \mathbb{R}$ be functions that are differentiable at $c \in (a, b)$. Then:*

(1) *The sum $f + g : (a, b) \rightarrow \mathbb{R}$ is differentiable at c, and*

$$(f + g)'(c) = f'(c) + g'(c).$$

(2) (Product Rule or the Leibniz Rule) *The product $fg : (a, b) \rightarrow \mathbb{R}$ is differentiable at c, and*

$$(fg)'(c) = f'(c)g(c) + f(c)g'(c).$$

(3) *If $f(x) \neq 0$ for $x \in (a, b)$, then $\frac{1}{f} : (a, b) \rightarrow \mathbb{R}$ is differentiable at c, and*

$$\left(\frac{1}{f}\right)'(c) = -\frac{f'(c)}{(f(c))^2}.$$

Proof. These claims follow from the Algebra of Limits, namely Theorem 3.15. Indeed,

$$\lim_{x \to c} \frac{(f + g)(x) - (f + g)(c)}{x - c} = \lim_{x \to c} \frac{f(x) + g(x) - f(c) - g(c)}{x - c}$$

$$= \lim_{x \to c} \frac{f(x) - f(c) + g(x) - g(c)}{x - c}$$

$$= \lim_{x \to c} \frac{f(x) - f(c)}{x - c} + \lim_{x \to c} \frac{g(x) - g(c)}{x - c}$$

$$= f'(c) + g'(c),$$

which proves (1). Also, (2) follows from the following:

$$\lim_{x \to c} \frac{(fg)(x) - (fg)(c)}{x - c} = \lim_{x \to c} \frac{f(x)g(x) - f(c)g(c)}{x - c}$$

$$= \lim_{x \to c} \frac{f(x)g(x) - f(c)g(x) + f(c)g(x) - f(c)g(c)}{x - c}$$

$$= \lim_{x \to c} \frac{f(x) - f(c)}{x - c} \cdot g(x) + \lim_{x \to c} f(c) \cdot \frac{g(x) - g(c)}{x - c}$$

$$= \lim_{x \to c} \frac{f(x) - f(c)}{x - c} \cdot \lim_{x \to c} g(x) + f(c) \lim_{x \to c} \frac{g(x) - g(c)}{x - c}$$

$$= f'(c)g(c) + f(c)g'(c),$$

where we have used the continuity of g at c (thanks to the fact that g is differentiable at c):

$$\lim_{x \to c} g(x) = g(c).$$

This completes the proof of (2).

Finally, let us show (3). We have

$$\lim_{x \to c} \frac{\left(\frac{1}{f}\right)(x) - \left(\frac{1}{f}\right)(c)}{x - c} = \lim_{x \to c} \frac{\frac{1}{f(x)} - \frac{1}{f(c)}}{x - c}$$

$$= \lim_{x \to c} -\frac{f(x) - f(c)}{x - c} \cdot \frac{1}{f(x)} \cdot \frac{1}{f(c)}$$

$$= -f'(c) \cdot \frac{1}{f(c)} \cdot \frac{1}{f(c)} = -\frac{f'(c)}{(f(c))^2}.$$

This completes the proof. □

Example 4.4.

(1) A repeated application of the Product Rule shows that the power function $x \mapsto x^n : \mathbb{R} \to \mathbb{R}$ ($n \in \mathbb{N}$) is differentiable and has the derivative $x \mapsto nx^{n-1}$:

$n = 1 :$ $(x^1)' = 1.$

$n = 2 :$ $(x^2)' = (x \cdot x)' = x' \cdot x + x \cdot x' = 1 \cdot x + x \cdot 1 = 2x.$

$n = 3 :$ $(x^3)' = (x^2 \cdot x)' = (x^2)' \cdot x + x^2 \cdot x' = 2x \cdot x + x^2 \cdot 1 = 3x^2.$

$$\cdots$$

$$(x^{n+1})' = (x^n \cdot x)' = (x^n)' \cdot x + x^n \cdot x' = nx^{n-1} \cdot x + x^n \cdot 1 = (n+1)x^n.$$

(2) A special case of the Product Rule is when $g \equiv c$ (constant). Then

$$(cf)' = c' \cdot f + c \cdot f' = 0 \cdot f + c \cdot f' = c \cdot f'.$$

(3) All polynomial functions are differentiable: if p is given by

$$p(x) = c_0 + c_1 x + c_2 x^2 + c_3 x^3 + \cdots + c_d x^d, \quad x \in \mathbb{R},$$

then $p'(x) = c_1 + 2c_2 x + 3c_3 x^2 + \cdots + dc_d x^{d-1}$, $x \in \mathbb{R}$. So the derivative of a polynomial of degree $d \in \mathbb{N}$ is a polynomial of degree $d - 1$.

(4) The function $x \mapsto 1/x : \mathbb{R} \setminus \{0\} \to \mathbb{R}$ is differentiable for all $x \neq 0$:

$$\frac{d}{dx}\left(\frac{1}{x}\right) = -\frac{x'}{x^2} = -\frac{1}{x^2}, \quad x \neq 0.$$

More generally, if $n \in \mathbb{N}$, then $\dfrac{d}{dx}\left(\dfrac{1}{x^n}\right) = -\dfrac{nx^{n-1}}{x^{2n}} = -\dfrac{n}{x^{n+1}}$, for $x \neq 0$. Thus

$$\frac{d}{dx} x^n = nx^{n-1} \tag{4.3}$$

also for $n < 0$. And clearly this also holds for $n = 0$. Hence (4.3) holds for all $n \in \mathbb{Z}$ and for all $x \in \mathbb{R} \setminus \{0\}$. ◇

Exercise 4.10. If $f, g, h : (a, b) \to \mathbb{R}$ are all differentiable at $c \in (a, b)$, then show that

$$(fgh)'(c) = f'(c)g(c)h(c) + f(c)g'(c)h(c) + f(c)g(c)h'(c).$$

Exercise 4.11. Let $a, b, c, d : \mathbb{R} \to \mathbb{R}$ be differentiable. Let $W : \mathbb{R} \to \mathbb{R}$ be defined by

$$W(x) = \det \begin{bmatrix} a(x) & b(x) \\ c(x) & d(x) \end{bmatrix}, \quad x \in \mathbb{R}.$$

Show that

$$W'(x) = \det \begin{bmatrix} a'(x) & b(x) \\ c'(x) & d(x) \end{bmatrix} + \det \begin{bmatrix} a(x) & b'(x) \\ c(x) & d'(x) \end{bmatrix}, \quad x \in \mathbb{R}.$$

Exercise 4.12. If $f(x) = x^3/3 - x^2/2 - 2x$, find

(1) $\{x \in \mathbb{R} : f'(x) = 0\}$.

(2) $\{x \in \mathbb{R} : f(x) > 0\}$.

Plot the graphs of f, f' in the interval $[-3, 3]$ using the computer. What do you observe?

Exercise 4.13. Let p be the polynomial given by

$$p(x) = 1 + \frac{x}{1!} + \frac{x^2}{2!} + \cdots + \frac{x^n}{n!}, \quad x \in \mathbb{R}.$$

Show that p can't have a repeated real root. (A polynomial p is said to have a repeated root α if $p(x) = (x - \alpha)^2 q(x)$ for some polynomial q.)

Corollary 4.1 (Quotient Rule). *Let*

(1) $f, g : (a, b) \to \mathbb{R}$ *be differentiable at* $c \in (a, b)$,

(2) $g(x) \neq 0$ *for all* $x \in (a, b)$, *and*

(3) $\dfrac{f}{g} : (a, b) \to \mathbb{R}$ *be defined by* $f \cdot \dfrac{1}{g}$, *that is,* $\left(\dfrac{f}{g}\right)(x) = \dfrac{f(x)}{g(x)}$, $x \in (a, b)$.

Then $\dfrac{f}{g}$ *is differentiable at* c, *and* $\left(\dfrac{f}{g}\right)'(c) = \dfrac{f'(c)g(c) - f(c)g'(c)}{(g(c))^2}$.

Proof. Using the Product Rule, we obtain

$$\left(\frac{f}{g}\right)'(c) = f'(c) \cdot \frac{1}{g(c)} + f(c) \cdot \left(\frac{1}{g}\right)'(c) = \frac{f'(c)g(c)}{(g(c))^2} + f(c) \cdot \left(-\frac{g'(c)}{(g(c))^2}\right)$$

$$= \frac{f'(c)g(c) - f(c)g'(c)}{(g(c))^2}.$$
□

Example 4.5.

(1) $\left(\dfrac{x^3}{1+x^2}\right)' = \dfrac{3x^2 \cdot (1+x^2) - x^3 \cdot (2x)}{(1+x^2)^2} = \dfrac{3x^2 + 3x^4 - 2x^4}{(1+x^2)^2} = \dfrac{3x^2 + x^4}{(1+x^2)^2}.$

(2) We will prove later on that $\sin' = \cos$ and $\cos' = -\sin$. We will accept these facts for now. For real x such that $\cos x \neq 0$, we have

$$\tan'x = \left(\frac{\sin}{\cos}\right)'(x) = \frac{(\sin'x)(\cos x) - (\sin x)(\cos'x)}{(\cos x)^2}$$

$$= \frac{(\cos x)(\cos x) - (\sin x)(-\sin x)}{(\cos x)^2} = \frac{(\cos x)^2 + (\sin x)^2}{(\cos x)^2}$$

$$= \frac{1}{(\cos x)^2} = (\sec x)^2.$$
◇

Exercise 4.14. Show that

$$\cot'x = -(\operatorname{cosec} x)^2,$$

$$\sec'x = (\tan x)(\sec x),$$

$$\operatorname{cosec}'x = -(\cot x)(\operatorname{cosec} x).$$

Exercise 4.15. Differentiate $\dfrac{4\sin x}{2x + \cos x}$.

4.1   Differentiable Inverse Theorem

Let $f : (a, b) \to \mathbb{R}$ be one-to-one on (a, b). Then we know that we can define its inverse $f^{-1} : f((a, b)) \to (a, b)$.

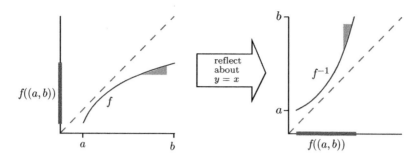

By looking at the fate of the little triangle when we reflect in the 45° line, we can guess what happens to the derivatives:

$$(f^{-1})'(f(c)) = \frac{1}{f'(c)}.$$

This is the content of the[1] Differentiable Inverse Theorem.

Theorem 4.2 (Differentiable Inverse Theorem). *If $f(a, b) \to \mathbb{R}$ is such that*

(1) *f is strictly increasing (or strictly decreasing),*

(2) *f is continuous,*

(3) *f is differentiable at $c \in (a, b)$, and*

(4) *$f'(c) \neq 0$,*

then $f^{-1} : f((a, b)) \to (a, b)$ is differentiable at $f(c)$ and $(f^{-1})'(f(c)) = \frac{1}{f'(c)}.$

Proof. We want to show that

$$\lim_{y \to f(c)} \frac{f^{-1}(y) - f^{-1}(f(c))}{y - f(c)} = \frac{1}{f'(c)}.$$

We will use Theorem 3.13 to show this. Let $(y_n)_{n \in \mathbb{N}}$ be any sequence with terms belonging to $f((a, b)) \backslash \{f(c)\}$, that converges to $f(c)$. Then we have that $y_n = f(x_n)$ for some sequence $(x_n)_{n \in \mathbb{N}} \in (a, b) \backslash \{c\}$. We want to show that

$$\lim_{n \to \infty} \frac{f^{-1}(y_n) - f^{-1}(f(c))}{y_n - f(c)} = \lim_{n \to \infty} \frac{x_n - c}{f(x_n) - f(c)} = \frac{1}{f'(c)}.$$

But by the continuity of f^{-1}, we know that since $y_n \to f(c)$, it follows that

$$f^{-1}(y_n) \to f^{-1}(f(c)),$$

that is, $x_n \to c$.

Since

$$\lim_{x \to c} \frac{f(x) - f(c)}{x - c} = f'(c),$$

---

[1] very important!

and as $(x_n)_{n\in\mathbb{N}} \in (a,b)\setminus\{c\}$ converges to c, we obtain (by Theorem 3.13 again) that

$$\lim_{n\to\infty} \frac{f(x_n) - f(c)}{x_n - c} = f'(c),$$

that is, $\lim_{n\to\infty} \underbrace{\frac{y_n - f(c)}{x_n - c}}_{\neq 0} = f'(c) \neq 0$, and so $\lim_{n\to\infty} \frac{x_n - c}{f(x_n) - f(c)} = \frac{1}{f'(c)}$. Done!   □

Example 4.6 (Derivative of the nth root function).

Let $n \in \mathbb{N}$. Then the function $x \xmapsto{f} x^n : (0,\infty) \to (0,\infty)$ is

(1) strictly increasing,

(2) continuous,

(3) differentiable on $(0,\infty)$, and

(4) $f'(x) = nx^{n-1} \neq 0$ for all $x \in (0,\infty)$.

Thus by the Differentiable Inverse Theorem, we conclude that the inverse of f,

$$y \xmapsto{f^{-1}} \sqrt[n]{y} : (0,\infty) \to (0,\infty),$$

is differentiable on $(0,\infty)$, and if $y = x^n$, where $x = \sqrt[n]{y} \in (0,\infty)$, then

$$(\sqrt[n]{y})' = \frac{1}{f'(x)} = \frac{1}{nx^{n-1}} = \frac{1}{n(\sqrt[n]{y})^{n-1}} = \frac{1}{ny^{1-\frac{1}{n}}} = \frac{1}{n}y^{\frac{1}{n}-1},$$

for all $y > 0$. So $(y^{\frac{1}{n}})' = \frac{1}{n}y^{\frac{1}{n}-1}$ for all $y > 0$.   ◇

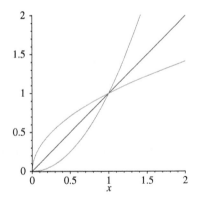

Example 4.7 (Derivative of the arcsin function).

The function $\sin : \left(-\frac{\pi}{2}, \frac{\pi}{2}\right) \to (-1,1)$ is

(1) strictly increasing,

(2) continuous,

(3) differentiable on $(-\pi/2, \pi/2)$, and

(4) $\sin'x = \cos x \neq 0$ for all $x \in (-\pi/2, \pi/2)$.

By the Differentiable Inverse Theorem, the inverse function

$$\sin^{-1} : (-1, 1) \rightarrow \left(-\frac{\pi}{2}, \frac{\pi}{2}\right)$$

is differentiable and for $y = \sin x \in (-1, 1)$, where $x \in \left(-\frac{\pi}{2}, \frac{\pi}{2}\right)$ we have

$$(\sin^{-1})'(y) = \frac{1}{\sin'x} = \frac{1}{\cos x} = \frac{1}{\sqrt{1 - (\sin x)^2}} = \frac{1}{\sqrt{1 - y^2}}.$$

◇

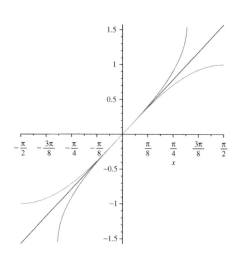

Example 4.8 (In the Differentiable Inverse Theorem, the condition $f'(c) \neq 0$ is not superfluous). The function $x \xmapsto{f} x^3 : \mathbb{R} \rightarrow \mathbb{R}$ is

(1) strictly increasing,

(2) continuous,

(3) differentiable on $\mathbb{R}$.

Note that

$$f'(0) = 3 \cdot 0^2 = 0.$$

Now let us show that f^{-1} is *not* differentiable at $f(0) = 0$.

Indeed, for $x > 0$,

$$\frac{f^{-1}(x) - f^{-1}(0)}{x - 0} = \frac{\sqrt[3]{x} - 0}{x - 0} = \frac{1}{x^{\frac{2}{3}}},$$

and if

$$x_n := \frac{1}{n}, \quad n \in \mathbb{N},$$

then

$$\frac{f^{-1}(x_n) - f^{-1}(0)}{x_n - 0} = n^{\frac{2}{3}},$$

and $(n^{\frac{2}{3}})_{n \in \mathbb{N}}$ does *not* converge. This shows that there is no real number L such that

$$\lim_{x \to 0} \frac{f^{-1}(x) - f^{-1}(0)}{x - 0} = L.$$

Hence f^{-1} is not differentiable at $f(0) = 0$.

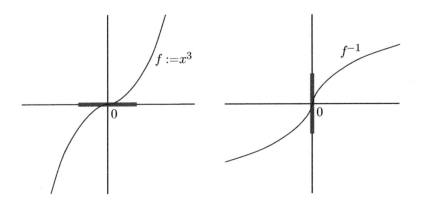

Exercise 4.16. Let $f : (0, \infty) \to \mathbb{R}$ be the strictly decreasing function given by

$$f(x) = \frac{1}{1 + x^2}, \quad x \in (0, \infty).$$

From Exercise 3.33, it follows that $f((0, \infty)) = (0, 1)$.

Show that $f^{-1} : (0, 1) \to (0, \infty)$ is differentiable, and find the derivative $(f^{-1})'$.

Exercise 4.17. Let $f : \left(-\frac{\pi}{2}, \frac{\pi}{2}\right) \to \mathbb{R}$ be the strictly increasing function given by

$$f(x) = x + \sin x, \quad x \in \left(-\frac{\pi}{2}, \frac{\pi}{2}\right).$$

Find the derivative $(f^{-1})'(0)$.

4.2   The Chain Rule

We will now learn how to differentiate the composition of functions, and the pertinent rule is called the 'Chain Rule'. This is one of the powerful tools that lends Calculus its name, that is, it is a rule, which is conveniently applied, and results in great calculational ease. It will make differentiation of complicated functions rather simple, by using this mechanical procedure. We will see this soon enough in the examples, but first let us state the Chain Rule.

Theorem 4.5 (Chain Rule). *Suppose that*

(1) $f : (a, b) \to \mathbb{R}$,
(2) $g : (A, B) \to \mathbb{R}$,
(3) $f((a, b)) \subset (A, B)$,
(4) f *is differentiable at* $c \in (a, b)$, *and*
(5) g *is differentiable at* $f(c) \in (A, B)$.

Then $g \circ f : (a, b) \to \mathbb{R}$ *is differentiable at* c, *and* $(g \circ f)'(c) = g'(f(c)) \cdot f'(c)$.

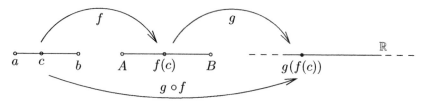

Example 4.9. We will see later on that $\sin' x = \cos x, x \in \mathbb{R}$. Using the Chain Rule, we obtain

$$\frac{d}{dx}\left(\sin(x^2)\right) = \left(\cos(x^2)\right) \cdot 2x, \quad x \in \mathbb{R}, \text{ and}$$

$$\frac{d}{dx}\left(\sin\frac{1}{x}\right) = \left(\cos\frac{1}{x}\right) \cdot \left(-\frac{1}{x^2}\right), \quad x \neq 0.$$

The Chain Rule can be applied repeatedly (forming a 'chain' of derivatives—hence the name of the rule!): for example,

$$\frac{d}{dx}\sin\left(\sin(x^2)\right) = \left(\cos\left(\sin(x^2)\right)\right) \cdot \left(\cos(x^2)\right) \cdot 2x, \quad x \in \mathbb{R}.$$

The above examples show the great power associated with this rule.   ◇

Proof of the Chain Rule. For $x \neq c$, we have

$$\frac{(g \circ f)(x) - (g \circ f)(c)}{x - c} = \frac{g(f(x)) - g(f(c))}{x - c},$$

and we would like to show that as $x \to c$, the above converges to $g'(f(c)) \cdot f'(c)$. It is tempting to multiply and divide by $f(x) - f(c)$ and write

$$\frac{g(f(x)) - g(f(c))}{x - c} = \frac{g(f(x)) - g(f(c))}{f(x) - f(c)} \cdot \frac{f(x) - f(c)}{x - c}$$

and say

'as $x \to c$, $\dfrac{f(x) - f(c)}{x - c} \to f'(c)$,

$f(x) \to f(c)$, and

$\dfrac{g(f(x)) - g(f(c))}{f(x) - f(c)} \to g'(f(c))$, completing the proof'.

However, even for $x \neq c$, and near c, it may happen that $f(x) - f(c) = 0$ (for example when $f \equiv f(c)$), and so division by $f(x) - f(c)$ is not possible then, rendering the above invalid. So instead, we proceed as follows.

Since g is differentiable at $f(c)$,

$$\lim_{y \to f(c)} \left(\frac{g(y) - g(f(c))}{y - f(c)} - g'(f(c)) \right) = 0.$$

So if we define $\varphi : (A, B) \to \mathbb{R}$ by

$$\varphi(y) = \begin{cases} \dfrac{g(y) - g(f(c))}{y - f(c)} - g'(f(c)) & \text{if } y \neq f(c), \\ 0 & \text{if } y = f(c), \end{cases}$$

then φ is continuous at $f(c)$ (in fact on (A, B)). Note that for $y \neq f(c)$,

$$g(y) - g(f(c)) = g'(f(c)) \cdot (y - f(c)) + \varphi(y)(y - f(c)). \tag{4.4}$$

Also if $y = f(c)$, then the left hand side of (4.4) is 0, and so is the right hand side. So (4.4) holds for *all* $y \in (A, B)$. In particular, for $x \in (a, b) \setminus \{c\}$, and with $y := f(x) \in (A, B)$, we have

$$g(f(x)) - g(f(c)) = g'(f(c)) \cdot (f(x) - f(c)) + \varphi(f(x))(f(x) - f(c)). \tag{4.5}$$

Dividing throughout by $x - c$, we obtain

$$\frac{g(f(x)) - g(f(c))}{x - c} = g'(f(c)) \cdot \frac{f(x) - f(c)}{x - c} + \varphi(f(x)) \cdot \frac{f(x) - f(c)}{x - c}.$$

But as $x \to c$, we have that $f(x) \to f(c)$, and so $\varphi(f(x)) \to \varphi(f(c)) = 0$. Hence

$$\lim_{x \to c} \frac{g(f(x)) - g(f(c))}{x - c} = g'(f(c)) \cdot f'(c) + 0 \cdot f'(c) = g'(f(c)) \cdot f'(c).$$

This completes the proof. □

Example 4.10. The function $x \mapsto \sqrt{1 - x^2} : (-1, 1) \to \mathbb{R}$ is the composition $g \circ f$, where

$$x \xmapsto{f} 1 - x^2 : (-1, 1) \to (0, \infty),$$

$$x \xmapsto{g} \sqrt{x} : (0, \infty) \to \mathbb{R}.$$

Thus, by the Chain Rule,

$$\frac{d}{dx} \sqrt{1 - x^2} = (g \circ f)'(x) = g'(f(x)) \cdot f'(x) = \frac{1}{2\sqrt{1 - x^2}} \cdot (-2x) = \frac{-x}{\sqrt{1 - x^2}},$$

for $x \in (-1, 1)$. ◇

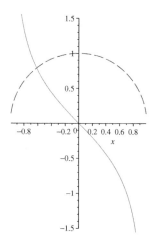

Example 4.11 (Derivative of rational powers).

Suppose that $r \in \mathbb{Q}$. We write

$$r = \frac{m}{n},$$

where $m \in \mathbb{Z}$ and $n \in \mathbb{N}$. We claim that $x \mapsto x^r : (0, \infty) \to (0, \infty)$ is differentiable, and that

$$\frac{d}{dx} x^r = r x^{r-1}, \quad x > 0.$$

Indeed, $x^r = x^{\frac{m}{n}} = (x^{\frac{1}{n}})^m$ is the composition $g \circ f$, where

$$x \stackrel{f}{\mapsto} \sqrt[n]{x} : (0, \infty) \to (0, \infty),$$

$$x \stackrel{g}{\mapsto} x^m : \mathbb{R} \to \mathbb{R}.$$

Thus by the Chain Rule,

$$\frac{d}{dx} x^r = m\left(x^{\frac{1}{n}}\right)^{m-1} \cdot \frac{1}{n} x^{\frac{1}{n}-1} = \frac{m}{n} x^{\frac{m}{n}-1} = r x^{r-1}, \quad x > 0.$$

For example, $\dfrac{d}{dx} x^{\frac{3}{2}} = \dfrac{3}{2}\sqrt{x}$, and $\dfrac{d}{dx} x^{-\frac{3}{2}} = -\dfrac{3}{2} x^{-\frac{5}{2}}$ for $x > 0$. $\diamondsuit$

Exercise 4.18. Find $f'(x)$ if $f(x) := \sin(\cos(1 + x^2))$, $x \in \mathbb{R}$. (You may use the fact that $\sin' = \cos$ and $\cos' = -\sin$.)

Exercise 4.19. (∗) Evaluate $\displaystyle\sum_{k=1}^{n} \frac{k^2}{2^k}$ using the function S given by $S(x) = \displaystyle\sum_{k=1}^{n} x^k$.

Exercise 4.20. Using the binomial expansion $(1 + x)^n = \displaystyle\sum_{k=0}^{n} \binom{n}{k} x^k$, find expressions for

(1) $\displaystyle\sum_{k=1}^{n} 3^k \binom{n}{k}$.

(2) $\displaystyle\sum_{k=1}^{n} k^2 \binom{n}{k}$.

(3) $\displaystyle\sum_{k=1}^{n} (2k+1) \binom{n}{k}$.

Exercise 4.21. Let $f : (-a, a) \to \mathbb{R}$ be differentiable, and let f be an odd function, that is, for all $x \in (-a, a)$, $f(-x) = -f(x)$. Show that f' is an even function, that is, $f'(-x) = f'(x)$ for all $x \in (-a, a)$, by calculating the derivative of $x \mapsto f(x) + f(-x) = 0$.

Exercise 4.22. Complete the following table:

x	$f(x)$	$f'(x)$	$g(x)$	$g'(x)$	$(f \circ g)(x)$	$(f \circ g)'(x)$	$(g \circ f)(x)$	$(g \circ f)'(x)$
0	1	1	3	3				
1	2	-9	2	9				
2	0	7	1	9				
3	3	6	0	-3				

Exercise 4.23. Complete the following table:

f	f'	$f' \circ f$	$f \circ f'$	$(f \circ f)'$
$1/x^3$				
$\cos x$				
x^3				
3				
$3x$				

Exercise 4.24. Let $f, g : (0, \infty) \to \mathbb{R}$ be differentiable. Find g' in terms of f, f' in each of the cases below, assuming that the given relation holds for all $x > 0$.

(1) $g(x) = f(x^2)$.

(2) $g(x) = (f(x))^2$.

(3) $g(x) = f(f(x))$.

(4) $g(x^2) = f(x)$.

4.3 Higher order derivatives and derivatives at boundary points

Successive or higher order derivatives

Differentiation applied successively leads to the notion of higher order derivatives.

Definition 4.2 (Higher order derivatives). Suppose that $f : (a, b) \to \mathbb{R}$ is differentiable. So we have its derivative function $f' : (a, b) \to \mathbb{R}$. If f' is itself differentiable at $c \in (a, b)$, we say that f is *twice differentiable at c*, and denote the derivative of f' at c by $f''(c)$. $f''(c)$ is called the *second order derivative of f at c*.

If f' is differentiable at every point of (a, b), then use the notation f'' for the mapping $x \mapsto f''(x) : (a, b) \to \mathbb{R}$.

If f'' is also differentiable at c, then we say that f is *thrice differentiable at c*, and denote the derivative of f'' at c by $f'''(c)$. Similarly, one defines n *times differentiable at c*, and the *nth order derivative $f^{(n)}$* for any $n \in \mathbb{N}$.

Instead of the notation

$$f''(c), \quad f'''(c), \quad f^{(n)}(c),$$

sometimes we use

$$\frac{d^2f}{dx^2}(c), \quad \frac{d^3f}{dx^3}(c), \quad \frac{d^nf}{dx^n}(c).$$

If for all $n \in \mathbb{N}$, f is n times differentiable at c, then we say that f is *infinitely (many times) differentiable at c*.

If f is infinitely differentiable at each point of an open interval (a, b), then we write

$$f \in C^\infty(a, b).$$

Elements of $C^\infty(a, b)$ are 'very smooth'. For example, all polynomials belong to $C^\infty(\mathbb{R})$. Differentiating a polynomial gives again a polynomial (of degree one less than the original one), which is in turn differentiable. Differentiating a polynomial of degree d $d + 1$ times gives the zero function. Later on, we will see other important examples: $\sin, \cos, \exp \in C^\infty(\mathbb{R})$, and $\log \in C^\infty(0, \infty)$.

Example 4.12. Let $k \in \mathbb{N}$, $c_k, a \in \mathbb{R}$ be fixed. Consider the polynomial p_k given by

$$p_k(x) = c_k(x - a)^k, \quad x \in \mathbb{R}.$$

Then we have

$p_k(x) = c_k \cdot (x - a)^k$	$p_k(a) = 0$
$p_k'(x) = c_k \cdot k \cdot (x - a)^{k-1}$	$p_k'(a) = 0$
$p_k''(x) = c_k \cdot k \cdot (k - 1) \cdot (x - a)^{k-2}$	$p_k''(a) = 0$
$p_k'''(x) = c_k \cdot k \cdot (k - 1) \cdot (k - 2) \cdot (x - a)^{k-3}$	$p_k'''(a) = 0$
$\cdots$	$\cdots$
$p_k^{(k-1)}(x) = c_k k(k - 1)(k - 2) \cdots (k - (k - 2))(x - a)^{k-(k-1)}$ $= c_k k(k - 1)(k - 2) \cdots 2(x - a)$	$p_k^{(k-1)}(a) = 0$
$p_k^{(k)}(x) = c_k k(k - 1)(k - 2) \cdots 2 \cdot 1 = c_k k!$	$p_k^{(k)}(a) = c_k k!$
$p_k^{(\ell)}(x) = 0$ for all $\ell > k$	$p_k^{(\ell)}(a) = 0, \ell > k.$

Conclusion: $p_k^{(\ell)}(a) = c_k \cdot k! \cdot \delta_{k\ell}$, where $\delta_{k\ell}$ is the *Kronecker delta*, given by

$$\delta_{k\ell} = \begin{cases} 1 & \text{if } k = \ell \\ 0 & \text{if } k \neq \ell. \end{cases}$$

We will use this later on when we discuss Taylor's Formula.   ◇

Exercise 4.25. Let $f : (0, \infty) \to \mathbb{R}$ be such that for all $x, y \in (0, \infty)$, $f(xy) = f(x) + f(y)$. If f is differentiable at 1, then show that f is differentiable at every $c \in (0, \infty)$ and that $f'(c) = f'(1)/c$. Conclude that f is infinitely differentiable. If $f'(1) = 2$, then find $f^{(n)}(3)$, $n \in \mathbb{N}$.

Exercise 4.26.

(1) Show that if f, g are twice differentiable on an open interval I, then

$$(fg)''(x) = f''(x) \cdot g(x) + 2 \cdot f'(x) \cdot g'(x) + f(x) \cdot g''(x), \quad x \in I.$$

(2) (∗) Prove that if f, g are infinitely differentiable on an open interval I, then

$$(fg)^{(n)}(x) = \sum_{k=0}^{n} \binom{n}{k} f^{(k)}(x) g^{(n-k)}(x), \quad x \in I.$$

(3) (∗) For a rational x and n a nonnegative integer, define $x^{[n]} := x(x-1)\cdots(x-n+1)$.

Show that if $x, y \in \mathbb{Q}$, then $(x+y)^{[n]} = \sum_{k=0}^{n} \binom{n}{k} x^{[k]} y^{[n-k]}$.

Hint: Differentiate t^{x+y} n times with respect to $t \in I := (0, \infty)$.

(In fact, after we have defined logarithms, and how to take real exponents, that is, the map $t \mapsto t^x : (0, \infty) \to \mathbb{R}$, where $x \in \mathbb{R}$, one can see that the same result holds even when $x, y \in \mathbb{R}$.)

Exercise 4.27. Let $f, g : \mathbb{R} \to \mathbb{R}$ be twice differentiable. The second order derivative of the map $x \mapsto (f \circ g)(x) = f(g(x)) : \mathbb{R} \to \mathbb{R}$ at the point c is

☐ (A) $f''(g(c)) \cdot g''(c)$.

☐ (B) $f''(g(c)) \cdot f'(g'(c)) \cdot g''(c)$.

☐ (C) $f''(g(c)) \cdot f'(g'(c)) \cdot f(g''(c))$.

☐ (D) $f''(g(c)) \cdot (g'(c))^2 + f'(g(c)) \cdot g''(c)$.

Exercise 4.28. Let $k \in \mathbb{N}$. What is $f^{(k)}(x)$ if f is given by

(1) $\dfrac{1}{(x-a)^n}$, $x \in \mathbb{R}\setminus\{a\}$. (Here $n \in \mathbb{N}$ and $a \in \mathbb{R}$ are fixed.)

(2) $\dfrac{1}{x^2 - 1}$, $x \in \mathbb{R}\setminus\{-1, 1\}$.

Exercise 4.29 (Differentiable, but not twice). Let $f : \mathbb{R} \to \mathbb{R}$ be given by $f(x) = x|x|$, $x \in \mathbb{R}$. Show that f is differentiable on $\mathbb{R}$, and find the derivative function f'. Is f' differentiable at 0? Write down a formula for a function that is n times differentiable everywhere, but such that $f^{(n)}$ is not differentiable at 0.

Derivatives at boundary points

Let $f : [a, b] \to \mathbb{R}$. We know what it means for f to be differentiable at an *interior* point of $[a, b]$, that is, at any point in the *open* interval (a, b).

$$a \qquad c \qquad\qquad b$$

But what about differentiability at the *endpoints* a and b? To take care of these hitherto omitted definitions, we now introduce the notions of the *left* derivative and the *right* derivative.

Definition 4.3 (Differentiability at a point; derivative). Let $f : [a, b] \to \mathbb{R}$.

(1) We say f *is differentiable at a* if its *right derivative $f'_+(a)$ at a* exists, that is, there exists a number $f'_+(a) \in \mathbb{R}$ such that

$$\lim_{x \to a+} \frac{f(x) - f(a)}{x - a} = f'_+(a).$$

In other words, for every $\epsilon > 0$, there exists a $\delta > 0$ such that whenever $0 < x - a < \delta$, we have

$$\left| \frac{f(x) - f(a)}{x - a} - f'_+(a) \right| < \epsilon.$$

(2) We say f *is differentiable at b* if its *left derivative $f'_-(b)$ at b* exists, that is, there exists a number $f'_-(b) \in \mathbb{R}$ such that

$$\lim_{x \nearrow b} \frac{f(x) - f(a)}{x - a} = f'_-(a).$$

In other words, for every $\epsilon > 0$, there exists a $\delta > 0$ such that whenever $0 < b - x < \delta$, we have

$$\left| \frac{f(x) - f(b)}{x - b} - f'_-(b) \right| < \epsilon.$$

(3) We say f *is differentiable* (on $[a, b]$) *iff is differentiable at each $x \in [a, b]$, and we call the map $f' : [a, b] \to \mathbb{R}$ given by*

$$f'(x) = \begin{cases} f'_+(a) & \text{if } x = a, \\ f'(x) & \text{if } x \in (a, b), \\ f'_-(b) & \text{if } x = b, \end{cases}$$

the derivative of f.

(4) *If f' is differentiable on $[a, b]$, then f is called* twice differentiable *and we let f'' denote the derivative of f', and so on.*

Just like $C[a, b] := \{f : [a, b] \to \mathbb{R} : f$ is continuous on $[a, b]\}$, one defines

$$C^1[a, b] := \{f : [a, b] \to \mathbb{R} : f \text{ is differentiable on } [a, b] \text{ and } f' \in C[a, b]\},$$

and more generally, for $n \geq 2$,

$$C^n[a, b] := \{f : [a, b] \to \mathbb{R} : f \text{ is n times differentiable on } [a, b] \text{ and } f^{(n)} \in C[a, b]\}.$$

Note that $C^1[a, b] \subsetneq C[a, b]$. (Why is there *strict* inclusion?)

4.4   Equations of tangent and normal lines to a curve

We can consider the graph of a function $f : (a, b) \to \mathbb{R}$ as a 'curve' in $\mathbb{R}^2$:

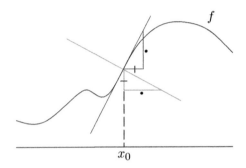

Then since $f'(x_0)$ is the slope of the tangent line to the graph of f at $(x_0, f(x_0))$, the *equation of the tangent line to the graph of f at $(x_0, f(x_0))$* is given by

$$\frac{y - f(x_0)}{x - x_0} = f'(x_0).$$

By the congruence of the two triangles shown in the picture above we see that the slope of the *normal line* at $(x_0, f(x_0))$ is, assuming $f'(x_0) \neq 0$,

$$-\frac{1}{f'(x_0)},$$

and so the *equation of the normal line to the graph of f at $(x_0, f(x_0))$* is given by

$$\frac{y - f(x_0)}{x - x_0} = -\frac{1}{f'(x_0)}.$$

Example 4.13. Consider the function f given by $f(x) = x^3 - 6x^2 + 8x$, $x \in \mathbb{R}$. What is the tangent line to the graph of f at the point $(3, f(3))$?

We have $f(3) = -3$, and $f'(x) = 3x^2 - 12x + 8$, so that $f'(3) = -1$. Hence the tangent line is given by

$$\frac{y - (-3)}{x - 3} = -1,$$

that is, $y = -x$. Similarly the normal line to the graph of f at $(3, -3)$ is

$$\frac{y - (-3)}{x - 3} = -\frac{1}{-1} = 1,$$

that is, $y = x - 6$.

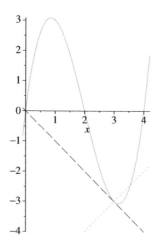

A *curve* γ is a function $t \mapsto \gamma(t) = (x(t), y(t)) : (a, b) \to \mathbb{R}^2$, where the maps $x, y : (a, b) \to \mathbb{R}$ are continuous.

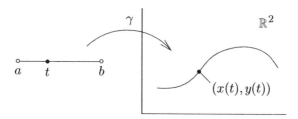

If the curve is 'smooth' (that is, if x, y are smooth, say continuously differentiable), then let us determine the equations of the tangent line and the normal line to the curve at a point $(x(t_0), y(t_0))$ on the curve, assuming that $x'(t_0)$ and $y'(t_0)$ are nonzero.

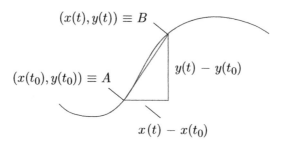

If we look at the picture above, we see that the slope of the chord AB is (assuming that $x(t) \neq x(t_0)$ for $0 < |t - t_0|$ small enough and $x'(t_0) \neq 0$)

$$\frac{y(t) - y(t_0)}{x(t) - x(t_0)} = \frac{\dfrac{y(t) - y(t_0)}{t - t_0}}{\dfrac{x(t) - x(t_0)}{t - t_0}} \xrightarrow{t \to t_0} \frac{y'(t_0)}{x'(t_0)},$$

and so the equation of the tangent line to the curve at the point $(x(t_0), y(t_0))$ is given by

$$\frac{y - y(t_0)}{x - x(t_0)} = \frac{y'(t_0)}{x'(t_0)}.$$

Similarly, the equation of the normal line to the curve at the point $(x(t_0), y(t_0))$ is (assuming that $y'(t_0) \neq 0$)

$$\frac{y - y(t_0)}{x - x(t_0)} = -\frac{x'(t_0)}{y'(t_0)}.$$

Example 4.14. Let $\gamma : \mathbb{R} \to \mathbb{R}^2$ be the curve given by $\gamma(t) = (\cos t, \sin t)$, $t \in \mathbb{R}$. Let $t_0 := \pi/4 \in \mathbb{R}$. What are the equations of the tangent line and the normal line to the curve at the point $(\cos t_0, \sin t_0) = (1/\sqrt{2}, 1/\sqrt{2})$ on the curve?

The tangent line is given by

$$\frac{y - \sin t_0}{x - \cos t_0} = \frac{\cos t_0}{-\sin t_0},$$

that is, $x \cos t_0 + y \sin t_0 = 1$, that is, (with $t_0 = \pi/4$) $x + y = \sqrt{2}$.

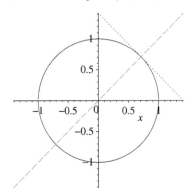

The normal line is given by

$$\frac{y - \sin t_0}{x - \cos t_0} = -\frac{-\sin t_0}{\cos t_0},$$

that is, $y = x \tan t_0$, that is, (with $t_0 = \pi/4$) $y = x$. Note that the normal line passes through the origin $(0,0)$, and this is expected since we know from elementary Euclidean geometry that for a circle, a line joining the center to a point on the circle is perpendicular to the tangent line at that point.   ◇

Remark 4.2. The graph of a function f is a curve where

$$x(t) := t,$$

$$y(t) := f(t).$$

Then the equation of the tangent line at $(x_0, f(x_0))$ is

$$\frac{y - f(x_0)}{x - x_0} = \frac{f'(x_0)}{1},$$

which matches what we had obtained earlier.

Exercise 4.30. Find the values of the constants a, b, c for which the graphs of f, g given by

$$f(x) := x^2 + ax + b,$$

$$g(x) := x^3 - c,$$

$x \in \mathbb{R}$, intersect at the point $(1, 2)$ and have the same tangent there.

Exercise 4.31. Find the equation of the tangent at $(\frac{1}{9}, 5)$ to the curve $t \mapsto (x(t), y(t))$, where

$$x(t) := \frac{1}{t^2},$$

$$y(t) := \sqrt{t^2 + 16},$$

for $t \in (0, \infty)$.

Exercise 4.32. Which of the following statements is always true about the tangent line to the graph of a differentiable function $f : \mathbb{R} \to \mathbb{R}$ at the point $(c, f(c))$?

☐ (A) The tangent line intersects the curve at precisely one point, namely at $(c, f(c))$.

☐ (B) The tangent line intersects the x-axis.

☐ (C) The tangent line intersects the y-axis.

☐ (D) Any line through $(c, f(c))$ other than the tangent line intersects the graph of f at at least one other point.

Implicitly defined curves

There are plane curves given by an equation of the form

$$F(x, y) = 0,$$

where $F : \mathbb{R}^2 \to \mathbb{R}$ is a nice function. Here we imagine solving for y, given a value of the variable x, to obtain $x \overset{y}{\mapsto} y(x)$ for some suitable function y. The curve is then $x \mapsto (x, y(x))$.

For example, if $F(x, y) := y - x^2 - 1$, then $y(x) = x^2 + 1$, and the curve is a parabola:

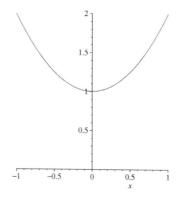

Remark 4.3. An explicit formula for an implicitly defined curve may not be available; for example

$$F(x, y) := y^5 + 2y^4 - 7y^3 + 3y^2 - 6y - x.$$

Given x, finding $y(x)$ here entails solving a quintic equation:

$$y^5 + 2y^4 - 7y^3 + 3y^2 - 6y - x = 0.$$

And even if a formula is available, it might only be 'locally valid'. For example, if F is given by $F(x, y) := x^2 + y^2 - 1$ (so that $F(x, y) = 0$ describes a circle), then

$$y(x) = \begin{cases} \sqrt{1 - x^2} & \text{if } y \geq 0, \\ -\sqrt{1 - x^2} & \text{if } y < 0. \end{cases}$$

To find the equation for the tangent and normal line to an implicitly defined curve, we need to find $y'(x)$, but this *can* be done without first explicitly finding y. The process, based on the Chain Rule, is called *Implicit Differentiation*, and the best way to see this is by looking at an example.

Example 4.15 (Tangents and normals to the circle using implicit differentiation). Let $F(x, y) := x^2 + y^2 - 1$. Then the point (x, y) on the curve described implicitly by F satisfies

$$x^2 + y^2 - 1 = 0.$$

Bearing in mind that y is a function of x, and viewing both sides of the above equation as functions of x, we obtain by differentiating with respect to x that (suppressing the argument of y—that is, writing y instead of $y(x)$ everywhere below)

$$2x + 2y \cdot \frac{dy}{dx} - 0 = 0,$$

so that if $y(x) \neq 0$, then

$$\frac{dy}{dx} = -\frac{x}{y}.$$

So at a point (x_0, y_0) on the circle, the equation of the tangent line is

$$\frac{y - y_0}{x - x_0} = -\frac{x_0}{y_0},$$

and using $x_0^2 + y_0^2 = 1$, this can be simplified to $yy_0 + xx_0 = 1$.

Similarly, the equation of the normal line is (assuming $x_0 \neq 0$)

$$\frac{y - y_0}{x - x_0} = \frac{y_0}{x_0},$$

that is, $y = \dfrac{y_0}{x_0}x.$ ◇

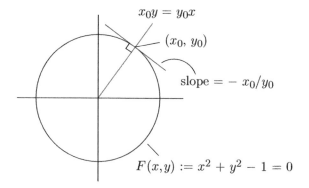

$$x_0 y = y_0 x$$

(x_0, y_0)

$$\text{slope} = -\, x_0/y_0$$

$$F(x, y) := x^2 + y^2 - 1 = 0$$

Example 4.16. Let $F(x, y) := y^5 + x^5 - y - 3x + 2$. Then $F(1, 1) = 0$. What is the equation of the tangent line to the curve implicitly defined by F at the point $(1, 1)$? We have

$$5y^4 \frac{dy}{dx} + 5x^4 - \frac{dy}{dx} - 3 = 0$$

and so $\dfrac{dy}{dx} = \dfrac{3 - 5x^4}{5y^4 - 1}$. So the slope of the tangent line at $(1, 1)$ is

$$\frac{dy}{dx}\bigg|_{(x,y)=(1,1)} = \frac{3 - 5x^4}{5y^4 - 1}\bigg|_{(x,y)=(1,1)} = \frac{3 - 5}{5 - 1} = -\frac{1}{2}.$$

Hence the equation of the tangent line is $\dfrac{y - 1}{x - 1} = -\dfrac{1}{2}$, that is, $x + 2y = 3$.   ◇

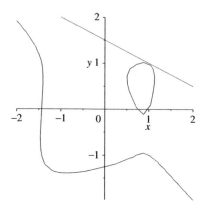

Exercise 4.33. Find the points on the curve given implicitly by $x^2 + xy + y^2 = 9$ at which

(1) the tangent is parallel to the x-axis

(2) the tangent is parallel to the y-axis.

Exercise 4.34. Find the tangents to the implicitly defined curve $x\sin(xy - y^2) = x^2 - 1$ at the point $(1, 1)$.

Exercise 4.35. Consider the implicitly defined $x^{2/3} + y^{2/3} = 2$, where $y = y(x)$ is a function of x locally around $(x, y) = (1, 1)$.

(1) Compute $\dfrac{dy}{dx}$ and $\dfrac{d^2y}{dx^2}$ when $x = 1$.

(2) Use a computer package such as Maple to plot the implicitly defined curve for positive values of x and y. Also in the same picture draw the line at $(1, 1)$ with slope $\dfrac{dy}{dx}\Big|_{x=1}$ found above. Do you observe tangency at $(1, 1)$?

(3) From the plots obtained in the previous part, can you explain the sign of $\dfrac{d^2y}{dx^2}\Big|_{x=1}$?

Exercise 4.36. Show that the curve defined implicitly by the equation $xy^3 + x^3y = 4$ has no horizontal tangent.

The Newton–Raphson Method for solving $f(x) = 0$ numerically

Given a nice function $f : (a, b) \to \mathbb{R}$, consider the problem of finding x such that $f(x) = 0$. Even for a relatively simple f, this may not be solvable 'analytically'; for example, if $f(x) := x - \cos x$.

When finding an *exact* solution is hopeless, one might settle for an *approximate* one. We discuss one such method given by Newton, which has a simple geometric idea behind it: for ξ near the desired point x_* we seek such that $f(x_*) = 0$, we may approximate the function f by its linear approximation at ξ (given by $f(x) \approx f(\xi) + f'(\xi)(x - \xi)$), and get an approximate value for x_* as

$$x_* \approx \xi - \frac{f(\xi)}{f'(\xi)}.$$

Thus we replace the graph of f by its tangent line, and find out where it intersects the x-axis, as shown below.

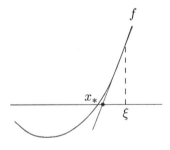

This process may be repeated (by replacing the old value of ξ by new approximate value of x_* just discovered), and one obtains the following algorithm:

Want $x_* : f(x_*) = 0$.

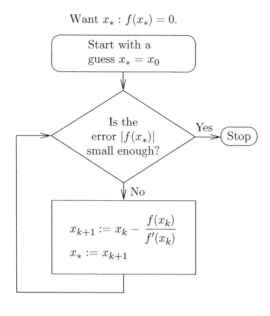

Step 1. Pick any point $(x_0, f(x_0))$ on the graph of f. This x_0 is an 'initial guess'. Draw the tangent line to the graph of f at $(x_0, f(x_0))$. Let x_1 be the x-coordinate of the point where the tangent line meets the x-axis.

Step 2. Take $(x_1, f(x_1))$ and repeat Step 1, giving a new point x_2, and so on.

Update equation: The equation of the tangent line at $(x_n, f(x_n))$ is

$$\frac{y - f(x_n)}{x - x_n} = f'(x_n),$$

and so when $y = 0$, we have $\dfrac{0 - f(x_n)}{x_{n+1} - x_n} = f'(x_n)$, that is,

$$x_{n+1} = x_n - \frac{f(x_n)}{f'(x_n)}.$$

The hope is that the sequence $x_1, x_2, x_3, \cdots$ converges to the desired solution x_* that satisfies $f(x_*) = 0$, but it might not.

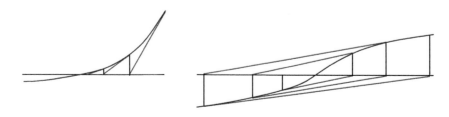

Example 4.17. Let us carry out the Newton–Raphson Method for the function f defined by $f(x) := x - \cos x$, $x \in \mathbb{R}$. Then $f'(x) = 1 + \sin x$, and so the update equation is

$$x_{n+1} = x_n - \frac{f(x_n)}{f'(x_n)} = x_n - \frac{x_n - \cos x_n}{1 + \sin x_n}, \quad n \geq 0.$$

Let us start with an initial guess of $x_0 := 2$. The table below shows successive iterations in the Newton–Raphson Method.

n	x_n (to 4 decimal places)	$f(x_n) = x_n - \cos x_n$ (to 4 decimal places)
0	2	2.4161
1	0.7345	−0.0077
2	0.7391	0
3	0.7391	0

Thus we see that the approximate value of x_* such that $f(x_*) = 0$ to 4 decimal places is given by $x_* \approx 0.7391$. ◇

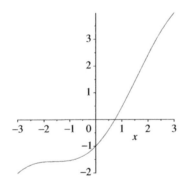

Exercise 4.37. Let f be given by $f(x) = x^2 - 2$. Suppose that the Newton–Raphson Method converges with the initial guess $x_0 := 1$. Generate a few rational approximations to $\sqrt{2}$.

Exercise 4.38. Show that $x^4 - x^3 - 75 = 0$ for some x lying between 3 and 4. Use the Newton–Raphson Method to find an approximate value of this x, starting with an initial guess of $x_0 := (3 + 4)/2 = 3.5$.

Exercise 4.39. Let $f : \mathbb{R} \to \mathbb{R}$ be defined by

$$f(x) = \begin{cases} \sqrt{x} & \text{if } x \geq 0, \\ -\sqrt{-x} & \text{if } x < 0. \end{cases}$$

Show that if the 'initial guess' $x_0 \neq 0$, then the corresponding sequence of iterates, generated by the Newton–Raphson Method for the equation $f(x) = 0$, does not converge.

4.5   Local minimisers and derivatives

Intuitively, we expect that when a function $f : (a, b) \to \mathbb{R}$ has a local bump or a local trough, then at the highest or lowest point x_* of the bump/trough, the tangent line should be horizontal, that is, the slope $f'(x_*) = 0$. The aim of this section is to prove this result.

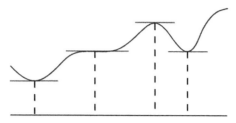

Definition 4.4 (Local minimum). We say that $f : (a, b) \to \mathbb{R}$ has a *local minimum* at $c \in (a, b)$ if there exists a $\delta > 0$ such that whenever $x \in (a, b)$ satisfies $|x - c| < \delta$, we have $f(x) \geq f(c)$.

In other words, 'locally' around c, the value assumed by f at c is the smallest. See Figure 4.1.

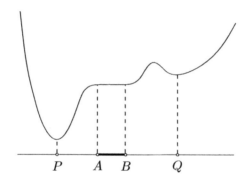

Figure 4.1   The points P, Q and all points in the interior of the line segment AB are all local minimisers.

Local maximisers are defined likewise.

Theorem 4.6. *If*

(1) $f : (a, b) \to \mathbb{R}$ *has a local minimum at* $x_* \in (a, b)$, *and*

(2) f *is differentiable at* x_*,

then $f'(x_*) = 0$.

Roughly,

$$\boxed{x_* \text{ is an interior local minimiser of } f} \Rightarrow \boxed{f'(x_*) = 0}.$$

An analogous result holds for a local maximiser. (Just consider $-f$, and note that x_* is a local maximiser of f if and only if x_* is a local minimiser of $-f$.)

Proof. Let $\epsilon > 0$. Then we know that there exists a $\delta > 0$ such that whenever $x \in (a, b)$ satisfies $0 < |x - x_*| < \delta$, we have

$$\left| \frac{f(x) - f(x_*)}{x - x_*} - f'(x_*) \right| < \epsilon.$$

But for all x near x_*, we also have $f(x) \geq f(x_*)$. So for such x satisfying also $0 < |x - x_*| < \delta$, we have

$$\frac{f(x) - f(x_*)}{x - x_*} - f'(x_*) < \epsilon, \tag{4.6}$$

$$-\frac{f(x) - f(x_*)}{x - x_*} + f'(x_*) < \epsilon, \tag{4.7}$$

$$f(x) \geq f(x_*). \tag{4.8}$$

Next, among these x, look at the ones $> x_*$. Then (4.6) gives:

$$\epsilon > \frac{f(x) - f(x_*)}{x - x_*} - f'(x_*) \geq 0 - f'(x_*), \text{ that is, } -f'(x_*) < \epsilon. \tag{4.9}$$

Also, by considering x for which (4.7) holds and $x < x_*$, we obtain:

$$\epsilon > -\frac{f(x) - f(x_*)}{x - x_*} + f'(x_*) \geq 0 + f'(x_*), \text{ that is, } f'(x_*) < \epsilon. \tag{4.10}$$

(4.9) and (4.10) imply $|f'(x_*)| < \epsilon$. But the choice of $\epsilon > 0$ was arbitrary, and hence we conclude that $f'(x_*) = 0$. $\qquad \square$

Example 4.18 (Tin manufacturing company). What is the smallest area of a tin can of cylindrical shape whose volume is specified to be $2\pi \cdot 1000 \approx 1845 \text{ cm}^3$?

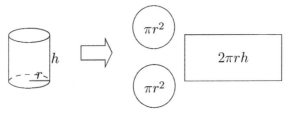

Let the radius of the tin can be r and the height be h. Its volume is $V := \pi r^2 h = 2\pi \cdot 1000$, so that

$$h = 2 \cdot \frac{1000}{r^2}.$$

Its surface area is

$$2 \cdot \pi r^2 + 2\pi r \cdot h = 2\pi r^2 + 2\pi r \cdot 2 \cdot \frac{1000}{r^2} = 2\pi r^2 + 2\pi \cdot 2 \cdot \frac{1000}{r} =: S(r).$$

Thus the function to be minimised is $r \xmapsto{S} S(r) : (0, \infty) \to \mathbb{R}$. If r_* is a minimiser, then we must have $S'(r_*) = 0$, that is,

$$S'(r_*) = 2\pi \cdot 2r_* - 2\pi \cdot 2 \cdot \frac{1000}{r_*^2} = 0,$$

and so $r_*^3 = 1000$. Hence $r_* = 10$ cm. The corresponding height is

$$h_* := 2 \cdot \frac{1000}{10^2} = 2 \cdot 10 = 20 \text{ cm.}$$

$\diamondsuit$

Exercise 4.40. Find the shortest distance from a given point $(0, b)$ on the y-axis with $b > 0$, to the parabola $y = x^2$.

Exercise 4.41. Does f given by $f(x) = (\sin x - \cos x)^2, x \in \mathbb{R}$, have a maximum value? If so, find it.

Exercise 4.42 (Snell's Law of Refraction via Fermat's Principle of Least Time). Consider the following example from the study of optics, where a light ray passes from air into glass, as shown in the picture below. The light ray bends or 'refracts' in a manner governed by Snell's Law (1621):

$$\frac{\sin \theta_a}{\sin \theta_b} = \mu,$$

where μ is the 'refractive index of glass with respect to air'. In 1662, Fermat derived Snell's Law on the basis of his Principle of Least Time, which says that light takes the path for which travel time is the smallest. Supposing that the speed of light in air is 1, and in glass, it is $1/\mu$, derive Snell's Law. See Figure 4.2.

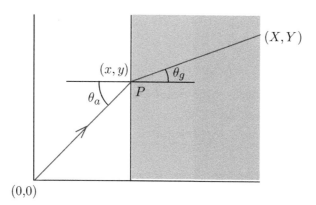

Figure 4.2 The y coordinate of the point P is considered to be a variable; x, X, Y are fixed and $x < X$. Fermat's Principle of Least Time says that, light is in a hurry, and the actual path taken by the light ray is the one corresponding to that special y for which the travel time from $(0, 0)$ to (X, Y) is the smallest.

4.6 Mean Value, Rolle's, Cauchy's Theorem

Now we will learn about three *equivalent* results, which will be used extensively in Calculus. For example, we will soon use them to prove Taylor's Formula, and later on to prove the most important result in Calculus, the Fundamental Theorem of Calculus, which will link the two disparate worlds of integration and differentiation.

Theorem 4.7 (Mean-Value Theorem).

If $f : [a, b] \rightarrow \mathbb{R}$ is

(1) *continuous on $[a, b]$, and*

(2) *differentiable on (a, b),*

then there exists a $c \in (a, b)$ such that $\dfrac{f(b) - f(a)}{b - a} = f'(c)$.

Theorem 4.8 (Rolle's Theorem).

If $f : [a, b] \rightarrow \mathbb{R}$

(1) *is continuous on $[a, b]$,*

(2) *is differentiable on (a, b), and*

(3) *$f(a) = f(b)$,*

then there exists a $c \in (a, b)$ such that $f'(c) = 0$.

Theorem 4.9 (Cauchy's (Generalised Mean Value) Theorem).

If $f, g : [a, b] \rightarrow \mathbb{R}$ are

(1) *continuous on $[a, b]$, and*

(2) *differentiable on (a, b),*

then there exists a $c \in (a, b)$ such that $(f(b) - f(a))g'(c) = (g(b) - g(a))f'(c)$.

Let us first note that if in Cauchy's Theorem, we take $g = x$, then we obtain the existence of a c such that $(f(b) - f(a)) \cdot 1 = (b - a)f'(c)$, and upon rearranging, we get

$$\frac{f(b) - f(a)}{b - a} = f'(c),$$

which is the conclusion in the Mean Value Theorem. So the Cauchy Theorem implies the Mean Value Theorem.

On the other hand, if in the Mean Value Theorem, we have that f also satisfies $f(a) = f(b)$, then the result tells us that there is a $c \in (a, b)$ such that

$$f'(c) = \frac{f(b) - f(a)}{b - a} = \frac{0}{b - a} = 0,$$

which is the conclusion in Rolle's Theorem. So the Mean Value Theorem implies Rolle's Theorem.

Thus:

$$\boxed{\text{Cauchy's Theorem}} \quad \Rightarrow \quad \boxed{\text{Mean Value Theorem}} \quad \Rightarrow \quad \boxed{\text{Rolle's Theorem}}.$$

But now we will prove Rolle's Theorem, and prove that each of the reverse implications hold too. Before doing so, let us explain where the Mean Value Theorem gets its name from.

Why 'Mean Value'?

If we think of $[a, b]$ as a time interval and $f(t)$ as being the position at time t of a particle moving along the real line, then

$$\frac{f(b) - f(a)}{b - a} = \frac{\text{total displacement}}{\text{time taken}} = \text{average or mean speed over } [a, b].$$

At some time instances, the instantaneous speed could have been *more* than this mean speed, while at other times *less* than the mean speed. The Mean Value Theorem says that some time instance c, the instantaneous speed $f'(c)$ was exactly *equal* to the mean speed![2]

Geometrically the Mean Value Theorem is intuitively expected, since if we look at the chord AB in the plane that joins the end points $A \equiv (a, f(a))$ and $B \equiv (b, f(b))$ of the graph of f, then the Mean Value Theorem is telling us that there is a point $c \in (a, b)$, where the tangent to f at the point $C \equiv (c, f(c))$ is parallel to the chord AB.

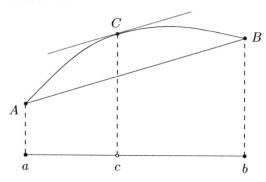

Exercise 4.43. Show that on the graph of any quadratic polynomial $x \overset{p}{\mapsto} p(x)$ (that is, polynomial of degree 2), the chord joining the points for which $x = a$ and $x = b$ is parallel to the tangent line at the midpoint $x = (a + b)/2$.

Proofs of the three theorems

Before we show Rolle's Theorem, let us remark that Rolle's Theorem is also visually obvious: if we think of $f(a) = f(b) = 0$ as specifying the horizontal seal level and f as describing the landscape profile, there must be a highest/lowest point where f' must vanish!

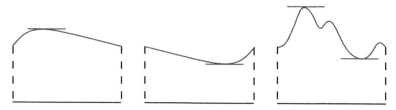

---

[2] After learning about the Fundamental Theorem of Calculus, we will also see that

$$\frac{f(b) - f(a)}{b - a} = \frac{1}{b - a} \int_a^b f'(t)dt,$$

and we may view the right hand side as an average/mean of all instantaneous speeds $f'(t)$ for t in $[a, b]$.

Proof of Rolle's Theorem (Theorem 4.8). If f is constant, then $f(x) = f(a)$ for all $x \in [a, b]$, and so $f' \equiv 0$. So any $c \in (a, b)$ does the job!

Now suppose that f is not constant. We have two cases:

1° There exists an $x \in [a, b]$ such that $f(x) > f(a) = f(b)$. Then clearly this x can't be a or b, and so $x \in (a, b)$. By the Extreme Value Theorem, there exists a $c \in [a, b]$ such that $f(c) \geq f(\xi)$ for all $\xi \in [a, b]$. Again, this c can't be a or b (since we already know that there are values (x!) where f takes bigger values than $f(a) = f(b)$). So $c \in (a, b)$ is such that for all $\xi \in (a, b), f(c) \geq f(\xi)$, that is, c is a maximiser of $f : (a, b) \to \mathbb{R}$. Consequently, $f'(c) = 0$.

2° There exists an $x \in [a, b]$ such that $f(x) < f(a) = f(b)$. This is analogous to 1°. (Or follows from 1° by looking at $-f$.)

This completes the proof of Rolle's Theorem. □

Example 4.19. The polynomial $p := 6x^5 + 13x + 1$ has *exactly one* real root.

By the Intermediate Value Theorem, it has *at least one* real root, because

$$p(0) = 6 \cdot 0^5 + 13 \cdot 0 + 1 = 1 > 0, \text{ and}$$

$$p(-1) = 6 \cdot (-1)^5 + 13 \cdot (-1) + 1 = -18 < 0.$$

But now by Rolle's Theorem, we also know *there can't be more than one zero*: For if $p(a) = p(b) = 0$ for $a < b$, then there would exist a $c \in (a, b)$ such that $p'(c) = 0$, but $p'(c) = 30c^4 + 13 \geq 13 > 0$, a contradiction! ◇

Proof of the Mean Value Theorem (Theorem 4.7). Define $\varphi : [a, b] \to \mathbb{R}$ by

$$\varphi(x) = (f(b) - f(a))x - (b - a)f(x), \quad x \in (a, b).$$

Then φ is continuous on $[a, b]$, differentiable on (a, b) and

$$\varphi(a) = (f(b) - f(a))a - (b - a)f(a)$$

$$= f(b)a - bf(a)$$

$$= (f(b) - f(a))b - (b - a)f(b)$$

$$= \varphi(b).$$

Moreover, for $x \in (a, b)$, we have $\varphi'(x) = f(b) - f(a) - (b - a)f'(x)$. By Rolle's Theorem, $\varphi'(c) = 0$ for some $c \in (a, b)$. Rearranging, we obtain

$$\frac{f(b) - f(a)}{b - a} = f'(c),$$

and this completes the proof of the Mean Value Theorem. □

Example 4.20. For all $x > 0$, $\sqrt{1+x} < 1 + \dfrac{1}{2}x$.

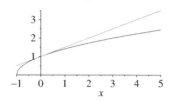

Consider $f : [0, \infty) \to \mathbb{R}$ defined by $f(x) = \sqrt{1+x}$ for $x \geq 0$. Then f is continuous on $[0, \infty)$, differentiable on $[0, \infty)$ and

$$f'(x) = \frac{1}{2\sqrt{1+x}}, \quad x \geq 0.$$

Fix $x > 0$. Applying the Mean Value Theorem to f on $[0, x]$, it follows that for some c satisfying $0 < c < x$, we have

$$\frac{\sqrt{1+x} - 1}{x} = \frac{f(x) - f(0)}{x - 0} = f'(c) = \frac{1}{2\sqrt{1+c}} < \frac{1}{2\sqrt{1+0}} = \frac{1}{2},$$

and so $\sqrt{1+x} < 1 + \dfrac{1}{2}x.$ ◇

Proof of Cauchy's Theorem (Theorem 4.9). Let $\varphi : [a, b] \to \mathbb{R}$ be defined by

$$\varphi(x) = \det \begin{bmatrix} f(x) & g(x) & 1 \\ f(a) & g(a) & 1 \\ f(b) & g(b) & 1 \end{bmatrix}, \quad x \in [a, b].$$

We have

$$\varphi(a) = \det \begin{bmatrix} f(a) & g(a) & 1 \\ f(a) & g(a) & 1 \\ f(b) & g(b) & 1 \end{bmatrix} = 0$$

since the first and second rows of the matrix are linearly dependent. Similarly,

$$\varphi(b) = \det \begin{bmatrix} f(b) & g(b) & 1 \\ f(a) & g(a) & 1 \\ f(b) & g(b) & 1 \end{bmatrix} = 0.$$

Also, for $x \in [a, b]$,

$$\varphi(x) = \det \begin{bmatrix} f(x) & g(x) & 1 \\ f(a) & g(a) & 1 \\ f(b) & g(b) & 1 \end{bmatrix}$$

$$= f(x)(g(a) - g(b)) - g(x)(f(a) - f(b)) + f(a)g(b) - f(b)g(a),$$

and so we see that φ is continuous on $[a, b]$, differentiable on (a, b) and

$$\varphi'(x) = f'(x)(g(a) - g(b)) - g'(x)(f(a) - f(b)) \quad (x \in (a, b)).$$

As $\varphi(a) = \varphi(b)$, it follows from Rolle's Theorem that there is a $c \in (a, b)$ for which $\varphi'(c) = 0$, that is, $f'(c)(g(a) - g(b)) - g'(c)(f(a) - f(b)) = 0$. Rearranging, we obtain the desired equality, completing the proof Cauchy's Theorem.   □

Exercise 4.44. Which of the following is equivalent to the statement of the Mean Value Theorem?

□ (A) If $f : [a, b] \to \mathbb{R}$ is continuous on $[a, b]$ and differentiable on (a, b), then there exists a $c \in (a, b)$ such that $f'(c) = 0$.

□ (B) If $f : [a, b] \to \mathbb{R}$ is continuous on $[a, b]$ and differentiable on (a, b), then there exists a $c \in (a, b)$ such that $(b - a)f'(c) = f(b) - f(a)$.

□ (C) If $f : [a, b] \to \mathbb{R}$ is continuous on $[a, b]$ and differentiable on (a, b), then there exists a unique $c \in (a, b)$ such that $f'(c) = \frac{f(b) - f(a)}{b - a}$.

□ (D) If $f : [a, b] \to \mathbb{R}$ is continuous on $[a, b]$ and differentiable on (a, b), then there exists a $c \in \{a, b\}$ such that $f'(c) = \frac{f(b) - f(a)}{b - a}$.

Exercise 4.45. Show that for every real $a, b \in \mathbb{R}$, $|\cos a - \cos b| \le |a - b|$.

Exercise 4.46. (∗) Suppose that $f : \mathbb{R} \to \mathbb{R}$ is differentiable, $|f'(x)| \le 1$ for all $x \in \mathbb{R}$, and that there exists an $a > 0$ such that $f(-a) = -a, f(a) = a$. Show that $f(0) = 0$.

Exercise 4.47. If $f : \mathbb{R} \to \mathbb{R}$ is differentiable and there exist $L, L' \in \mathbb{R}$ such that

$$\lim_{x \to \infty} f(x) = L, \quad \text{and} \quad \lim_{x \to \infty} f'(x) = L',$$

then prove that $L' = 0$.

Exercise 4.48. Let $c \in (a, b)$, and let $f : (a, b) \to \mathbb{R}$ be such that f is

(1) differentiable on $(a, b) \backslash \{c\}$,

(2) continuous on (a, b), and

(3) $\lim_{x \to c} f'(x)$ exists.

Then f is differentiable at c, and $f'(c) = \lim_{x \to c} f'(x)$.

Contrast this situation with the case of the function $x \mapsto |x|$ with $c = 0$.

Exercise 4.49. Let $f : (a, b) \to \mathbb{R}$ be differentiable on (a, b) and suppose that there is a number M such that for all $x \in (a, b)$, $|f'(x)| \le M$. Show that f is uniformly continuous on (a, b).

Exercise 4.50. Prove that if $c_0, \cdots, c_d$ are any real numbers satisfying

$$\frac{c_0}{1} + \frac{c_1}{2} + \cdots + \frac{c_d}{d + 1} = 0,$$

then the polynomial $c_0 + c_1 x + \cdots + c_d x^d$ has a zero in $(0, 1)$.

Exercise 4.51. Suppose that f is n times differentiable and that $f(x) = 0$ for $n + 1$ distinct x. Prove that $f^{(n)}(x) = 0$ for some x.

Exercise 4.52. Show that there are exactly two real values of x such that

$$x^2 = x \sin x + \cos x$$

and that they lie in $\left(-\frac{\pi}{2}, \frac{\pi}{2}\right)$.

Exercise 4.53. Find a function $f : \mathbb{R} \to \mathbb{R}$ such that $f'(-1) = 1/2, f'(0) = 0$ and $f''(x) > 0$ for all $x \in \mathbb{R}$, or prove that such a function cannot exist.

Corollary 4.10. *Suppose that $f : (a, b) \to \mathbb{R}$ is differentiable on (a, b). Then:*

(1) *If $f'(x) > 0$ for all $x \in (a, b)$, then f is strictly increasing.*

(2) *If f is strictly increasing, then $f'(x) \geq 0$ for all $x \in (a, b)$.*

(3) *$f'(x) \geq 0$ for all $x \in (a, b)$ if and only if f is increasing.*

(4) *$f'(x) = 0$ for all $x \in (a, b)$ if and only if f is constant.*

Proof. (1) For each pair of numbers $x_1, x_2 \in (a, b)$, with $x_1 < x_2$, it follows by the Mean Value Theorem, that

$$f(x_2) - f(x_1) = \underbrace{f'(x)}_{>0} \underbrace{(x_2 - x_1)}_{>0}$$

for some x between x_1 and x_2, and so $f(x_2) > f(x_1)$. Hence f is strictly increasing.

(2) Let $c \in (a, b)$. Then

$$f'(c) = \lim_{x \to c} \frac{f(x) - f(c)}{x - c} = \lim_{x \searrow c} \frac{f(x) - f(c)}{x - c} \geq 0,$$

where the last equality follows from the fact that for all $x > c$, $f(x) - f(c) > 0$ and $x - c > 0$. As the choice of $c \in (a, b)$ was arbitrary, the claim follows.

(3) The proof is analogous to (1) and (2).

(4) If f is constant, then clearly $f' \equiv 0$. Vice versa, if $f' \equiv 0$, then for any pair of numbers $x_1, x_2 \in (a, b)$, it follows by the Mean Value Theorem, that

$$f(x_2) - f(x_1) = \underbrace{f'(x)}_{=0} (x_2 - x_1) = 0,$$

so that $f(x_2) = f(x_1)$. Hence f is constant on (a, b). $\qquad\square$

Note that in (2), it may happen that f' is zero at some points, and it may fail to be positive. For example, consider the function x^3 on $\mathbb{R}$. It is strictly increasing, but

$$\frac{d}{dx}x^3\bigg|_{x=0} = 3 \cdot 0^2 = 0.$$

A similar version holds with 'decreasing' instead of 'increasing':

Corollary 4.11. *Suppose that $f : (a,b) \to \mathbb{R}$ is differentiable on (a,b). Then:*

(1) *If $f'(x) < 0$ for all $x \in (a,b)$, then f is strictly decreasing.*

(2) *If f is strictly decreasing, then $f'(x) \leq 0$ for all $x \in (a,b)$.*

(3) *$f'(x) \leq 0$ for all $x \in (a,b)$ if and only if f is decreasing.*

Exercise 4.54. Let $f : \mathbb{R} \to \mathbb{R}$ be such that for all $x, y \in \mathbb{R}$, $|f(x) - f(y)| \leq (x-y)^2$. Prove that f is constant.

Exercise 4.55. ($*$) Find all functions $f : \mathbb{R} \to \mathbb{R}$ such that f is differentiable on $\mathbb{R}$ and for all $x \in \mathbb{R}$ and all $n \in \mathbb{N}$,

$$f'(x) = \frac{f(x+n) - f(x)}{n}.$$

Hint: Conclude that f must be twice differentiable and calculate $f''(x)$.

Exercise 4.56 (Simple Harmonic Oscillator $y'' + y = 0$).

(1) If $y = y(x)$ is a solution of the differential equation

$$y'' + y = 0, \tag{4.11}$$

then show that $y^2 + (y')^2$ is constant.

(2) Use part (1) to show that every solution to (4.11) has the form

$$y(x) = A\cos x + B\sin x$$

for suitable constants A, B. Proceed as follows: It is easy to show that all functions of the above form satisfy (4.11). Let y be a solution. For it to have the form $A\cos x + B\sin x$, it is necessary that $A = y(0)$ and $B = y'(0)$. Now consider

$$f(x) := y(x) - y(0)\cos x - y'(0)\sin x,$$

and apply (1) to f, making use of the fact that $f(0) = f'(0) = 0$.

(3) Use part (2) to prove the trigonometric addition formulae for $\alpha, \beta \in \mathbb{R}$:

$$\sin(\alpha + \beta) = (\sin\alpha)(\cos\beta) + (\cos\alpha)(\sin\beta),$$
$$\cos(\alpha + \beta) = (\cos\alpha)(\cos\beta) - (\sin\alpha)(\sin\beta).$$

Exercise 4.57. Let $f : \mathbb{R} \to \mathbb{R}$. We call $x \in \mathbb{R}$ a *fixed point* of f if $f(x) = x$.

(1) If f is differentiable, and for all $x \in \mathbb{R}$, $f'(x) \neq 1$, then prove that f has at most one fixed point.

(2) (*) Let the sequence $(x_n)_{n \in \mathbb{N}}$ be generated by taking an arbitrary real x_1, and setting $x_{n+1} = f(x_n)$ for $n \in \mathbb{N}$. Show that if there is an $M < 1$ such that for all $x \in \mathbb{R}$, $|f'(x)| \leq M$, then there is a fixed point x_* of f, and that $x_* = \lim_{n \to \infty} x_n$.

(3) Visualise the process described in the part (2) above via the zigzag/cobweb path

$$(x_1, x_2) \to (x_2, x_2) \to (x_2, x_3) \to (x_3, x_3) \to (x_3, x_4) \to \cdots .$$

(4) Prove that the function $f : \mathbb{R} \to \mathbb{R}$ defined by

$$f(x) = x + \frac{1}{1 + e^x} \quad (x \in \mathbb{R})$$

has no fixed point, although $0 < f'(x) < 1$ for all $x \in \mathbb{R}$. Is this a contradiction to the result in part (2) above? Explain.

4.7   Taylor's Formula

We had seen earlier that for a polynomial p_k given by

$$p_k(x) = c_k(x - a)^k,$$

we have

$$p_k(a) = 0, \quad p_k'(a) = 0, \cdots, \quad p_k^{(k-1)}(a) = 0, \quad p_k^{(k)}(a) = c_k k!, \quad p_k^{(k+1)}(a) = 0, \cdots .$$

Thus if we have a polynomial p that is a linear combination of such terms,

$$p(x) = c_0 + c_1(x - a) + c_2(x - a)^2 + \cdots + c_d(x - a)^d,$$

then we have

$$p(a) = c_0,$$
$$p'(a) = c_1 \cdot 1!,$$
$$p''(a) = c_2 \cdot 2!,$$
$$\vdots$$
$$p^{(d)}(a) = c_d \cdot d!,$$
$$p^{(d+1)}(a) = 0,$$
$$\vdots$$

So there is a special relationship between the coefficients c_k and the successive derivatives of p at a:

$$c_k = \frac{p^{(k)}(a)}{k!}, \quad 0 \le k \le d.$$

Now, suppose that we *start* with a smooth enough function $f : \mathbb{R} \to \mathbb{R}$ and form a related d degree polynomial p given by

$$p(x) := f(a) + \frac{f'(a)}{1!}(x-1) + \cdots + \frac{f^{(d)}(a)}{d!}(x-a)^d, \quad x \in \mathbb{R}.$$

Then:

(1) p is a polynomial,

(2) p is related to f,

(3) in fact, from what we have just learnt, the coefficients of p are related to the derivatives of p at a, and so

$$\frac{p^{(k)}(a)}{k!} = \frac{f^{(k)}(a)}{k!}, \quad 0 \le k \le d,$$

that is,

$$p(a) = f(a),$$
$$p'(a) = f'(a),$$
$$p''(a) = f''(a),$$
$$\vdots$$
$$p^{(d)}(a) = f^{(d)}(a).$$

So *at a, p matches very well with f.

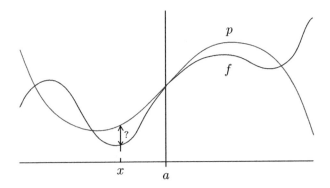

It is thus natural to ask:

How big is the error $f(x) - p(x)$ when $x \ne a$?

Taylor's Formula answers this question.

Theorem 4.12 (Taylor's Formula). *Let* $f, f', \cdots, f^{(d+1)}$ *all exist on an open interval I. If a and x are any two points of I, then there is a point ξ between them (and distinct from them) such that*

$$f(x) = \underbrace{f(a) + \frac{f'(a)}{1!}(x - a) + \cdots + \frac{f^{(d)}(a)}{d!}(x - a)^d}_{p(x)} + \underbrace{\frac{f^{(d+1)}(\xi)}{(d + 1)!}(x - a)^{d+1}}_{\text{error}}.$$

Remark 4.4.

(1) If we have an estimate on how big $|f^{(d+1)}|$ is on I, then we have a handle on the error term, and how big that is. This allows one to get polynomial approximations to f.

(2) The name of the result is after the English mathematician Brook Taylor, who stated a version of it in 1712, but there were also other mathematicians involved in its genesis such as Lagrange and Gregory.

Proof. We use induction on d. When $d = 0$, the result follows from the Mean Value Theorem: there exists a ξ between a and x, and not equal to a and to ξ, such that

$$\frac{f(x) - f(a)}{x - a} = f'(\xi),$$

which upon rearrangement gives $f(x) = f(a) + f'(\xi)(x - a)$, and so the claim is true.

Suppose that the result is true for some d. Now we show it for $d + 1$. Consider an f such that $f', \cdots, f^{d+2}$ exist. Let

$$p(x) := f(a) + \frac{f'(a)}{1!}(x - a) + \cdots + \frac{f^{(d+1)}(a)}{(d + 1)!}(x - a)^{d+1}. \tag{4.12}$$

Let $b \in I$. We want to show that

$$f(b) - p(b) = \frac{f^{(d+2)}(\xi)}{(d + 2)!}(b - a)^{d+2}$$

for some ξ between a and b. To achieve this, we will essentially use Taylor's Formula for f' (induction hypothesis!) and use the Cauchy version of the Mean Value Theorem to get Taylor's Formula for f. To this end, let us define F and G by

$$F := f - p,$$
$$G := (x - a)^{d+2}.$$

Then $F(a) = f(a) - p(a) = 0$ and $G(a) = 0$. By Cauchy's Theorem, there exists a c between a and b, which is different from either of them, such that

$$(F(b) - \underbrace{F(a)}_{=0})G'(c) = (G(b) - \underbrace{G(a)}_{=0})F'(c).$$

Thus $(f(b) - p(b))(d + 2)(c - a)^{d+1} = (b - a)^{d+2} \cdot (f'(c) - p'(c))$, and upon rearranging,

$$f(b) - p(b) = \frac{(b - a)^{d+2}}{(c - a)^{d+1}(d + 2)} \cdot (f'(c) - p'(c)). \tag{4.13}$$

But we will now see that p' serves as a Taylor approximating polynomial for f'! Indeed, it follows from (4.12) that

$$p'(x) = 0 + \frac{f'(a)}{1!} \cdot 1 + \frac{f''(a)}{2!} \cdot 2(x - a) \cdots + \frac{f^{(d+1)}(a)}{(d + 1)!} \cdot (d + 1)(x - a)^d$$

$$= f'(a) + \frac{(f')'(a)}{1!}(x - a) + \frac{(f')''(a)}{2!}(x - a)^2 + \cdots + \frac{(f')^{(d)}(a)}{d!} \cdot (x - a)^d.$$

Hence by the induction hypothesis (applied to f', whose derivatives up to order $d + 1$ exist on I), we have that there exists a ξ between a and c (and this ξ is then also between a and b) such that

$$f'(c) = \underbrace{f'(a) + \frac{(f')'(a)}{1!}(c - a) + \cdots + \frac{(f')^{(d)}(a)}{d!} \cdot (c - a)^d}_{p'(c)} + \frac{(f')^{(d+1)}(\xi)}{(d + 1)!}(c - a)^{d+1}.$$

Thus

$$f'(c) - p'(c) = \frac{f^{(d+2)}(\xi)}{(d + 1)!}(c - a)^{d+1}.$$

Substituting this in (4.13), we obtain

$$f(b) - p(b) = \frac{(b - a)^{d+2}}{(c - a)^{d+1}(d + 2)} \cdot \frac{f^{(d+2)}(\xi)}{(d + 1)!}(c - a)^{d+1} = \frac{f^{(d+2)}(\xi)}{(d + 2)!}(b - a)^{d+2}.$$

This completes the proof. $\qquad\qquad\qquad\qquad\qquad\qquad\qquad\qquad\qquad\qquad\qquad\qquad\qquad\square$

Example 4.21. What is sin 1 to three decimal places?

We will learn later on that $\sin' = \cos$, $\cos' = -\sin$, $|\cos x| \leq 1$ for all $x \in \mathbb{R}$, $\sin 0 = 0$ and $\cos 0 = 1$. Let us take these facts for granted right now. Thus with $f := \sin$, we have

$$f' = \cos, \quad f'' = -\sin, \quad f''' = -\cos, \quad f^{(4)} = \sin, \quad \cdots .$$

So we observe that for any $n \in \mathbb{N}$,

$$f^{(2n+1)} = (-1)^n \cos,$$

$$f^{(2n)} = (-1)^n \sin .$$

By Taylor's Formula,

$$\sin 1 = \sin 0 + \frac{\cos 0}{1!} \cdot 1 - \frac{\sin 0}{2!} \cdot 1 - \frac{\cos 0}{3!} \cdot 1 + \cdots + \frac{(-1)^n \sin 0}{(2n)!} + \frac{(-1)^n \cos c}{(2n + 1)!}$$

for some $c \in (0, 1)$. Hence

$$\sin 1 = \frac{1}{1!} - \frac{1}{3!} + \frac{1}{5!} - + \cdots + \frac{(-1)^n}{(2n)!} + \frac{(-1)^n \cos c}{(2n+1)!}.$$

In order to find $\sin 1$ to three decimal places, we take n large enough so that

$$\left| \frac{(-1)^n \cos c}{(2n+1)!} \right| < 10^{-4}.$$

As $|\cos c| \le 1$, this is guaranteed if $(2n+1)! > 10^4$. We have $4! = 24$, $5! = 120$, $6! = 720$, $7! = 5040$, $8! > 10^4$. So we take $n = 4$. Then up to 3 decimal places,

$$\sin 1 \approx \frac{1}{1!} - \frac{1}{3!} + \frac{1}{5!} - \frac{1}{7!} = 1 - \frac{1}{6} + \frac{1}{120} - \frac{1}{5040} = \frac{5040 - 840 + 42 - 1}{5040} = \frac{4241}{5040}$$

$$= 0.8415 \quad \text{(to three decimal places)}.$$

(A scientific calculator gives 0.8414709848.) $\diamondsuit$

Exercise 4.58. Use Taylor's Formula to show that $\lim\limits_{x \to 0} \dfrac{\sin x - x}{x^3} = -\dfrac{1}{6}$.

Exercise 4.59 (*o*-notation). A special notation introduced by Landau in 1909 is particularly suited to Taylor's formula. Given functions f, g in an open interval I containing a such that g is nonzero in I, we write

$$\text{`}f(x) = o(g(x)) \text{ as } x \to a \text{'}$$

if

$$\lim_{x \to a} \frac{f(x)}{g(x)} = 0.$$

The symbol $f(x) = o(g(x))$ is read '$f(x)$ is little-oh of $g(x)$' or '$f(x)$ is of smaller order than $g(x)$', and is intended to convey the idea that for x near a, $f(x)$ is small compared to $g(x)$. For example,

$$f(x) = o(1) \text{ as } x \to a \text{ means that } \lim_{x \to a} f(x) = 0,$$

and

$$f(x) = o(x) \text{ as } x \to a \text{ means that } \lim_{x \to a} \frac{f(x)}{x} = 0.$$

Also, if h is a function on I, then we write '$f(x) = g(x) + o(h(x))$ as $x \to a$' to mean that '$f(x) - g(x) = o(h(x))$ as $x \to a$'.

(1) If f has a continuous $(n+1)$st derivative in some open interval containing the compact interval $[a - \epsilon, a + \epsilon]$, and if $M := \max\{|f^{n+1}(x)| : |x - a| \le \epsilon\}$, then it follows from Taylor's Formula that

$$f(x) = \sum_{k=0}^{n} \frac{f^{(k)}(a)}{k!} (x - a)^k + o((x - a)^n) \text{ as } x \to a.$$

(2) Show that $\tan x = x + \dfrac{x^3}{3} + o(x^3)$ as $x \to 0$.

(3) Show that $\lim\limits_{x \to 0} \dfrac{\tan x - x}{\sin x - x \cos x} = 1$.

4.8   Convexity

'Convex' functions play an important role in optimisation. We will learn in this section:

(1) What a convex function is (that is a function whose graph lies below all possible chords).

(2) For a twice differentiable function f, convexity is equivalent to the condition that $f'' \geq 0$ pointwise.

(3) What the role of convex functions is in optimisation (the necessity condition we learnt about earlier, of vanishing derivative for minimisers, becomes sufficient if the function at hand happens to be convex).

What is a convex function?

Definition 4.5 (Convex function). Let I be an interval. A function $f : I \to \mathbb{R}$ is said to be *convex* if for all $x, y \in I$ and all $t \in (0, 1)$,

$$f((1 - t)x + ty) \leq (1 - t)f(x) + tf(y).$$

Geometric meaning of convexity. 'The graph of f lies below all possible chords'.

The point $(1 - t)x + ty$ on the x-axis divides the segment joining the points x to y in the ratio $t : 1 - t$. The convexity inequality says that the corresponding value of f, namely $f((1 - t)x + ty)$ is at most $(1 - t)f(x) + tf(y)$, and so the point

$$((1 - t)x + ty, f((1 - t)x + ty))$$

on the graph of f lies below the point

$$((1 - t)x + ty, (1 - t)f(x) + tf(y))$$

on the 'chord' joining the points $(x, f(x))$ and $(y, f(y))$.

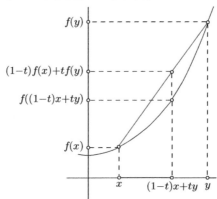

Exercise 4.60. Show that $x \mapsto |x| : \mathbb{R} \to \mathbb{R}$ is convex.

Exercise 4.61. Let $f : I \to \mathbb{R}$ be a function on an interval $I \subset \mathbb{R}$. We define the *epigraph* of f by

$$U(f) := \bigcup_{x \in I}\{(x,y) : y \geq f(x)\} \subset I \times \mathbb{R}.$$

In other words, $U(f)$ is the 'region above and on the graph of f'. A subset $C \subset \mathbb{R}^2$ is called a *convex set* if for all $v_1, v_2 \in C$ and for all $t \in (0,1)$, $(1-t) \cdot v_1 + t \cdot v_2 \in C$. Show that f is a convex function if and only if $U(f)$ is a convex set.

Example 4.22. Consider the function $f : \mathbb{R} \to \mathbb{R}$ given by $f(x) = x^2$, $x \in \mathbb{R}$. We claim that this function is convex.

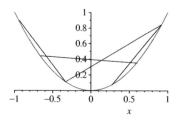

Indeed, we note that for $x_1, x_2 \in \mathbb{R}$ and $t \in (0,1)$,

$$f((1-t)x_1 + tx_2)$$
$$= ((1-t)x_1 + tx_2)^2 = (1-t)^2x_1^2 + 2t(1-t)x_1x_2 + t^2x_2^2$$
$$= (1-t)x_1^2 + tx_2^2 + ((1-t)^2 - (1-t))x_1^2 + (t^2 - t)x_2^2 + 2t(1-t)x_1x_2$$
$$= (1-t)x_1^2 + tx_2^2 - t(1-t)(x_1^2 + x_2^2 - 2x_1x_2)$$
$$= (1-t)x_1^2 + tx_2^2 - \underbrace{t(1-t)(x_1 - x_2)^2}_{\geq 0} \leq (1-t)x_1^2 + tx_2^2 = (1-t)f(x_1) + tf(x_2).$$

Consequently, f is convex.   ◇

So checking convexity using the definition can be cumbersome. However, the following result makes it very easy.

Theorem 4.13. *Let $f : I \to \mathbb{R}$ be twice differentiable on an interval I. Then f is convex on I if and only if for all $x \in I$, $f''(x) \geq 0$.*

The convexity of $x \mapsto x^2$ is now immediate, as $\dfrac{d^2}{dx^2}x^2 = 2 > 0$, $x \in \mathbb{R}$.

Example 4.23. We have

$$\frac{d^2}{dx^2}e^x = e^x > 0, \quad x \in \mathbb{R},$$

and so $x \mapsto e^x$ is convex. So, for all $x_1, x_2 \in \mathbb{R}$, $t \in (0,1)$, $e^{(1-t)x_1 + tx_2} \leq (1-t)e^{x_1} + te^{x_2}$.

Here is one more example. The function $x \xmapsto{g} \sqrt{1 + x^2}$ is convex on $\mathbb{R}$. Indeed,

$$g'(x) = \frac{x}{\sqrt{1 + x^2}},$$

and so

$$g''(x) = \frac{1}{\sqrt{1 + x^2}} + x \cdot \left(-\frac{1}{2}\right) \cdot \frac{1}{(1 + x^2)^{3/2}} \cdot 2x$$

$$= \frac{1}{\sqrt{1 + x^2}} - \frac{x^2}{(1 + x^2)\sqrt{1 + x^2}} = \frac{1}{(1 + x^2)^{3/2}} > 0$$

for all $x \in \mathbb{R}$. ◇

Proof of Theorem 4.13.

(1) Suppose that $f''(x) \geq 0$ for all $x \in I$. Let $x, y \in I$ be such that $x < y$, and let $t \in (0, 1)$. Applying the Mean Value Theorem to f on the interval $[x, (1 - t)x + ty]$ gives

$$\frac{f((1 - t)x + ty) - f(x)}{t(y - x)} = f'(c_1)$$

for some c_1 such that $x < c_1 < (1 - t)x + ty$. Also, by the Mean Value Theorem for f on $[(1 - t)x + ty, y]$, we have

$$\frac{f(y) - f((1 - t)x + ty)}{(1 - t)(y - x)} = f'(c_2)$$

for some c_2 such that $(1 - t)x + ty < c_2 < y$.

As $f''(\xi) \geq 0$ for all $\xi \in I$, we know that f' is increasing on I, and since $c_1 < c_2$, it follows that $f'(c_1) \leq f'(c_2)$, that is,

$$\frac{f((1 - t)x + ty) - f(x)}{t(y - x)} \leq \frac{f(y) - f((1 - t)x + ty)}{(1 - t)(y - x)}.$$

Upon rearranging, we obtain $f((1 - t)x + ty) \leq tf(y) + (1 - t)f(x)$. So f is convex. This proves the 'if' part.

(2) Now suppose that f is convex. Let $x, u, y \in I$ be such that $x < u < y$. If

$$t := \frac{u - x}{y - x},$$

then $t \in (0, 1)$, and

$$1 - t = \frac{y - u}{y - x}.$$

From the convexity of f, we obtain

$$\frac{y - u}{y - x}f(x) + \frac{u - x}{y - x}f(y) \geq f\left(\frac{y - u}{y - x}x + \frac{u - x}{y - x}y\right) = f(u),$$

that is,

$$(y - x)f(u) \leq (u - x)f(y) + (y - u)f(x). \tag{4.14}$$

From (4.14), we obtain $(y - x)f(u) \leq (u - x)f(y) + (y - x + x - u)f(x)$, that is,

$$(y - x)f(u) - (y - x)f(x) \leq (u - x)f(y) - (u - x)f(x),$$

and so

$$\frac{f(u) - f(x)}{u - x} \leq \frac{f(y) - f(x)}{y - x}.$$

Passing the limit as $u \searrow x$, we obtain

$$f'(x) \leq \frac{f(y) - f(x)}{y - x}. \tag{4.15}$$

From (4.14), we also have that $(y - x)f(u) \leq (u - y + y - x)f(y) + (y - u)f(x)$, that is, $(y - x)f(u) - (y - x)f(y) \leq (u - y)f(y) - (u - y)f(x)$, and so

$$\frac{f(y) - f(x)}{y - x} \leq \frac{f(y) - f(u)}{y - u}.$$

Passing the limit as $u \nearrow y$, we obtain

$$\frac{f(y) - f(x)}{y - x} \leq f'(y). \tag{4.16}$$

From (4.15) and (4.16), we obtain $f'(x) \leq f'(y)$. So f' is increasing on I. Hence $f''(\xi) \geq 0$ for all $\xi \in I$ since

$$f''(\xi) = \lim_{x \searrow \xi} \frac{f'(x) - f'(\xi)}{x - \xi} \geq 0,$$

where the last inequality follows since both the numerator is always nonnegative and denominator is always positive. This completes the proof of the 'only if' part.   □

Remark 4.5. Actually in part (2) above, we have shown the following (without the assumption of *twice* differentiability):

Proposition 4.14. *If* $f : I \to \mathbb{R}$ *is convex and differentiable on* I, *then* f' *is increasing on* I.

Here is an example of a function that is not convex. Consider $x \overset{f}{\mapsto} \sqrt{x} : (0, \infty) \to \mathbb{R}$.
Then $f' = \dfrac{1}{2\sqrt{x}}$ and so $f'' = -\dfrac{1}{4x^{3/2}} < 0$.

So f is not convex on $(0, \infty)$. On the other hand, $-f$ *is* convex. Functions $f : I \to \mathbb{R}$ on an interval I for which $-f$ is convex, are called *concave*.[3] For example, we will see later on that also the logarithm function $\log : (0, \infty) \to \mathbb{R}$ is concave.

Exercise 4.62. Which of the following statements is/are always true?

☐ (A) $x \mapsto 1/x : (0, \infty) \to \mathbb{R}$ is convex.

☐ (B) $x \mapsto -\sin x : (0, \pi/2) \to \mathbb{R}$ is convex.

☐ (C) If $f : \mathbb{R} \to [0, \infty)$ is convex, then $x \mapsto \sqrt{f(x)}$ is convex.

☐ (D) If $f : \mathbb{R} \to [0, \infty)$ is convex, then $x \mapsto (f(x))^2$ is convex.

What does convexity have to do with optimisation?

We had seen the following necessary condition for minimisers in Theorem 4.6:

If $f : (a, b) \to \mathbb{R}$ has a local minimiser $x_* \in (a, b)$ and f is differentiable at x_*, then $f'(x_*) = 0$.

However, the vanishing of the derivative at a point is *not sufficient* in general for that point to be a local minimiser.

Example 4.24. If we look at $f : \mathbb{R} \to \mathbb{R}$ given by $f(x) = x^3$, $x \in \mathbb{R}$, then with $x_* := 0$, we have that $f'(x_*) = 3x_*^2 = 3 \cdot 0^2 = 0$, but clearly $x_* = 0$ is not a local minimiser of f, since $f(-\epsilon) = -\epsilon^3 < 0 = f(0)$ for all $\epsilon > 0$, no matter how small.

---

[3] Some books refer to our convex functions as 'concave up', and our concave functions as 'concave down'. One can justify this alternative terminology as follows. Imagine a convex function, and think of a thin glass placed along its graph. Then the *upper* portion of this glass looks like the cross section of a concave lens, and so it makes sense to call a convex function concave *up*. On the other hand, if we place a thin glass along the graph of a (in our sense) concave function, then the *lower* surface of this glass looks like a concave lens, and so it makes sense to call this function concave *down*.

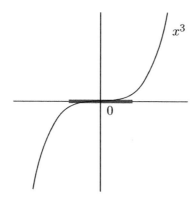

$$f'(0) = 0, \text{ but } 0 \text{ is not a minimiser for the map } f.$$

$\diamondsuit$

But for *convex* functions, the vanishing derivative condition is *sufficient*!

Theorem 4.15. *Let* $f : (a,b) \to \mathbb{R}$ *and* $x_* \in (a,b)$. *If*

    (1) f *is convex on* (a,b),

    (2) f *is differentiable on* (a,b), *and*

    (3) $f'(x_*) = 0$,

then x_* *is a minimiser of* f.

Proof. Suppose that $x_0 \in (a,b)$ is such that $f(x_0) < f(x_*)$. We have only two possible cases:

    1° $x_0 > x_*$. By the Mean Value Theorem for f on $[x_*, x_0]$, there exists a $c \in (x_*, x_0)$ such that

$$f'(x_*) = 0 > \boxed{\frac{\boxed{f(x_0) - f(x_*)}^{}}{\boxed{x_0 - x_*}_+}} = f'(c),$$

    contradicting the convexity of f.

    2° $x_* > x_0$. By the Mean Value Theorem for f on $[x_0, x_*]$, there exists a $c \in (x_0, x_*)$ such that

$$f'(x_*) = 0 < \frac{\boxed{f(x_*) - f(x_0)}^{+}}{\boxed{x_* - x_0}_+} = f'(c),$$

    again contradicting the convexity of f.

So $f(x) \geq f(x_*)$ for all $x \in I$. Hence x_* is a minimiser. $\qquad\qquad\square$

Example 4.25. Let us revisit Example 4.18, where we considered the minimisation of the function $S : (0, \infty) \to \mathbb{R}$ given by

$$S(r) = 2\pi r^2 + 2\pi \cdot 2 \cdot \frac{1000}{r}, \quad r > 0.$$

We have $S'(r) = 4\pi r - 4\pi \cdot \dfrac{1000}{r^2}$, and

$$S''(r) = 4\pi + 8\pi \cdot \frac{1000}{r^3} > 0$$

for all $r \in (0, \infty)$. Thus S is convex. We had found out earlier that $S(r) = 0$ if and only if $r = 10$ cm. Hence S has a unique minimiser, given by $r_* = 10$ cm. $\qquad \diamondsuit$

Exercise 4.63. Let $a, b > 0$. In the Cartesian plane, a straight line path is drawn from the point $(0, a)$ to the horizontal x-axis, and then to $(1, b)$ as shown:

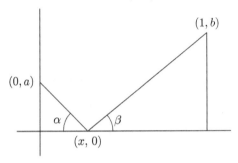

Show, using Calculus, that the total length is shortest when $\alpha = \beta$. Can you find a geometric proof of this? *Hint:* Think of the x-axis as a reflecting surface and the path as a ray of light.

Exercise 4.64. If $a_1, \cdots , a_n \in \mathbb{R}$, then find the minimum value of $\displaystyle\sum_{k=1}^{n} (x - a_k)^2$, $x \in \mathbb{R}$.

Exercise 4.65 (The Cauchy–Schwarz Inequality).

(1) Let $a > 0$, $b, c \in \mathbb{R}$ and consider the function $f : \mathbb{R} \to \mathbb{R}$ given by

$$f(t) = at^2 + bt + c, \quad t \in \mathbb{R}.$$

Show that f has a minimiser and that the minimum value of f is $-\dfrac{b^2 - 4ac}{4a}$.

(2) Show that for all $n \in \mathbb{N}$ and all $a_1, \cdots , a_n, b_1, \cdots , b_n \in \mathbb{R}$, there holds that

$$(a_1^2 + \cdots + a_n^2)(b_1^2 + \cdots + b_n^2) \geq (a_1 b_1 + \cdots + a_n b_n)^2.$$

Hint: Consider $\displaystyle\sum_{k=1}^{n} (ta_k - b_k)^2$.

Exercise 4.66 (The Arithmetic Mean-Geometric Mean Inequality).

(1) Suppose that $f : I \to \mathbb{R}$ is a convex function on an interval $I \subset \mathbb{R}$. If $n \in \mathbb{N}$, and $x_1, \cdots , x_n \in I$, then show that

$$f\left(\frac{x_1 + \cdots + x_n}{n}\right) \leq \frac{f(x_1) + \cdots + f(x_n)}{n}.$$

(2) Show that $-\log : (0, \infty) \to \mathbb{R}$ is convex.

(3) Prove the Arithmetic Mean-Geometric Mean Inequality: for nonnegative real numbers $a_1, \cdots, a_n$, there holds that

$$\frac{a_1 + \cdots + a_n}{n} \geq \sqrt[n]{a_1 \cdots a_n}.$$

(The left hand side above is called the *arithmetic mean of* $a_1, \cdots, a_n$, while the right hand side is called their *geometric mean*.)

Exercise 4.67. Consider all the rectangles with perimeter equal to a fixed length $p > 0$. Which of the following is true for the unique rectangle that is a square, compared to the other rectangles?

☐ (A) It has the largest area and the largest length of diagonal.

☐ (B) It has the largest area and the smallest length of diagonal.

☐ (C) It has the smallest area and the largest length of diagonal.

☐ (D) It has the smallest area and the smallest length of diagonal.

Exercise 4.68. Sketch the curve given by $y = 2x^3 + 2x^2 - 2x - 1$ after locating intervals of increase/decrease, intervals of convexity/concavity,[4] points of local maxima/minima, and points of inflection (places where $f'' = 0$). How many times, and approximately where does the curve cross the x-axis?

Exercise 4.69. A wire of length ℓ is cut into two pieces, one being bent to form a square, and the other to form an equilateral triangle. How should the wire be cut if the sum of the two areas is to be minimised?

Exercise 4.70. Consider the graph of the derivative f' of a smooth $(C^\infty) f : (-9, 9) \to \mathbb{R}$ on the interval $I := (-9, 9)$ shown in the following picture.

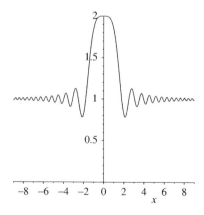

---

[4] A function is *concave* if $-f$ is convex.

Determine if each of the following statements is true, false or if there isn't enough information to determine whether the statement is true or false.

(1) f is increasing on I.

(2) f is convex on I.

(3) f is zero at most once on I.

(4) $f'''(0) \leq 0$.

(5) $f'''(0) < 0$.

(6) f'' is convex on I.

(7) f has a local maximum at 0.

4.9 $\frac{0}{0}$ form of l'Hôpital's Rule

l'Hôpital's Rule is a useful result for finding limits involving a ratio of two functions by knowing what the limit is for the ratio of the *derivatives* of the two functions. It is named after the French mathematician l'Hôpital (pronounced 'low-pee-taal') who wrote the first textbook on Calculus around 1700!

Roughly, the rule says that if $f(a) = g(a) = 0$ (the reason behind calling it the '$\frac{0}{0}$ form'[5]) and

$$\lim_{x \to a} \frac{f'(x)}{g'(x)} = \ell,$$

then

$$\lim_{x \to a} \frac{f(x)}{g(x)} = \ell$$

as well.

It is enough to consider the 'right-sided' version, where $x \searrow a$. Then we have a corresponding 'left-sided' version, and the two can be put together.

Theorem 4.16 ($\frac{0}{0}$ form of l'Hôpital's Rule). *If $f, g : [a, b] \to \mathbb{R}$ are such that*

(1) $f(a) = g(a) = 0$

(2) $g(x) \neq 0$ *for $x \in (a, b]$,*

(3) f, g *are continuous on $[a, b]$,*

(4) f, g *are differentiable on (a, b),*

(5) $g'(x) \neq 0$ *for all $x \in (a, b)$, and*

(6) $\lim\limits_{x \searrow a} \dfrac{f'(x)}{g'(x)} = \ell,$

then $\lim\limits_{x \to \infty} \dfrac{f(x)}{g(x)} = \ell.$

---

[5] Later on in Theorem 5.11, we will learn about a $\frac{\infty}{\infty}$ version of the l'Hôpital's Rule, where the numerator and denominator functions tend to ∞ as x approaches a.

Proof. Let $\epsilon > 0$. Then there exists a $\delta > 0$ such that for all $x \in [a, b]$ satisfying $0 < x - a < \delta$, we have

$$\left| \frac{f'(x)}{g'(x)} - \ell \right| < \epsilon.$$

By the Cauchy Mean Value Theorem, it follows that for such x, there exists a $c_x \in (a, x)$ such that

$$\frac{f(x)}{g(x)} = \frac{f(x) - f(a)}{g(x) - g(a)} = \frac{f'(c_x)}{g'(c_x)}.$$

But this c_x satisfies $0 < c_x - a < \delta$, and so

$$\left| \frac{f(x)}{g(x)} - \ell \right| = \left| \frac{f'(c_x)}{g'(c_x)} - \ell \right| < \epsilon.$$

Hence $\lim\limits_{x \to \infty} \dfrac{f(x)}{g(x)} = \ell$. □

Here are some examples.

Example 4.26.

(1) $\lim\limits_{x \to 1} \dfrac{1 - \sqrt[5]{x}}{1 - \sqrt[3]{x}} = \dfrac{3}{5}$.

With $f(x) := 1 - \sqrt[5]{x}$ and $g(x) := 1 - \sqrt[3]{x}$, we have $f(1) = g(1) = 0$. Also,

$$f'(x) = -\frac{1}{5} x^{-4/5},$$

$g'(x) = -\dfrac{1}{3} x^{-2/3} \neq 0$ for x near 1, and

$$\lim_{x \to 1} \frac{f'(x)}{g'(x)} = \lim_{x \to 1} \frac{-\frac{1}{5} x^{-4/5}}{-\frac{1}{3} x^{-2/3}} = \frac{-\frac{1}{5} \cdot 1}{-\frac{1}{3} \cdot 1} = \frac{3}{5}.$$

Hence by l'Hôpital's Rule, $\lim\limits_{x \to 1} \dfrac{f(x)}{g(x)} = \lim\limits_{x \to 1} \dfrac{1 - \sqrt[5]{x}}{1 - \sqrt[3]{x}} = \dfrac{3}{5}$ too.

(2) $\lim\limits_{x \to 0} \dfrac{\sin x}{x} = 1$.

With $f(x) := \sin x$ and $g(x) := x$, we have that $f(0) = g(0) = 0$. Also, $f'(x) = \cos x$, and $g'(x) = 1 > 0$ for all x. Thus

$$\lim_{x \to 0} \frac{f'(x)}{g'(x)} = \lim_{x \to 0} \frac{\cos x}{1} = \cos 0 = 1.$$

Hence by l'Hôpital's Rule, $\lim\limits_{x \to 0} \dfrac{f(x)}{g(x)} = \lim\limits_{x \to 0} \dfrac{\sin x}{x} = 1$ as well.

(3) $\lim\limits_{x \to 0} \dfrac{1 - \cos x}{x^2} = \dfrac{1}{2}$.

With $f(x) := 1 - \cos x$ and $g(x) := x^2$, we have $f(0) = g(0) = 0$. Also, $f'(x) = \sin x$, and $g'(x) = 2x \neq 0$ for $x \neq 0$. Thus

$$\lim_{x \to 1} \frac{f'(x)}{g'(x)} = \lim_{x \to 1} \frac{\sin x}{2x} = \frac{1}{2} \cdot \lim_{x \to 1} \frac{\sin x}{x} = \frac{1}{2} \cdot 1 = \frac{1}{2},$$

by (2) above. Hence by l'Hôpital's Rule, $\lim\limits_{x \to 0} \dfrac{f(x)}{g(x)} = \lim\limits_{x \to 0} \dfrac{1 - \cos x}{x^2} = \dfrac{1}{2}$ too. ◇

Example 4.27. One should be careful[6] when applying l'Hôpital's Rule and check that the hypotheses for its application are all actually satisfied. For example, consider the reckless application giving

$$\lim_{x \to 0+} \frac{x+1}{x} = \lim_{x \to 0} \frac{1}{1} = 1,$$

while in fact this is wrong, and the limit does not exist since

$$\lim_{x \to 0+} \frac{x+1}{x} = \lim_{x \to 0+} \left(1 + \frac{1}{x}\right) = \infty.$$

What went wrong? $(x+1)\big|_{x=0} = 1 \neq 0!$ ◇

Exercise 4.71. What is wrong with the following application of l'Hôpital's Rule?

$$\lim_{x \to 1} \frac{x^3 + x - 2}{x^2 - 3x + 2} = \lim_{x \to 1} \frac{3x^2 + 1}{2x - 3} = \lim_{x \to 1} \frac{6x}{2} = 3.$$

Show that the limit above is actually equal to -4.

Exercise 4.72. Revisit Exercise 4.58, but now use l'Hôpital's Rule to show that

$$\lim_{x \to 0} \frac{\sin x - x}{x^3} = -\frac{1}{6}.$$

Exercise 4.73. Compute

$$\lim_{x \to \sqrt{3}} \frac{\tan^{-1} x - \pi/3}{x - \sqrt{3}},$$

where $\tan^{-1} : \mathbb{R} \to (-\pi/2, \pi/2)$ denotes the inverse of $\tan : (-\pi/2, \pi/2) \to \mathbb{R}$.

Exercise 4.74. Find $\displaystyle\lim_{x \to 0+} \frac{\tan \sqrt{x}}{\sqrt{x}}$.

Exercise 4.75. Find $\displaystyle\lim_{x \to 1} \frac{(2x - x^4)^{1/2} - x^{1/3}}{1 - x^{3/4}}$.

Notes

Exercises 4.22, 4.23, are based on [**A**]. Exercises 4.31, 4.33, 4.35 stem from [**G**]. Exercise 4.57 is based on [**R**].

---

[6] As in one of the Spiderman comics, a narrative panel reads 'with great power there must also come—great responsibility!'

5

Integration

In this chapter, we study

(1) the definition of the integral

(2) some of its basic properties

(3) its applications.

The integral $\displaystyle\int_a^b f(x)dx$ will be the area under the graph of a function $f : [a, b] \rightarrow \mathbb{R}$.

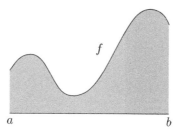

5.1 Towards a definition of the integral

Let $f : [a, b] \rightarrow \mathbb{R}$ be a 'nice' function and consider its graph:

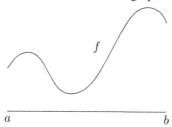

The How and Why of One Variable Calculus, First Edition. Amol Sasane.
© 2015 John Wiley & Sons, Ltd. Published 2015 by John Wiley & Sons, Ltd.

It is a basic problem in geometry to calculate the area under the graph of such a function f. Let us (for now) denote this area by $A(f)$. For example, when $f : [-r, r] \to \mathbb{R}$ is given by

$$f(x) = \sqrt{r^2 - x^2}, \quad -r \le x \le r,$$

then we would like to calculate the area $A(f)$ under the graph of f, which is the area of the semicircular region:

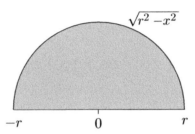

But what do we mean by 'area' and for which $f : [a, b] \to \mathbb{R}$ does $A(f)$ exist?

Consider first a very simple case, namely when $f : [a, b] \to \mathbb{R}$ is a constant function

$$f(x) = c, \quad x \in [a, b].$$

Then clearly the area $A(f)$ under the graph of f should be the area of the shaded rectangle, given by

$$A(f) = c \cdot (b - a) \quad \text{(the product of the length with the breadth of the rectangle)}.$$

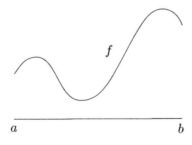

But what if f is not constant, and instead looks like this?

Well, if there are numbers M, m such that

$$m \leq f(x) \leq M \ \text{ for all } x \in [a,b],$$

then clearly we should have

$$m \cdot (b - a) \leq A(f) \leq M \cdot (b - a)$$

as illustrated by the pictures in Figure 5.1.

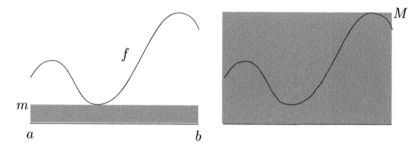

Figure 5.1 The area $A(f)$ under the graph of f is flanked by the areas of the two shaded rectangles, that is, it satisfies $m(b - a) \leq A(f) \leq M(b - a)$.

This gives us the idea that we can estimate the area $A(f)$ by considering little rectangles, as shown in Figure 5.2, and we anticipate that if we make the rectangles finer and finer, then we should be able to approximate $A(f)$ better and better.

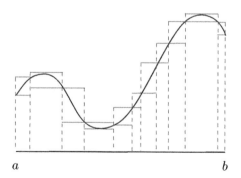

Figure 5.2 The area $A(f)$ under the graph of f satisfies $\underline{S} \leq A(f) \leq \overline{S}$, where $\underline{S}$ is the sum of all the areas of the rectangles shown above which lie *below* the graph of f and $\overline{S}$ is the sum of all the areas of the rectangles shown in the picture which lie *above* the graph of f.

In order to make this precise, we introduce the notions of

(1) a *partition P* of an interval $[a, b]$, and

(2) an *upper/lower sum* associated with a partition P of $[a, b]$ and a bounded function $f : [a, b] \rightarrow \mathbb{R}$.

Partition of an interval $[a, b]$.

Definition 5.1 (Partition of an interval). A *partition* (of an interval $[a, b] \subset \mathbb{R}$) is a finite set $P = \{x_0, x_1, \cdots, x_{n-1}, x_n\}$ such that

$$x_0 := a < x_1 < x_2 < x_3 < \cdots < x_{n-1} < b =: x_n.$$

See Figure 5.3. The collection of all partitions of $[a, b]$ is denoted by $\mathcal{P}_{[a,b]}$.

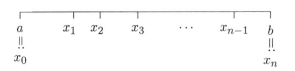

Figure 5.3   A partition P of $[a, b]$.

Example 5.1. The sets

$$\{a, b\},$$

$$\left\{ a, \frac{a+b}{2}, b \right\},$$

$$\left\{ a, a + \frac{b-a}{3}, b \right\},$$

$$P_n := \left\{ a, a + \frac{b-a}{n}, a + 2\frac{b-a}{n}, \cdots, a + (n-1)\frac{b-a}{n}, b \right\} \quad (n \in \mathbb{N}),$$

are examples of partitions of $[a, b]$, and all of these belong to $\mathcal{P}_{[a,b]}$.   ◇

Exercise 5.1. Which of the following statements is true?

(1) $\{0, 1, 1/2, 1/3, \cdots\}$ is a partition of $[0, 1]$.

(2) Every interval $[a, b]$ has an infinite number of partitions.

(3) $\{0, 1, 2, 3\}$ is a partition of $[0, \infty)$.

(4) $\{1/3, 1/2, 3/4\}$ is a partition of $[0, 1]$.

Bounded functions.

Definition 5.2 (Bounded function). A function $f : [a, b] \to \mathbb{R}$ is said to be *bounded* if there exist M, m such that for all $x \in [a, b]$, $m \leq f(x) \leq M$. See Figure 5.4.

This is equivalent to each of the following:

(1) The range of f, namely the set $\{f(x) : x \in [a, b]\}$, is a bounded set.

(2) There exists an $M \geq 0$ such that for all $x \in [a, b]$, $|f(x)| \leq M$.

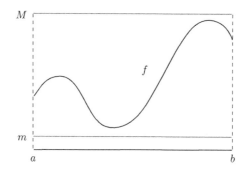

Figure 5.4   A bounded function $f : [a, b] \to \mathbb{R}$.

The equivalence of the boundedness of f with (1) is obvious from the definitions, while the equivalence of (1) and (2) follows immediately from Exercise 1.20.

Pictorially, if we imagine a light source at '$x = +\infty$', sending parallel light rays to the left, then the 'shadow of the graph of f on the y-axis' is a bounded set.

Example 5.2. The function $f : [0, 1] \to \mathbb{R}$ given by $f(x) = x^2, x \in \mathbb{R}$, is bounded. Indeed, for all $x \in [0, 1]$, we have $m := 0 \leq f(x) = x^2 \leq 1 =: M$.

On the other hand, the function $g : [0, 1] \to \mathbb{R}$ given by

$$g(x) = \begin{cases} 1/x & \text{if } x \in (0, 1], \\ 0 & \text{if } x = 0, \end{cases}$$

is not bounded, since if there exists an $M \in \mathbb{R}$ such that $g(x) \leq M$ for all $x \in [0, 1]$, then in particular, for all $n \in \mathbb{N}$, with $x := 1/n \in [0, 1]$, we would have

$$g(x) = \frac{1}{1/n} = n \leq M, \quad n \in \mathbb{N},$$

which is impossible by the Archimedean Property of $\mathbb{R}$.   ◇

Upper sum $\overline{S}(f, P)$ of f associated with a partition P.

Definition 5.3 (Upper sum). Let $f : [a, b] \to \mathbb{R}$ be a bounded function and P be a partition of $[a, b]$. The *upper sum* $\overline{S}(f, P)$ of f associated with a partition P is

$$\overline{S}(f, P) := \sum_{k=0}^{n-1} M_k \cdot (x_{k+1} - x_k),$$

where $M_k := \sup_{x \in [x_k, x_{k+1}]} f(x), k = 0, 1, \cdots, n - 1$.

The set $\{f(x) : x \in [x_k, x_{k+1}]\}$, namely, the range of f restricted to the subinterval $[x_k, x_{k+1}]$ of $[a, b]$, is nonempty and bounded above (by any upper bound for the range of f on $[a, b]$). So, M_k above makes sense for all ks.

The upper sum is formed by the addition of the various terms $M_k \cdot (x_{k+1} - x_k)$ for the different ks. Each one of such terms is just the area of the rectangle with base as the interval $[x_k, x_{k+1}]$ and height M_k for the various ks. See Figure 5.5.

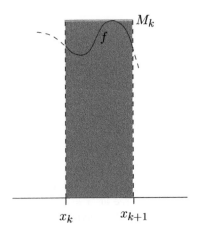

Figure 5.5 The area of the shortest rectangle that lies above the graph of f in the interval $[x_k, x_{k+1}]$.

Thus the term $M_k \cdot (x_{k+1} - x_k)$ is just the area of the shortest rectangle lying above the graph of f in the interval $[x_k, x_{k+1}]$.

The rationale behind the notation $\overline{S}(f, P)$ is that S is for 'sum' (of areas of rectangles), the ⁻ reminds us that the rectangles have their upper edges lying *above* the graph of f, and the (f, P) tells us *which* function f and partition P of $[a, b]$ we are forming the upper sum for. The picture in Figure 5.6 shows how a particular upper sum is formed:

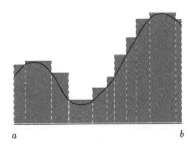

Figure 5.6 The upper sum $\overline{S}(f, P)$ associated with f and the partition P is the sum of the areas of the shortest rectangles that lie above the graph of f in each of the intervals $[x_k, x_{k+1}]$, $k = 0, 1, \cdots, n - 1$, of the partition.

Example 5.3. For $n \in \mathbb{N}$, let P_n be the partition

$$P_n := \left\{ 0, \frac{1}{n}, \frac{2}{n}, \frac{3}{n}, \cdots, \frac{n-1}{n}, 1 \right\}$$

of the interval $[0, 1]$, and let $f : [0, 1] \to \mathbb{R}$ be the squaring function

$$f(x) = x^2, \quad x \in [0, 1].$$

As f is increasing, it is clear that

$$M_k := \sup_{x \in [\frac{k}{n}, \frac{k+1}{n}]} f(x) = \frac{(k+1)^2}{n^2}.$$

Thus the upper sum $\overline{S}(f, P_n)$ associated with f and P_n is given by

$$\overline{S}(f, P_n) = \sum_{k=0}^{n-1} M_k \cdot \left(\frac{k+1}{n} - \frac{k}{n} \right) = \sum_{k=0}^{n-1} M_k \cdot \frac{1}{n}$$

$$= \sum_{k=0}^{n-1} \frac{(k+1)^2}{n^2} \cdot \frac{1}{n} = \frac{1}{n^3} \sum_{k=0}^{n-1} (k+1)^2$$

$$= \frac{1}{n^3} (1^2 + 2^2 + 3^2 + \cdots + n^2) \overset{(*)}{=} \frac{1}{n^3} \cdot \frac{n(n+1)(2n+1)}{6}$$

$$= \frac{1}{6} \left(1 + \frac{1}{n} \right) \left(2 + \frac{1}{n} \right).$$

Here in the step $(*)$, we have used the fact that for all $n \in \mathbb{N}$, the sum of the first n squares is

$$1^2 + 2^2 + 3^2 + \cdots + n^2 = \frac{n(n+1)(2n+1)}{6},$$

which the interested student may justify using induction on n. ◇

Lower sum $\underline{S}(f, P)$ of f associated with a partition P.

Definition 5.4 (Lower sum). Let $f : [a, b] \to \mathbb{R}$ be a bounded function and P be a partition of $[a, b]$. The *lower sum* $\underline{S}(f, P)$ of f associated with a partition P is

$$\underline{S}(f, P) := \sum_{k=0}^{n-1} m_k \cdot (x_{k+1} - x_k),$$

where $m_k := \inf_{x \in [x_k, x_{k+1}]} f(x), k = 0, 1, \cdots, n - 1.$

The set $\{f(x) : x \in [x_k, x_{k+1}]\}$, namely, the range of f restricted to the subinterval $[x_k, x_{k+1}]$ of $[a, b]$, is nonempty and bounded below (by any lower bound for the range of f on $[a, b]$). So, m_k above makes sense for all ks.

The lower sum is obtained by adding the various terms $m_k \cdot (x_{k+1} - x_k)$ for the different ks. Each such term is just the area of the rectangle with base as the interval $[x_k, x_{k+1}]$ and height m_k. See Figure 5.7.

Thus the term $m_k \cdot (x_{k+1} - x_k)$ is just the area of the tallest rectangle lying below the graph of f in the interval $[x_k, x_{k+1}]$.

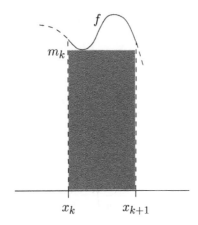

Figure 5.7   The area of the tallest rectangle that lies below the graph of f in the interval $[x_k, x_{k+1}]$.

The rationale behind the notation $\underline{S}(f, P)$ is that S is for 'sum' (of areas of rectangles), the $\underline{\ }$ reminds us that the rectangles have their upper edges lying *below* the graph of f, and the (f, P) tells us *which* function f and partition P of $[a, b]$ we are forming the lower sum for. The picture in Figure 5.8 shows how a particular lower sum is formed:

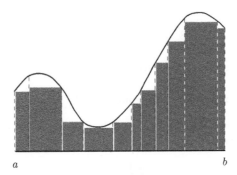

Figure 5.8   The lower sum $\underline{S}(f, P)$ associated with f and the partition P is the sum of the areas of the tallest rectangles that lie below the graph of f in each of the intervals $[x_k, x_{k+1}]$, $k = 0, 1, \cdots, n - 1$, of the partition.

Example 5.4. For $n \in \mathbb{N}$, let P_n be the partition

$$P_n := \left\{ 0, \frac{1}{n}, \frac{2}{n}, \frac{3}{n}, \cdots, \frac{n-1}{n}, 1 \right\}$$

of the interval $[0, 1]$ and let $f : [0, 1] \to \mathbb{R}$ be the squaring function

$$f(x) = x^2, \quad x \in [0, 1].$$

As f is increasing, it is clear that

$$m_k := \sup_{x \in [\frac{k}{n}, \frac{k+1}{n}]} f(x) = \frac{k^2}{n^2}.$$

Thus the lower sum $\underline{S}(f, P_n)$ associated with f and P_n is given by

$$\underline{S}(f, P_n) = \sum_{k=0}^{n-1} m_k \cdot \left(\frac{k+1}{n} - \frac{k}{n} \right) = \sum_{k=0}^{n-1} m_k \cdot \frac{1}{n}$$

$$= \sum_{k=0}^{n-1} \frac{k^2}{n^2} \cdot \frac{1}{n} = \frac{1}{n^3} \sum_{k=0}^{n-1} k^2$$

$$= \frac{1}{n^3} (0^2 + 1^2 + 2^2 + \cdots + (n-1)^2) = \frac{1}{n^3} \cdot \frac{(n-1)n(2n-1)}{6}$$

$$= \frac{1}{6} \left(1 - \frac{1}{n} \right) \left(2 - \frac{1}{n} \right).$$

◇

In order to arrive at a sensible definition of the integral of $f : [a, b] \to \mathbb{R}$, that is, of the area $A(f)$ under the graph of $f : [a, b] \to \mathbb{R}$, we first make the following observations, which will help us to formulate this sought after definition:

(1) Clearly, we expect the area $A(f)$ under the graph of $f : [a, b] \to \mathbb{R}$ to satisfy

$$A(f) \leq \overline{S}(f, P)$$

for *any* partition P, and so the number $A(f)$ should be a lower bound for the set of all upper sums $\overline{S}(f, P)$, where P belongs to the collection $\mathcal{P}_{[a,b]}$ of all partitions of $[a, b]$. Thus

$$A(f) \leq \overline{S}(f) := \inf_{P \in \mathcal{P}_{[a,b]}} \overline{S}(f, P). \tag{5.1}$$

(2) Similarly, we expect the area $A(f)$ under the graph of $f : [a, b] \to \mathbb{R}$ to satisfy

$$\underline{S}(f, P) \leq A(f)$$

for *any* partition P, and so the number $A(f)$ should be an upper bound for the set of all lower sums $\underline{S}(f, P)$, where P belongs to the collection $\mathcal{P}_{[a,b]}$ of all partitions of $[a, b]$. Thus

$$\sup_{P \in \mathcal{P}_{[a,b]}} \underline{S}(f, P) =: \underline{S}(f) \leq A(f). \tag{5.2}$$

(3) Putting (5.1) and (5.2) together, we see that our notion of the integral must satisfy

$$\underline{S}(f) \leq A(f) \leq \overline{S}(f).$$

See Figure 5.9. Also, as our partitions P get finer, we expect that for nice functions f (for which we *can* define the area under its graph), $\underline{S}(f, P) \approx \overline{S}(f, P)$, and so for such nice functions, we would then expect that $\underline{S}(f) = A(f) = \overline{S}(f)$. And this motivates the following definition.

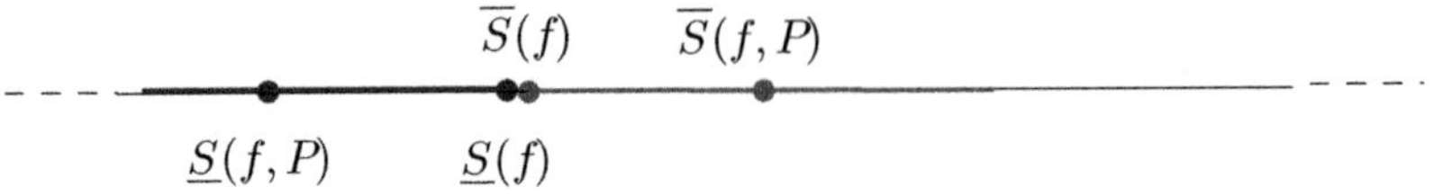

$$\overline{S}(f) \qquad \overline{S}(f, P)$$

$$\underline{S}(f, P) \qquad \underline{S}(f)$$

Figure 5.9 The integral $A(f)$ is flanked by $\underline{S}(f)$ and $\overline{S}(f)$. For fine partitions, we expect $\overline{S}(f, P)$ to be close to $\underline{S}(f, P)$, and this makes $\overline{S}(f)$ and $\underline{S}(f)$ to be equal to each other. Thus for nice functions f, $A(f) = \underline{S}(f) = \overline{S}(f)$.

Definition 5.5 (Riemann integral of a Riemann integrable function). Let $\mathcal{P}_{[a,b]}$ be the collection of all partitions of $[a, b]$ and let $f : [a, b] \to \mathbb{R}$ be bounded. Then f is said to be *Riemann integrable* (on $[a, b]$) if

$$\underline{S}(f) = \overline{S}(f),$$

and the *Riemann integral*, denoted by $\int_a^b f(x)dx$, is defined to be this common value:

$$\int_a^b f(x)dx = \underline{S}(f) = \overline{S}(f).$$

The set of all Riemann integrable functions on $[a, b]$ is denoted by $RI[a, b]$.

In the notation $\int_a^b f(x)dx$, the

$$\int$$

symbol is really an elongated S from 'sum', and the '$f(x)dx$' reminds us that in the upper and lower sums, we have areas of little rectangles, whose base length is an elemental change dx in x, and height is $f(x)$. The a and b at the bottom and top simply indicate what interval $[a, b]$ we are working with. The function f is often referred to as the *integrand*.

We will soon show that in general for *any* bounded function $f : [a, b] \to \mathbb{R}$ (Riemann integrable or not), we have

$$\overline{S}(f) \geq \underline{S}(f).$$

For non-Riemann integrable functions, one has a strict inequality above, and for Riemann integrable functions, one has an equality.

In order to prove the inequality above, we will need to investigate what happens to upper and lower sums when points are added to a partition. The new partition obtained by the process of adding extra points is called a refinement.

Definition 5.6 (Refinement of a partition). If P, P_* are partitions of $[a, b]$ such that $P \subset P_*$, then P_* is called a *refinement of* P.

When a partition is refined, one can imagine that the approximations to the area under the graph of f becomes better, and so lower sums ought to increase, and upper sums ought to decrease. This is exactly what happens, and this is the content of the next result.

Lemma 5.1 (Refinement Lemma). *If* P, P_* *are partitions of* $[a, b]$ *with* $P \subset P_*$, *and* $f : [a, b] \to \mathbb{R}$ *is bounded, then* $\overline{S}(f, P_*) \leq \overline{S}(f, P)$, *and* $\underline{S}(f, P_*) \geq \underline{S}(f, P)$.

Proof. Let $P = \{x_0, x_1, \cdots, x_{n-1}, x_n\}$. First suppose that P_* has just *one* extra point x_*, occurring in some subinterval $[x_k, x_{k+1}]$. See Figure 5.10.

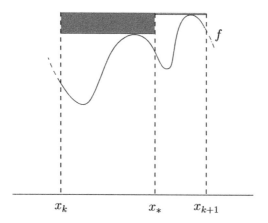

Figure 5.10 $\overline{S}(f, P) - \overline{S}(f, P_*)$ essentially is the nonnegative area of the shaded rectangle.

If we compare $\overline{S}(f, P)$ with $\overline{S}(f, P_*)$, we notice that most of the terms in the two sums are identical, except for the terms involving the interval $[x_k, x_{k+1}]$. We have

$$\overline{S}(f, P) - \overline{S}(f, P_*)$$

$$= \left(\sup_{x \in [x_k, x_{k+1}]} f(x) \right) \cdot (x_{k+1} - x_k)$$

$$- \left(\sup_{x \in [x_k, x_*]} f(x) \right) \cdot (x_* - x_k) - \left(\sup_{x \in [x_*, x_{k+1}]} f(x) \right) \cdot (x_{k+1} - x_*)$$

$$= \left(\sup_{x \in [x_k, x_{k+1}]} f(x) \right) \cdot (x_{k+1} - x_* + x_* - x_k)$$

$$- \left(\sup_{x \in [x_k, x_*]} f(x) \right) \cdot (x_* - x_k) - \left(\sup_{x \in [x_*, x_{k+1}]} f(x) \right) \cdot (x_{k+1} - x_*)$$

$$= \left(\sup_{x \in [x_k, x_{k+1}]} f(x) - \sup_{x \in [x_k, x_*]} f(x) \right) \cdot (x_* - x_k)$$

$$+ \left(\sup_{x \in [x_k, x_{k+1}]} f(x) - \sup_{x \in [x_*, x_{k+1}]} f(x) \right) \cdot (x_{k+1} - x_*)$$

$$\geq 0 + 0 = 0.$$

If P_* has several additional points (instead of just one additional point), then we repeat the argument several times, considering one extra point in each step to obtain

$$\overline{S}(f, P_*) \ \leq \ \cdots \ \leq \ \overline{S}(f, P_2) \ \leq \ \overline{S}(f, P_1) \ \leq \ \overline{S}(f, P)$$

where

P_1 is a refinement of P having *one* more point than P,

P_2 is a refinement of P_1 having *one* more point than P_1,

and *two* extra points than P,

$\cdots$ and so on.

Thus $\overline{S}(f, P_*) \leq \overline{S}(f, P)$.

The proof of $\underline{S}(f, P_*) \geq \underline{S}(f, P)$ is analogous.   □

Corollary 5.2. *If* $f : [a, b] \to \mathbb{R}$ *is bounded then* $\overline{S}(f) \geq \underline{S}(f)$.

Proof. If P, P' are any two refinements of $[a, b]$, then $P \cup P'$ is a refinement of P as well as P', and so by the Refinement Lemma, we have

$$\overline{S}(f, P) \geq \overline{S}(f, P \cup P') \geq \underline{S}(f, P \cup P') \geq \underline{S}(f, P').$$

Thus $\overline{S}(f, P) \geq \underline{S}(f, P')$, for *any* two partitions P, P'. (So **any** upper sum is always bigger than **any** lower sum!)

Let P be a fixed partition. For any partition $P' \in \mathcal{P}_{[a,b]}$, $\overline{S}(f, P) \geq \underline{S}(f, P')$. Thus

$$\overline{S}(f, P) \geq \sup_{P' \in \mathcal{P}_{[a,b]}} \underline{S}(f, P') = \underline{S}(f).$$

As the choice of $P \in \mathcal{P}_{[a,b]}$ was arbitrary, $\inf_{P \in \mathcal{P}_{[a,b]}} \overline{S}(f, P) = \overline{S}(f) \geq \underline{S}(f)$.   □

Let us now show that the squaring function is Riemann integrable on $[0, 1]$, and let us calculate its value.

Example 5.5. Consider the bounded function $f : [0, 1] \to \mathbb{R}$ given by

$$f(x) = x^2, \quad x \in [0, 1].$$

We will show that $f \in RI[0, 1]$ and that $\displaystyle\int_0^1 x^2 dx = \frac{1}{3}$.

Rather than considering *all* partitions, it turns out that we can be efficient and consider the special partitions

$$P_n = \left\{ 0, \frac{1}{n}, \frac{2}{n}, \frac{3}{n}, \cdots, \frac{n-1}{n}, 1 \right\}, \quad n \in \mathbb{N}.$$

We had found out earlier in Examples 5.3 and 5.4 that

$$\overline{S}(f, P_n) = \frac{1}{6} \left(1 + \frac{1}{n} \right) \left(1 + \frac{2}{n} \right) \quad \text{and} \quad \underline{S}(f, P_n) = \frac{1}{6} \left(1 - \frac{1}{n} \right) \left(1 - \frac{2}{n} \right).$$

Thus

$$\overline{S}(f) = \inf_{P \in \mathcal{P}_{[0,1]}} \overline{S}(f,P) \le \inf_{n \in \mathbb{N}} \overline{S}(f,P_n) = \inf_{n \in \mathbb{N}} \frac{1}{6}\left(1 + \frac{1}{n}\right)\left(2 + \frac{1}{n}\right) \overset{(*)}{=} \frac{1}{3}, \text{ and}$$

$$\underline{S}(f) = \sup_{P \in \mathcal{P}_{[0,1]}} \underline{S}(f,P) \ge \sup_{n \in \mathbb{N}} \underline{S}(f,P_n) = \sup_{n \in \mathbb{N}} \frac{1}{6}\left(1 - \frac{1}{n}\right)\left(2 - \frac{1}{n}\right) \overset{(**)}{=} \frac{1}{3}.$$

For the justification of $(*)$ and $(**)$, note that the sequence with n term

$$\frac{1}{6}\left(1 + \frac{1}{n}\right)\left(2 + \frac{1}{n}\right)$$

is decreasing and bounded below by 0, and hence convergent to

$$\inf_{n \in \mathbb{N}} \frac{1}{6}\left(1 + \frac{1}{n}\right)\left(2 + \frac{1}{n}\right).$$

On the other hand, from the Algebra of Limits, it is easy to see that the limit must be

$$\frac{1}{6}\left(1 + \lim_{n \to 0} \frac{1}{n}\right)\left(2 + \lim_{n \to 0} \frac{1}{n}\right) = \frac{1}{6}(1 + 0)(2 + 0) = \frac{1}{3}.$$

The proof of $(**)$ is analogous.

Hence $\frac{1}{3} \ge \overline{S}(f) \ge \underline{S}(f) \ge \frac{1}{3}$, and so $\overline{S}(f) = \underline{S}(f) = \frac{1}{3}$.

Thus $f \in RI[0,1]$ and $\int_0^1 x^2 dx = \frac{1}{3}$.   ◇

In the example above, we had to work rather hard to find the integral of a simple function. But we will soon learn about the Fundamental Theorem of Calculus, which will enable us to avoid such complicated calculations with partitions, lower and upper sums, infimums and supremums etc. Indeed, the Fundamental Theorem of Calculus says that if the integrand f is the derivative of a function F, then

$$\int_a^b f(x)dx = F(b) - F(a) \ !$$

In light of this result, we can now easily evaluate our previous example for the squaring function. Indeed, we simply note that the integrand $f := x^2$ is the derivative of $F := x^3/3$, and so

$$\int_0^1 x^2 dx = \frac{1^3}{3} - \frac{0^3}{3} = \frac{1}{3}.$$

But before we establish the Fundamental Theorem of Calculus, we will first learn about a few basic, but important properties of the Riemann integral in the next section.

Are *all* bounded functions $f : [a,b] \to \mathbb{R}$ Riemann integrable? The answer is no, and here is an example.

Example 5.6. ($1_{\mathbb{Q}} \notin RI[0, 1]$.) Consider the indicator function[1] $1_{\mathbb{Q}}$ of the rationals:

$$1_{\mathbb{Q}}(x) = \begin{cases} 1 & \text{if } x \in \mathbb{Q}, \\ 0 & \text{if } x \notin \mathbb{Q}. \end{cases}$$

Clearly $1_{\mathbb{Q}}$ is bounded: for all x, $0 \le 1_{\mathbb{Q}}(x) \le 1$.

We will show that (the restriction of) $1_{\mathbb{Q}}$ on $[0, 1]$ is not Riemann integrable on $[0, 1]$ by showing that

$$\overline{S}(1_{\mathbb{Q}}) \ge 1 > 0 \ge \underline{S}(1_{\mathbb{Q}}).$$

Let $P = \{x_0 = 0, x_1, \cdots, x_{n-1}, x_n = 1\}$ be any partition of $[0, 1]$. Then each $[x_k, x_{k+1}]$ contains a rational number, say $\alpha_k \in \mathbb{Q}$, and an irrational number, say $\beta_k \notin \mathbb{Q}$. Thus

$$M_k := \sup_{x \in [x_k, x_{k+1}]} f(x) \ge f(\alpha_k) = 1, \quad \text{and}$$

$$m_k := \inf_{x \in [x_k, x_{k+1}]} f(x) \le f(\beta_k) = 0.$$

Hence

$$\overline{S}(1_{\mathbb{Q}}, P) = \sum_{k=0}^{n-1} M_k \cdot (x_{k+1} - x_k) \ge \sum_{k=0}^{n-1} 1 \cdot (x_{k+1} - x_k)$$

$$= (x_1 - x_0) + (x_2 - x_1) + \cdots + (x_n - x_{n-1}) = x_n - x_0 = 1 - 0 = 1.$$

Similarly

$$\underline{S}(1_{\mathbb{Q}}, P) = \sum_{k=0}^{n-1} m_k \cdot (x_{k+1} - x_k) \le \sum_{k=0}^{n-1} 0 \cdot (x_{k+1} - x_k) = 0.$$

So $\overline{S}(1_{\mathbb{Q}}) = \inf_{P \in \mathcal{P}_{[a,b]}} \overline{S}(1_{\mathbb{Q}}, P) = 1 > 0 = \sup_{P \in \mathcal{P}_{[a,b]}} \overline{S}(1_{\mathbb{Q}}, P) = \underline{S}(1_{\mathbb{Q}})$, and $1_{\mathbb{Q}} \notin RI[0, 1]$.   ◇

Hence we have $RI[a, b] \subsetneq B[a, b]$, where $B[a, b]$ denotes the set of all bounded functions on $[a, b]$.

Exercise 5.2. (∗) Is the following function f Riemann integrable on $[0, 1]$?

$$f(x) = \begin{cases} 0 & \text{if } x \in [0, 1] \backslash \mathbb{Q}, \\ x & \text{if } x \in [0, 1] \cap \mathbb{Q}. \end{cases}$$

Hint: For any partition $P = \{x_0 = 0 < x_1 < \cdots < x_{n-1} < x_n = 1\}$,

$$x_{k+1} \ge \frac{x_{k+1} + x_k}{2}, \quad k = 0, \cdots, n - 1.$$

Use this to find a positive lower bound on upper sums.

Let us now show that there is an ample supply of Riemann integrable functions: all continuous functions are Riemann integrable, that is, $C[a, b] \subset RI[a, b]$.

---

[1] If S is a subset of $\mathbb{R}$, then the *indicator function* 1_S is defined by $1_S(x) = 1$ if $x \in S$ and 0 if $x \notin S$.

Theorem 5.3. *Every continuous function on $[a,b]$ is Riemann integrable on $[a,b]$.*

Proof. As f is continuous on $[a,b]$ and since $[a,b]$ is a compact interval, f is also *uniformly continuous* on $[a,b]$. Let $\epsilon > 0$. Then there exists a $\delta > 0$ such that whenever $x,y \in [a,b]$ satisfy $|x-y| < \delta$, we have $|f(x)-f(y)| < \epsilon$. Consider any partition $P_* = \{x_0, x_1, \cdots, x_{n-1}, x_n\}$ such that

$$\max_{k \in \{0,1,\cdots,n-1\}} |x_{k+1} - x_k| < \delta.$$

Then with $M_k := \sup_{x \in [x_k, x_{k+1}]} f(x)$ and $m_k := \inf_{x \in [x_k, x_{k+1}]} f(x)$, we have

$$\overline{S}(f, P_*) - \underline{S}(f, P_*) = \sum_{k=0}^{n-1} (M_k - m_k)(x_{k+1} - x_k)$$

$$\leq \sum_{k=0}^{n-1} \epsilon(x_{k+1} - x_k) = \epsilon(b - a).$$

Thus

$$0 \leq \overline{S}(f) - \underline{S}(f) \leq \overline{S}(f, P_*) - \underline{S}(f, P_*) \leq \epsilon(b - a).$$

Since $\epsilon > 0$ was arbitrary, it follows that $\overline{S}(f) = \underline{S}(f)$, that is, $f \in RI[a,b]$. □

Example 5.7. All polynomial functions, being continuous, are Riemann integrable on every compact interval $[a,b]$. ◇

Example 5.8 (Definition of π). Consider the continuous function $f : [-1,1] \to \mathbb{R}$ defined by $f(x) = \sqrt{1-x^2}, x \in [-1,1]$.

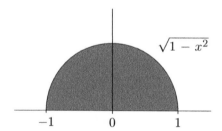

Figure 5.11   The definition of the number $\pi := 2 \int_{-1}^{1} \sqrt{1-x^2}dx$.

Then f is Riemann integrable. We *define* the number $\pi \in \mathbb{R}$ by

$$\pi := 2 \int_{-1}^{1} \sqrt{1-x^2}dx = \text{two times the area of the semicircular disk of radius 1.}$$

See Figure 5.11. (Later on, we will *prove* that this is the same π one meets in high school, namely, for a circle of radius r, its circumference will be $2\pi r$, and the area enclosed by it will be equal to πr^2.) ◇

Exercise 5.3. (The aim of this exercise is twofold: first, to show that $C[a,b] \subsetneq RI[a,b]$, and secondly, to point out that what we have been discussing is *signed* area under the graph of f, so that if the graph of a function lies below the x-axis, then that portion of the area is attributed a *negative* sign.) Let $f : [0, 2] \to \mathbb{R}$ be given by

$$f(x) = \begin{cases} 1 & \text{if } x \in [0, 1], \\ -1 & \text{if } x \in (1, 2]. \end{cases}$$

Show that $f \in RI[0, 2]$ and that $\int_0^2 f(x)dx = 0$.

Hint: Consider the partitions $P_n := \left\{0, 1, 1 + \dfrac{1}{n}, 2\right\}$, $n \in \mathbb{N}$.

Exercise 5.4. Which of the following statements is/are true?

□ (A) Continuous functions $f : [a, b] \to \mathbb{R}$ are Riemann integrable on $[a, b]$.

□ (B) Riemann integrable functions $f : [a, b] \to \mathbb{R}$ are continuous on $[a, b]$.

□ (C) Differentiable functions $f : [a, b] \to \mathbb{R}$ are Riemann integrable on $[a, b]$.

□ (D) Bounded functions $f : [a, b] \to \mathbb{R}$ are Riemann integrable on $[a, b]$.

Exercise 5.5. For a partition $P = \{x_0 = a, x_1, \cdots, x_{n-1}, x_n = b\}$ of $[a, b]$, with

$$x_k < x_{k+1}, \quad k = 0, \cdots, n - 1,$$

define

$$\Phi(P) := \max\{x_{k+1} - x_k : k = 0, \cdots, n - 1\}.$$

Which of the following is always true for any bounded function $f : [a, b] \to \mathbb{R}$?

□ (A) If P_2 is a refinement of P_1 (that is, $P_1 \subset P_2$), then $\Phi(P_2) \leq \Phi(P_1)$.

□ (B) If $\Phi(P_2) \leq \Phi(P_1)$, then $\overline{S}(f, P_2) \leq \overline{S}(f, P_1)$.

□ (C) If $\Phi(P_2) \leq \Phi(P_1)$, then $\underline{S}(f, P_2) \leq \underline{S}(f, P_1)$.

□ (D) If $\Phi(P_2) \leq \Phi(P_1)$, then $\underline{S}(f, P_2) \leq \overline{S}(f, P_1)$.

5.2   Properties of the Riemann integral

Theorem 5.4.[2] *If* $f, g \in RI[a, b]$ *and* $\alpha \in \mathbb{R}$, *then* $f + g \in RI[a, b]$, $\alpha \cdot f \in RI[a, b]$,

$$\int_a^b (f(x) + g(x))dx = \int_a^b f(x)dx + \int_a^b g(x)dx, \quad and$$

$$\int_a^b \alpha \cdot f(x)dx = \alpha \cdot \int_a^b f(x)dx.$$

Proof. Let $\epsilon > 0$. Then there exist partitions P_f, P_g of $[a, b]$ such that

$$\overline{S}(f, P_f) < \overline{S}(f) + \epsilon/2,$$

$$\overline{S}(g, P_g) < \overline{S}(g) + \epsilon/2.$$

---

[2] The content of this result can be expressed in linear algebraic language by saying that $RI[a, b]$ forms a *vector space* with operations of vector addition and scalar multiplication defined in a pointwise manner, and that the map $f \mapsto \int_a^b f(x)dx : RI[a, b] \to \mathbb{R}$ is a *linear transformation*.

Then $P := P_f \bigcup P_g =: \{x_0, x_1, \cdots, x_{n-1}, x_n\}$ is a refinement of P_f and P_g, and

$$\overline{S}(f+g) \leq \overline{S}(f+g, P) = \sum_{k=0}^{n-1} \left(\sup_{x \in [x_k, x_{k+1}]} (f(x) + g(x)) \right) (x_{k+1} - x_k)$$

$$\leq \sum_{k=0}^{n-1} \left(\sup_{x \in [x_k, x_{k+1}]} f(x) + \sup_{x \in [x_k, x_{k+1}]} g(x) \right) (x_{k+1} - x_k)$$

$$= \overline{S}(f, P) + \overline{S}(g, P) \leq \overline{S}(f, P_f) + \overline{S}(g, P_g)$$

$$< \overline{S}(f) + \epsilon/2 + \overline{S}(g) + \epsilon/2 = \overline{S}(f) + \overline{S}(g) + \epsilon.$$

As $\epsilon > 0$ was arbitrary, it follows that

$$\overline{S}(f+g) \leq \overline{S}(f) + \overline{S}(g). \tag{5.3}$$

In a similar manner, we can show that

$$\underline{S}(f+g) \geq \underline{S}(f) + \underline{S}(g). \tag{5.4}$$

Here are the details. Let $\epsilon > 0$. Then there are partitions P_f, P_g of $[a, b]$ so that

$$\underline{S}(f, P_f) > \underline{S}(f) - \epsilon/2,$$
$$\underline{S}(g, P_g) > \underline{S}(g) - \epsilon/2.$$

Then $P := P_f \bigcup P_g =: \{x_0, x_1, \cdots, x_{n-1}, x_n\}$ is a refinement of P_f, P_g, and

$$\underline{S}(f+g) \geq \overline{S}(f+g, P) = \sum_{k=0}^{n-1} \left(\inf_{x \in [x_k, x_{k+1}]} (f(x) + g(x)) \right) (x_{k+1} - x_k)$$

$$\geq \sum_{k=0}^{n-1} \left(\inf_{x \in [x_k, x_{k+1}]} f(x) + \inf_{x \in [x_k, x_{k+1}]} g(x) \right) (x_{k+1} - x_k)$$

$$= \underline{S}(f, P) + \underline{S}(g, P) \geq \underline{S}(f, P_f) + \underline{S}(g, P_g)$$

$$> \underline{S}(f) - \epsilon/2 + \underline{S}(g) - \epsilon/2 = \underline{S}(f) + \underline{S}(f) - \epsilon.$$

As $\epsilon > 0$ was arbitrary, we obtain (5.4).

From (5.3) and (5.4), we have

$$\underline{S}(f) + \underline{S}(g) \leq \underline{S}(f+g) \leq \overline{S}(f+g) \leq \overline{S}(f) + \overline{S}(g). \tag{5.5}$$

Since $f, g \in RI[a, b]$, we have $\underline{S}(f) = \overline{S}(f)$ and $\underline{S}(g) = \overline{S}(g)$. Thus the first and last terms in (5.5) are equal. Consequently, $\underline{S}(f + g) = \overline{S}(f + g)$, that is, $f + g \in RI[a, b]$. Moreover,

$$\int_a^b (f(x) + g(x))dx = \overline{S}(f + g) = \overline{S}(f) + \overline{S}(g) = \int_a^b f(x)dx + \int_a^b g(x)dx.$$

To show the second claim, we consider the three possible cases that $\alpha > 0$, $\alpha = 0$, and $\alpha < 0$ separately.

$\underline{1°}$ $\alpha > 0$. We will use the result from Exercise 1.8. For every partition P of $[a, b]$, we have

$$\overline{S}(\alpha \cdot f, P) = \sum_{k=0}^{n-1} \left(\sup_{x \in [x_k, x_{k+1}]} (\alpha f(x)) \right) (x_{k+1} - x_k)$$

$$= \sum_{k=0}^{n-1} \alpha \left(\sup_{x \in [x_k, x_{k+1}]} f(x) \right) (x_{k+1} - x_k) = \alpha \cdot \overline{S}(f, P),$$

$$\underline{S}(\alpha \cdot f, P) = \sum_{k=0}^{n-1} \left(\inf_{x \in [x_k, x_{k+1}]} (\alpha f(x)) \right) (x_{k+1} - x_k)$$

$$= \sum_{k=0}^{n-1} \alpha \left(\inf_{x \in [x_k, x_{k+1}]} f(x) \right) (x_{k+1} - x_k) = \alpha \cdot \underline{S}(f, P).$$

Thus

$$\overline{S}(\alpha \cdot f) = \inf_{P \in \mathcal{P}_{[a,b]}} \overline{S}(\alpha \cdot f, P) = \inf_{P \in \mathcal{P}_{[a,b]}} \alpha \cdot \overline{S}(f, P) = \alpha \inf_{P \in \mathcal{P}_{[a,b]}} \overline{S}(f, P)$$

$$= \alpha \overline{S}(f) = \alpha \underline{S}(f)$$

$$= \alpha \sup_{P \in \mathcal{P}_{[a,b]}} \underline{S}(f, P) = \sup_{P \in \mathcal{P}_{[a,b]}} \alpha \underline{S}(f, P) = \sup_{P \in \mathcal{P}_{[a,b]}} \underline{S}(\alpha \cdot f, P) = \underline{S}(\alpha \cdot f).$$

Hence $\alpha \cdot f \in RI[a, b]$ and $\int_a^b \alpha \cdot f(x)dx = \alpha \cdot \int_a^b f(x)dx$.

$\underline{2°}$ $\alpha = 0$. Then we have $\alpha f(x) = 0$ for all $x \in [a, b]$, and so for every partition P of $[a, b]$, $\overline{S}(\alpha \cdot f, P) = 0 = \underline{S}(\alpha \cdot f, P)$, so that $\overline{S}(\alpha \cdot f) = 0 = \underline{S}(\alpha \cdot f)$. Hence $\alpha \cdot f \in RI[a, b]$ and

$$\int_a^b \alpha \cdot f(x)dx = 0 = 0 \cdot \int_a^b f(x)dx = \alpha \cdot \int_a^b f(x)dx.$$

$\underline{3°}$ $\alpha < 0$. Let $P = \{x_0, x_1, \cdots, x_{n-1}, x_n\}$ be any partition of $[a, b]$. First let $\alpha = -1$. Then

$$\overline{S}(-f, P) = \sum_{k=0}^{n-1} \left(\sup_{x \in [x_k, x_{k+1}]} -f(x) \right) (x_{k+1} - x_k)$$

$$= \sum_{k=0}^{n-1} \left(- \inf_{x \in [x_k, x_{k+1}]} f(x) \right) (x_{k+1} - x_k) = -\underline{S}(f, P).$$

By replacing f by $-f$, we also obtain from the above that $\underline{S}(-f,P) = -\overline{S}(f,P)$. Hence we have

$$\overline{S}(-f) = \inf_{P \in \mathcal{P}_{[a,b]}} \overline{S}(-f,P) = \inf_{P \in \mathcal{P}_{[a,b]}} -\underline{S}(f,P) = -\sup_{P \in \mathcal{P}_{[a,b]}} \underline{S}(f,P) = -\underline{S}(f)$$

$$\underline{S}(-f) = \sup_{P \in \mathcal{P}_{[a,b]}} \underline{S}(-f,P) = \sup_{P \in \mathcal{P}_{[a,b]}} -\overline{S}(f,P) = -\inf_{P \in \mathcal{P}_{[a,b]}} \overline{S}(f,P) = -\overline{S}(f).$$

Thus

$$\overline{S}(-f) = -\underline{S}(f) = -\overline{S}(f) = \underline{S}(-f),$$

and so $-f \in RI[a,b]$, and $\int_a^b -f(x)dx = \overline{S}(-f) = -\overline{S}(f) = -\int_a^b f(x)dx.$

For general $\alpha < 0$, we have $\alpha = -|\alpha|$, and as $f \in RI[a,b]$, it follows from $1°$ that $|\alpha| \cdot f \in RI[a,b]$. From the above, we now obtain that $-|\alpha| \cdot f \in RI[a,b]$, that is, $\alpha \cdot f \in RI[a,b]$. Also,

$$\int_a^b \alpha f(x)dx = \int_a^b -|\alpha|f(x)dx = -\int_a^b |\alpha|f(x)dx = -|\alpha| \int_a^b f(x)dx = \alpha \int_a^b f(x)dx.$$

This completes the proof. □

Example 5.9. For $n \in \mathbb{N}$, $x \mapsto x^n : \mathbb{R} \to \mathbb{R}$ is continuous and so $x^n \in RI[a,b]$ for all a,b. So the polynomial $p := c_0 \cdot 1 + c_1 \cdot x + \cdots + c_d \cdot x^d \in RI[a,b]$, and moreover,

$$\int_a^b p(x)dx = c_0 \int_a^b 1dx + c_1 \int_a^b x\,dx + \cdots + c_d \int_a^b x^d dx.$$

After learning the Fundamental Theorem of Calculus, we will know that

$$\int_a^b x^n dx = \int_a^b \frac{d}{dx}\frac{x^{n+1}}{n+1}\,dx = \frac{b^{n+1} - a^{n+1}}{n+1}, \quad n = 0, 1, 2, 3, \cdots.$$

So $\int_a^b p(x)dx = c_0(b-a) + c_1 \frac{b^2 - a^2}{2} + \cdots + c_d \frac{b^{d+1} - a^{d+1}}{d+1}.$ ◇

The following result will play an important role in the sequel.

Theorem 5.5 (Riemann Condition). *Let $f : [a,b] \to \mathbb{R}$ be bounded. Then*

$$f \in RI[a,b] \Leftrightarrow \boxed{\begin{array}{l} \textit{for all } \epsilon > 0, \textit{ there exists a partition } P_\epsilon \in \mathcal{P}_{[a,b]} \textit{ such that} \\ \overline{S}(f,P_\epsilon) - \underline{S}(f,P_\epsilon) < \epsilon. \end{array}}$$

Proof.

($\Leftarrow$:) For all $\epsilon > 0$, $0 \le \overline{S}(f) - \underline{S}(f) \le \overline{S}(f,P_\epsilon) - \underline{S}(f,P_\epsilon) < \epsilon$ and so $\overline{S}(f) = \underline{S}(f)$.

($\Rightarrow$:) Suppose $f \in RI[a,b]$. Let $\epsilon > 0$. Then there exists a partition P_1 such that

$$\overline{S}(f,P_1) < \overline{S}(f) + \frac{\epsilon}{2}.$$

Similarly, there exists a partition P_2 such that

$$\underline{S}(f,P_2) > \underline{S}(f) - \frac{\epsilon}{2}.$$

Let P_ϵ be the refinement of P_1 and P_2; $P_\epsilon := P_1 \cup P_2$. Then

$$\overline{S}(f,P_\epsilon) \leq \overline{S}(f,P_1) < \overline{S}(f) + \frac{\epsilon}{2}, \quad \text{and}$$

$$\underline{S}(f,P_\epsilon) \geq \underline{S}(f,P_2) > \underline{S}(f) - \frac{\epsilon}{2}.$$

Consequently, $0 \leq \overline{S}(f,P_\epsilon) - \underline{S}(f,P_\epsilon) < \underbrace{\overline{S}(f) - \underline{S}(f)}_{=0} + \epsilon = \epsilon.$ ☐

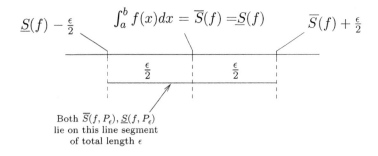

$\underline{S}(f) - \frac{\epsilon}{2}$ $\int_a^b f(x)dx = \overline{S}(f) = \underline{S}(f)$ $\overline{S}(f) + \frac{\epsilon}{2}$

$\frac{\epsilon}{2}$ $\frac{\epsilon}{2}$

Both $\overline{S}(f,P_\epsilon), \underline{S}(f,P_\epsilon)$
lie on this line segment
of total length ϵ

Let us now show that restrictions of Riemann integrable functions are Riemann integrable.

Theorem 5.6. *If* $[c,d] \subset [a,b]$ *and* $f \in RI[a,b]$, *then* $f \in RI[c,d]$.

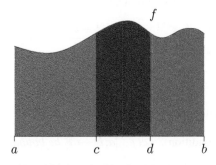

f

a c d b

Proof. We will use the Riemann Condition to show this. Let $\epsilon > 0$. Since f belongs to $RI[a,b]$, there exists a partition P_ϵ of $[a,b]$ such that $\overline{S}(f,P_\epsilon) - \underline{S}(f,P_\epsilon) < \epsilon$. Let $P'_\epsilon := P_\epsilon \cup \{c,d\} = P_{[a,c]} \cup P_{[c,d]} \cup P_{[d,b]}$, where

$$P_{[a,c]} \quad \text{is a partition of } [a,c],$$

$$P_{[c,d]} \quad \text{is a partition of } [c,d],$$

$$P_{[d,b]} \quad \text{is a partition of } [d,b].$$

We know that

$$\overline{S}(f,P_\epsilon) \geq \overline{S}(f,P'_\epsilon) = \overline{S}(f,P_{[a,c]}) + \overline{S}(f,P_{[c,d]}) + \overline{S}(f,P_{[d,b]}),$$
$$\underline{S}(f,P_\epsilon) \leq \underline{S}(f,P'_\epsilon) = \underline{S}(f,P_{[a,c]}) + \underline{S}(f,P_{[c,d]}) + \underline{S}(f,P_{[d,b]}).$$

Thus

$$\epsilon > \overline{S}(f,P_\epsilon) - \underline{S}(P,\epsilon)$$
$$\geq \overline{S}(f,P_{[a,c]}) - \underline{S}(f,P_{[a,c]}) + \overline{S}(f,P_{[c,d]}) - \underline{S}(f,P_{[c,d]}) + \overline{S}(f,P_{[d,b]}) - \underline{S}(f,P_{[d,b]})$$
$$\geq 0 + \overline{S}(f,P_{[c,d]}) - \underline{S}(f,P_{[c,d]}) + 0 = \overline{S}(f,P_{[c,d]}) - \underline{S}(f,P_{[c,d]}).$$

Hence by the Riemann Condition, $f \in RI[c,d]$. $\qquad\qquad\square$

Exercise 5.6. Let $f : [a,b] \to \mathbb{R}$, $a < c < b$, $f \in RI[a,c]$ and $f \in RI[c,b]$. Then $f \in RI[a,b]$ and moreover

$$\int_a^b f(x)dx = \int_a^c f(x)dx + \int_c^b f(x)dx.$$

Exercise 5.7. Let $f : [a,b] \to \mathbb{R}$ be a bounded function, such that f has only one discontinuity at $c \in (a,b)$. Show that $f \in RI[a,b]$. Extend the result to a finite number of discontinuities of f in (a,b).

Theorem 5.7. If $f, g \in RI[a,b]$, then $f \cdot g \in RI[a,b]$.

Proof. Let $\epsilon > 0$. Let $M_f, M_g > 0$ be such that $|f(x)| < M_f$ and $|g(x)| < M_g$ for all $x \in [a,b]$. Since $f \in RI[a,b]$, there exists a partition P_f of $[a,b]$ such that

$$\overline{S}(f,P_f) - \underline{S}(f,P_f) < \frac{\epsilon}{2M_g}.$$

Also, since $g \in RI[a,b]$, there exists a partition P_g of $[a,b]$ such that

$$\overline{S}(f,P_g) - \underline{S}(f,P_g) < \frac{\epsilon}{2M_f}.$$

Let $P = P_f \bigcup P_g =: \{x_0, x_1, \cdots, x_{n-1}, x_n\}$ be the refinement of P_f and P_g. For a bounded function φ on $[a,b]$ and a $k \in \{0, 1, \cdots, n-1\}$, we use the notation

$$M_{\varphi,k} := \sup_{x \in [x_k, x_{k+1}]} \varphi(x), \quad \text{and}$$

$$m_{\varphi,k} := \inf_{x \in [x_k, x_{k+1}]} \varphi(x).$$

Then for $x, y \in [x_k, x_{k+1}]$,

$$(f \cdot g)(x) - (f \cdot g)(y) = f(x)g(x) - f(x)g(y) + f(x)g(y) - f(y)g(y)$$

$$= f(x)(g(x) - g(y)) + (f(x) - f(y))g(y)$$

$$\le |f(x)||g(x) - g(y)| + |g(y)||f(x) - f(y)|$$

$$\le M_f(M_{g,k} - m_{g,k}) + M_g(M_{f,k} - m_{f,k}).$$

As $x, y \in [x_k, x_{k+1}]$ were arbitrary, it follows from the above that

$$M_{f \cdot g,k} - m_{f \cdot g,k} \le M_f(M_{g,k} - m_{g,k}) + M_g(M_{f,k} - m_{f,k}).$$

Thus

$$\overline{S}(f \cdot g) - \underline{S}(f \cdot g) \le \overline{S}(f \cdot g, P) - \underline{S}(f \cdot g, P)$$

$$\le M_f(\overline{S}(g, P) - \underline{S}(g, P)) + M_g(\overline{S}(f, P) - \underline{S}(f, P))$$

$$\le M_f(\overline{S}(g, P_g) - \underline{S}(g, P_g)) + M_g(\overline{S}(f, P_g) - \underline{S}(f, P_g))$$

$$\le M_f \frac{\epsilon}{2M_f} + M_g \frac{\epsilon}{2M_g} = \epsilon.$$

By the Riemann Condition, we conclude that $f \cdot g \in RI[a, b]$.   □

Some conventions

When defining $\int_a^b f(x)dx$, we assumed that $a < b$.

To simplify matters in what is to follow, we will adopt the following new definitions:

(1) If $a = b$, then *every* $f : [a, b] \to \mathbb{R}$ is Riemann integrable, and we define

$$\int_a^a f(x)dx := 0.$$

(2) If $a > b$ and $f : [b, a] \to \mathbb{R}$ is Riemann integrable, then we define

$$\int_a^b f(x)dx := -\int_b^a f(x)dx.$$

Theorem 5.8 (Domain additivity). *Suppose that $f \in RI[a, b]$ and let c lie between a and b.*
Then

$$\int_a^b f(x)dx = \int_a^c f(x)dx + \int_c^b f(x)dx.$$

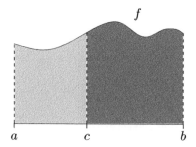

f

a c b

Proof. Since restrictions of Riemann integrable functions are Riemann integrable, we know that $f \in RI[a, c]$ and $f \in RI[c, b]$. The claim now follows immediately from Exercise 5.6. $\square$

Some useful inequalities associated with Riemann integration

Theorem 5.9. *Let $f, g \in RI[a, b]$. Then*

(1) *If for all $x \in [a, b]$, $f(x) \geq 0$, then $\displaystyle\int_a^b f(x)dx \geq 0$.*

(2) *If for all $x \in [a, b]$, $f(x) \geq g(x)$, then $\displaystyle\int_a^b f(x)dx \geq \int_a^b g(x)dx$.*

(3) *$|f| \in RI[a, b]$ and $\left| \displaystyle\int_a^b f(x)dx \right| \leq \int_a^b |f(x)|dx$.*

(4) *Let $f \in C[a, b]$ and for all $x \in [a, b]$ $f(x) \geq 0$. If $\displaystyle\int_a^b f(x)dx = 0$, then $f \equiv 0$ on $[a, b]$, that is, f is identically zero on $[a, b]$.*

Proof.

(1) We have

$$\int_a^b f(x)dx = \underline{S}(f) = \sup_{P \in \mathcal{P}_{[a,b]}} \underline{S}(f, P) \geq \underline{S}(f, \{a, b\}) = \underbrace{\left(\inf_{x \in [a,b]} f(x) \right)}_{\geq 0} (b - a) \geq 0.$$

(2) We just apply (1) to the function $h := f - g$. Then since $h(x) \geq 0$ for all $x \in [a, b]$ and $f - g \in RI[a, b]$, we obtain

$$\int_a^b f(x)dx - \int_a^b g(x)dx = \int_a^b (f(x) - g(x))dx = \int_a^b h(x)dx \geq 0.$$

Thus $\displaystyle\int_a^b f(x)dx \geq \int_a^b g(x)dx$.

(3) Let $\epsilon > 0$. By the Riemann Condition, there exists a partition

$$P_\epsilon = \{x_0, x_1, \cdots, x_{n-1}, x_n\}$$

of $[a, b]$ such that $\overline{S}(f, P_\epsilon) - \underline{S}(f, P_\epsilon) < \epsilon$.

Claim: $\overline{S}(|f|, P_\epsilon) - \underline{S}(|f|, P_\epsilon) < \epsilon$.

For any fixed $k \in \{0, 1, \cdots, n-1\}$, let $x, y \in [x_k, x_{k+1}]$. With

$$M_k := \sup_{x \in [x_k, x_{k+1}]} f(x) \quad \text{and}$$

$$m_k := \inf_{x \in [x_k, x_{k+1}]} f(x),$$

we have that $f(x) - f(y) \le M_k - m_k$, and $f(y) - f(x) \le M_k - m_k$. Consequently, we have that $|f(x) - f(y)| \le M_k - m_k$, and so

$$||f|(x) - |f|(y)| = |f(x)| - |f(y)| \le |f(x) - f(y)| \le M_k - m_k.$$

Thus

$$\sup_{x \in [x_k, x_{k+1}]} |f|(x) - \inf_{y \in [x_k, x_{k+1}]} |f|(y) \le M_k - m_k = \sup_{x \in [x_k, x_{k+1}]} f(x) - \inf_{y \in [x_k, x_{k+1}]} f(y).$$

Consequently, $\overline{S}(|f|, P_\epsilon) - \underline{S}(|f|, P_\epsilon) \le \overline{S}(f, P_\epsilon) - \underline{S}(f, P_\epsilon) < \epsilon$. This completes the proof of the claim.

By the Riemann Condition, $|f| \in RI[a, b]$.

Moreover, for all $x \in [a, b], f(x) \le |f(x)|$ and $-f(x) \le |f(x)|$, and so it follows that

$$\int_a^b f(x)dx \le \int_a^b |f(x)|dx \quad \text{and} \quad -\int_a^b f(x)dx \le \int_a^b |f(x)|dx.$$

Thus $\left| \int_a^b f(x)dx \right| \le \int_a^b |f(x)|dx.$

(4) Suppose that $\neg \left(f \equiv 0 \text{ on } [a, b] \right)$.

Then there exists a $c \in [a, b]$ such that $f(c) \ne 0$. As $f \ge 0, f(c) > 0$.

Take $\epsilon := f(c)/2 > 0$. Since f is continuous at c, there exists a $\delta > 0$ such that whenever $x \in [a, b]$ satisfies $|x - c| < \delta$,

$$|f(x) - f(c)| < \epsilon = \frac{f(c)}{2},$$

and so $f(c) - f(x) \le |f(c) - f(x)| = |f(x) - f(c)| < \frac{f(c)}{2}$. Hence

$$f(x) > f(c) - \frac{f(c)}{2} = \frac{f(c)}{2} > 0 \text{ for } x \in [a, b] \cup (c - \delta, c + \delta).$$

(This also shows that if $c = a$ or $c = b$, then there are other values of c where f is positive. So there is no loss of generality in assuming that $c \in (a, b)$. Also, by shrinking δ if necessary, we may assume that $a < c - \delta < c + \delta < b$.)

With $P_* := \{a, c - \delta, c + \delta, b\}$, we have

$$\int_a^b f(x)dx = \underline{S}(f) \ge \underline{S}(f, P_*)$$

$$= \left(\inf_{x \in [a, c-\delta]} f(x) \right)(c - \delta - a) + \left(\inf_{x \in [c-\delta, c+\delta]} f(x) \right) 2\delta + \left(\inf_{x \in [c-\delta, b]} f(x) \right)(b - c - \delta)$$

$$\ge 0 + \frac{f(c)}{2} \cdot (2\delta) + 0 = \delta \cdot f(c) > 0,$$

a contradiction. □

Exercise 5.8.

(1) Let $f, g \in RI[a, b]$. Show that $\max\{f, g\}$ and $\min\{f, g\}$ also belong to $RI[a, b]$, where

$$\max\{f, g\} := \max\{f(x), g(x)\},$$
$$\min\{f, g\} := \min\{f(x), g(x)\},$$

for all $x \in [a, b]$. *Hint:* $\max\{a, b\} = \frac{a+b+|a-b|}{2}$ for $a, b \in \mathbb{R}$.

(2) The aim of this exercise is twofold: firstly, to show that the pointwise supremum of a sequence of Riemann integrable functions need not be Riemann integrable, and secondly, to demonstrate that the pointwise limit of Riemann integrable functions need not be Riemann integrable.

Let $r_1, r_2, r_3, \cdots$ be an enumeration of the rationals in $[0, 1]$. Define $f_n : [0, 1] \to \mathbb{R}$ by

$$f_n(x) = \begin{cases} 1 & \text{if } x \in \{r_1, \cdots, r_n\}, \\ 0 & \text{otherwise} \end{cases}$$

Is each $f_n \in RI[0, 1]$? Let $\sup\limits_{n \in \mathbb{N}} f_n : [0, 1] \to \mathbb{R}$ be the function defined by

$$\left(\sup_{n \in \mathbb{N}} f_n \right)(x) = \sup_{n \in \mathbb{N}} f_n(x), \quad x \in [0, 1].$$

Is $\sup\limits_{n \in \mathbb{N}} f_n$ Riemann integrable?

Exercise 5.9. We have seen in Theorems 5.4, 5.7, and 5.9(3) that if $f, g \in RI[a, b]$, then so is their pointwise sum, product and their respective modulus. Give examples of bounded functions $f, g : [0, 1] \to \mathbb{R}$ that are *not* Riemann integrable, but for which $|f|, f + g, fg$ are all Riemann integrable on $[0, 1]$.

Exercise 5.10 (An integral mean value result). (∗) Let $f \in C[a, b]$, $\varphi \in RI[a, b]$, and let ρ be pointwise nonnegative. (We may interpret ρ as the 'mass density' of a rod, along the interval $[a, b]$, made of a possibly inhomogeneous material. If $\rho \equiv c$, a constant, then the rod has uniform density along its length.) Use the Intermediate Value Theorem for f to show that there is a $c \in [a, b]$ such that

$$\int_a^b f(x)\rho(x)dx = f(c) \int_a^b \rho(x)dx.$$

In particular, if $\rho \equiv 1$, then we obtain $\dfrac{1}{b-a} \displaystyle\int_a^b f(x)dx = f(c)$.

(If $f(x) = x$, then we can interpret the position c as the 'center of mass/gravity' of the horizontal (inhomogeneous) rod, namely the place about which if the rod is pivoted, it will remain balanced, since the moments about that point due to the weight of the constituent particles of the rod add up to 0. If the rod is homogeneous, then we see that the center of mass c is given by

$$\frac{b^2 - a^2}{2} = \int_a^b x \cdot 1 \, dx = c \cdot \int_a^b 1 \, dx = c \cdot (b - a),$$

that is, $c = \dfrac{a + b}{2}$, as expected based on our physical intuition.)

Give an example to show that the assumption $f \in C[a, b]$ cannot be dropped for the conclusion to hold. Moreover, provide an example to show that the nonnegativity of ρ is also a necessary condition.

Exercise 5.11 (Cantor set). (∗) The Cantor set is constructed as follows. Let $F_1 := [0, 1]$ and delete from F_1 the open interval $(\frac{1}{3}, \frac{2}{3})$ which is its middle third, and denote the remaining set by F_2. Thus we have that $F_2 = [0, \frac{1}{3}] \cup [\frac{2}{3}, 1]$. Next, delete from F_2 the middle thirds of its two pieces, namely the open intervals $(\frac{1}{9}, \frac{2}{9})$ and $(\frac{7}{9}, \frac{8}{9})$, and denote the remaining set by F_3. It can be checked that $F_3 = [0, \frac{1}{9}] \cup [\frac{2}{9}, \frac{1}{3}] \cup [\frac{2}{3}, \frac{7}{9}] \cup [\frac{8}{9}, 1]$. Continuing this process, that is, at each step deleting the open middle third of each interval remaining from the previous step, we obtain a sequence $F_1, F_2, F_3, \cdots$ of sets, each member of which contains all the subsequent members.

The Cantor set C is defined by $C = \bigcap\limits_{n=1}^{\infty} F_n$.

C is contained in $[0, 1]$ and consists of all those points in the interval $[0, 1]$, which are 'eventually left over' after the removal of all the open intervals $(\frac{1}{3}, \frac{2}{3})$, $(\frac{1}{9}, \frac{2}{9})$, $(\frac{7}{9}, \frac{8}{9})$, $\cdots$. What are these points? Clearly the end points of the intervals making up F_n do remain, and so C contains these:

$$0, 1, \frac{1}{3}, \frac{2}{3}, \frac{1}{9}, \frac{2}{9}, \frac{7}{9}, \frac{8}{9}, \cdots.$$

Are there any other points in C? In fact, C contains many more points than the above list of end points. After all, the above list of endpoints is countable, but it can be shown that C is uncountable, as follows.

We will prove that there is a one-to-one correspondence between points of C and the points of $[0, 1]$. First note that any point x in C is associated with a sequence of letters 'L' or 'R' as follows. Indeed, let $x \in C$. Then for any n, $x \in F_n$, and when the middle thirds of each subinterval in F_n is removed, x is present in either the left part or the right part of the subinterval, and the nth term in the sequence is letters is L or R accordingly. For example, the points

$$0 \equiv \text{L,L,L,L,L,L}, \cdots,$$

$$1 \equiv \text{R,R,R,R,R,R}, \cdots,$$

$$\frac{1}{3} \equiv \text{L,R,R,R,R,R}, \cdots,$$

$$\frac{2}{9} \equiv \text{L,R,L,L,L,L}, \cdots,$$

$$\frac{20}{27} \equiv \text{R,L,R,L,L,L}, \cdots.$$

But points in $[0, 1]$ are also in one-to-one correspondence with such sequences. Indeed,

$$[0, 1] = \left[0, \tfrac{1}{2}\right] \cup \left(\tfrac{1}{2}, 1\right]$$

$$= \left[0, \tfrac{1}{4}\right] \cup \left(\tfrac{1}{4}, \tfrac{1}{2}\right] \cup \left(\tfrac{1}{2}, \tfrac{3}{4}\right] \cup \left(\tfrac{3}{4}, 1\right]$$

$$= \left[0, \tfrac{1}{8}\right] \cup \left(\tfrac{1}{8}, \tfrac{1}{4}\right] \cup \left(\tfrac{1}{4}, \tfrac{3}{8}\right] \cup \left(\tfrac{3}{8}, \tfrac{1}{2}\right] \cup \left(\tfrac{1}{2}, \tfrac{5}{8}\right] \cup \left(\tfrac{5}{8}, \tfrac{3}{4}\right] \cup \left(\tfrac{3}{4}, \tfrac{7}{8}\right] \cup \left(\tfrac{7}{8}, 1\right]$$

$$\cdots.$$

If $x \in [0, 1]$, then for each n, we can look at the nth equality, and see if x falls in the left or the right part when each subinterval in the right hand side of the nth equality is divided into two parts, and this gives the $(n + 1)$st term of the sequence of Ls and Rs associated with x: for example,

$$0 \equiv \text{L,L,L,L,L,L,} \cdots,$$

$$1 \equiv \text{R,R,R,R,R,R,} \cdots,$$

$$\frac{1}{2} \equiv \text{L,R,R,R,R,R,} \cdots.$$

As $[0, 1]$ is uncountable, it follows that so is C.

It turns out that the Cantor set is an important set, as it is often a source of interesting examples/counterexamples in Analysis. (For example, as the sum of the lengths of the intervals removed is

$$\frac{1}{3} + 2\frac{1}{3^2} + 4\frac{1}{3^3} + \cdots = 1,$$

(we can factor out $1/3$ to obtain a geometric series, which can be summed), the '(Lebesgue length) measure' of F is $1 - 1 = 0$. So this is an example of an uncountable set with 'Lebesgue measure' 0.)

The aim of this exercise is to show that there exist Riemann integrable functions that have infinitely many points of discontinuity. Indeed, we will show that the indicator function $\mathbf{1}_C$ of the Cantor set is Riemann integrable on $[0, 1]$. Proceed as follows.

(1) As $C \subset F_n$, clearly $\mathbf{1}_C \le \mathbf{1}_{F_n}$. Since $\mathbf{1}_{F_n}$ has only finitely many discontinuities, it follows that

$$\int_0^1 \mathbf{1}_{F_n}(x)\,dx = \text{length of the intervals in } F_n = \left(\frac{2}{3}\right)^n.$$

Show this. Conclude that if $\epsilon > 0$, then there exists a partition P of $[0, 1]$ such that

$$\overline{S}(\mathbf{1}_{F_n}, P) < \left(\frac{2}{3}\right)^n + \epsilon.$$

Deduce that $\overline{S}(\mathbf{1}_C) \le 0$.

(2) As $\mathbf{1}_C \ge 0$, it is clear that $\underline{S}(\mathbf{1}_C, P) \ge 0$ for all partitions P of $[0, 1]$, and so $\underline{S}(\mathbf{1}_C) \ge 0$.

(3) Conclude from Parts (1) and (2) that $\mathbf{1}_C \in RI[0, 1]$, and that $\int_0^1 \mathbf{1}_C(x)\,dx = 0$.

Exercise 5.12. Can the assumption that $f \in C[0,1]$ in Theorem 5.9 (4) be replaced by the condition that $f \in RI[0,1]$?

Exercise 5.13 (The Dirac δ function). For doing quantum mechanical computations, the physicist P.A.M. Dirac introduced the δ 'function' (as eigenstates of the position operator). The aim of this exercise is to show that a classical such function does not exist[3]. Show that there is no function $\delta : \mathbb{R} \to \mathbb{R}$, which has the property that for all $a > 0$,

(1) $\delta \in RI[-a,a]$,

(2) for every $\varphi \in C[-a,a]$, $\displaystyle\int_{-a}^{a} \delta(x)\varphi(x)dx = \varphi(0)$.

5.3   Fundamental Theorem of Calculus

Calculus has two components:

Differentiation	Integration
Local process: derivative at a point depends only on values of the function near the point.	Global process: takes into account values of the function in the entire interval.

But now we will learn about a bridge between these two seemingly different worlds of differentiation and integration, namely the Fundamental Theorem of Calculus, which says, roughly that the two processes of differentiation and integration are inverses of each other.

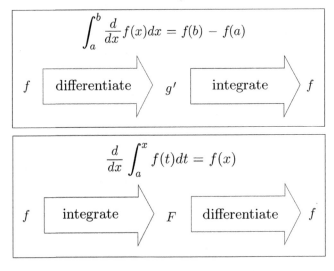

$$\int_a^b \frac{d}{dx} f(x)dx = f(b) - f(a)$$

f differentiate g' integrate f

$$\frac{d}{dx} \int_a^x f(t)dt = f(x)$$

f integrate F differentiate f

---

[3] However, the mathematician Laurent Schwartz later gave a mathematical foundation for the Dirac δ-function by viewing it as a 'generalised function' or 'distribution', in which one thinks of δ as a (linear) map $\delta : C_0^\infty(\mathbb{R}) \to \mathbb{R}$, which sends $\varphi \in C_0^\infty(\mathbb{R})$ to the number $\varphi(0)$. Here $C_0^\infty(\mathbb{R})$ denotes the set of all functions $\varphi : \mathbb{R} \to \mathbb{R}$, which are infinitely many times differentiable (that is, in $C^\infty(\mathbb{R})$) and vanish outside some compact interval (which may depend on φ). Laurent Schwartz was awarded the Fields Medal in 1950 for his work on the theory of distributions, which also plays a fundamental role in the study of partial differential equations.

Before stating the Fundamental Theorem of Calculus, we give the following definition.

Definition 5.7 (Primitive of a function). Let $f : [a, b] \to \mathbb{R}$. Then a function $F : [a, b] \to \mathbb{R}$ is called a *primitive of f* if

(1) F is differentiable on $[a, b]$ and

(2) for every $x \in [a, b]$, $F'(x) = f(x)$.

Example 5.10 (Primitives are not unique). Both the functions

$$\frac{x^2}{2} \quad \text{and} \quad \frac{x^2}{2} - 399$$

are primitives of x. In fact, any function $x^2 + C$, where C is an arbitrary constant, is a primitive of x. ◇

The above example shows that primitives are not unique. But we will show later on that they are 'unique up to additive constants', that is, for any two primitives $F, \widetilde{F}$ of f, there is a constant C (depending on the pair $F, \widetilde{F}$) such that $\widetilde{F} = F + C$ on $[a, b]$.

Theorem 5.10 (Fundamental Theorem of Calculus). *Let $f \in RI[a, b]$. Then:*

(1) *If f has a primitive F, then $\displaystyle\int_a^x f(t)dt = F(x) - F(a)$ for all $x \in [a, b]$.*

(2) *Let $F : [a, b] \to \mathbb{R}$ be defined by*

$$F(x) := \int_a^x f(t)dt, \quad x \in [a, b].$$

If f is continuous at $c \in [a, b]$, then F is differentiable at c and

$$F'(c) = f(c),$$

In particular, if $f \in C[a, b]$, then F is a primitive of f.

Proof. (of Part (1):)

(If $x = a$, then both the left hand side and right hand side are 0, and so the result holds. So let us assume that $x > a$.) Let $P = \{x_0, x_1, \cdots, x_{n-1}, x_n\}$ be *any* partition of $[a, x]$. By the Mean Value Theorem,

$$\frac{F(x_{k+1}) - F(x_k)}{x_{k+1} - x_k} = f(c_k),$$

for some $c_k \in (x_k, x_{k+1})$. Thus

$$\overline{S}(f, P) = \sum_{k=0}^{n-1} \left(\sup_{x \in [x_k, x_{k+1}]} f(x) \right) (x_{k+1} - x_k)$$

$$\geq \sum_{k=0}^{n-1} f(c_k)(x_{k+1} - x_k) = \sum_{k=0}^{n-1} (F(x_{k+1}) - F(x_k))$$

$$= F(x_1) - F(x_0) + F(x_2) - F(x_1) + \cdots + F(x_n) - F(x_{n-1})$$

$$= F(x_n) - F(x_0) = F(x) - F(a),$$

that is, for any partition P of $[a,x]$, $\overline{S}(f,P) \geq F(x) - F(a)$, and so

$$\overline{S}(f) \geq F(x) - F(a). \tag{5.6}$$

Similarly,

$$\underline{S}(f,P) = \sum_{k=0}^{n-1} \left(\inf_{x \in [x_k, x_{k+1}]} f(x) \right) (x_{k+1} - x_k)$$

$$\leq \sum_{k=0}^{n-1} f(c_k)(x_{k+1} - x_k) = \sum_{k=0}^{n-1} (F(x_{k+1}) - F(x_k)) = F(x) - F(a),$$

that is, for any partition P, $\underline{S}(f,P) \leq F(x) - F(a)$, and so

$$\underline{S}(f) \leq F(x) - F(a). \tag{5.7}$$

From (5.6) and (5.7), we obtain

$$\int_a^x f(t)dt = \underline{S}(f) \leq F(x) - F(a) \leq \overline{S}(f) = \int_a^x f(t)dt.$$

Consequently, $F(x) - F(a) = \int_a^x f(t)dt$. This finishes the proof of Part (1). $\qquad \square$

Before moving on to the proof of Part (2), here is an example illustrating Part (1).

Example 5.11. With $F := x^3/3$ and $f := x^2$, we have $F' = f$ on $\mathbb{R}$. Since we have that $f \in C[0,1] \subset RI[0,1]$, it follows from the above and the Fundamental Theorem of Calculus that

$$\int_0^1 x^2 dx = \frac{1^3}{3} - \frac{0^3}{3} = \frac{1}{3}.$$

Note the remarkable simplicity now obtained (as opposed to the calculation done earlier in Example 5.5), thanks to the Fundamental Theorem of Calculus. $\qquad \diamond$

Now let us continue with the proof of Part (2) of the Fundamental Theorem of Calculus.

Proof. (of Part (2)): Let $\epsilon > 0$. As f is continuous at c, there exists a $\delta > 0$ such that whenever $t \in [a,b]$ satisfies $|t - c| \leq \delta$, $|f(t) - f(c)| < \epsilon$. Let $x \in [a,b] \setminus \{c\}$. Then by the definition of F and the result on Domain Additivity, we obtain

$$\frac{F(x) - F(c)}{x - c} = \frac{1}{x - c} \left(\int_a^x f(t)dt - \int_a^c f(t)dt \right)$$

$$= \frac{1}{x - c} \int_c^x f(t)dt. \tag{5.8}$$

Also, by Part (1) of the Fundamental Theorem of Calculus,

$$\int_c^x f(c)dt = \int_c^x (f(c) \cdot t)'dt$$

$$= f(c) \cdot x - f(c) \cdot c$$

$$= f(c) \cdot (x - c),$$

and so for $x \in [a,b] \backslash \{c\}$,

$$f(c) = \frac{1}{x-c} \int_c^x f(c)dt. \tag{5.9}$$

From (5.8) and (5.9),

$$\left| \frac{F(x) - F(c)}{x - c} - f(c) \right| = \left| \frac{1}{x-c} \int_c^x f(t)dt - \frac{1}{x-c} \int_c^x f(c)dt \right|$$

$$= \frac{1}{|x-c|} \cdot \left| \int_c^x (f(t) - f(c))dt \right|$$

for all $x \in [a,b] \backslash \{c\}$. So for $x \in [a,b]$ satisfying $0 < |x - c| < \delta$, we have

$$\left| \frac{F(x) - F(c)}{x - c} - f(c) \right| = \frac{1}{|x-c|} \cdot \left| \int_c^x (f(t) - f(c))dt \right|$$

$$\leq \frac{1}{|x-c|} \int_c^x |f(t) - f(c)|dt$$

$$\leq \overline{S}(|f(\cdot) - f(c)|, \{c, x\})$$

$$\leq \frac{1}{|x-c|} \cdot \epsilon \cdot |x - c| = \epsilon.$$

Consequently, $F'(c) = f(c)$. $\qquad \square$

Geometric interpretation of the Fundamental Theorem of Calculus

The plausibility of Part (2) of the Fundamental Theorem of Calculus can be illustrated geometrically. See the following figure, in which we have depicted the graph of a Riemann integrable function f.

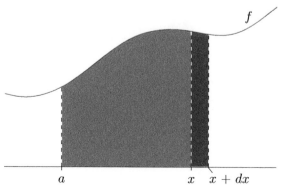

Let F be defined by

$$F(x) = \int_a^x f(t)dt, \quad x \geq a.$$

Consider an $x \geq a$, and imagine increasing x by a tiny amount dx. Then $F(x)$ is the area under the graph of f from a to x. Thus the area of the little strip is

$$F(x + dx) - F(x) \approx f(x) \cdot dx,$$

and dividing throughout by dx, we obtain

$$F'(x) \approx \frac{F(x + dx) - F(x)}{dx} \approx f(x).$$

Example 5.12. For $n \in \mathbb{Z}\setminus\{-1\}$,

$$\left(\frac{x^{n+1}}{n+1}\right)' = x^n, \quad x \neq 0.$$

If $b > a > 0$, then by the Fundamental Theorem of Calculus,

$$\int_a^b x^n dx = \frac{x^{n+1}}{n+1}\bigg|_a^b := \frac{b^{n+1} - a^{n+1}}{n+1}.$$

(The notation $F(x)\big|_a^b$ means $F(b) - F(a)$.)

What if $n = -1$? We will soon define the 'logarithm' function $\log : (0, \infty) \to \mathbb{R}$ by

$$\log x := \int_1^x \frac{1}{t}\, dt, \quad x > 0.$$

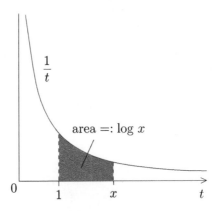

By the Fundamental Theorem of Calculus, $(\log x)' = \left(\int_1^x \frac{1}{t}\, dt\right)' = \frac{1}{x}, x > 0.$ ◇

Example 5.13. Later on, we will learn about trigonometric functions

$$\sin, \cos : \mathbb{R} \to \mathbb{R},$$

and we will prove that $\sin' = \cos$ and $\cos' = -\sin$.

So by the Fundamental Theorem of Calculus,

$$\int_a^b \sin x\, dx = -\cos x \Big|_a^b = -\cos b - (-\cos a) = \cos a - \cos b \text{ and}$$

$$\int_a^b \cos x\, dx = \sin x \Big|_a^b = \sin b - \sin a.$$ ◇

Example 5.14. Another important function we will meet soon is the exponential function $\exp : \mathbb{R} \to \mathbb{R}$, which has the property that $\exp' = \exp$, and so by the Fundamental Theorem of Calculus,

$$\int_a^b \exp x\, dx = \exp x \Big|_a^b = \exp b - \exp a.$$ ◇

Exercise 5.14 (Leibniz's Rule for Integrals). If $f \in C[a, b]$ and u, v are differentiable on $[c, d]$ and $u([c, d]) \subset [a, b]$, $v([c, d]) \subset [a, b]$, then

$$\frac{d}{dx} \int_{u(x)}^{v(x)} f(t)dt = f(v(x)) \cdot v'(x) - f(u(x)) \cdot u'(x), \quad x \in [c, d].$$

Exercise 5.15. For $x \in \mathbb{R}$, define

$$F(x) := \int_0^{2x} \sin(t^2)dt, \text{ and } G(x) := \int_0^{x^2} \sin(\sqrt{|t|})dt.$$

Find F' and G'.

Exercise 5.16. Let $f : [0, \infty) \to \mathbb{R}$ be continuous. Find f in each of the cases below if the given equation is known to hold for all $x \geq 0$, or if no such f exists, justify why not.

(1) $\int_0^{x^2} f(t)dt = \exp(-x^2)$.

(2) $\int_0^{f(x)} t^2 dt = \exp(-x^2)$.

(3) $\int_0^{\exp(-x^2)} f(t)dt = x^2$.

Exercise 5.17. Let f be a continuous function on $\mathbb{R}$ and $\lambda \neq 0$. Consider

$$y(x) = \frac{1}{\lambda} \int_0^x f(t) \sin(\lambda(x - t)) dt \text{ for } x \in \mathbb{R}.$$

Show that y is a solution to the inhomogeneous differential equation $y''(x) + \lambda^2 y(x) = f(x)$ for all $x \in \mathbb{R}$ and with the initial conditions $y(0) = 0$ and $y'(0) = 0$.

Exercise 5.18. Let $V(q)$ denote the voltage required to place a charge q on the plates of a capacitor. The elemental work done to place a small charge dq on a capacitor with charge q is $V(q) \cdot dq$. Thus the work required to charge a capacitor from $q = a$ to $q = b$ is the area under the graph of $q \mapsto V(q)$ on the interval $[a, b]$, that is, the integral

$$\int_a^b V(q) dq.$$

Show that if the voltage is proportional to the charge, then the work done to place a charge Q on an uncharged capacitor is

$$\frac{1}{2} QV(Q).$$

Exercise 5.19. Using $(1 + x)^n = \sum_{k=0}^{n} \binom{n}{k} x^k$, find $\sum_{k=1}^{n} \frac{1}{k+1} \binom{n}{k}$.

Exercise 5.20. Let $f : (1, 2) \to \mathbb{R}$ be given by

$$f(x) = \int_1^{x^2} \frac{(\cos \sqrt{t})(\sin \sqrt{t})}{\sqrt{t}} dt$$

for $x \in (1, 2)$. Does f have a local maximiser on $(1, 2)$? If so, where?

As an application of the Fundamental Theorem of Calculus, we will now show the $\frac{\infty}{\infty}$ form of the l'Hôpital Rule.

$\frac{\infty}{\infty}$ form of l'Hôpital's Rule

Theorem 5.11 ($\frac{\infty}{\infty}$ form of l'Hôpital's Rule). *If*

(1) $f, g : (a, \infty) \to \mathbb{R}$ *are differentiable,*

(2) $g, g' > 0$ *on* (a, ∞),

(3) $\lim_{x \to \infty} g(x) = \infty$, *and*

(4) $\lim_{x \to \infty} \dfrac{f'(x)}{g'(x)} = \ell \in \mathbb{R}$,

then $\lim_{x \to \infty} \dfrac{f(x)}{g(x)} = \ell$.

Proof.

$\underline{1°}$ First suppose that $\ell = 0$. Let $\epsilon > 0$. Let $c > 0$ be such that for $x \geq c$,

$$-\frac{\epsilon}{2} < \frac{f'(x)}{g'(x)} < \frac{\epsilon}{2},$$

and since $g' > 0$,

$$-\frac{\epsilon}{2}g'(x) < f'(x) < \frac{\epsilon}{2}g'(x).$$

Integrating from c to x $(> c)$, we obtain

$$-\frac{\epsilon}{2}(g(x) - g(c)) < f(x) - f(c) < \frac{\epsilon}{2}(g(x) - g(c)).$$

Dividing by $g(x) > 0$, it follows that

$$-\frac{\epsilon}{2}\left(1 - \frac{g(c)}{g(x)}\right) < \frac{f(x)}{g(x)} - \frac{f(c)}{g(x)} < \frac{\epsilon}{2}\left(1 - \frac{g(c)}{g(x)}\right).$$

Let $d_1 > c$ be such that for $x > d_1$,

$$-\frac{\epsilon}{4} < \frac{f(c)}{g(x)} < \frac{\epsilon}{4}.$$

Also, let $d_2 > d_1$ $(> c)$ be such that for all $x > d_2$,

$$-\frac{1}{2} < \frac{g(c)}{g(x)} < \frac{1}{2}.$$

Hence for $x > d_2$, it follows from the above that

$$-\epsilon = -\frac{\epsilon}{2}\left(1 + \frac{1}{2}\right) - \frac{\epsilon}{4} = -\epsilon < \frac{f(x)}{g(x)} < \epsilon = \frac{\epsilon}{4} + \frac{\epsilon}{2}\left(1 + \frac{1}{2}\right) = \epsilon.$$

Consequently, $\lim\limits_{x \to \infty} \dfrac{f(x)}{g(x)} = 0 \, (= \ell)$.

$\underline{2°}$ *General ℓ.* Consider

$$F := f - \ell \cdot g,$$

$$G := g.$$

Then F, G are differentiable, $G, G' > 0$, and

$$\frac{F'}{G'} = \frac{f' - \ell \cdot g'}{g'} = \frac{f'}{g'} - \ell \xrightarrow{x \to \infty} \ell - \ell = 0.$$

By 1°, $\dfrac{F}{G} \xrightarrow{x \to \infty} 0$, that is, $\dfrac{f}{g} - \ell \xrightarrow{x \to \infty} 0$, and so $\lim\limits_{x \to \infty} \dfrac{f(x)}{g(x)} = \ell$. $\qquad \square$

Example 5.15 ($\lim\limits_{x\to\infty}\dfrac{\log x}{x}=0$). Indeed,

(1) $f := \log x$ and $g := x$ are differentiable on $(0,\infty)$,

(2) $g = x$ and $g' = 1$ are > 0 on $(0,\infty)$,

(3) $\lim\limits_{x\to\infty} g(x) = \lim\limits_{x\to\infty} x = \infty$,

(4) $\lim\limits_{x\to\infty}\dfrac{f'(x)}{g'(x)} = \lim\limits_{x\to\infty}\dfrac{1/x}{1} = 0$,

and so by l'Hôpital's Rule, $\lim\limits_{x\to\infty}\dfrac{f(x)}{g(x)} = \lim\limits_{x\to\infty}\dfrac{\log x}{x} = 0$.   ◇

One can also show a similar result with $\lim\limits_{x\to\infty}$ being replaced by $\lim\limits_{x\searrow 0}$.

Corollary 5.3 ($\frac{\infty}{\infty}$ form of l'Hôpital's Rule). *If*

(1) $f, g : (0, a) \to \mathbb{R}$ *are differentiable,*

(2) $g > 0$ *and* $g' < 0$ *on* $(0, a)$,

(3) $\lim\limits_{x\searrow 0} g(x) = \infty$, *and*

(4) $\lim\limits_{x\searrow 0}\dfrac{f'(x)}{g'(x)} = \ell \in \mathbb{R}$,

then $\lim\limits_{x\searrow 0}\dfrac{f(x)}{g(x)} = \ell$.

Proof. Let $F, G : (\frac{1}{a}, \infty) \to \mathbb{R}$ be defined by

$$F(x) := f\left(\frac{1}{x}\right), \quad \text{and } G(x) := g\left(\frac{1}{x}\right)$$

for $x > \frac{1}{a}$. Note that $G > 0$ and $\lim\limits_{x\to\infty} G(x) = \lim\limits_{x\searrow 0} g(x) = \infty$. Moreover,

$$F'(x) := f'\left(\frac{1}{x}\right)\cdot\left(-\frac{1}{x^2}\right),$$

$$G'(x) := g'\left(\frac{1}{x}\right)\cdot\left(-\frac{1}{x^2}\right)$$

for $x > \frac{1}{a}$. Note that from the second expression for G', using the hypothesis that $g' < 0$, we obtain $G' > 0$. Finally,

$$\frac{F'(x)}{G'(x)} = \frac{f'\left(\frac{1}{x}\right)\cdot\left(-\frac{1}{x^2}\right)}{g'\left(\frac{1}{x}\right)\cdot\left(-\frac{1}{x^2}\right)} = \frac{f'\left(\frac{1}{x}\right)}{g'\left(\frac{1}{x}\right)} \xrightarrow{x\to\infty} \ell.$$

Hence by the previous version of the $\frac{\infty}{\infty}$ version of l'Hôpital's Rule, we have

$$\frac{F(x)}{G(x)} = \frac{f\left(\frac{1}{x}\right)}{g\left(\frac{1}{x}\right)} \xrightarrow{x \to \infty} \ell,$$

that is, $\lim\limits_{x \searrow 0} \dfrac{f(x)}{g(x)} = \ell.$ □

Example 5.16 ($\lim\limits_{x \to 0+} x \log x = 0$). We write

$$x \log x = \frac{\log x}{1/x} = \frac{f(x)}{g(x)},$$

where $f := \log x$ and $g := \dfrac{1}{x}$. Then

(1) $f = \log x$, $g = \dfrac{1}{x}$ are differentiable on $(0, \infty)$,

(2) $g = \dfrac{1}{x} > 0$ and $g' = -\dfrac{1}{x^2} < 0$ on $(0, \infty)$,

(3) $\lim\limits_{x \to 0+} g(x) = \lim\limits_{x \to 0+} \dfrac{1}{x} = \infty$, and

(4) $\lim\limits_{x \to 0+} \dfrac{f'(x)}{g'(x)} = \lim\limits_{x \to 0+} \dfrac{1/x}{-1/x^2} = \lim\limits_{x \to 0+} -x = 0.$

So by l'Hôpital's Rule, $\lim\limits_{x \to 0+} \dfrac{f(x)}{g(x)} = \lim\limits_{x \to 0+} \dfrac{\log x}{1/x} = \lim\limits_{x \to 0+} x \log x = 0.$ ◇

Exercise 5.21. Evaluate $\lim\limits_{x \to 0+} \dfrac{1}{x^3} \displaystyle\int_0^x \dfrac{t^2}{t^6 + 1}\, dt.$

We will now learn about two of the most important and powerful techniques of calculating integrals, which are both based on the Fundamental Theorem of Calculus:

(1) Integration by Parts, and

(2) Integration by a Change of Variables or by Substitution.

Integration by parts

Recall the Leibniz Rule for differentiation:

$$(fg)' = f'g + fg'.$$

If this is integrated from a to b, then the result is

$$\int_a^b f'(x)g(x)\,dx + \int_a^b f(x)g'(x)\,dx = f(b)g(b) - f(a)g(a).$$

This formula is useful when the integrand can be written as a product in such a way that one factor ($f'(x)$) can be integrated, while the other ($g(x)$) can be differentiated with the net effect (new integrand= $f(x)g'(x)$) that is good (that is,

$$\int_a^b f(x)g'(x)dx$$

is easy to find.) The best way to see this is by considering examples. But first we state the result.

Theorem 5.13 (Integration by Parts).

Let

    (1) $f \in C^1[a, b]$,

    (2) $g = G' \in C[a, b]$.

Then $\displaystyle\int_a^b f(x)g(x)dx = f(x)G(x)\Big|_a^b - \int_a^b f'(x)G(x)dx.$

Proof. $(f\,G)' = f \cdot g + f' \cdot G$, and so

$$f(x)G(x)\Big|_a^b - \int_a^b f'(x)G(x)dx = \int_a^b f(x)g(x)dx$$

by the Fundamental Theorem of Calculus.   □

Example 5.17. Consider the integral $\displaystyle\int_0^1 x\sqrt{1-x}\,dx.$

Since $\left((1-x)^{3/2}\right)' = \dfrac{3}{2}(1-x)^{1/2}(-1)$, $-\dfrac{2}{3}(1-x)^{3/2}$ is a primitive of $\sqrt{1-x}$. Thus

$$\int_0^1 x\sqrt{1-x}\,dx = x \cdot \left(-\frac{2}{3}\right)(1-x)^{\frac{3}{2}}\Big|_0^1 - \int_0^1 -\frac{2}{3}(1-x)^{\frac{3}{2}}dx$$

$$= 0 + \frac{2}{3}\int_0^1 (1-x)^{\frac{3}{2}}dx = \frac{2}{3}\frac{1}{\frac{3}{2}+1}(1-x)^{\frac{3}{2}+1}(-1)\Big|_0^1$$

$$= \frac{4}{15} \cdot (0-1) \cdot (-1) = \frac{4}{15}.$$

◇

In the previous example, how did we decide upon *integrating* $\sqrt{1-x}$ and *differentiating* x when using Integration by Parts? There is no fixed algorithm for this, but a general rule of thumb is to follow the following scheme:

'L   I   A   T   E'

$\longleftarrow$ $\longrightarrow$

differentiate   integrate

where the letters L, I, A, T, E stand for the following classes of functions (some of which we will define later on in this chapter):

L: Logarithmic

I: Inverse trigonometric $(\sin^{-1}, \cos^{-1}, \tan^{-1}, \cdots)$

A: Algebraic $(x^n, \sqrt{1 - x^m}, \cdots)$

T: Trigonometric $(\sin, \cos, \cdots)$

E: Exponential $(e^x, e^{-x}, \cdots)$.

Here are a few examples.

Example 5.18. Consider the integral

$$\int_1^3 \log x \, dx.$$

We view the integrand $\log x$ as the product of the two functions 1 (algebraic) and $\log x$ (logarithmic). The LIATE Rule of Thumb tells us that we ought to try differentiating $\log x$ and integrating 1. Thus

$$\int_1^3 \log x \, dx = \int_1^3 \log x \cdot 1 \, dx = (\log x) \cdot x \Big|_1^3 - \int_1^3 \frac{1}{x} \cdot x \, dx$$

$$= 3 \log 3 - \underbrace{(\log 1)}_{=0} \cdot 1 - \int_1^3 1 \, dx = 3 \log 3 - (3 - 1) = 3 \log 3 - 2. \qquad \Diamond$$

Example 5.19. We have

$$\int_a^b x \sin x \, dx = x(-\cos x) \Big|_a^b - \int_a^b 1 \cdot (-\cos x) \, dx$$

$$= -b \cos b + a \cos a + \int_a^b \cos x \, dx$$

$$= a \cos a - b \cos b + \sin x \Big|_a^b$$

$$= a \cos a - b \cos b + \sin b - \sin a. \qquad \Diamond$$

Sometimes in order to evaluate an integral, one might have to use Integration by Parts a couple of times, and the following example illustrates this.

Example 5.20. Consider the integral $\int_0^x (\exp t)(\cos t) dt$.

(We will define the exponential function $\exp : \mathbb{R} \to \mathbb{R}$ later on, but right now, all we need to know now is that $\exp' = \exp$ and $\exp 0 = 1$.) The trick is to integrate by parts *twice*. Things don't look good after the first use, but they get better after the second when the original integral appears again:

$$\int_0^x \exp t \cdot \cos t \, dt = \exp t \cdot \sin t \Big|_0^x - \int_0^x \exp t \cdot \sin t \, dt$$

$$= \exp x \cdot \sin x - \exp 0 \cdot \sin 0 - \int_0^x \exp t \cdot \sin t \, dt$$

$$= \exp x \cdot \sin x - \exp t \cdot (-\cos t) \Big|_0^x + \int_0^x \exp t \cdot (-\cos t) \, dt$$

$$= \exp x \cdot \sin x + \exp x \cdot \cos x - \exp 0 \cdot \cos 0 - \int_0^x \exp t \cdot \cos t \, dt,$$

and we're back to square one. So

$$\int_0^x \exp t \cdot \cos t \, dt = \frac{(\exp x)(\cos x + \sin x) - 1}{2}.$$

$\diamondsuit$

Exercise 5.22. Evaluate $\displaystyle\int_1^2 x \log x \, dx$.

Exercise 5.23. Let m, n be nonnegative integers. Show that

$$\int_0^1 x^m (1-x)^n dx = \frac{m!n!}{(m+n+1)!}.$$

Hint: If $I(m, n)$ is the integral, then show the 'recurrence relation'

$$I(m, n) = \frac{n}{m+1} I(m+1, n-1),$$

for all nonnegative integers m and all $n \in \mathbb{N}$.

Exercise 5.24. For $f \in C[a, b]$, show that for all $x \in [a, b]$,

$$\int_a^x \left(\int_a^u f(t) dt \right) du = \int_a^x (x-u)f(u) du.$$

Hint: Start with the right hand side.

Remark 5.1. More generally, one can show, using induction, that

$$\int_a^x \left(\int_a^{u_n} \left(\cdots \left(\int_a^{u_1} f(t) dt \right) du_1 \right) \cdots \right) du_n = \int_a^x \frac{f(u) \cdot (x-u)^n}{n!} du.$$

Exercise 5.25 (Taylor's Formula with Integral Remainder). Let n be a nonnegative integer and $f \in C^{n+1}[a, b]$. Show that

$$f(b) = f(a) + f'(a)(b-a) + \cdots + \frac{f^{(n)}(a)}{n!}(b-a)^n + \frac{1}{n!} \int_a^b (b-t)^n f^{(n+1)}(t) dt.$$

(Note that as opposed to the Taylor's Formula we have met before, where the error term contained an undetermined c, now the 'integral remainder' does not involve such an undetermined number c.)

Integration by Substitution/Change of Variables

Theorem 5.14 (Integration by Substitution/Change of Variables). *If*

(1) $\varphi \in C^1[\alpha, \beta]$ *and* $\varphi([\alpha, \beta]) = [a, b]$, *and*

(2) $f \in C[a, b]$,

then $\displaystyle\int_{\varphi(\alpha)}^{\varphi(\beta)} f(x)dx = \int_{\alpha}^{\beta} f(\varphi(t)) \cdot \varphi'(t)dt.$

Proof. Define $F : [a, b] \to \mathbb{R}$ by $F(x) = \displaystyle\int_a^x f(t)dt$, for $x \in [a, b]$.

If $H : [\alpha, \beta] \to \mathbb{R}$ is defined by $H = F \circ \varphi$, then by the Chain Rule, we have

$$H'(t) = F'(\varphi(t)) \cdot \varphi'(t) = f(\varphi(t)) \cdot \varphi'(t), \quad t \in [\alpha, \beta].$$

Thus

$$\int_{\alpha}^{\beta} f(\varphi(t)) \cdot \varphi'(t) = \int_{\alpha}^{\beta} H'(t)dt = H(\beta) - H(\alpha) = F(\varphi(\beta)) - F(\varphi(\alpha))$$

$$= \int_a^{\varphi(\beta)} f(x)dx - \int_a^{\varphi(\alpha)} f(x)dx = \int_{\varphi(\alpha)}^{\varphi(\beta)} f(x)dx,$$

where the last equality follows by Domain Additivity. □

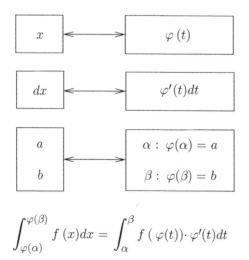

$$\int_{\varphi(\alpha)}^{\varphi(\beta)} f(x)dx = \int_{\alpha}^{\beta} f(\varphi(t)) \cdot \varphi'(t)dt$$

Figure 5.12   How to remember the Change of Variables/Substitution Method.

Example 5.21. Consider the integral $\int_0^1 t\sqrt{1-t^2}\,dt$.

We make the substitution $x = \varphi(t) = 1 - t^2, t \in [0,1]$, and take $f(x) := \sqrt{x}, x \in [0,1]$:

$$dx = -2t\,dt, \qquad t\,dt = -\frac{1}{2}\,dx;$$

$$t = 0 \quad \Rightarrow \quad x = 1,$$

$$t = 1 \quad \Rightarrow \quad x = 0.$$

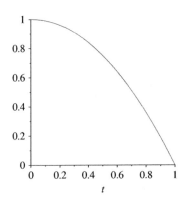

Thus

$$\int_0^1 t\sqrt{1-t^2}\,dt = \int_1^0 \sqrt{x}\left(-\frac{1}{2}\right)dx = \frac{1}{2}\int_0^1 \sqrt{x}\,dx$$

$$= \frac{1}{2} \cdot \frac{1}{1 + \frac{1}{2}} x^{1+\frac{1}{2}}\Big|_0^1 = \frac{1}{2} \cdot \frac{2}{3} \cdot (1^{3/2} - 0^{3/2}) = \frac{1}{3}.$$

$\diamondsuit$

Example 5.22. Consider the integral $\int_0^{\pi/2} (\sin t)^5 \cos t\,dt$.

We use the substitution $x = \varphi(t) = \sin t, t \in [0, \frac{\pi}{2}]$, and take $f(x) := x^5, x \in [0,1]$:

$$dx \quad = \quad \cos t\,dt;$$

$$t = 0 \quad \Rightarrow \quad x = 0,$$

$$t = \frac{\pi}{2} \quad \Rightarrow \quad x = 1.$$

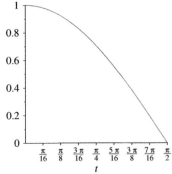

Thus $\displaystyle\int_0^{\pi/2} (\sin t)^5 \cos t\, dt = \int_0^1 x^5 dx = \frac{1}{6} x^6 \Big|_0^1 = \frac{1}{6}.$ ◇

Example 5.23. Consider the integral $\displaystyle\int_2^5 \frac{1}{t \log t}\, dt.$

We make the substitution $x = \varphi(t) = \log t$, $t \in [2,5]$, and take f given by $f(x) := 1/x$ for x in the interval $[\log 2, \log 5]$:

$$dx = \frac{1}{t}\, dt; \quad t = 2 \Rightarrow x = \log 2, \text{ and } t = 5 \Rightarrow x = \log 5.$$

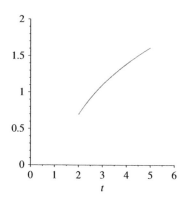

Thus $\displaystyle\int_2^5 \frac{1}{t \log t}\, dt = \int_{\log 2}^{\log 5} \frac{1}{x}\, dx = \log x \Big|_{\log 2}^{\log 5} = \log(\log 5) - \log(\log 2).$ ◇

Example 5.24. If $f \in C[-a, a]$ is an odd function, then $\displaystyle\int_{-a}^a f(x)dx = 0.$

Using the substitution $t = -x$ (so that $dt = -dx$, $x = a$ when $t = -a$, and $x = 0$ when $t = 0$), we obtain

$$\int_{-a}^0 f(x)dx = \int_a^0 f(-t)(-1)dt = \int_0^a f(-t)dt = \int_0^a -f(t)dt = -\int_0^a f(t)dt.$$

So $\displaystyle\int_{-a}^a f(x)dx = \int_{-a}^0 f(x)dx + \int_0^a f(x)dx = -\int_0^a f(t)dt + \int_0^a f(x)dx = 0.$

For example,

$$\int_{-\pi/2}^{\pi/2} (\sin x)^5 (\cos x)\, dx = 0, \text{ and } \int_{-1/2}^{1/2} \log \frac{1+x}{1-x}\, dx = 0.$$

What happens if $f \in C[-a, a]$ is an *even* function? Proceeding in the same manner as above, we see that

$$\int_{-a}^a f(x)\, dx = 2\int_0^a f(x)\, dx.$$ ◇

Exercise 5.26. Evaluate the following integrals without doing any computations!

(1) $\int_{-1}^{1} x^3 \sqrt{1 - x^2} \, dx.$

(2) $\int_{-1}^{1} (x^3 + 9)\sqrt{1 - x^2} \, dx.$

Exercise 5.27. Evaluate $\int_{0}^{\frac{1}{4}} \dfrac{x}{\sqrt{1 - 4x^2}} \, dx.$

Exercise 5.28. Evaluate $\int_{0}^{16} \sqrt[4]{x} \cdot \sqrt{\sqrt[4]{x^5} + 1} \, dx.$

Exercise 5.29. Let $T > 0$, and $f : \mathbb{R} \to \mathbb{R}$ be a continuous function, which is T-periodic, that is, $f(x + T) = f(x)$ for all $x \in \mathbb{R}$. Show that the integral

$$\int_{a}^{a+T} f(x) \, dx$$

has the same value for all $a \in \mathbb{R}$.

Exercise 5.30. For $0 \le x < \dfrac{\pi}{2}$, find $\int_{0}^{x} \tan t \, dt.$

Exercise 5.31. Find $\int_{0}^{\pi/2} (\sin x) \exp(\cos x) \, dx.$

Exercise 5.32. Find $\int_{0}^{1} 2x \exp(3x^2) \, dx.$

5.4 Riemann sums

It turns out that some integrals, such as

$$\int_{0}^{x} \exp(-t^2) dt \ \text{ for } x > 0 \quad \text{or} \quad \int_{0}^{\pi/2} \sqrt{1 - k^2(\sin\theta)^2} \, d\theta \ \text{ for } 0 < k < 1,$$

although they exist, can't be expressed in terms of elementary functions. In such cases, from the point of view of applications (engineering, physics, etc.), one might be willing to settle for a value that is an approximation of the true value within a certain accuracy level. So the need arises for having a numerical scheme for finding the Riemann integral. In this section, we learn the heart of such a numerical recipe.

We have seen that

(1) If $f \in RI[a, b]$, then for *any* partition P of $[a, b]$,

$$\underline{S}(f, P) \le \int_{a}^{b} f(x) dx \le \overline{S}(f, P).$$

(2) Given $\epsilon > 0$, for *certain* partitions P_ϵ of $[a, b]$, we can approximate

$$\int_a^b f(x)dx$$

to within an accuracy of ϵ by $\underline{S}(f, P_\epsilon)$ and $\overline{S}(f, P_\epsilon)$ because

$$\overline{S}(f, P_\epsilon) - \underline{S}(f, P_\epsilon) < \epsilon.$$

We will now learn that

(3) No matter what accuracy level is specified, by taking *any* 'sufficiently fine' partition P, the lower/upper sums $\underline{S}(f, P)$ and $\overline{S}(f, P)$ approximate

$$\int_a^b f(x)dx$$

to within the accuracy level.

Before stating this result precisely, we need to understand what we mean by the 'fineness' of a partition.

Definition 5.8 (Fineness of a partition). Let $P = \{x_0, x_1, \cdots, x_{n-1}, x_n\}$ be a partition of $[a, b]$. The *fineness* $\Phi(P)$ *of P* is defined by

$$\Phi(P) := \max\{x_{k+1} - x_k : 0 \le k \le n - 1\}.$$

small $\Phi(P)$ large $\Phi(P)$

Theorem 5.15. *Let $f \in RI[a, b]$. Then:*

> *For every $\epsilon > 0$, there exists a $\delta > 0$ such that for every $P \in \mathcal{P}_{[a,b]}$ with $\Phi(P) < \delta$,*
> $\overline{S}(f, P) - \underline{S}(f, P) < \epsilon.$

Proof. Let $\epsilon > 0$. Since $f \in RI[a, b]$, there exists a partition P_ϵ of $[a, b]$ such that

$$\overline{S}(f, P_\epsilon) - \underline{S}(f, P_\epsilon) < \frac{\epsilon}{2}.$$

Let

n_ϵ be the number of points in P_ϵ,

δ_ϵ be the length of the *shortest* subinterval in P_ϵ,

$M := \sup_{x \in [a,b]} |f(x)|.$

Let $P = \{x_0, x_1, \cdots, x_{n-1}, x_n\}$ be any partition of $[a, b]$ such that

$$\Phi(P) < \delta := \min\left\{\delta_\epsilon, \frac{\epsilon}{8Mn_\epsilon}\right\}.$$

We claim that $\overline{S}(f, P) - \underline{S}(f, P) < \epsilon$.

Let $P_* := P \bigcup P_\epsilon$. Then $\overline{S}(f, P_*) \leq \overline{S}(f, P_\epsilon)$ and $\underline{S}(f, P_*) \geq \underline{S}(f, P_\epsilon)$, and so we have that $\overline{S}(f, P_*) - \underline{S}(f, P_*) \leq \overline{S}(f, P_\epsilon) - \underline{S}(f, P_\epsilon) < \epsilon$. Now let us compare the upper/lower sums corresponding to the partitions P_* and P.

The terms in $\overline{S}(f, P_*)$ and $\overline{S}(f, P)$ are mostly the same, except for the following situation. If $y \in [x_k, x_{k+1}]$ is a point of P_* that is not in P, then $[x_k, x_{k+1}]$ do not contain any point of P_* other than y. (This is because otherwise if y' is another point in $[x_k, x_{k+1}]$ that also belongs to $P_* \backslash P$, we would have $y' \in P_\epsilon$ and so $\delta_\epsilon \leq |y - y'| \leq x_{x+1} - x_k \leq \Phi(P) < \delta \leq \delta_\epsilon$, a contradiction!) So $\overline{S}(f, P)$ contains the term

$$\left(\sup_{x \in [x_k, x_{k+1}]} f(x)\right) \cdot (x_{k+1} - x_k),$$

while $\overline{S}(f, P_*)$ contains the term

$$\left(\sup_{x \in [x_k, y]} f(x)\right) \cdot (y - x_k) + \left(\sup_{x \in [y, x_{k+1}]} f(x)\right) \cdot (x_{k+1} - y).$$

Thus the error committed in replacing the single term by the sum of the two terms is at most $2 \cdot M \cdot \delta$. But there are at most n_ϵ points of P_ϵ, and so

$$\overline{S}(f, P) < \overline{S}(f, P_*) + 2 \cdot M \cdot \delta \cdot n_\epsilon.$$

Similarly, $\underline{S}(f, P) > \underline{S}(f, P_*) - 2 \cdot M \cdot \delta \cdot n_\epsilon$. Hence

$$\overline{S}(f, P) - \underline{S}(f, P) < \overline{S}(f, P_*) - \underline{S}(f, P_*) + 4M\delta n_\epsilon < \frac{\epsilon}{2} + 4M\frac{\epsilon}{8Mn_\epsilon}n_\epsilon = \epsilon.$$

This completes the proof.   □

Riemann sums

There are *other* sums associated with a partition P (besides the *upper* sum and *lower* sum) that lie between $\overline{S}(f, P)$ and $\underline{S}(f, P)$, and so for sufficiently fine partitions, these other sums will also approximate the integral

$$\int_a^b f(x)dx.$$

These other sums are called *Riemann sums* and have the form

$$S(f, P) := \sum_{k=0}^{n-1} f(\xi_k)(x_{k+1} - x_k),$$

where for each $k = 0, 1, \cdots, n - 1$, ξ_k is any point in $[x_k, x_{k+1}]$.

Clearly $\underline{S}(f,P) \leq S(f,P) \leq \overline{S}(f,P)$. See Figure 5.13.

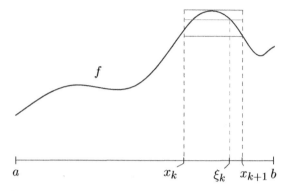

Figure 5.13 $\displaystyle\inf_{x\in[x_k,x_{k+1}]} f(x) \leq f(\xi_k) \leq \sup_{x\in[x_k,x_{k+1}]} f(x).$

The advantage of using Riemann sums over using $\overline{S}(f,P)$ and $\underline{S}(f,P)$ is that these latter numbers might be harder to determine, since we have to mess about with infs and sups, while in order to find the Riemann sum $S(f,P)$, we just need to evaluate the function at the points ξ_ks.

Corollary 5.16. *If $f \in RI[a,b]$ and*

$$P_n := \left\{ a,\ a + \frac{b-a}{n},\ a + 2\cdot\frac{b-a}{n},\cdots,a+(n-1)\cdot\frac{b-a}{n},\ b \right\}, \quad n \in \mathbb{N},$$

then $\displaystyle\int_a^b f(x)dx = \lim_{n\to\infty} S(f,P_n).$

Proof. $\displaystyle\Phi(P_n) = \frac{b-a}{n} \xrightarrow{n\to\infty} 0.$ $\qquad\qquad\square$

Example 5.25. We will show that

$$\lim_{n\to\infty}\left(\frac{1}{n}+\frac{1}{n+1}+\cdots+\frac{1}{2n}\right) = \log 2,$$

by viewing the sum as a Riemann sum as follows. We have

$$\frac{1}{n}+\frac{1}{n+1}+\cdots+\frac{1}{2n-1} = \frac{1}{n(1+\frac{0}{n})}+\frac{1}{n(1+\frac{1}{n})}+\frac{1}{n(1+\frac{2}{n})}+\cdots+\frac{1}{n(1+\frac{n-1}{n})}$$

$$= \sum_{k=0}^{n-1} f(\xi_k)\cdot(x_{k+1}-x_k) = S\left(\frac{1}{x},P_n\right),$$

where

$$P_n := \left\{ a = 1,\ 1 + \frac{1}{n},\ 1 + 2 \cdot \frac{1}{n}, \cdots, 1 + (n-1) \cdot \frac{1}{n},\ b = 2 \right\}, \quad n \in \mathbb{N},$$

$$f := \frac{1}{x} \in C[1,2] \subset RI[1,2],$$

$$\xi_k := x_k := 1 + k\frac{1}{n}, \quad k = 0, 1, \cdots, n-1.$$

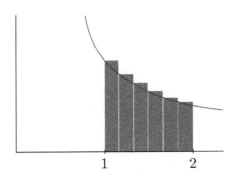

By definition,

$$\int_1^2 \frac{1}{x}\, dx = \log 2.$$

Hence

$$\lim_{n \to \infty} \left(\frac{1}{n} + \frac{1}{n+1} + \cdots + \frac{1}{2n} \right) = \lim_{n \to \infty} \left(\frac{1}{n} + \frac{1}{n+1} + \cdots + \frac{1}{2n-1} \right) + \lim_{n \to \infty} \frac{1}{2n}$$

$$= \log 2 + 0 = \log 2. \qquad \Diamond$$

Example 5.26. What is $1^{1/3} + 2^{1/3} + 3^{1/3} + \cdots + 1000^{1/3}$ approximately?

Taking

$$P_n := \left\{ a = 0,\ \frac{1}{n},\ 2 \cdot \frac{1}{n}, \cdots, (n-1) \cdot \frac{1}{n},\ b = 1 \right\}, \quad n \in \mathbb{N},$$

$$f := x^{1/3} \in C[0,1] \subset RI[0,1],$$

$$\xi_k := x_{k+1} := \frac{k+1}{n}, \quad k = 0, 1, \cdots, n-1,$$

we obtain

$$S(f, P_n) = \sum_{k=0}^{n-1} f(x_{k+1}) \cdot (x_{k+1} - x_k) = \sum_{k=0}^{n-1} \left(\frac{k+1}{n} \right)^{1/3} \cdot \frac{1}{n}$$

$$= \frac{1^{1/3} + 2^{1/3} + 3^{1/3} + \cdots + n^{1/3}}{n^{4/3}}.$$

As

$$\lim_{n\to\infty} S(f, P_n) = \int_0^1 f(x)dx = \int_0^1 x^{1/3}\, dx = \frac{3}{4},$$

we have

$$S(f, 1000) \approx \frac{3}{4},$$

that is,

$$\frac{1^{1/3} + 2^{1/3} + 3^{1/3} + \cdots + 1000^{1/3}}{1000^{4/3}} \approx \frac{3}{4}.$$

Hence

$$1^{1/3} + 2^{1/3} + 3^{1/3} + \cdots + 1000^{1/3} \approx \frac{3}{4} \cdot 1000^{4/3} = \frac{3}{4} \cdot 10000 = 7500.$$

(With the help of a calculator, $1^{1/3} + 2^{1/3} + 3^{1/3} + \cdots + 1000^{1/3} = 7504.723$ to three decimal places.)   ◇

Exercise 5.33. Show that

$$\lim_{n\to\infty} \left(\frac{1}{\sqrt{1^2 + n^2}} + \frac{1}{\sqrt{2^2 + n^2}} + \frac{1}{\sqrt{3^2 + n^2}} + \cdots + \frac{1}{\sqrt{n^2 + n^2}} \right) = \int_0^1 \frac{1}{\sqrt{x^2 + 1}}\, dx.$$

(We will show later on that this latter value is $\log(1 + \sqrt{2})$.)

Exercise 5.34. Find $\displaystyle \lim_{n\to\infty} \sum_{k=0}^{n-1} \frac{n}{n^2 + k^2}$.

Exercise 5.35. Find $\displaystyle \lim_{n\to\infty} \sum_{k=1}^{n} \frac{1}{\sqrt{n^2 + kn}}$.

Exercise 5.36. Using the computer, write a program to find

$$\int_0^{10} e^{-x^2}\, dx$$

approximately using a Riemann sum. For example, take the partition

$$P_n := \left\{ a = 0,\ \frac{10}{n},\ 2 \cdot \frac{10}{n}, \cdots, (n-1) \cdot \frac{10}{n},\ b = 10 \right\}$$

with n, say, 10000. It can be shown that the 'improper integral' (something we will study in the following section)

$$\int_0^{\infty} e^{-x^2}\, dx = \frac{\sqrt{\pi}}{2}.$$

As e^{-x^2} decreases rapidly with increasing x, and since $e^{-10^2} = e^{-100}$ has order of magnitude[4] 10^{-44}, it follows that the tail is

$$\int_{10}^{\infty} e^{-x^2} dx = \int_0^{\infty} e^{-(u+10)^2} du \quad \text{(with } u = x - 10\text{)}$$

$$= \int_0^{\infty} e^{-100} e^{-u^2} e^{-20u} du$$

$$\leq e^{-100} \int_0^{\infty} e^{-u^2} \cdot 1 du = e^{-100} \frac{\sqrt{\pi}}{2} \sim 10^{-44},$$

and so it can be neglected. Hence, our Riemann sum should give a reasonably good approximation to $\frac{\sqrt{\pi}}{2}$, and in turn, we can find an approximation for π. What approximate value for π do you obtain based on your computer program?

5.5   Improper integrals

So far, we have defined

$$\int_a^b f(x)dx$$

where $-\infty < a < b < \infty$ (that is, a and b are finite) and f is Riemann integrable (and in particular f is bounded).

Now we give meaning to the integral when either the domain of integration is unbounded or the function to be integrated (the integrand) becomes unbounded.

We will do this using the limits of integrals of the 'nice' type (with bounded integrand and a compact interval $[a, b]$), provided the relevant limits exist.

Definition 5.9 (Convergence/Divergence of improper integrals). Suppose that a is a real number, and let the function $f : [a, \infty) \to \mathbb{R}$ be such that for all $y \in (a, \infty), f \in RI[a, y]$. If

$$\lim_{y \to \infty} \int_a^y f(x) \, dx$$

exists (that is, there is a real number that is this limit), then we define

$$\int_a^{\infty} f(x) \, dx := \lim_{y \to \infty} \int_a^y f(x) \, dx,$$

and we say *the improper integral* $\int_a^{\infty} f(x) \, dx$ *exists* or $\int_a^{\infty} f(x) \, dx$ *converges*.

If $\int_a^{\infty} f(x) \, dx$ doesn't converge, then we say $\int_a^{\infty} f(x) \, dx$ *diverges/doesn't exist*.

---

[4] This phrase is used to indicate the rough size of a number, just like in everyday conversation, for example when one says 'She has a six figure salary'.

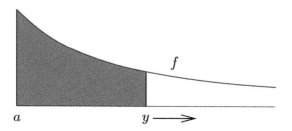

f

a $y \longrightarrow$

Example 5.27 (For $r < -1$, $\int_1^\infty x^r \, dx$ exists).

We have $\epsilon := -(r+1) > 0$. Thus

$$\left(\frac{x^{r+1}}{r+1}\right)' = (r+1) \cdot \frac{x^r}{r+1} = x^r,$$

and so

$$\int_1^y x^r \, dx = \frac{y^{r+1} - 1}{r+1}.$$

Hence

$$\lim_{y\to\infty} \int_1^y x^r \, dx = \lim_{y\to\infty} \frac{y^{r+1} - 1}{r+1}$$

$$= \lim_{y\to\infty} \frac{y^{-\epsilon} - 1}{r+1}$$

$$= \frac{0-1}{r+1} = -\frac{1}{r+1}.$$

So, if $r < -1$, then the improper integral $\int_1^\infty x^r \, dx$ exists and

$$\int_1^\infty x^r \, dx = -\frac{1}{r+1}. \qquad \diamondsuit$$

Example 5.28 ($\int_1^\infty \frac{1}{x} \, dx$ diverges).

Let P be the partition $\{1, 2, \cdots, n-1, n\}$ of the interval $[1, n]$ for $n \geq 2$. Then

$$\int_1^n \frac{1}{x} \, dx \geq \underline{S}\left(\frac{1}{x}, P\right) = \frac{1}{2} + \frac{1}{3} + \cdots + \frac{1}{n}.$$

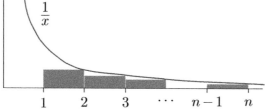

$\frac{1}{x}$

$1 \quad 2 \quad 3 \quad \cdots \quad n-1 \quad n$

But from Example 2.16, we know that the increasing sequence

$$\left(\frac{1}{2} + \frac{1}{3} + \cdots + \frac{1}{n} \right)_{n \in \mathbb{N}}$$

diverges, and so it can't be bounded. As

$$y \mapsto \int_1^y \frac{1}{x}\, dx : (1, \infty) \to \mathbb{R}$$

is increasing (indeed the integrand $1/x > 0$ for all $x \geq 1$), it follows from the above that

$$\lim_{y \to \infty} \int_1^y \frac{1}{x}\, dx = \infty.$$

Hence the improper integral $\int_1^\infty \frac{1}{x}\, dx$ diverges.   ◇

Theorem 5.16. *The improper integral* $\int_1^\infty x^r\, dx$ *converges* $\Leftrightarrow$ $\boxed{r < -1}$.

Proof. We have already seen that

$$\text{if } r < -1, \text{ then } \int_1^\infty x^r\, dx \text{ converges, and}$$

$$\text{if } r = -1, \text{ then } \int_1^\infty x^r\, dx \text{ diverges.}$$

Now suppose that $r > -1$. Then $r + 1 > 0$, and if $x \geq 1$, then $x^{r+1} \geq 1$. (We will learn this later when we define real exponents of nonnegative numbers: so far we at least know this for rational rs.) Thus

$$x^r \geq \frac{1}{x}.$$

So for $y > 1$,

$$\int_1^y x^r\, dx \geq \int_1^y \frac{1}{x}\, dx.$$

Since $\lim\limits_{y \to \infty} \int_1^y \frac{1}{x}\, dx = \infty$, it follows from the above that also

$$\lim_{y \to \infty} \int_1^y x^r\, dx = \infty,$$

and so the improper integral $\int_1^\infty \frac{1}{x}\, dx$ diverges if $r > -1$.   □

For example, $\int_1^\infty \frac{1}{\sqrt{x}}\, dx$ diverges, while $\int_1^\infty \frac{1}{x^2}\, dx$ converges.

Example 5.29. The improper integral $\int_0^\infty \frac{1}{1 + x^2}\, dx$ converges. For $y > 1$,

$$\int_0^y \frac{1}{1 + x^2}\, dx = \int_0^1 \frac{1}{1 + x^2}\, dx + \int_1^y \frac{1}{1 + x^2}\, dx$$

$$\leq \int_0^1 \frac{1}{1 + x^2}\, dx + \int_1^y \frac{1}{x^2}\, dx$$

$$\leq \int_0^1 \frac{1}{1 + x^2}\, dx + \int_1^y \frac{1}{x^2}\, dx \quad (\text{since } r = -2 < -1).$$

Thus $y \mapsto \int_0^y \dfrac{1}{1+x^2}\, dx$ is bounded above, and moreover, it is increasing. So

$$\lim_{y \to \infty} \int_0^\infty \frac{1}{1+x^2}\, dx$$

exists.

Definition 5.10 (Convergence/Divergence of improper integrals). Similar to our earlier definition, the improper integral

$$\int_{-\infty}^a f(x)\, dx$$

converges if $\displaystyle \lim_{y \to \infty} \int_{-y}^a f(x)\, dx$ *exists, and* $\displaystyle \int_{-\infty}^a f(x)\, dx := \lim_{y \to \infty} \int_{-y}^a f(x)\, dx$.

If $\displaystyle \int_{-\infty}^a f(x)\, dx$ does not converge, we say it *diverges*.

Definition 5.11 (Convergence of improper integrals). The improper integral

$$\int_{-\infty}^\infty f(x)\, dx$$

converges if both $\displaystyle \int_{-\infty}^0 f(x)\, dx$ and $\displaystyle \int_0^\infty f(x)\, dx$ exist, and we define

$$\int_{-\infty}^\infty f(x)\, dx := \int_{-\infty}^0 f(x)\, dx + \int_0^\infty f(x)\, dx.$$

Example 5.30. $\displaystyle \int_{-\infty}^\infty \frac{1}{1+x^2}\, dx$ converges. We have seen that

$$\int_0^\infty \frac{1}{1+x^2}\, dx$$

converges. For $y > 0$ (using the substitution $u = -x$, so that $du = -dx$, $u = 0$ when $x = 0$, and $u = y$ when $x = -y$), we have

$$\int_{-y}^0 \frac{1}{1+x^2}\, dx = \int_y^0 \frac{1}{1+(-u)^2}(-1)\, du = \int_0^y \frac{1}{1+u^2}\, du.$$

Since $\displaystyle \lim_{y \to \infty} \int_0^y \frac{1}{1+u^2}\, du$ exists, it follows from the above that

$$\lim_{y \to \infty} \int_{-y}^0 \frac{1}{1+x^2}\, dx$$

exists, that is $\displaystyle \int_{-\infty}^0 \frac{1}{1+x^2}\, dx$ converges. Since both

$$\int_{-\infty}^0 \frac{1}{1+x^2}\, dx \quad \text{and} \quad \int_0^\infty \frac{1}{1+x^2}\, dx$$

converge, $\displaystyle \int_{-\infty}^\infty \frac{1}{1+x^2}\, dx$ converges.

Example 5.31 ($\lim\limits_{y\to\infty} \int_{-y}^{y} f(x)\,dx$ may exist, but not $\int_{-\infty}^{\infty} f(x)\,dx$). We have

$$\int_{-y}^{y} x\,dx = \frac{x^2}{2}\Big|_{-y}^{y} = \frac{y^2}{2} - \frac{y^2}{2} = 0,$$

and so $\lim\limits_{y\to\infty} \int_{-y}^{y} f(x)\,dx = 0$. But,

$$\int_{0}^{y} x\,dx = \frac{x^2}{2}\Big|_{0}^{y} = \frac{y^2}{2} - 0 = \frac{y^2}{2},$$

and so $\lim\limits_{y\to\infty} \int_{0}^{y} f(x)\,dx$ does not exist. Consequently $\int_{-\infty}^{\infty} f(x)\,dx$ diverges.   ◇

Theorem 5.18. *If* $\int_{-\infty}^{\infty} f(x)\,dx$ *converges, then* $\lim\limits_{y\to\infty} \int_{-y}^{y} f(x)\,dx$ *exists, and*

$$\int_{-\infty}^{\infty} f(x)\,dx = \lim_{y\to\infty} \int_{-y}^{y} f(x)\,dx.$$

Proof. Indeed,

$$\int_{-\infty}^{\infty} f(x)\,dx = \int_{-\infty}^{0} f(x)\,dx + \int_{0}^{\infty} f(x)\,dx = \lim_{y\to\infty} \int_{-y}^{0} f(x)\,dx + \lim_{y\to\infty} \int_{0}^{y} f(x)\,dx$$

$$= \lim_{y\to\infty} \left(\int_{-y}^{0} f(x)\,dx + \int_{0}^{y} f(x)\,dx \right) = \lim_{y\to\infty} \int_{-y}^{y} f(x)\,dx.$$ □

Definition 5.12 (Absolutely convergent improper integral).

$\int_{a}^{\infty} f(x)\,dx$ is *absolutely convergent* if $\int_{a}^{\infty} |f(x)|\,dx$ is convergent.

The terminology used earlier suggests that absolutely convergent improper integrals ought to be first of all convergent: after all if we call a child a 'good boy', the child should be first of all a boy! The following result gives the needed justification.

Theorem 5.19.

If $\int_{a}^{\infty} f(x)\,dx$ *is absolutely convergent, then* $\int_{a}^{\infty} f(x)\,dx$ *converges.*

Proof. Set $f_+(x) = \dfrac{|f(x)| + f(x)}{2}, f_-(x) = \dfrac{|f(x)| - f(x)}{2}$, for $x \in [a, \infty)$. Then

$$0 \le f_+(x) = \frac{|f(x)| + f(x)}{2} \le \frac{|f(x)| + |f(x)|}{2} = |f(x)|,$$

$$0 \le f_-(x) = \frac{|f(x)| - f(x)}{2} \le \frac{|f(x)| + |f(x)|}{2} = |f(x)|.$$

Since $\displaystyle\int_a^\infty |f(x)|\,dx = \lim_{y\to\infty}\int_a^y |f(x)|\,dx$ exists, it follows from the above that

$$\int_a^\infty f_+(x)\,dx = \lim_{y\to\infty}\int_a^y f_+(x)\,dx \text{ and } \int_a^\infty f_-(x)\,dx = \lim_{y\to\infty}\int_a^y f_-(x)\,dx$$

exist. Since $f = f_+ - f_-$, we have

$$\lim_{y\to\infty}\int_a^y f(x)\,dx = \lim_{y\to\infty}\int_a^y (f_+(x) - f_-(x))\,dx$$

$$= \lim_{y\to\infty}\int_a^y f_+(x)\,dx - \lim_{y\to\infty}\int_a^y f_-(x)\,dx$$

exists. Thus $\displaystyle\int_a^\infty f(x)\,dx$ converges. $\qquad\Box$

Example 5.32. We will show $\displaystyle\int_1^\infty \frac{\sin x}{x}\,dx$ converges. We have $\left|\dfrac{\sin x}{x}\right| \le \dfrac{1}{x}$, but

$$\int_1^\infty \frac{1}{x}\,dx$$

diverges. So such a simplistic idea doesn't work. We will 'increase the power of x in the denominator' using Integration by Parts:

$$\int_1^y \frac{\sin x}{x}\,dx = \frac{1}{x}(-\cos x)\Big|_1^y - \int_1^y -\frac{1}{x^2}(-\cos x)\,dx = -\frac{\cos y}{y} + \frac{\cos 1}{1} - \int_1^y \frac{\cos x}{x^2}\,dx.$$
$$(5.10)$$

Since $\left|\dfrac{\cos x}{x^2}\right| \le \dfrac{1}{x^2}$, and since $\displaystyle\int_1^\infty \frac{1}{x^2}\,dx$ converges, it follows that

$$\int_1^\infty \frac{\cos x}{x^2}\,dx$$

is absolutely convergent, and so, convergent. Also, $\displaystyle\lim_{y\to\infty}\frac{-\cos y}{y} = 0$. From (5.10),

$$\lim_{y\to\infty}\int_1^y \frac{\sin x}{x}\,dx$$

exists, that is, $\displaystyle\int_1^\infty \frac{\sin x}{x}\,dx$ converges. $\qquad\Diamond$

There is another kind of improper integral, in which the domain of the integrand is bounded, but the integrand is unbounded.

Definition 5.13 (Convergence of an improper integral). Let $a, b \in \mathbb{R}$ with $a < b$.

(1) $f : (a, b] \to \mathbb{R}$ be such that for every $y > a, f \in RI[y, b]$. If

$$\lim_{y\searrow a}\int_y^b f(x)\,dx$$

exists, then we say that the improper integral $\int_{a+}^{b} f(x)\,dx$ *converges*, and

$$\int_{a+}^{b} f(x)\,dx := \lim_{y \searrow a} \int_{y}^{b} f(x)\,dx.$$

Sometimes, we denote $\int_{a+}^{b} f(x)\,dx$ simply by $\int_{a}^{b} f(x)\,dx$.

(2) The improper integral

$$\int_{a}^{b-} f(x)\,dx$$

is defined analogously for a function $f : [a,b) \to \mathbb{R}$ be such that for every $y < b$, $f \in RI[a,y]$.

(3) Finally, consider an unbounded function $f : (a,b) \to \mathbb{R}$, which is Riemann integrable on $[x,y]$ for all $x, y \in (a,b)$ with $x < y$. Set $c := \frac{a+b}{2}$.

If $\int_{a+}^{c} f(x)\,dx$ converges and $\int_{c}^{b-} f(x)\,dx$ converges, then we say

$$\int_{a+}^{b-} f(x)\,dx$$

converges, and define $\int_{a+}^{b-} f(x)\,dx := \int_{a+}^{c} f(x)\,dx + \int_{c}^{b-} f(x)\,dx$.

Theorem 5.20. $\int_{0}^{1} x^{r}\,dx$ converges $\Leftrightarrow \boxed{r > -1}$.

Proof. For $r \neq -1$, we have $\int_{y}^{1} x^{r}\,dx = \frac{x^{r+1}}{r+1}\Big|_{y}^{1} = \frac{1 - y^{r+1}}{r+1}$.

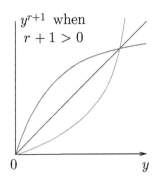

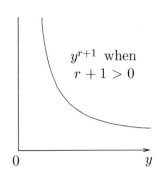

$\underline{1°}$ $r > -1$. Then $r + 1 > 0$, and $\lim_{\substack{y \to 0 \\ y > 0}} y^{r+1} = 0$. Thus

$$\lim_{\substack{y \to 0 \\ y > 0}} \int_y^1 x^r \, dx = \lim_{\substack{y \to 0 \\ y > 0}} \frac{1 - y^{r+1}}{r + 1} = \frac{1 - 0}{r + 1} = \frac{1}{r + 1}.$$

Hence $\int_0^1 x^r \, dx$ converges.

$\underline{2°}$ $r < -1$. Then $r + 1 < 0$, and $\lim_{\substack{y \to 0 \\ y > 0}} y^{r+1} = \infty$. Thus

$$\lim_{\substack{y \to 0 \\ y > 0}} \int_y^1 x^r \, dx = \lim_{\substack{y \to 0 \\ y > 0}} \frac{1 - y^{r+1}}{r + 1} = \infty.$$

Hence $\int_0^1 x^r \, dx$ diverges.

$\underline{3°}$ $r = -1$. Then for $n \geq 2$, by considering the partition

$$\left\{ \frac{1}{n}, \frac{2}{n}, \frac{3}{n}, \cdots, \frac{n-1}{n}, \frac{n}{n} \right\} \in \mathcal{P}_{[\frac{1}{n}, 1]}$$

and the function $1/x \in C[\frac{1}{n}, 1] \subset RI[\frac{1}{n}, 1]$, we obtain the inequality

$$\int_{1/n}^1 \frac{1}{x} \, dx \geq S(1/x, P) = \frac{1}{n} \cdot \frac{1}{2/n} + \frac{1}{n} \cdot \frac{1}{3/n} + \cdots + \frac{1}{n} \cdot \frac{1}{n/n}$$

$$= \frac{1}{2} + \frac{1}{3} + \cdots + \frac{1}{n} \xrightarrow{n \to \infty} \infty.$$

Hence $\lim_{\substack{y \to 0 \\ y > 0}} \int_y^1 x^r \, dx$ does not exist, and so $\int_0^1 \frac{1}{x} \, dx$ diverges.   □

$$\int_{0+}^1 x^r \, dx \text{ converges} \qquad \int_1^\infty x^r \, dx \text{ converges}$$
$$\Leftrightarrow r > -1. \qquad \qquad \Leftrightarrow r < -1.$$

One can also consider 'mixed' combinations of improper integrals such as

$$\int_{a+}^b f(x) \, dx + \int_b^\infty f(x) \, dx$$

written as $\int_{a+}^\infty f(x) \, dx$. Here is an example.

Example 5.33 (The Gamma function Γ). The simplest functions one meets in applications are the *algebraic functions*, which are the polynomials, rational functions, and the nth root function. Loosely speaking if a function is not a combination of these, then one calls it *transcendental*. Among the transcendental functions are the logarithm function log, the exponential function exp, the trigonometric functions sin, cos, and so on. We will soon define these elementary transcendental functions and their properties in the subsequent section. Besides the elementary transcendental functions, there are functions that also appear frequently enough (typically in specific subdisciplines such as statistics, quantum mechanics, number theory, etc.) that they warrant their own special symbols. Such functions are called *special functions*. Arguably, the most common special function or the least special of the special functions is the Gamma function Γ, which transcends multiple subdisciplines such as quantum mechanics, analytic number theory, statistics, and so on. The Γ function is defined in terms of an improper integral, and the aim of this exercise is to study the 'well definedness' of Gamma function for positive values of the argument. In other words, we want to prove the following:

Claim: If $s > 0$, then the improper integral $\Gamma(s) := \int_{0+}^{\infty} e^{-t} \cdot t^{s-1}\, dt$ converges.

(Here we use e^{-t} for $\exp(-t)$. We will use several properties of the exponential function in the following, which can be accepted on faith now, but will be proved in the following section when we study the exponential function.)

The improper integral is interpreted as a sum

$$\underbrace{\int_{0+}^{1} e^{-t} t^{s-1}\, dt}_{\text{I}} + \underbrace{\int_{1}^{\infty} e^{-t} t^{s-1}\, dt}_{\text{II}}.$$

(Note that for values of t near 1, the integrand is well behaved, but for t near 0, for $s < 1$, the integrand is unbounded.)

Let us study the convergence of I and II.

I: We know that

$$\int_{0+}^{1} t^{s-1}\, dt$$

converges for $s - 1 > -1$, that is, for $s > 0$. As $0 \le e^{-t} t^{s-1} \le 1 \cdot t^{s-1}$ for $t \in (0, 1]$, it follows that

$$\int_{0+}^{1} e^{-t} t^{s-1}\, dt$$

converges.

II: Let $n \in \mathbb{N}$ be such that $n > s$. Then for $t \ge 1$,

$$e^t = 1 + \frac{t}{1!} + \frac{t^2}{2!} + \cdots \ge \frac{t^n}{n!},$$

and so $\dfrac{e^{-t} t^{s-1} t^{n-s+1}}{n!} \le 1$, that is,

$$e^{-t} t^{s-1} \le \frac{n!}{t^{1+n-s}}.$$

As $\displaystyle\int_1^\infty t^r \, dt$ converges for $r < -1$, it follows that

$$\int_1^\infty e^{-t} t^{s-1} \, dt$$

converges.

We remark that the Gamma function was introduced by Euler in 1729. It is a solution to the following *interpolation problem*:

Find a smooth curve that connects the points on the graph of the factorial function $n \mapsto n! : \mathbb{N} \to \mathbb{N}$.

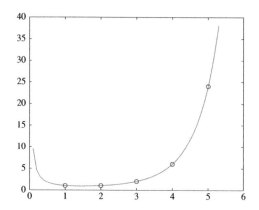

Of course there are infinitely many such functions, but the Gamma function is somewhat special. There is a result, called the Bohr[5]-Mollerup Theorem, stating that the Gamma function is the *unique* function f, which is

(1) positive,

(2) logarithmically convex (that is, $\log \circ \, \Gamma$ is a convex function[6])

(3) for all $x > 0$, $f(x+1) = xf(x)$.   ◇

Exercise 5.37. Find $\displaystyle\int_9^\infty \frac{1}{(x-3)^2} \, dx.$

Exercise 5.38. Determine whether or not the following improper integrals exist:

(1) $\displaystyle\int_0^\infty \frac{1}{\sqrt{1+x^3}} \, dx.$

(2) $\displaystyle\int_0^\infty \frac{x}{1+\sqrt{x^3}} \, dx.$

---

[5] Incidentally, this is *Harald* Bohr, the mathematician brother of the physicist *Niels* Bohr. Harald Bohr made fundamental contributions to the theory of almost periodic functions, and also won a silver medal in the 1908 Olympics in football.

[6] For example, e^{x^2} is convex. Logarithmically convex functions are convex, but not vice versa. x^2 is convex, but $\log(x^2) = 2\log|x|$ is not convex.

Exercise 5.39 (Properties of the Gamma function Γ).

    (1) Show that $\Gamma(1) = 1$.

    (2) (∗) For $s > 0$, $\Gamma(s+1) = s \cdot \Gamma(s)$.

    (3) Show that for all $n \in \mathbb{N}$, $\Gamma(n+1) = n!$.

Exercise 5.40 (∗) Suppose that $f : [0, \infty) \to [0, \infty)$ is such that

$$\int_0^\infty f(x)dx \tag{5.11}$$

exists. Intuitively, we expect that the area under the graph of the nonnegative f in intervals $[x, \infty)$ to become smaller and smaller as x becomes larger and larger, and so one is tempted to conclude that f itself must have limit 0 as $x \to \infty$. Show with an example that this needn't be the case.

    Suppose now that we know that $f : [0, \infty) \to [0, \infty)$ is such that, besides having that the improper integral (5.11) exists, also f is differentiable and

$$\int_0^\infty f'(x)dx$$

exists. Show that $\lim_{x\to\infty} f(x) = 0$.

Exercise 5.41 (Convolution). For $f, g : \mathbb{R} \to \mathbb{R}$, which are both zero outside some compact interval, we define the *convolution* $f * g : \mathbb{R} \to \mathbb{R}$ by

$$(f * g)(t) = \int_{-\infty}^\infty f(\tau)g(t - \tau)d\tau, \quad t \in \mathbb{R},$$

assuming that the integral exists for each t.

(1) Note that the graph of $g(-\cdot)$ is obtained by reflecting the graph of g about the y-axis, and for a fixed t, the graph of $g(t - \cdot)$ is a shifted version of the graph of $g(-\cdot)$. So in order to find out the value $(f * g)(t)$, one may proceed as follows.

    (a) Draw the graph of f and g.

    (b) Reflect the graph of g about the y-axis.

    (c) Translate the graph of $g(-\cdot)$ by $|t|$ units to the left if $t < 0$ and to the right if $t \geq 0$.

    (d) Multiply the functions f and $g(t - \cdot)$ pointwise, and find the area under the graph of this pointwise product.

Use this procedure to graphically determine the convolution $1_{[0,1]} * 1_{[0,1]}$, where $1_{[0,1]}$ is the indicator function of the interval $[0, 1]$:

$$1_{[0,1]}(x) = \begin{cases} 1 & \text{if } x \in [0, 1], \\ 0 & \text{if } x \in \mathbb{R}\backslash[0, 1]. \end{cases}$$

(2) Show that $f * g = g * f$. (That is the convolution operation $*$ is 'commutative'.)

Exercise 5.42 (Differentiation under the integral sign[7]). While we don't learn the theory behind this, let us try to use this tool formally, that is, without paying attention to rigour. To find

$$\int_0^\infty \frac{\sin x}{x}\, dx,$$

we consider the more general integral

$$I(\alpha) = \int_0^\infty e^{-\alpha x} \frac{\sin x}{x}\, dx,$$

by introducing the 'parameter' α. (So our integral of interest corresponds to $\alpha = 0$.) Then by differentiating with respect to α, we obtain

$$I'(\alpha) = \int_0^\infty (-x) e^{-\alpha x} \frac{\sin x}{x}\, dx$$

$$= \int_0^\infty -e^{-\alpha x} \sin x\, dx = -\frac{1}{\alpha^2 + 1} e^{-\alpha x}((-\alpha)\sin x - \cos x)\Big|_0^\infty$$

$$= -\frac{1}{\alpha^2 + 1},$$

where we have used Integration by Parts (twice), as in Example 5.20. Integrating (from α to ∞), we have $0 - I(\alpha) = I(\infty) - I(\alpha) = -\frac{\pi}{2} + \tan^{-1}\alpha$, and so

$$\int_0^\infty \frac{\sin x}{x}\, dx = I(0) = \frac{\pi}{2}.$$

Try your hand at formally finding the values of the following integrals.

(1) $\displaystyle\int_0^\infty \frac{(\sin x)^2}{x^2}\, dx$ by considering $I(\alpha) = \displaystyle\int_0^\infty \frac{(\sin(\alpha x))^2}{x^2}\, dx$.

(2) $\displaystyle\int_0^\infty e^{-x} \frac{\sin x}{x}\, dx$ by considering $I(\alpha) = \displaystyle\int_0^\infty e^{-x} \frac{\sin(\alpha x)}{x}\, dx$.

(3) $\displaystyle\int_0^1 \frac{x - 1}{\log x}\, dx$ by considering $I(\alpha) = \displaystyle\int_0^1 \frac{x^\alpha - 1}{\log x}\, dx$.

(4) $\displaystyle\int_0^\infty \frac{\tan^{-1}(\pi x) - \tan^{-1}x}{x}\, dx$ by considering $I(\alpha) = \displaystyle\int_0^\infty \frac{\tan^{-1}(\alpha x) - \tan^{-1}x}{x}\, dx$.

Exercise 5.43 (Gravitational potential energy and escape velocity).

(1) The gravitational potential energy (due to the gravitational pull of the Earth) at distance an R from the center of the earth can be thought to be the amount of work

---

[7] Differentiation under the integral sign is mentioned in Feynman's memoir *Surely You're Joking, Mr. Feynman!* in the chapter 'A Different Box of Tools', where it is narrated that he learnt it from a Calculus book by Woods while in high school, and using this tool for doing integrals (where standard methods such as contour integration or a simple series expansion had failed), he built a reputation for doing integrals.

done to bring an object from separation R to far away ('at infinity $R = \infty$'). By Newton's Law of Gravitation, the force experienced by a mass m at a separation r from the center of the Earth is given by

$$F = \frac{GMm}{r^2},$$

where G is the universal gravitation constant and M is the mass of the Earth. The work done to move a mass m at r through a small distance dr is given by

$$\text{Work done} = (\text{Force}) \cdot (\text{Displacement}) = \frac{GMm}{r^2} dr,$$

where we have assumed that the force is constant over $[r, r + dr]$ for small dr. So, the potential energy $V(R)$ at R is given by the (improper) integral

$$V(R) := \int_R^\infty \frac{GMm}{r^2} dr.$$

Find an explicit expression for $V(R)$.

(2) The *escape velocity* $v_e(R)$ at a separation R from the center of the earth is defined as to be the one that imparts enough kinetic energy to the object in order to overcome the gravitation potential energy $V(R)$. Show that

$$v_e = \sqrt{\frac{2GM}{R}}.$$

Using the following values, determine the escape velocity of a rocket on the surface of the earth:

$$\text{Radius of the Earth } R_\oplus = 6,371 \text{ km},$$

$$\text{Mass of the Earth } M_\oplus = 5.97219 \times 10^{24} \text{ kg},$$

$$\text{Universal gravitational constant } G = 6.67384 \times 10^{-11} \text{ m}^3 \text{ kg}^{-1} \text{ s}^{-2}.$$

(3) Assuming that M is the mass of a star (say the Sun, whose mass is estimated to be $M_\odot = 1.99 \times 10^{30}$ kg), for what radius does the escape velocity equal the speed of light, $c = 3 \times 10^8 \text{ ms}^{-1}$? This radius, r_s, is called the *Schwarzchild radius* (of the black hole).

Exercise 5.44. If λ is a positive real number, then show that $\int_0^\infty e^{-\lambda x} dx = \frac{1}{\lambda}$.

Exercise 5.45. Does the improper integral $\int_0^\infty e^{-x^2} dx$ converge?

Exercise 5.46. Does $\int_2^\infty \frac{1}{x \log x} dx$ converge? What about $\int_2^\infty \frac{1}{x(\log x)^2} dx$?

5.6   Elementary transcendental functions

We will use the theory of Riemann integration to introduce and define the elementary transcendental functions:

(1) logarithm,

(2) exponential,

(3) trigonometric functions such as $\sin, \cos, \tan,$ etc.

We will give formal definitions of these and derive their several interesting properties based on our definitions and the Calculus tools we have been developing. En route we will also formally define e, π. (π we have met before in Example 5.8.) These things are of great practical value, since they form the foundations of concepts such as angle polar coordinates, and so on.

The logarithm function

Q. What does a drowning Calculus teacher say?
A. log, loglog, logloglog, $\cdots$.

$\dfrac{1}{t} : (0, \infty) \to \mathbb{R}$ is continuous, and so for any $[a, b] \subset (0, \infty)$, $\dfrac{1}{t} \in RI[a, b]$.

Definition 5.14 (Logarithm function). The *logarithm* $\log : (0, \infty) \to \mathbb{R}$ is defined by

$$\log x := \int_1^x \frac{1}{t}\, dt, \quad x > 0.$$

As $\dfrac{1}{t} > 0$ for all $t \in (0, \infty)$, we have $\log x \geq 0$ if $x \geq 1$.

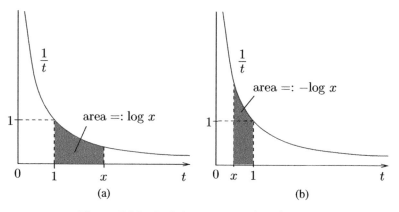

(a)   (b)

Figure 5.14   Definition of $\log : (0, \infty) \to \mathbb{R}$.

We note that owing to our convention with the Riemann integral and the limits of integration described on page 204, $\log x < 0$ for $0 < x < 1$. See Figure 5.14.

Moreover,

$$\log 1 = \int_1^1 \frac{1}{t}\, dt = 0.$$

Theorem 5.21 (Properties of the logarithm). $\log : (0, \infty) \to \mathbb{R}$ *is*

(1) *differentiable, and* $\log' x = \dfrac{1}{x}$ *for all* $x \in (0, \infty)$;

(2) *strictly increasing, and concave;*

(3) *onto,* $\lim_{x \to \infty} \log x = \infty$, $\lim_{x \to 0+} \log x = -\infty$.

Proof.

(1) By the Fundamental Theorem of Calculus, $\dfrac{d}{dx} \displaystyle\int_1^x \dfrac{1}{t}\, dt = \dfrac{1}{x}$.

(2) Since $\log' x = \dfrac{1}{x} > 0$ for all $x \in (0, \infty)$, it follows that log is strictly increasing. Moreover, $\log'' x = -\dfrac{1}{x^2} < 0$ for all $x \in (0, \infty)$, and so log is concave.

(3) Let $n \geq 2$. Then

$$\log n = \int_1^n \frac{1}{t}\, dt = \sum_{k=2}^n \int_{k-1}^k \frac{1}{t}\, dt \geq \sum_{k=2}^n \int_{k-1}^k \frac{1}{k}\, dt = \sum_{k=2}^n \frac{1}{k}.$$

Since 'the Harmonic Series diverges', $\left(\displaystyle\sum_{k=2}^n \frac{1}{k} \right)_{n \in \mathbb{N}}$ is not bounded above. So we obtain that for all $y > 0$, there exists an $n \geq 2$ such that

$$\log n \geq \sum_{k=2}^n \frac{1}{k} \geq y \geq 0 = \log 1.$$

Thus by the Intermediate Value Theorem applied to the continuous[8] function log on $[1, n]$, there exists an $x \in [1, n]$ such that $\log x = y$.

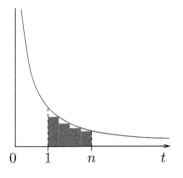 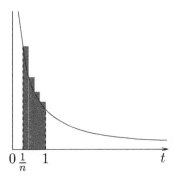

Also,

$$\log \frac{1}{n} = -\int_{1/n}^1 \frac{1}{t}\, dt = -\sum_{k=2}^n \int_{\frac{1}{k}}^{\frac{1}{k-1}} \frac{1}{t}\, dt \leq -\sum_{k=2}^n \int_{\frac{1}{k}}^{\frac{1}{k-1}} (k-1)\, dt = -\sum_{k=2}^n \frac{1}{k}.$$

---

[8] we know that log is differentiable on $(0, \infty)$ and hence continuous there

Since

$$\left(\sum_{k=2}^{n} \frac{1}{k} \right)_{n \in \mathbb{N}}$$

is not bounded above, we obtain that for all $y \le 0$, there exists an $n \ge 2$ such that

$$\log \frac{1}{n} \le - \sum_{k=2}^{n} \frac{1}{k} < y \le 0 = \log 1.$$

So by the Intermediate Value Theorem applied to the continuous function log on $[\frac{1}{n}, 1]$, there exists an $x \in [\frac{1}{n}, 1]$ such that $\log x = y$. Hence log is onto.

Finally, we will show that $\lim_{x \to \infty} \log x = \infty$ and $\lim_{x \to 0+} \log x = -\infty$.

Let $M \in \mathbb{R}$. Choose $x_0 \in (0, \infty)$ such that $\log x_0 = M$. As log is increasing, we have that for $x > x_0$, $\log x \ge \log x_0 = M$. Hence

$$\lim_{x \to \infty} \log x = \infty.$$

Also, log is increasing, and so for all $x \in (0, \infty)$ such that $0 < x < x_0$ (with the above choice of x_0), $\log x \le \log x_0 = M$. Thus

$$\lim_{x \to 0+} \log x = -\infty.$$

This completes the proof. $\qquad\qquad\qquad\qquad\qquad\qquad\qquad\qquad\qquad\qquad$ □

On the basis of the above properties, one can sketch the graph of log. See Figure 5.15.

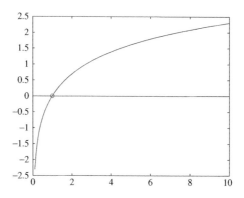

Figure 5.15   Graph of log.

Definition of the number e

The properties (2), (3) in Theorem 5.21 of the logarithm function say in particular that $\log : (0, \infty) \to \mathbb{R}$ is one-to-one (since it is *strictly* increasing) and onto, that is, it is *bijective*. Thus there exists a *unique* number $e \in (0, \infty)$ such that $\log e = 1$. This is called *Euler's number* (explaining the choice of the letter) or sometimes also as *Napier's constant*. See Figure 5.16.

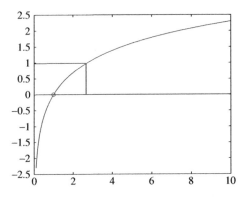

Figure 5.16   Definition of *e*: the horizontal line through 1 on the *y*-axis intersects the graph of log at a point, whose *x*-coordinate is the number *e*.

Here are some crude estimates for *e*: $2 \le e \le 4$. Indeed,

$$\log 2 = \int_1^2 \frac{1}{t}\, dt \le \int_1^2 1\, dt = 1,$$

$$\log 4 = \int_1^4 \frac{1}{t}\, dt = \int_1^2 \frac{1}{t}\, dt + \int_2^4 \frac{1}{t}\, dt \ge \int_1^2 \frac{1}{2}\, dt + \int_2^4 \frac{1}{4}\, dt = \frac{1}{2} + \frac{1}{2} = 1.$$

Since log is strictly increasing, $2 \le e \le 4$. (If $e < 2$, then $1 < \log 2 \le 1$, a contradiction, and if $e > 4$, then $1 > 4 \ge 1$, again a contradiction.)

We will soon show that

$$e = \sum_{n=0}^{\infty} \frac{1}{n!} := \lim_{N \to \infty} \sum_{n=0}^{N} \frac{1}{n!}.$$

This allows us to show that up to three decimal places, $e = 2.718 \cdots$. We will also show that $e \notin \mathbb{Q}$, that is, *e* is irrational.

Remark 5.2.   One can in fact also show that *e* is not 'algebraic', that is, it is 'transcendental'. What do we mean by this? Here are the relevant definitions. A real number α is called *algebraic* if there is a nonzero polynomial *p* with rational coefficients such that $p(\alpha) = 0$. Of course all rational numbers are algebraic since if $r = \frac{n}{d}$, where $n, d \in \mathbb{Z}$ and $d \neq 0$, then with $p := dx - n$, we obtain

$$p(r) = d \cdot \frac{n}{d} - n = 0.$$

But the set $\mathbb{A}$ of algebraic numbers strictly contains $\mathbb{Q}$: for example, $\sqrt{2} \in \mathbb{A} \backslash \mathbb{Q}$, since with $p := x^2 - 2$, we have $p(\sqrt{2}) = 0$. It can be shown that $\mathbb{A}$ is a *field* (with the same operations of addition and multiplication inherited from $\mathbb{R}$). A real number τ is called *transcendental* if $\tau \notin \mathbb{A}$. It can be proved that $e, \pi \in \mathbb{R} \backslash \mathbb{A}$; see **[S2]**. Thus $\mathbb{Q} \subsetneq \mathbb{A} \subsetneq \mathbb{R}$.

Let us now show another important property of the logarithm.

Theorem 5.22. *For $a, b \in (0, \infty)$, $\log(ab) = \log a + \log b$.*

Adders multiply on log tables.

Proof. Fix $b \in (0, \infty)$. Define $f : (0, \infty) \to \mathbb{R}$ by $f(x) := \log(x \cdot b) - \log x$, $x > 0$. Then

$$f'(x) = \frac{1}{x \cdot b} \cdot b - \frac{1}{x} = 0, \quad x > 0.$$

Hence f is constant. In particular, $f(a) = f(1)$, for $a > 0$, that is,

$$\log(ab) - \log a = \log(1 \cdot b) - \log 1 = \log b - 0 = \log b.$$

Rearranging, we obtain the desired equality.   □

Exercise 5.47 (Euler's constant γ). Another important number named after Euler is Euler's constant γ, which plays a role, among other things, in Number Theory.

(1) Prove that $\dfrac{1}{n} \geq \log(n + 1) - \log n \geq \dfrac{1}{n+1}$ for all $n \in \mathbb{N}$.

(2) If

$$a_n := 1 + \frac{1}{2} + \frac{1}{3} + \cdots + \frac{1}{n} - \log n,$$

then show that $(a_n)_{n \in \mathbb{N}}$ is decreasing and that $a_n \geq 0$ for all $n \in \mathbb{N}$. Conclude that there is a number

$$\gamma := \lim_{n \to \infty} \left(1 + \frac{1}{2} + \frac{1}{3} + \cdots + \frac{1}{n} - \log n \right).$$

(Approximately, $\gamma = 0.5772156649 \cdots$, but it is not known whether γ is rational or irrational. The reason behind the choice of the symbol is that γ is closely related to the Γ function. For example, $\gamma = -\Gamma'(1)$.)

Exercise 5.48. Show that for all $x \geq 0$, $x - \dfrac{x^2}{2} \leq \log(1 + x) \leq x - \dfrac{x^2}{2} + \dfrac{x^3}{3}$.

Exercise 5.49. In 'hyperbolic geometry' of the unit disk $\mathbb{D} := \{(x, y) \in \mathbb{R}^2 : x^2 + y^2 < 1\}$ in the plane, straight lines in $\mathbb{D}$ are circular arcs that are orthogonal to the bounding circle $\mathbb{T}$. See Figure 5.17.

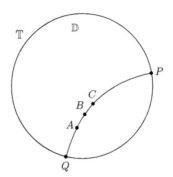

Figure 5.17   A straight line passing through A, B, C.

The distance between two points A, B is then taken as

$$d(A, B) := \log \left(\frac{\overset{\frown}{AP}}{\overset{\frown}{AQ}} \cdot \frac{\overset{\frown}{BQ}}{\overset{\frown}{BP}} \right),$$

where $\overset{\frown}{AP}$ denotes the circular Euclidean arc length of the circular arc AP, etc.

Note that since $\overset{\frown}{AP} > \overset{\frown}{BP}$ and $\overset{\frown}{BQ} > \overset{\frown}{AQ}$, $d(A, B)$ are always nonnegative. Also, one can check that $d(A, B) = d(B, A)$. Show that for three points A, B, C lying on such a line, in that order, $d(A, B) + d(B, C) = d(A, C)$.

Exercise 5.50 (Stirling's Formula).

(1) Compute the improper integral $\int_0^1 \log x \, dx$.

(2) Based on the result in the previous part, try giving a formal nonrigorous explanation of Stirling's Formula: for large n, $\log n! \approx n \log n - n$ (implying the approximation $n! \approx e^{n \log n - n} = n^n e^{-n}$).

Exercise 5.51. Show that for all $x \in \mathbb{R}$,

$$\int_0^x \frac{1}{\sqrt{1 + t^2}} \, dt = \log(x + \sqrt{1 + x^2}),$$

$$\int_0^x \sqrt{1 + t^2} \, dt = \frac{x\sqrt{1 + x^2} + \log(x + \sqrt{1 + x^2})}{2}.$$

Exercise 5.52. Find $\lim\limits_{x \to \infty} \dfrac{\log(\log x)}{\log x}$.

The exponential function

Definition 5.15 (Exponential function exp). $\exp : \mathbb{R} \to (0, \infty)$ is the inverse of (the bijective function) $\log : (0, \infty) \to \mathbb{R}$. Thus for $x \in \mathbb{R}$ and $y \in (0, \infty)$,

$$\exp x = y \quad \Leftrightarrow \quad \log y = x.$$

See Figure 5.18.

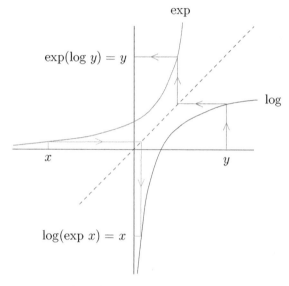

Figure 5.18 $\exp : \mathbb{R} \to (0, \infty)$ and $\log : (0, \infty) \to \mathbb{R}$ are inverses of each other.

By definition,

$$\exp x > 0 \qquad \text{for all } x \in \mathbb{R},$$
$$\exp 0 = 1 \qquad \text{since } \log 1 = 0,$$
$$\exp 1 = e \qquad \text{since } \log e = 1.$$

Here are further properties of the exponential function.

Theorem 5.23 (Properties of exp). $\exp : \mathbb{R} \to (0, \infty)$ *is such that:*

(1) $\exp$ *is differentiable, and* $\exp' x = \exp x$ *for all* $x \in \mathbb{R}$;

(2) $\exp$ *is strictly increasing, and convex;*

(3) *for all* $\alpha, \beta \in \mathbb{R}$, $\exp(\alpha + \beta) = (\exp \alpha) \cdot (\exp \beta)$;

(4) $\lim\limits_{x \to \infty} \exp x = \infty$, $\lim\limits_{x \to -\infty} \exp x = 0$.

Proof.

(1) Let $x \in \mathbb{R}$ and $c \in (0, \infty)$ be such that $\log c = x$. So by the Differentiable Inverse Theorem (applied to $\log : (0, \infty) \to \mathbb{R}$, which is strictly increasing, continuous, differentiable at $x = \log c$ and $\log' c = 1/c \neq 0$), it follows that $\exp : \mathbb{R} \to (0, \infty)$ is differentiable at $x = \log c$ and that

$$\exp' x = \exp'(\log c) = \frac{1}{\log' c} = \frac{1}{1/c} = c = \exp x.$$

(2) Since $\exp' x = \exp x > 0$ for all $x \in \mathbb{R}$, it follows that exp is strictly increasing. Also, exp is convex because $\exp'' x = \exp x > 0$ for all $x \in \mathbb{R}$.

(3) Let $a := \exp \alpha$ and $b := \exp \beta$. Then we have that $\log a = \alpha$ and $\log b = \beta$, and thus $\log(ab) = \log a + \log b = \alpha + \beta$. So $(\exp \alpha)(\exp \beta) = ab = \exp(\log ab) = \exp(\alpha + \beta)$.

(4) Let $M > 0$. Set $x_0 := \log M$. Then $\exp x_0 = M$. As exp is increasing, we have for all $x \geq x_0$ that $\exp x \geq \exp x_0 = M$. Hence

$$\lim_{x \to \infty} \exp x = \infty.$$

Let $\epsilon > 0$. Then for all $x < \log \epsilon$, we have $\exp x < \exp(\log \epsilon) = \epsilon$, and since $\exp x > 0$, $|\exp x - 0| = \exp x < \epsilon$. Hence $\lim_{x \to -\infty} \exp x = 0$. $\square$

Taylor's formula for exp

For all $k \geq 0$ and all $x \in \mathbb{R}$, $\exp^{(k)} x = \exp x$. So the Taylor polynomial at 0 of degree n for $f := \exp$ is

$$p_n(x) := f(0) + \frac{f'(0)}{1!}x + \cdots + \frac{f^{(n)}(0)}{n!}x^n$$

$$= \exp 0 + \frac{\exp 0}{1!}x + \cdots + \frac{\exp 0}{n!}x^n$$

$$= 1 + \frac{1}{1!}x + \frac{1}{2!}x^2 + \frac{1}{3!}x^3 + \cdots + \frac{1}{n!}x^n.$$

Thus Taylor's Formula gives the existence of a c_x between 0 and x such that

$$\exp x = 1 + \frac{1}{1!}x + \frac{1}{2!}x^2 + \frac{1}{3!}x^3 + \cdots + \frac{1}{n!}x^n + \frac{\exp c_x}{(n+1)!}x^{n+1}.$$

In particular,

$$e = \exp 1 = 1 + \frac{1}{1!} + \frac{1}{2!} + \frac{1}{3!} + \cdots + \frac{1}{n!} + \frac{\exp c}{(n+1)!}$$

for some $c \in (0, 1)$. But $0 < \exp c < \exp 1 = e < 4$, and so we obtain

$$0 < e - \left(1 + \frac{1}{1!} + \frac{1}{2!} + \frac{1}{3!} + \cdots + \frac{1}{n!}\right) < \frac{4}{(n+1)!}. \tag{5.12}$$

Thus the sequence (of rational numbers)

$$\left(1 + \frac{1}{1!} + \frac{1}{2!} + \frac{1}{3!} + \cdots + \frac{1}{n!}\right)_{n \in \mathbb{N}}$$

converges to e, and we write $e = \sum_{n=0}^{\infty} \frac{1}{n!}$.

We can use (5.12) to find e to an arbitrary number of decimal places. The following table shows the terms of the above sequence for $n = 1$ to 9.

n	$1 + \frac{1}{1!} + \frac{1}{2!} + \frac{1}{3!} + \cdots + \frac{1}{n!}$	$\frac{4}{(n+1)!}$
1	2	2
2	2.5	$0.666\cdots$
3	$2.666\cdots$	$0.1666\cdots$
4	$2.708333\cdots$	$0.0333\cdots$
5	$2.71666\cdots$	$0.00555\cdots$
6	$2.7180555\cdots$	$0.0007936\cdots$
7	$2.718253968\cdots$	$0.0000992\cdots$
8	$2.71827877\cdots$	$0.000011\cdots$
9	$2.718281525\cdots$	$0.0000011\cdots$

Theorem 5.24. $e \notin \mathbb{Q}$.

Proof. Suppose that $e = \frac{p}{q}$, where $p, q \in \mathbb{N}$. We recall (5.12), which says that for all $n \in \mathbb{N}$,

$$0 < e - \left(1 + \frac{1}{1!} + \frac{1}{2!} + \frac{1}{3!} + \cdots + \frac{1}{n!}\right) \leq \frac{4}{(n+1)!}.$$

Let us multiply throughout by $n!$, where $n > \max\{4, q\}$. Then

$$0 < \text{ (an integer) } < \frac{4}{n+1} < 1,$$

which is impossible!   □

We will now learn about some useful consequences of the properties of log and exp.

Corollary 5.25.

(1) $\lim_{h \to 0} \dfrac{\log(1 + h)}{h} = 1.$

(2) $\lim_{h \to 0} \dfrac{(\exp h) - 1}{h} = 1.$

(1) and (2) imply that for small h, $\log(1 + h) \approx h$, and $\exp h \approx 1 + h$.

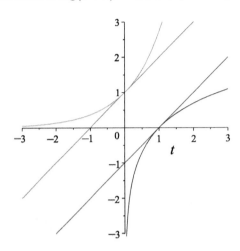

Proof. We have

$$\lim_{h \to 0} \frac{\log(1 + h)}{h} = \lim_{h \to 0} \frac{\log(1 + h) - \log 1}{(1 + h) - 1} = \log' 1 = \frac{1}{1} = 1,$$

$$\lim_{h \to 0} \frac{(\exp h) - 1}{h} = \lim_{h \to 0} \frac{(\exp h) - \exp 0}{h - 0} = \exp' 0 = \exp 0 = 1.$$

$\square$

Corollary 5.26. *For all $n \in \mathbb{N}$,*

(1) $\displaystyle \lim_{x \to \infty} \frac{\log x}{\sqrt[n]{x}} = 0;$

(2) $\displaystyle \lim_{x \to \infty} \frac{x^n}{\exp x} = 0.$

(1) says that any nth roots grows faster than $\log x$. (2) says that $\exp x$ grows faster than any polynomial!

Proof. Let $f := \log x$ and $g := \sqrt[n]{x}$. Then f, g are differentiable, and

$$f' = \frac{1}{x},$$

$$g' = \frac{1}{n} x^{\frac{1}{n} - 1}.$$

So $g, g' > 0$ on $(0, \infty)$. We also have $\displaystyle \lim_{x \to \infty} g(x) = \lim_{x \to \infty} x^{\frac{1}{n}} = \infty$, and

$$\lim_{x \to \infty} \frac{f'(x)}{g'(x)} = \lim_{x \to \infty} \frac{1/x}{\frac{1}{n} x^{\frac{1}{n} - 1}} = \lim_{x \to \infty} \frac{n}{x^{\frac{1}{n}}} = 0.$$

So by the $\frac{\infty}{\infty}$ form of l'Hôpital's Rule, $0 = \lim\limits_{x \to \infty} \dfrac{f(x)}{g(x)} = \lim\limits_{x \to \infty} \dfrac{\log x}{\sqrt[n]{x}}$.

With $g := \exp x$, we have $g' = \exp x$, and so $g, g' > 0$ on $\mathbb{R}$. Also, $\lim\limits_{x \to \infty} g(x) = \infty$.

So a repeated application of the $\frac{\infty}{\infty}$ form of l'Hôpital's Rule gives

$$\lim_{x \to \infty} \frac{x^n}{\exp x} = \lim_{x \to \infty} \frac{n x^{n-1}}{\exp x} = \lim_{x \to \infty} \frac{n \cdot (n-1) x^{n-2}}{\exp x} = \cdots = \lim_{x \to \infty} \frac{n!}{\exp x} = 0. \qquad \square$$

Real powers of positive real numbers

The logarithm and exponential functions allow us to define a^b, where $a > 0$ and $b \in \mathbb{R}$, in a manner which is consistent with whatever notions of powers we have developed so far (and we will show this below).

Definition 5.16 (Real powers of positive reals). If $a > 0$ and $b \in \mathbb{R}$, then we define a^b by $a^b := \exp(b \log a)$.

Recall that, earlier, we had defined a^r for $a > 0$ and $r \in \mathbb{Q}$. Let us first check that our new definition matches with the old one in the special case when b happens to be a rational number r. To this end, define $f : (0, \infty) \to \mathbb{R}$ by $f(x) = (\log x^r) - r \log x$, $x > 0$. Then

$$f'(x) = \frac{1}{x^r} \cdot r x^{r-1} - r \cdot \frac{1}{x} = 0.$$

So f is constant on $(0, \infty)$. Hence

$$(\log x^r) - r \log x = f(x) = f(1) = (\log 1^r) - r \log 1 = 0.$$

Consequently, $\log x^r = r \log x$, that is, $x^r = \exp(r \log x)$. In fact, this motivates the definition above, since the right hand side, namely $\exp(r \log x)$, makes sense for $r \notin \mathbb{Q}$ too!

Let $a > 0$ and consider the function $x \mapsto a^x : \mathbb{R} \to (0, \infty)$. Here are sketches of the graphs of $x \mapsto 2^x$ and $x \mapsto (1/2)^x$.

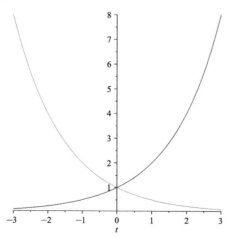

Theorem 5.27. *Let $a > 0$. Then $\dfrac{d}{dx} a^x = (\log a) \cdot a^x$ for all $x \in \mathbb{R}$.*

Proof. We have $a^x = \exp(x \log a)$, and so

$$\frac{d}{dx} a^x = \frac{d}{dx} \exp(x \log a) = \exp(x \log a) \cdot \frac{d}{dx}(x \log a) = a^x \cdot \log a$$

for all $x \in \mathbb{R}$.   □

Theorem 5.28. *$e^x = \exp x$ for all $x \in \mathbb{R}$.*

Proof. If $a = e > 0$, $b = x \in \mathbb{R}$, then $a^b = e^x = \exp(x \log e) = \exp(x \cdot 1) = \exp x$.   □

Based on the above result, from now on, we will write e^x instead of $\exp x$.

We had seen earlier that $e = \lim\limits_{n\to\infty} \left(1 + \dfrac{1}{1!} + \dfrac{1}{2!} + \dfrac{1}{3!} + \cdots + \dfrac{1}{n!}\right)$. We now show:

Theorem 5.29. *If $x \neq 0$, then $e^x = \lim\limits_{n\to\infty} \left(1 + \dfrac{x}{n}\right)^n$. In particular,*

$$e = \lim_{n\to\infty} \left(1 + \frac{1}{n}\right)^n.$$

Proof. We know that $\lim\limits_{h\to 0} \dfrac{\log(1 + h)}{h} = 1$. Thus for $x \neq 0$, we obtain

$$\lim_{n\to\infty} \frac{\log\left(1 + \frac{x}{n}\right)}{\frac{x}{n}} = 1,$$

that is, $\lim\limits_{n\to\infty} n \log\left(1 + \dfrac{x}{n}\right) = x$. As exp is continuous, it follows from here that

$$\lim_{n\to\infty} \exp\left(n \log\left(1 + \frac{x}{n}\right)\right) = \exp x = e^x,$$

that is, $\lim\limits_{n\to\infty} \left(1 + \dfrac{x}{n}\right)^n = e^x$. Putting $x = 1$, we obtain $e = \lim\limits_{n\to\infty} \left(1 + \dfrac{1}{n}\right)^n$.   □

Theorem 5.30. *For all $x \in \mathbb{R}$, $e^x = \lim\limits_{n\to\infty} \left(1 + \dfrac{x}{1!} + \dfrac{x^2}{2!} + \dfrac{x^3}{3!} + \cdots + \dfrac{x^n}{n!}\right)$.*

Proof. Fix $x \in \mathbb{R}$. For $n \in \mathbb{N}$, there exists a $c_{x,n}$ between 0 and x such that

$$e^x = 1 + \frac{x}{1!} + \frac{x^2}{2!} + \frac{x^3}{3!} + \cdots + \frac{x^n}{n!} + \frac{e^{c_{x,n}}}{(n+1)!} x^{n+1}.$$

So

$$\left| e^x - \left(1 + \frac{x}{1!} + \frac{x^2}{2!} + \frac{x^3}{3!} + \cdots + \frac{x^n}{n!}\right) \right| = \frac{e^{c_{x,n}}}{(n+1)!} |x|^{n+1}.$$

But $c_{x,n} \leq |x|$, and so $e^{c_{x,n}} \leq e^{|x|}$. Hence

$$\left| e^x - \left(1 + \frac{x}{1!} + \frac{x^2}{2!} + \frac{x^3}{3!} + \cdots + \frac{x^n}{n!} \right) \right| \leq \frac{e^{|x|}}{(n+1)!} |x|^{n+1} =: a_{n+1}.$$

We have

$$a_{n+1} = \frac{e^{|x|}}{(n+1)!} |x|^{n+1} = \frac{|x|}{n+1} \cdot \frac{e^{|x|}}{n!} |x|^n = \frac{|x|}{n+1} \cdot a_n < a_n,$$

where the last inequality holds for all $n > |x|$. So, the sequence $(a_n)_{n \in \mathbb{N}}$ is eventually decreasing and bounded below by 0. So it is convergent, with limit, say, L. But from the above, we see that

$$L = \lim_{n \to \infty} a_{n+1} = \lim_{n \to \infty} \frac{|x|}{n+1} \cdot a_n = 0 \cdot L = 0.$$

Hence $e^x = \lim_{n \to \infty} \left(1 + \frac{x}{1!} + \frac{x^2}{2!} + \frac{x^3}{3!} + \cdots + \frac{x^n}{n!} \right).$ $\square$

Exercise 5.53. If $a, b > 0$, $b \neq 1$, then we define 'the logarithm of a to the base b', by

$$\log_b a := \frac{\log a}{\log b}.$$

(1) Show that $b^{\log_b a} = a$.

(2) Show that if $a > 0$ and $b, c \in \mathbb{R}$, then $(a^b)^c = a^{(b \cdot c)}$.

(3) Prove that $\log_2 3$ is irrational.

(4) Show that there are irrational numbers a, b such that a^b is rational.

(5) Find the flaw in the following argument given for the claim that $1 > 2$.
'We know that $4 > 2$. As the logarithm is a strictly increasing function, taking $\log_{1/2}$ of both sides, we obtain $-2 = \log_{1/2} 4 > \log_{1/2} 2 = -1$, and so $1 > 2$'.

(6) Sketch the graphs of $x \mapsto \log x$, $\log_{1/2} x$, $\log_{10} x$ in the same picture.

Exercise 5.54. Find the derivative of $f : (0, \infty) \to \mathbb{R}$, where $f(x) = \dfrac{\log x}{x}$, $x > 0$.
Which is bigger: e^π or π^e? (You may use the estimates $e < 3 < \pi$.)

Exercise 5.55 (A potpourri of first-order differential equations).

(1) (Homogeneous linear.) Let a be a continuous function on the open interval I. Let $x_0 \in I$, $y_0 \in \mathbb{R}$. Show that the equation

$$f'(x) = a(x)f(x), \quad x \in I,$$

satisfying the condition

$$f(x_0) = y_0,$$

has the unique solution

$$f(x) = y_0 \exp\left(\int_{x_0}^{x} a(\xi)d\xi\right), \quad x \in I.$$

Hint: For uniqueness, differentiate $\exp\left(-\int_{x_0}^{x} a(\xi)d\xi\right) f(x)$, where f is any solution.
(In particular, when $a(x) \equiv a$ is a constant, then the 'initial value problem'

$$\begin{cases} f'(x) = af(x), \\ f(0) = y_0, \end{cases}$$

has the unique solution given by $f(x) = e^{ax}x_0, x \geq 0$.)

(2) (Inhomogeneous linear.) Let a, b be continuous functions on the open interval I. Show that the equation

$$f'(x) = a(x)f(x) + b(x), \quad x \in I,$$

satisfying the condition

$$f(x_0) = y_0,$$

has the unique solution

$$f(x) = y_0 e^{A(x)} + e^{A(x)} \int_{x_0}^{x} b(\xi)e^{-A(\xi)}d\xi, \quad x \in I,$$

where $A(x) := \int_{x_0}^{x} a(\xi)d\xi, x \in I$.

(3) (Separable.) If p, q are functions on $\mathbb{R}$, then a differential equation of the form

$$y'(x) = \frac{p(x)}{q(y(x))}$$

is called *separable*. One can then formally write

$$q(y)dy = p(x)dx$$

and 'integrate both sides' to solve the differential equation. This nonsense can be justified as follows: Suppose that y is a solution and that Q is a primitive for the function

$$x \mapsto q(y(x)) \cdot y'(x),$$

and P is a primitive for

$$x \mapsto p(x).$$

Then clearly $(P - Q)' = 0$, and so $P - Q$ is a constant, from which one can hope to find y. Carry out this procedure in the special case of the differential equation

$$xy' + y = y^2,$$

assuming that $x > 0$, and that $y(x) > 1$ for all $x > 0$.

Exercise 5.56 (Newton's Law of Cooling). Newton's law of cooling states that an object cools at a rate proportional to the difference of its temperature and the temperature of the surrounding medium. Find the temperature $\Theta(t)$ of an object at time t, in terms of its temperature Θ_0 at time 0, assuming that the temperature of the surrounding medium is kept at a constant, M. What happens as $t \to \infty$?

Exercise 5.57 (Radioactive decay). A radioactive substance diminishes at a rate proportional to the amount present. If $A(t)$ is the amount at time t, this means that $A'(t) = -cA(t)$ for some $c > 0$.

(1) Find $A(t)$ in terms of the amount $A(0) = A_0$ present at time 0.

(2) Show that there is a number τ (the *half-life* of the radioactive element) with the property that $A(t + \tau) = A(t)/2$ for all t.

Exercise 5.58 (Compound interest). An amount P is deposited in a bank that pays an interest at a rate r per year, compounded m times a year. Thus the total amount (of principal plus interest) at the end of n years is

$$A = P\left(1 + \frac{r}{m}\right)^{mn}.$$

If r, n are kept fixed, then show that this amount approaches the limit Pe^{rn} as $m \to \infty$. This motivates the following definition. We say that the money grows at an annual rate r when compounded continuously if the amount $A(t)$ after t years is Pe^{rt} for all $t \geq 0$. Approximately how long does it take for a bank account to double in value if it receives interest at an annual rate of 6% if

(1) compounded continuously?

(2) it is just a simple interest?

Exercise 5.59 (The hyperbolic trigonometric functions). The *hyperbolic sin* and *hyperbolic cos* functions $\sinh, \cosh : \mathbb{R} \to \mathbb{R}$ are defined as

$$\cosh x := \frac{e^x + e^{-x}}{2}, \quad \text{and} \quad \sinh x := \frac{e^x - e^{-x}}{2},$$

for $x \in \mathbb{R}$. (sinh, cosh are pronounced 'shine' and 'cosh', respectively.)

(1) Show that $\sinh 0 = 0$, $\cosh 0 = 1$, and

$$\sinh' x = \cosh x, \quad \cosh' x = \sinh x, \quad x \in \mathbb{R},$$

$$\cosh(x + y) = (\cosh x)(\cosh y) + (\sinh x)(\sinh y), \quad x, y \in \mathbb{R}.$$

(2) Sketch the graphs of sinh and cosh.

(3) Show that for any $t \in \mathbb{R}$, the point $(\cosh t, \sinh t)$ is on the hyperbola $x^2 - y^2 = 1$. Sketch the portion of the hyperbola obtained.

(4) Using Maple/Mathematica/Matlab or some other suitable computer package, sketch the graph of the *hyperbolic tan* function tanh defined by

$$\tanh x = \frac{\sinh x}{\cosh x}, \quad x \in \mathbb{R}.$$

(tanh is pronounced 'than'.)

Exercise 5.60. Show that $\lim_{x \to 0+} x^x = 1$. Conclude that $\lim_{n \to \infty} n^{1/n} = 1$.

Exercise 5.61. Show that the improper integral $\int_0^1 \frac{1}{x^x} \, dx$ exists.

Note that for $x > 0$, $\dfrac{1}{x^x} = e^{-x \log x} = \displaystyle\sum_{n=0}^{\infty} \frac{(-x \log x)^n}{n!}$.

Based on this, try giving a formal argument to justify the identity $\displaystyle\int_0^1 \frac{1}{x^x} \, dx = \sum_{n=1}^{\infty} \frac{1}{n^n}$.

Exercise 5.62. Find $\displaystyle\lim_{n \to \infty} \left(\left(1 + \frac{1}{n}\right)\left(1 + \frac{2}{n}\right) \cdots \left(1 + \frac{n}{n}\right) \right)^{\frac{1}{n}}$.

Exercise 5.63. Order the following functions from the fastest growing to the slowest:

$$2^x, \ e^x, \ x^x, \ (\log x)^x, \ e^{x/2}, \ x^{1/2}, \ \log_2 x, \ \log(\log x), \ (\log x)^2, \ x^e, \ x^2, \ \log x, \ (2x)^x, \ x^{2x}.$$

Exercise 5.64. Suppose that we know that the value of $\log_{10} 2 = 0.3010$ (to four decimal places). Using this find the number of digits in 2^{399}.

Exercise 5.65. Evaluate $\displaystyle\int_3^4 \frac{x^2 + 3x + 9}{(x+1)(x-2)^2} \, dx$.

Hint: See Exercise 3.51.

Exercise 5.66. (∗) Solve for positive real x, y:

$$3^x - 2^y = 23,$$

$$\log_3 x + \log_y 2 = 2.$$

Hint: First try to guess a solution. Using the fact that $x \mapsto 3^x, \log_3 x$ are strictly increasing, show that this solution is the only one.

Exercise 5.67. (∗) Consider the sequence $(a_n)_{n \in \mathbb{N}}$ whose terms are generated recursively as follows:

$$a_1 := 1,$$

$$a_n = n(1 + a_{n-1}), \quad n \ge 2.$$

Prove that $\displaystyle\lim_{n \to \infty} \left(1 + \frac{1}{a_1}\right)\left(1 + \frac{1}{a_2}\right)\left(1 + \frac{1}{a_3}\right) \cdots \left(1 + \frac{1}{a_n}\right) = e.$

Trigonometric functions

We begin by introducing the notion of *angle in radians*.

Definition 5.17 (Angle made with the positive real axis of a point in the upper half plane on the unit circle $\mathbb{T}$). Consider the unit circle[9] $\mathbb{T}$ with center $(0,1)$ and radius 1:

$$\mathbb{T} = \{(x,y) \in \mathbb{R}^2 : x^2 + y^2 = 1\}.$$

Let $A \equiv (1,0)$ and $O \equiv (0,0)$. If P is a point in $\mathbb{T}$, and P is in the upper half plane, then we define the *angle AOP in radians*, denoted by $\angle AOP$, to be twice the area of the shaded sector shown in Figure 5.19.

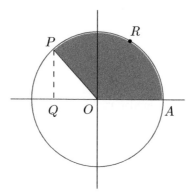

Figure 5.19   Definition of angle in radians.

Thus if P is the point $(-1, 0)$, then the angle AOP is

$$2 \int_{-1}^{1} \sqrt{1 - x^2} dx = \pi \text{ radians.}$$

Let us find an expression for $\angle AOP$ in terms of the x coordinate of the point P. If P is the point (x, y), then let us find the area of the shaded sector. We have the two cases $x \geq 0$ and $x \leq 0$ depicted below.

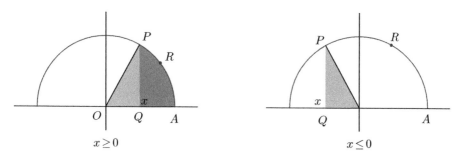

$$x \geq 0 \qquad\qquad\qquad x \leq 0$$

---

[9] The peculiar notation $\mathbb{T}$ is used, since it is the '(one-dimensional) *torus*'. A two-dimensional torus $\mathbb{T}^2$ is the surface of a doughnut.

In the case when we have $x \geq 0$, the angle $\Theta(x)$ of the point P having the coordinates $(x, y) = (x, \sqrt{1 - x^2})$ is

$$\Theta(x) = 2 \left(\text{area of the sector } AOP \right)$$

$$= 2 \left(\left(\text{area of the triangle } \Delta POQ \right) + \left(\text{area under the circular arc } \overset{\frown}{PRA} \right) \right)$$

$$= 2 \left(\frac{1}{2} x \sqrt{1 - x^2} + \int_x^1 \sqrt{1 - \xi^2} d\xi \right).$$

On the other hand, when we have $x \leq 0$, then the angle $\Theta(x)$ of the point P having coordinates $(x, y) = (x, \sqrt{1 - x^2})$ is

$$\Theta(x) = 2 \left(\text{area of the sector } OAP \right)$$

$$= 2 \left(\left(\text{area under the circular arc } \overset{\frown}{PRA} \right) - \left(\text{area of the triangle } \Delta POQ \right) \right)$$

$$= 2 \left(\int_x^1 \sqrt{1 - \xi^2} d\xi - \frac{1}{2} (-x) \sqrt{1 - x^2} \right)$$

$$= 2 \left(\frac{1}{2} x \sqrt{1 - x^2} + \int_x^1 \sqrt{1 - \xi^2} d\xi \right).$$

Thus for *all* $x \in [-1, 1]$, the angle of $P = (x, \sqrt{1 - x^2})$ in radians is given by

$$\Theta(x) = 2 \left(\frac{1}{2} x \sqrt{1 - x^2} + \int_x^1 \sqrt{1 - \xi^2} d\xi \right).$$

Clearly, $\Theta : [-1, 1] \to \mathbb{R}$ is continuous, since $x \mapsto x\sqrt{1 - x^2}$ is continuous and by the Fundamental Theorem of Calculus, since the integrand $\sqrt{1 - \xi^2}$ is continuous, the map

$$x \mapsto \int_x^1 \sqrt{1 - \xi^2} d\xi$$

is differentiable on $[-1, 1]$, and in particular, continuous there. The picture below shows a plot of the graph of Θ.

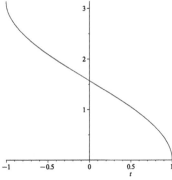

Also, for $x \in (-1, 1)$,

$$\Theta'(x) = 2\left(\frac{1}{2}\sqrt{1 - x^2} + \frac{1}{2}x \cdot \frac{(-2x)}{2\sqrt{1 - x^2}} - \sqrt{1 - x^2}\right)$$

$$= \left(-\frac{\sqrt{1 - x^2}}{2} - \frac{x^2}{2\sqrt{1 - x^2}}\right) = -\frac{1}{\sqrt{1 - x^2}} < 0.$$

Hence Θ is strictly decreasing in $(-1, 1)$, and thus also on $[-1, 1]$, thanks to its continuity on $[-1, 1]$. (See Exercise 3.8.) So the map $\Theta : [-1, 1] \rightarrow [0, \pi]$ is a strictly decreasing, continuous function on $[-1, 1]$. $\Theta(x)$ decreases from $\pi = \Theta(-1)$ to $0 = \Theta(1)$ as x increases from -1 to 1. By the Intermediate Value Theorem Θ assumes all values between π and 0. Consequently, $\Theta : [-1, 1] \rightarrow [0, \pi]$ is one-to-one and onto. In other words, it is a bijection. Hence the function Θ has an inverse $\Theta^{-1} : [0, \pi] \rightarrow [-1, 1]$, and this is the cosine function! So given any angle $\alpha \in [0, \pi]$, there is a unique $x \in [-1, 1]$ such that $\Theta(x) = \alpha$. This unique $x = \Theta^{-1}(\alpha)$ is called the cosine of the angle α.

Definition 5.18 (Cosine and sine of an angle). For $\alpha \in [0, \pi]$, we define

    (1) the *cosine* of α, denoted by $\cos \alpha$, by $\cos \alpha = \Theta^{-1}(\alpha)$.

    (2) the *sine* of α, denoted by $\sin \alpha$, by $\sin \alpha = \sqrt{1 - (\Theta^{-1}(\alpha))^2}$.

The reason we have given such precise definitions is because we want to derive all the familiar properties of the trigonometric functions from scratch. So right now these definitions might seem cumbersome, but once we have done what we want to do with them (that is prove all their familiar properties), the primary purpose of the above unwieldy definitions would have been served! In any case, things are not as complicated as they seem. The way to think about the above definitions is like this: if we choose any angle α in $[0, \pi]$, then that decides a sector of the unit circle with the positive x-axis, as shown below, whose area is half that of the angle, namely $\alpha/2$. This sector then determines a point P on the unit circle, which has an x coordinate, which is precisely $\Theta^{-1}(\alpha) = \cos \alpha$, and the y-coordinate of P is given by $\sqrt{1 - x^2} = \sqrt{1 - (\Theta^{-1}(\alpha))^2} = \sin \alpha$.

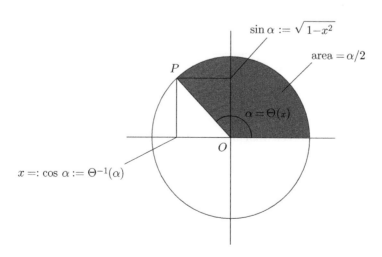

The graphs of sin, cos $: [0, \pi] \to [-1, 1]$ are shown below.

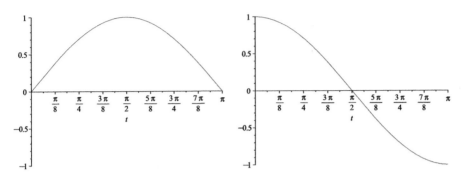

Note that since Θ is strictly decreasing and continuous, it follows that so is cos on $[0, \pi]$. Also, since

$$\sin \alpha = \sqrt{1 - (\cos \alpha)^2}, \quad \alpha \in [0, \pi],$$

the function sin is also continuous on $[0, \pi]$.

Next, we extend the definitions of sin and cos from $[0, \pi]$ to $[0, 2\pi]$ as follows.

Definition 5.19 (Cosine and sine of an angle (continued)). For $\alpha \in [\pi, 2\pi]$ (so that we have $\alpha - \pi \in [0, \pi]$), we define

(1) the *cosine* of α, denoted by $\cos \alpha$, by $\cos \alpha = -\cos(\alpha - \pi)$.

(2) the *sine* of α, denoted by $\sin \alpha$, by $\sin \alpha = -\sin(\alpha - \pi)$.

Since $\sin \pi = 0 = \sin 0$, from the above definition and the continuity of sin on $[0, \pi]$, it follows that sin is continuous on $[0, 2\pi]$. Similarly cos is continuous on $[0, 2\pi]$.

The graphs of sin, cos $: [0, 2\pi] \to [-1, 1]$ are shown below.

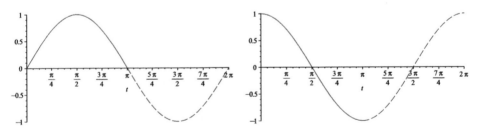

And finally, we extend sin, cos from $[0, 2\pi]$ to all of $\mathbb{R}$ by demanding 2π-periodicity.

Definition 5.20 (Cosine and sine of an angle (continued)). For $\alpha \in \mathbb{R} \setminus [0, 2\pi]$, we define

(1) the *cosine* of α, denoted by $\cos \alpha$, by $\cos \alpha = \cos \left(\alpha - 2\pi \left\lfloor \dfrac{\alpha}{2\pi} \right\rfloor \right)$.

(2) the *sine* of α, denoted by $\sin \alpha$, by $\sin \alpha = \sin \left(\alpha - 2\pi \left\lfloor \dfrac{\alpha}{2\pi} \right\rfloor \right)$.

Since $\sin 2\pi = 0 = \sin 0$, from the above definition and the continuity of $\sin$ on $[0, 2\pi]$, it follows that $\sin$ is continuous on $\mathbb{R}$. Similarly, $\cos$ is continuous on $\mathbb{R}$.

The picture below shows the graphs of $\sin : \mathbb{R} \to [-1, 1]$ (top) and $\cos : \mathbb{R} \to [-1, 1]$ (bottom).

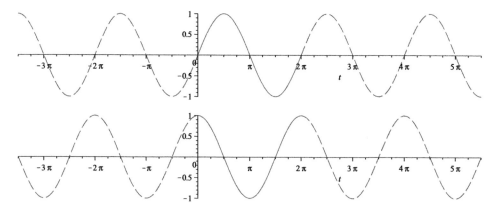

Let us now turn to the properties of $\sin$ and $\cos$. We already know from the definition that $\sin x = \sin(x + 2\pi)$, and $\cos x = \cos(x + 2\pi)$ for all $x \in \mathbb{R}$. Moreover, since we have $(\sin x)^2 + (\cos x)^2 = 1$ for $x \in [0, \pi]$, it follows that for $x \in [\pi, 2\pi]$,

$$(\sin x)^2 + (\cos x)^2 = (-\sin(x - \pi))^2 + (-\cos(x - \pi))^2$$
$$= (\sin(x - \pi))^2 + (\cos(x - \pi))^2$$
$$= 1.$$

Finally, by the 2π-periodicity, we see that $(\sin x)^2 + (\cos x)^2 = 1$ for *all* $x \in \mathbb{R}$.

Theorem 5.31. *For all* $x \in \mathbb{R}$, $\cos'x = -\sin x$ *and* $\sin'x = \cos x$.

Proof. We prove this result by considering the following cases:

$\underline{1°}$ $\alpha \in (0, \pi)$. Let x_α be such that $\Theta(x_\alpha) = \alpha$. Then by the Differentiable Inverse Theorem,

$$\cos'\alpha = (\Theta^{-1})'(\alpha) = \frac{1}{\Theta'(x_\alpha)} = \frac{1}{-\frac{1}{\sqrt{1-x_\alpha^2}}} = -\sqrt{1 - x_\alpha^2} = -\sin\alpha.$$

Also, $\sin'\alpha = \left(\sqrt{1 - (\cos\alpha)^2}\right)' = \dfrac{2(\cos\alpha)(\sin\alpha)}{2\sqrt{1 - (\cos\alpha)^2}} = \cos\alpha.$

$\underline{2°}$ $\alpha \in (\pi, 2\pi)$. Then by the chain rule,

$$\cos'\alpha = (-\cos(\alpha - \pi))' = \left(\sin(\alpha - \pi)\right) \cdot 1 = -\sin\alpha,$$
$$\sin'\alpha = (-\sin(\alpha - \pi))' = -\left(\cos(\alpha - \pi)\right) \cdot 1 = \cos\alpha.$$

<u>3°</u> $\alpha \in \mathbb{R}\backslash\pi\mathbb{Z}$. Then there exists an $n \in \mathbb{Z}$ such that $\alpha - n \cdot 2\pi \in (0, 2\pi)$. Thus by the 2π-periodicity of sin and cos,

$$\cos'\alpha = (\cos(\alpha - n \cdot 2\pi))' = -\left(\sin(\alpha - n \cdot 2\pi)\right) \cdot 1 = -\sin\alpha,$$

$$\sin'\alpha = (\sin(\alpha - n \cdot 2\pi))' = \left(\cos(\alpha - n \cdot 2\pi)\right) \cdot 1 = \cos\alpha.$$

<u>4°</u> Finally, suppose that $\alpha \in \pi\mathbb{Z}$. We will use Exercise 4.48. Recall that over there, we showed that if $c \in (a, b)$, and $f : (a, b) \to \mathbb{R}$ is such that f is
(1) differentiable on $(a, b)\backslash\{c\}$,
(2) continuous on (a, b), and
(3) $\lim_{x \to c} f'(x)$ exists,
then f is differentiable at c and $f'(c) = \lim_{x \to c} f'(x)$.

We apply this result with $c = n\pi$ with $n \in \mathbb{Z}$. Since we know that sin is differentiable on $(n\pi - \pi, n\pi + \pi)\backslash\{n\pi\}$, sin is continuous everywhere, and

$$\lim_{x \to n\pi} \sin'x = \lim_{x \to n\pi} \cos x = \cos(n\pi),$$

it follows from the above result that sin is differentiable at $n\pi$, and moreover,

$$\sin'(n\pi) = \lim_{x \to n\pi} \sin'x = \cos(n\pi).$$

Similarly, cos is differentiable at $n\pi$ and $\cos'(n\pi) = -\sin(n\pi)$.
This completes the proof. □

Corollary 5.32. $\lim_{x \to 0} \dfrac{\sin x}{x} = 1$.

Proof. We have $\lim_{x \to 0} \dfrac{\sin x}{x} = \lim_{x \to 0} \dfrac{\sin x - \sin 0}{x - 0} = \sin'0 = \cos 0 = 1$. □

Theorem 5.33. $(*)$ $\pi \notin \mathbb{Q}$.

Proof. [10] Suppose that $\pi \in \mathbb{Q}$, and let $\pi = \dfrac{n}{d}$, where $n, d \in \mathbb{N}$. Set

$$p(x) := \frac{x^m(n - d \cdot x)^m}{m!},$$

$$P(x) := p(x) - p^{(2)}(x) + p^{(4)}(x) - + \cdots + (-1)^m p^{(2m)}(x),$$

where we will specify the natural number m later on.

We will now show that p and all its derivatives evaluated at 0 are integers. To see this, we proceed as follows. Note that p only has terms of degree at least m and at most $2m$, and so

$$p(x) = \frac{x^m(n - d \cdot x)^m}{m!} = c_m x^m + c_{m+1}x^{m+1} + \cdots + c_{2m}x^{2m},$$

---
[10] based on [N5].

for some $c_m, \cdots, c_{2m}$. Since $m!p = x^m(n - d \cdot x)^m$ has integer coefficients, it follows that $m!c_m, m!c_{m+1}, \cdots, m!c_{2m} \in \mathbb{Z}$. Now $p^{(j)}(0) = 0$, for all $j = 0, 1, \cdots, m - 1$. Also, $p^{(j)}(0) = 0$ if $j > 2m$. What if $j = m, \cdots, 2m$? If $j = m$, then

$$p^{(j)}(x) = m!c_m + \text{ terms containing } x,$$

and so $p^{(j)}(0) = m!c_m \in \mathbb{Z}$ if $j = m$. Similarly, it can be seen that

$$p^{(j)}(0) = j!c_j, \quad j = m, \cdots, 2m.$$

Hence for $j = m, \cdots, 2m$, $p^{(j)}(0) = \underbrace{c_j m!}_{\in \mathbb{Z}} \cdot \left(\underbrace{(m+1) \cdot \ldots \cdot j}_{\text{an integer}} \right) \in \mathbb{Z}.$

So we conclude that $p^{(j)}(0)$ is an integer for all $j \geq 0$.

It also follows from the above that p and all its derivatives $p^{(j)}, j \geq 1$, have integer values for $x = \pi = n/d$, thanks to the (easily verified) relation

$$p(x) = p\left(\frac{n}{d} - x\right).$$

We have

$$\frac{d}{dx}\left(P'(x)\sin x - P(x)\cos x\right)$$

$$= P''(x)\sin x + P'(x)\cos x - P'(x)\cos x + P(x)\sin x = (P''(x) + P(x))\sin x$$

$$= \Big(\quad p^{(2)}(x) - p^{(4)}(x) + \cdots + (-1)^{m-1}p^{(2m)}(x) + (-1)^m p^{(2m+2)}(x)$$

$$+ p(x) - p^{(2)}(x) + p^{(4)}(x) - \cdots + (-1)^m p^{(2m)}(x) \Big)\sin x$$

$$= p(x)\sin x \ \ (\text{using } p^{(2m+2)}(x) \equiv 0),$$

and

$$\int_0^\pi p(x)\sin x \, dx = \left(P'(x)\sin x - P(x)\cos x\right)\Big|_0^\pi = P(\pi) + P(0). \qquad (5.13)$$

Now $P(\pi) + P(0)$ is a positive integer: integer because $p^{(j)}(\pi)$ and $p^{(j)}(0)$ are integers, and positive because the integrand is continuous on $[0, \pi]$ and positive on $(0, \pi)$. But we have for $0 < x < \pi = n/d$, that $0 < n - d \cdot x < n$ and

$$0 < p(x)\sin x \leq p(x) \cdot 1 = \frac{x^m(n - d \cdot x)^m}{m!} < \frac{x^m n^m}{m!} < \frac{\pi^m n^m}{m!} = \frac{(\pi n)^m}{m!}.$$

But for $m > \pi n$, we have

$$\frac{(\pi n)^m}{m!} = \frac{\pi n}{1} \cdot \frac{\pi n}{2} \cdots \frac{\pi n}{\lfloor \pi n \rfloor} \cdot \frac{\pi n}{\lfloor \pi n \rfloor + 1} \cdots \frac{\pi n}{m} \leq \frac{(\pi n)^{\lfloor \pi n \rfloor}}{(\lfloor \pi n \rfloor)!} \cdot 1 \cdots 1 \cdot \frac{\pi n}{m} \xrightarrow{m \to \infty} 0.$$

So we arrive at the contradiction that on the one hand, the integral in (5.13) is a positive integer, but the above inequality shows that this integral can be made arbitrarily small by taking m sufficiently large. Consequently π is irrational!   □

In Exercise 4.56, we had shown the following 'trigonometric addition formulae': for all $\alpha, \beta \in \mathbb{R}$,

$$\sin(\alpha + \beta) = (\sin \alpha)(\cos \beta) + (\cos \alpha)(\sin \beta), \quad \text{and}$$
$$\cos(\alpha + \beta) = (\cos \alpha)(\cos \beta) - (\sin \alpha)(\sin \beta).$$

Theorem 5.34. *For all* $\alpha \in \mathbb{R}$, $\cos \alpha = \cos(-\alpha)$, *and* $\sin(-\alpha) = -\sin \alpha$.

Proof. First, we note that for $x \in [-1, 1]$,

$$\Theta(x) = 2\left(\frac{1}{2}x\sqrt{1-x^2} + \int_x^1 \sqrt{1-\xi^2}d\xi\right). \tag{5.14}$$

Thus

$$\Theta(-x) = 2\left(-\frac{1}{2}x\sqrt{1-x^2} + \int_{-x}^1 \sqrt{1-\xi^2}d\xi\right)$$

$$= 2\left(-\frac{1}{2}x\sqrt{1-x^2} + \int_x^{-1} \sqrt{1-u^2}(-1)du\right) \quad \text{(substituting } u = -\xi)$$

$$= 2\left(-\frac{1}{2}x\sqrt{1-x^2} + \int_{-1}^x \sqrt{1-\xi^2}d\xi\right). \tag{5.15}$$

Adding (5.14) and (5.15), we obtain

$$\Theta(x) + \Theta(-x) = 2\left(\int_{-1}^x \sqrt{1-\xi^2}du + \int_x^1 \sqrt{1-\xi^2}d\xi\right) = 2\int_{-1}^1 \sqrt{1-\xi^2}d\xi = \pi.$$

So if $\alpha \in [0, \pi]$ is such that $\cos \alpha = x$, then

$$-x = \cos(\Theta(-x)) = \cos(\pi - \Theta(x)) = \cos(\pi - \alpha).$$

Consequently, for $\alpha \in [0, \pi]$, we have that $\cos \alpha = -\cos(\pi - \alpha)$. Hence for $\alpha \in [0, \pi]$, $\cos(-\alpha) = \cos(2\pi - \alpha) = \cos((\pi - \alpha) + \pi) = -\cos(\pi - \alpha) = -(-\cos \alpha) = \cos \alpha$. Also, for $\alpha \in [\pi, 2\pi]$,

$$\cos(-\alpha) = \cos(2\pi - \alpha) = \cos(\pi - (\alpha - \pi)) = -\cos(\alpha - \pi) = -(-\cos \alpha) = \cos \alpha.$$

Thus $\cos \alpha = \cos(-\alpha)$ for all $\alpha \in [0, 2\pi]$. For general $\alpha \in \mathbb{R}$, we first write $\alpha = \theta + n \cdot 2\pi$ for some $\theta \in [0, 2\pi)$ and some integer n, so that

$$\cos \alpha = \cos(\theta + n \cdot 2\pi) = \cos \theta = \cos(-\theta) = \cos(-\theta - n \cdot 2\pi) = \cos(-\alpha).$$

Finally as the derivative of a differentiable even function is odd (see Exercise 4.3), and since $\sin$ is the derivative of the even function $-\cos$, it follows that $\sin$ is odd. This completes the proof.

(An alternative, slicker proof can be given based along the same lines as in Exercise 4.56, as follows. First we note that the function $x \overset{y}{\mapsto} \cos(-x)$ satisfies the differential equation

$y'' + y = 0$. Indeed, we have

$$y'(x) = \frac{d}{dx}\cos(-x) = (-\sin(-x)) \cdot (-1) = \sin(-x),$$

$$y''(x) = \frac{d}{dx}\sin(-x) = (\cos(-x)) \cdot (-1) = -\cos(-x) = -y(x),$$

and so $y'' + y = 0$. Also, we have

$$y(0) = \cos(-0) = \cos 0 = 1,$$
$$y'(0) = \sin(-0) = \sin 0 = 0.$$

Thus by part (2) of Exercise 4.56, it follows, using the facts above, that

$$\cos(-x) = y(x) = y(0)\cos x + y'(0)\sin x = 1 \cdot \cos x + 0 \cdot \sin x = \cos x$$

for all $x \in \mathbb{R}$. Differentiating both sides with respect to x, we also get

$$(-\sin(-x)) \cdot (-1) = -\sin x,$$

that is, $\sin(-x) = -\sin x$ for all $x \in \mathbb{R}$.)   □

tan, cot, sec, cosec

The other four auxiliary trigonometric functions are defined in terms of sin and cos by the usual formulae

$$\tan x := \frac{\sin x}{\cos x},$$

$$\cot x := \frac{\cos x}{\sin x},$$

$$\sec x := \frac{1}{\cos x},$$

$$\csc x := \frac{1}{\sin x},$$

for those real x where the denominator in the respective expression is not zero, and the functions are called the *tangent, cotangent, secant*, and the *cosecant* of x, respectively. Thus these functions are defined for all real x except for certain isolated points. Since sin and cos are 2π-periodic, each of the above also inherit the periodicity property $f(x + 2\pi) = f(x)$ for all $x, x + 2\pi$ in the respective domains of definitions. tan and cot in fact have a smaller period π.

All other standard results about the auxiliary trigonometric functions can now be derived from the basic properties of sin and cos. For example, if $x \in (-\pi/2, \pi/2)$, then we have by the quotient rule that

$$\tan' x = \frac{d}{dx}\left(\frac{\sin x}{\cos x}\right) = \frac{(\cos x)(\cos x) - (\sin x)(-\sin x)}{(\cos x)^2} = \frac{1}{(\cos x)^2} > 0.$$

Hence $\tan : (-\pi/2, \pi/2) \to \mathbb{R}$ is strictly increasing, and continuous. Its graph is displayed below.

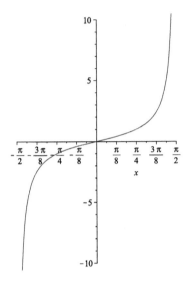

Also,

$$\lim_{x \nearrow \frac{\pi}{2}} \tan x = \infty$$

(as $\sin(\pi/2) = 1$ and $\cos x > 0 = \cos(\pi/2)$ for $0 < x < \pi/2$), and

$$\lim_{x \searrow -\frac{\pi}{2}} \tan x = -\infty$$

(as $\sin(-\pi/2) = -1$ and $\cos x > \cos(-\pi/2) = 0$ for $-\pi/2 < x < 0$). So, besides being injective, $\tan : (-\pi/2, \pi/2) \to \mathbb{R}$ is surjective as well.

Hence $\tan : (-\pi/2, \pi/2) \to \mathbb{R}$ is a bijection, and so it possesses an inverse function, $\tan^{-1} : \mathbb{R} \to (-\pi/2, \pi/2)$, which is called the *arctangent* function, sometimes denoted by arctan.

By the Differentiable Inverse Theorem, for $y = \tan x \in \mathbb{R}$, $x \in (-\pi/2, \pi/2)$,

$$(\tan^{-1})'(y) = (\tan^{-1})'(\tan x) = \frac{1}{\tan' x} = \frac{1}{\dfrac{1}{(\cos x)^2}} = \frac{1}{\dfrac{(\cos x)^2 + (\sin x)^2}{(\cos x)^2}}$$

$$= \frac{1}{1 + (\tan x)^2} = \frac{1}{1 + y^2}.$$

Exercise 5.68. Evaluate the following limits:

(1) $\displaystyle\lim_{x \to 0} \frac{3^{\sin x} - 1}{x}$.

(2) $\displaystyle\lim_{x\to 0} \frac{\sin x - x + x^3/6}{x^3}$.

(3) $\displaystyle\lim_{x\to 0} \frac{\cos x - 1 + x^2/2}{x^4}$.

(4) $\displaystyle\lim_{x\to 0} \left(\frac{1}{x} - \frac{1}{\sin x} \right)$.

Exercise 5.69. Evaluate $\displaystyle\int_{-\frac{1}{2}}^{\frac{1}{2}} (\cos x) \cdot \log \left(\frac{1-x}{1+x} \right) dx$.

Exercise 5.70. Prove that if $m, n \in \mathbb{N}$, then

$$\int_{-\pi}^{\pi} (\sin(mx))(\sin(nx))dx = \left\{ \begin{array}{ll} 0 & \text{if } m \neq n, \\ \pi & \text{if } m = n \end{array} \right\} = \int_{-\pi}^{\pi} (\cos(mx))(\cos(nx))dx,$$

$$\int_{-\pi}^{\pi} (\cos(mx))(\sin(nx))dx = 0.$$

Exercise 5.71 (Fourier series). Let $n \in \mathbb{N}$ and $a_0, a_1, \cdots, a_n$ and $b_1, \cdots, b_n$ be real numbers. Consider the function $f : \mathbb{R} \to \mathbb{R}$ defined by

$$f(x) := a_0 + \sum_{k=1}^{n} (a_k \cos(kx) + b_k \sin(kx)), \quad x \in \mathbb{R}.$$

(1) Show that f is 2π-periodic, that is, $f(x + 2\pi) = f(x)$ for all $x \in \mathbb{R}$.

(2) Prove that $a_0 = \dfrac{1}{2\pi} \displaystyle\int_{-\pi}^{\pi} f(x)dx$, and for $k = 1, \cdots, n$,

$$a_k = \frac{1}{\pi} \int_{-\pi}^{\pi} f(x) \cos(kx)dx, \quad \text{and}$$

$$b_k = \frac{1}{\pi} \int_{-\pi}^{\pi} f(x) \sin(kx)dx.$$

(3) Let $a_k := 0$ for $0 \leq k \leq n$, and

$$b_k := \left\{ \begin{array}{ll} \dfrac{4}{k\pi} & \text{if } k \text{ is even,} \\ 0 & \text{otherwise.} \end{array} \right.$$

Take $n = 3$, $n = 33$ and $n = 333$, successively, and in each case, plot the resulting f using a package such as Maple, Mathematica or Matlab on the computer. What do you observe?

Exercise 5.72 (Fixed points of sin and cos). (∗)

(1) Show that 0 is the only fixed point of $\sin : \mathbb{R} \to \mathbb{R}$.

(2) Prove that $\cos : \mathbb{R} \to \mathbb{R}$ has a unique fixed point, which we denote by c_*.

(3) Determine experimentally the value of c_* as follows. In your scientific calculator, enter any number and press the cos key repeatedly. After a while, the display stabilises. Explain by means of a picture, why this displayed value must be c_* (approximately).

Exercise 5.73 (Addition formula for tan).

(1) Show that for real x, y such that $x, y, x + y$ are not in $\pi\mathbb{Z} + \dfrac{\pi}{2}$, there holds that

$$\tan(x + y) = \frac{\tan x + \tan y}{1 - (\tan x)(\tan y)}.$$

(2) Prove that $\tan 1° \notin \mathbb{Q}$. (Here one *degree*, $1°$, is the angle measuring $\dfrac{\pi}{180}$ radians.)

Exercise 5.74. (∗)

(1) Show that $\displaystyle\lim_{m \to \infty} \lim_{n \to \infty} (\cos(2\pi m! x))^n = \begin{cases} 1 & \text{if } x \in \mathbb{Q}, \\ 0 & \text{if } x \notin \mathbb{Q}. \end{cases}$

(2) Using the fact that $e = \displaystyle\sum_{n=0}^{\infty} \frac{1}{n!}$ and the above result show that $e \notin \mathbb{Q}$.

Exercise 5.75 (Visual differentiation of tan). We have learnt that

$$\tan' x = \frac{1}{(\cos x)^2} = 1 + (\tan x)^2. \tag{5.16}$$

Here we offer a formal[11] geometric explanation. Consider the triangle shown below.

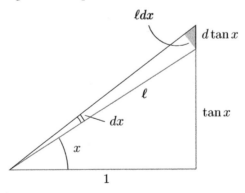

___

[11] The word 'formal' is often used in Mathematics in order to describe a situation where attention is focussed on the 'form/structure' as opposed to strict mathematical rigour. So the usage of 'formal' here is *not* intended to be the opposite of 'informal', and hence rigorous!

If we increase x by a small amount dx, then $\tan x$ increases by the amount $d\tan x$ shown in the figure. Show that (5.16) holds by imagining that in the limiting case when dx diminishes to 0, the little shaded triangle is 'ultimately' similar to the big one wih side lengths $1, \ell, \tan x$.

Exercise 5.76. Find $\displaystyle\int_0^1 \frac{1}{1+t^2}\,dt$.

Exercise 5.77. (∗) Show that $\displaystyle\int_0^{\frac{\pi}{2}} \frac{1}{1+(\tan x)^{\sqrt 3}}\,dx = \frac{\pi}{4}$.

(The integrand is taken as 1 when $x = 0$, and 0 when $x = \dfrac{\pi}{2}$.)

Hint: Consider the symmetry in the graph of $\dfrac{1}{1+(\tan x)^{\sqrt 3}}$ about $\left(\dfrac{\pi}{4}, \dfrac{1}{2}\right)$.

Exercise 5.78. For $x \in (-1,1)$, the inverse $\sin^{-1} : (-1,1) \to (-\pi/2, \pi/2)$ of the strictly increasing continuous function $\sin : (-\pi/2, \pi/2) \to (-1,1)$ exists.
For $y \in (-1,1)$, prove that
$$(\sin^{-1})'(y) = \frac{1}{\sqrt{1-y^2}},$$
and that
$$\int_0^y \frac{1}{\sqrt{1-t^2}}\,dt = \sin^{-1}y,$$
$$\int_0^y \sqrt{1-t^2}\,dt = \frac{y\sqrt{1-y^2} + \sin^{-1}y}{2}.$$

Exercise 5.79.

(1) Using the fact that $\cos t \le 1$ for all real t, and 'integrating both sides' from 0 to x, show that for all nonnegative real x, $\sin x \le x$.

(2) (∗) If x is a real number, then show that $\cos(\sin x) \ge \sin(\cos x)$.
Hint: First, let $x \in [0, \pi/2]$, and show that $\cos(\sin x) \ge |\cos x| \ge |\sin(\cos x)|$. Next extend this to $[-\pi/2, \pi/2]$, and finally to $\mathbb{R}$ using π-periodicity of each of the functions in this chain of inequalities.

Polar coordinates

For any point P with coordinates (x, y) on the unit circle $\mathbb{T}$ for which $y \ge 0$ (that is P is in the *upper* half of the Cartesian plane), we have a well-defined notion of the angle $\Theta(x)$ the point P makes with the positive real axis.

Definition 5.21 (Angle made with the positive real axis of a point in the lower half plane on the unit circle $\mathbb{T}$). For a point P with coordinates (x, y) on the unit circle $\mathbb{T}$ in the *lower* half of the Cartesian plane (that is, $y < 0$), we define the *angle that P makes with the positive real axis* as $-\Theta(x)$.

For example, the angle of $P = (0, -1)$ made with the positive real axis is $-\pi/2$ radians. Note that the angle that $(-1, 0)$ made with the positive real axis is defined to be π radians (and not $-\pi$ radians). Thus $(-1, 0)$ (by our convention) is taken to be a point in the *upper* (and not *lower*) half plane.

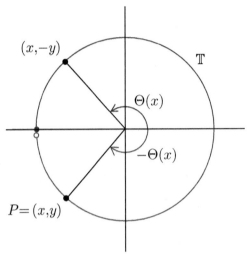

So any point *on the unit circle* can be specified by the angle that it makes with the positive real axis. What if the point does *not* lie on the unit circle $\mathbb{T}$? Well, if the point $P = (x, y)$ is not the origin, then we can draw a ray $\overrightarrow{OP}$ emanating from the origin $O = (0, 0)$, which passes through (x, y). This ray $\overrightarrow{OP}$ intersects the circle at some point P', with coordinates, say (x', y'), and we define the angle that P makes with the positive real axis as the angle that P' makes with the positive real axis (that is, it is $\Theta(x')$ if $y \geq 0$, $-\Theta(x')$ if $y < 0$). Before we describe this as a definition, let us look at the picture shown below.

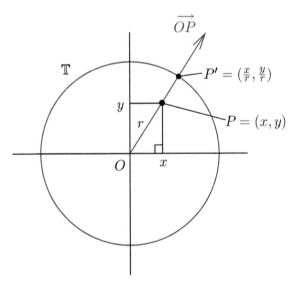

Let $r := \sqrt{x^2 + y^2}$. We note that $r \geq 0$ and in fact, $r \neq 0$ since (x, y) is assumed to be not the origin $O = (0, 0)$. So we know that $r > 0$. Then since $x^2 + y^2 = r^2$, we have

$$\left(\frac{x}{r}\right)^2 + \left(\frac{y}{r}\right)^2 = 1,$$

and so $\left(\frac{x}{r}, \frac{y}{r}\right)$ lies on the unit circle $\mathbb{T}$. Also,

$$O = (0, 0), \quad P = (x, y) \quad \text{and} \quad \left(\frac{x}{r}, \frac{y}{r}\right)$$

lie on a straight line, and since $r > 0$, we see that the ray $\overrightarrow{OP}$ intersects $\mathbb{T}$ at

$$P' := (x', y') := \left(\frac{x}{r}, \frac{y}{r}\right).$$

Hence the angle θ of P made with the positive real axis is

$$\theta = \begin{cases} \Theta\left(\dfrac{x}{r}\right) & \text{if } y \geq 0, \\ -\Theta\left(\dfrac{x}{r}\right) & \text{if } y < 0. \end{cases}$$

Recalling that $\cos : [0, \pi] \to [-1, 1]$ is the inverse of $\Theta : [-1, 1] \to [0, \pi]$, we have

$$\theta = \begin{cases} \cos^{-1}\left(\dfrac{x}{r}\right) & \text{if } y \geq 0, \\ -\cos^{-1}\left(\dfrac{x}{r}\right) & \text{if } y < 0. \end{cases}$$

Note that vice versa, if we know the angle made by a point $P \neq (0, 0)$ with the positive real axis, then we know where P' is on the unit circle, and this enables us to determine where P is by knowing also the distance r of P to the origin O: we simply look at the ray $\overrightarrow{OP'}$ (which has P on it somewhere), and P is precisely that point on the ray $\overrightarrow{OP'}$ which is at a distance $r > 0$ from O. The pair (r, θ) of real numbers (where $r > 0$ and $\theta \in (-\pi, \pi]$) are called the *polar coordinates* of P (which has the *Cartersian coordinates* (x, y)).

Definition 5.22 (Polar coordinates of a point in $\mathbb{R}^2 \backslash \{(0, 0)\}$). The *polar coordinates* of the point $P = (x, y) \in \mathbb{R}^2 \backslash \{(0, 0)\}$ are (r, θ), where

$$r := \sqrt{x^2 + y^2} \ (> 0),$$

$$\theta := \begin{cases} \cos^{-1}\left(\dfrac{x}{r}\right) & \text{if } y \geq 0, \\ -\cos^{-1}\left(\dfrac{x}{r}\right) & \text{if } y < 0, \end{cases}$$

where $\cos^{-1} : [-1, 1] \to [0, \pi]$ is the inverse of $\cos : [0, \pi] \to [-1, 1]$.

So given the rectangular coordinates, we can calculate the corresponding polar coordinates. How about going the other way around? It will be convenient to also have ready-made expressions for this. Suppose that a point has polar coordinates (r, θ). Then what are its

Cartesian coordinates in terms of (r, θ)? Since $\cos \theta = \cos(-\theta)$, it follows that $x = r \cos \theta$. As $x^2 + y^2 = r^2$, we see that $y = \pm r \sin \theta$. We have the two cases:

$\underline{1^\circ}$ $y \geq 0$. Then $\theta \in [0, \pi]$, and so $\sin \theta \geq 0$. So $y = r \sin \theta$.

$\underline{2^\circ}$ $y < 0$. Then $\theta \in (-\pi, 0]$, and so $\sin \theta = -\sin(-\theta) \leq 0$. So $y = r \sin \theta$.

So in either case ($y \geq 0$ or $y < 0$), $y = r \sin \theta$. Hence $(x, y) = (r \cos \theta, y \sin \theta)$.

We summarise these observations in the following table.

$$P \in \mathbb{R}^2 \backslash \{(0,0)\} :$$

Cartesian coordinates	Polar coordinates
$(x, y), x, y \in \mathbb{R}$	$(r, \theta), r > 0, \theta \in (-\pi, \pi]$
$x = r \cos \theta$ $y = r \sin \theta$	$r = \sqrt{x^2 + y^2}$ $\theta := \begin{cases} \cos^{-1}\left(\dfrac{x}{r}\right) & \text{if } y \geq 0, \\ -\cos^{-1}\left(\dfrac{x}{r}\right) & \text{if } y < 0. \end{cases}$

In the definition of the polar coordinates (or for that matter, the angle of a point on the unit circle in the lower half plane), we arranged things so that θ lies in the interval $(-\pi, \pi]$. But instead of this convention, one can give other possible definitions. For example, it is also usual practice for θ to belong to interval $[0, 2\pi)$. In this case the formula for the angle when $y < 0$ changes from

$$-\cos^{-1}\left(\frac{x}{r}\right) \quad \text{to} \quad 2\pi - \cos^{-1}\left(\frac{x}{r}\right).$$

Exercise 5.80. Find the polar coordinates of the following points given in Cartesian/ rectangular coordinates:

$$(1, 1) \quad (1, 0) \quad (0, 1) \quad (-1, 0) \quad (-1, -1) \quad (0, -1).$$

Curves in polar coordinates

Just like an equation in the Cartesian coordinates (x, y), such as $x^2 + y^2 = 1$, can describe a curve in $\mathbb{R}^2$, an equation in the polar coordinates (r, θ) can be used to describe a curve in $\mathbb{R}^2$. Some curves have simpler equations in polar coordinates. For example, the unit circle $\mathbb{T}$ is just described by $r = 1$! Indeed, these are all points with polar coordinates (r, θ), which are at a distance of 1 from the origin, that is, $r = 1$; the angle θ does not matter, that is, any $\theta \in (-\pi, \pi]$ is allowed. So

$$\mathbb{T} = \{ \text{ points with Cartesian coordinates } (x, y) \in \mathbb{R}^2 : x^2 + y^2 = 1 \}$$
$$= \{ \text{ points with polar coordinates } (r, \theta) \in (0, \infty) \times (-\pi, \pi] : r = 1 \}.$$

Here are a few other examples.

Example 5.34 (Circle again). Consider the equation $r = 2\sin\theta$ in polar coordinates (r, θ). We claim that the set of points described by it is the circle with center having Cartesian coordinates $(0, 1)$ and radius 1.

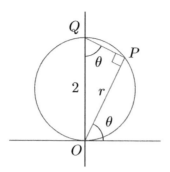

This is clear from the picture: the angle subtended by the diameter in a semicircle is $90°$, so that if P makes an angle[12] θ with the positive real axis, then $\angle PQO = \theta$ as well. So from the right angled triangle PQO, we see that $r = \ell(OP) = 2\sin\theta$.

Another way to see this is by proceeding analytically as follows. If P is a point whose polar coordinates satisfy $r = 2\sin\theta$, then its Cartesian coordinates (x, y) satisfy $x^2 + y^2 = r^2 = r \cdot r = r \cdot 2\sin\theta = 2r\sin\theta = 2y$, and so $x^2 + y^2 - 2y + 1 = 1$, that is, $(x - 0)^2 + (y - 1)^2 = 1^2$. This means that the point (x, y) is at a distance 1 from $(0, 1)$. So the given equation describes a circle with center having the Cartesian coordinates $(0, 1)$ and radius 1. ◇

Example 5.35 (Archimedean spiral). The equation $r = a + b\theta$ describes an *Archimedean spiral*. One can use Maple to plot curves given in polar coordinates. One should first invoke the plot package, and then use the command polarplot. Using the commands

```
> with(plots):
> polarplot(t, t=0 .. Pi, axis[radial]=[color="Blue"])
```

gives the leftmost plot in the following picture. ◇

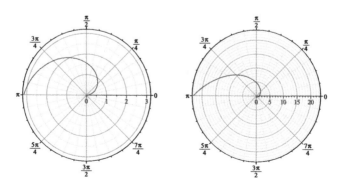

---

[12] assumed to be less than $\pi/2$ radians at the moment

Example 5.36 (Logarithmic spiral). The equation $r = e^\theta$ describes a *logarithmic spiral*, as shown in the rightmost plot of the previous figure.   ◇

Example 5.37 (Cardioid). The equation $r = 2(1 + \cos\theta)$ describes a curve called the *cardioid*, and is a 'heart-shaped' curve as shown in the leftmost plot below.   ◇

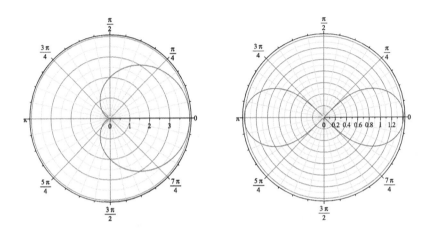

Example 5.38 (Lemniscate). The equation $r^2 = 2 \cdot \cos(2\theta)$ describes a curve called the *lemniscate* and is a '∞-shaped' curve, as shown in the rightmost plot above. Can you explain the shape of the curve based on the equation? (*Hint:* Look at the graph of $\cos(2\theta)$ on $(-\pi, \pi]$. When are the values positive?)   ◇

Exercise 5.81. Find the equation of the line $y = 3x + 1$ in polar coordinates. Use Maple to plot the resulting curve described in polar coordinates, and verify that you get what you expect to see.

Exercise 5.82. Describe the curve $r = (\tan\theta)(\sec\theta)$ for polar coordinates (r, θ) in terms of Cartesian coordinates.

Exercise 5.83. Find the equation in Cartesian coordinates for the curve described by the relation $r = (2 + \cos\theta)^{-1}$ for polar coordinates (r, θ). Use Maple to plot the resulting curve.

5.7   Applications of Riemann Integration

We will now see a few selected applications of the Riemann integral to problems in planar and solid geometry. We will learn to calculate

(1) the area of a region between two curves,

(2) volumes of bodies obtained by revolving planar regions,

(3) lengths of smooth curves, and

(4) the curved surface area of a surface obtained by revolving a curve.

Area of a region between two curves

Suppose that $\bar{f}, \underline{f} \colon [a,b] \to \mathbb{R}$ are two Riemann integrable functions on $[a,b]$ such that for all $x \in [a,b]$, we have that $\underline{f}(x) \leq \bar{f}(x)$.

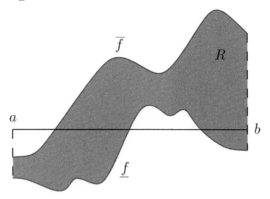

Then the area of the region R, given by

$$R := \{(x,y) \in \mathbb{R}^2 \colon \text{ for all } x \in [a,b], \; \underline{f}(x) \leq y \leq \bar{f}(x)\}$$

is defined to be

$$\text{Area}(R) := \int_a^b \left(\bar{f}(x) - \underline{f}(x) \right) dx.$$

Example 5.39. For $x \in [0,1]$, $\bar{f} := x^2 \geq x^3 =: \underline{f}$. Thus the area of the region between these two curves is

$$\int_0^1 (x^2 - x^3)\,dx = \left(\frac{x^3}{3} - \frac{x^4}{4} \right) \Big|_0^1 = \frac{1}{3} - \frac{1}{4} = \frac{1}{12}. \qquad \diamondsuit$$

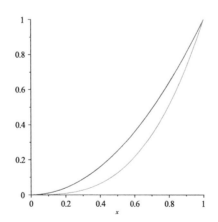

Example 5.40 (Area of a circle of radius r). Consider the circle given by

$$x^2 + y^2 = r^2,$$

where $r > 0$. The area of the circular disk enclosed by the circle is the area of the region between the graphs of the functions $\bar{f} := \sqrt{r^2 - x^2}$ and $\underline{f} := -\sqrt{r^2 - x^2}$.

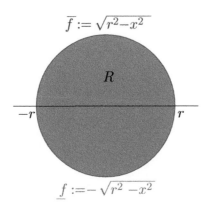

$$\bar{f} := \sqrt{r^2 - x^2}$$

$$R$$

$$-r \qquad \qquad r$$

$$\underline{f} := -\sqrt{r^2 - x^2}$$

Thus the area of the disk is

$$\text{Area}(R) = \int_{-r}^{r} \left(\sqrt{r^2 - x^2} - (-\sqrt{r^2 - x^2}) \right) dx = 2 \int_{-r}^{r} \sqrt{r^2 - x^2} dx$$

$$= 4 \int_{0}^{r} \sqrt{r^2 - x^2} dx,$$

where the last equality follows because $x \mapsto \sqrt{r^2 - x^2}$ is an even function.

We now use the substitution $x = r \cos \theta$, so that $dx = -r \sin \theta d\theta$, and when $x = 0$, we have $\theta = \pi/2$, while if $x = r$ then we have $\theta = 0$. So we obtain

$$\text{Area}(R) = 4 \int_{0}^{r} \sqrt{r^2 - x^2} dx = 4 \int_{\pi/2}^{0} \sqrt{r^2 - r^2(\cos \theta)^2} \cdot (-r \sin \theta) d\theta$$

$$= 4r^2 \int_{0}^{\pi/2} (\sin \theta)^2 d\theta = 2r^2 \int_{0}^{\pi/2} (1 - \cos(2\theta)) d\theta$$

$$= 2r^2 \left(\frac{\pi}{2} - \frac{\sin(2\theta)}{2} \Big|_{0}^{\pi/2} \right) = 2r^2 \left(\frac{\pi}{2} - 0 \right)$$

$$= \pi r^2.$$

Thus with $\pi :=$ area of a circular disk with radius 1 (as we had done earlier), we obtain that

$$\frac{\text{area of a circular disk with radius } r}{r^2} = \text{constant} = \pi.$$

(This proves a result taken for granted in high school.) $\qquad \qquad \diamond$

Example 5.41 (Area of a sector of radius r subtending an angle φ at the center). We will just consider the case when $\varphi \in [0, \pi/2)$, since the other cases can be done by successively adding areas of quarters of circular sectors (whose area we know is equal to $\pi r^2/4$). See the following picture.

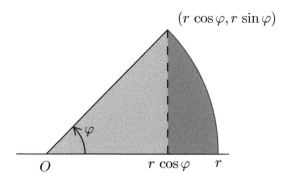

$(r\cos\varphi, r\sin\varphi)$

$r\cos\varphi$

O $r\cos\varphi$ r

The area is given by

$$\text{Area} = \frac{1}{2} \cdot (r\cos\varphi) \cdot (r\sin\varphi) + \int_{r\cos\varphi}^{r} \sqrt{r^2 - x^2}\,dx$$

$$= \frac{1}{4}r^2\sin(2\varphi) + \int_{\varphi}^{0} (r\sin\theta) \cdot (-r\sin\theta)d\theta \qquad \begin{array}{l}\text{(using the substitution} \\ x = r\cos\theta)\end{array}$$

$$= \frac{1}{4}r^2\sin(2\varphi) + \int_{0}^{\varphi} r^2(\sin\theta)^2 d\theta$$

$$= \frac{1}{4}r^2\sin(2\varphi) + \frac{r^2}{2}\int_{0}^{\varphi} (1 - \cos(2\theta))d\theta$$

$$= \frac{1}{4}r^2\sin(2\varphi) + \frac{\varphi r^2}{2} - r^2 \cdot \frac{\sin(2\theta)}{4}\Big|_{0}^{\varphi}$$

$$= \frac{1}{4}r^2\sin(2\varphi) + \frac{\varphi r^2}{2} - r^2 \cdot \frac{\sin(2\varphi)}{4} + r^2 \cdot 0$$

$$= \frac{\varphi r^2}{2}.$$

So the area of a sector of radius r subtending an angle $\varphi \in [0, \pi/2)$ is $\dfrac{\varphi r^2}{2}$.   ◇

Exercise 5.84. Find the area of the ellipse described by

$$\frac{x^2}{a^2} + \frac{y^2}{b^2} = 1,$$

where $a, b > 0$. What happens when $a = b$?

Exercise 5.85. (∗) The horizontal line $y = c$ intersects the curve $y = 2x - 3x^3$ in the first quadrant as shown in the following picture.

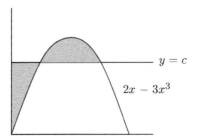

Determine c so that the areas of the two shaded regions are equal.

Exercise 5.86. The aim of this exercise is to complete the plausibility argument using the Method of Exhaustion mentioned in the Introduction.

(1) (Definition of π as in high school.) In elementary school, one learns that the ratio of the circumference C_d of a circle of diameter d to its diameter is constant, and this constant is defined to be the number π. Give an argument for this fact (which we will reprove in Example 5.49) based on the following diagram.

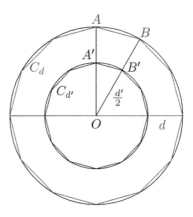

(2) (Area of a circle of radius r.) We inscribe a regular polygon with n sides inside the circle of radius r, and triangulate it by joining the center of the circle to the vertices of the polygon. By looking at the picture below, justify the expression πr^2 for the area of the circle.

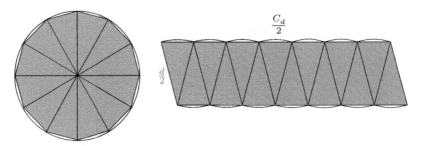

Area of a region between curves described in polar coordinates

The area of the sector of a disk with radius r and subtending an angle φ at the center is $r^2\varphi/2$, and this enables us to calculate the area of regions between curves described in polar coordinates.

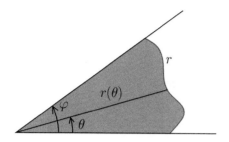 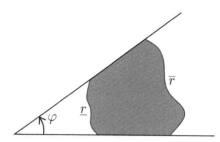

If $r : [0, \varphi] \to [0, \infty)$, then the area of the region

$$R := \{\, \text{points with polar coordinates } (\rho, \theta) : \rho = r(\theta),\ 0 \leq \theta \leq \varphi \}$$

is defined to be $\displaystyle\int_0^\varphi \frac{(r(\theta))^2}{2}\, d\theta.$

Similarly, if $\bar{r}, \underline{r} : [0, \varphi] \to [0, \infty)$, then the area of the region

$$R := \{\, \text{points with polar coordinates } (\rho, \theta) : \underline{r}(\theta) \leq \rho \leq \bar{r}(\theta),\ 0 \leq \theta \leq \varphi \}$$

is defined to be $\displaystyle\int_0^\varphi \frac{(\bar{r}(\theta))^2 - (\underline{r}(\theta))^2}{2}\, d\theta.$

The rationale behind these definitions is as follows. Suppose that we partition the interval $[0, \varphi]$ into

$$\theta_0 := 0 < \theta_1 < \theta_2 < \cdots < \theta_{n-1} < \theta_n = \varphi.$$

Then the region

$$R := \{\, \text{points with polar coordinates } (\rho, \theta) : \rho = r(\theta),\ 0 \leq \theta \leq \varphi \}$$

is subdivided into 'wedges', as shown in the picture below.

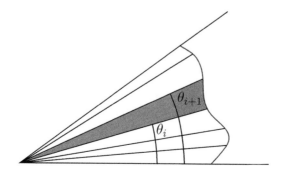

The area of the ith wedge is

$$\stackrel{(*)}{=} \text{ area of the sector of radius } r(\theta_i) \text{ subtending an angle } \theta_{i+1} - \theta_i + \text{ (an error)}_i.$$

So the area of the region is

$$= \sum_{i=0}^{n-1} \text{area of wedge } i$$

$$\stackrel{(**)}{=} \sum_{i=0}^{n-1} \frac{(r(\theta_i))^2}{2} \cdot (\theta_{i+1} - \theta_i) + \text{ error.}$$

As n grows, the error in the approximation $(*)$ diminishes, and the sum on the right hand side of $(**)$ approaches

$$\int_0^{\varphi} \frac{(r(\theta))^2}{2} d\theta.$$

Example 5.42 (Area enclosed by the cardioid). The cardioid is given in polar coordinates by $r = 2(1 + \cos\theta)$, and so the area enclosed by it is

$$\text{Area} = 2 \cdot \int_0^{\pi} \frac{(2(1 + \cos\theta))^2}{2} d\theta$$

$$= 4 \int_0^{\pi} \left(1 + 2\cos\theta + (\cos\theta)^2\right) d\theta$$

$$= 4 \left(\pi + 2\sin\theta\Big|_0^{\pi} + \int_0^{\pi} \frac{1 + \cos(2\theta)}{2} d\theta\right)$$

$$= 4 \left(\pi + 0 + \frac{\pi}{2} + 0\right)$$

$$= 4 \cdot \frac{3\pi}{2}$$

$$= 6\pi. \qquad \qquad \diamond$$

Exercise 5.87. Calculate the area enclosed by the two 'petals' of the lemniscate given in polar coordinates by $r^2 = 2\cos(2\theta)$.

Volumes of solids of revolution

Definition 5.23 (Solid of revolution). A subset of $\mathbb{R}^3$, which is obtained by revolving a planar region about an axis is called a *solid of revolution*.

Here are a few examples.

Planar region   Solid of revolution

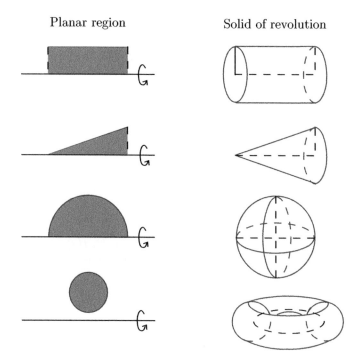

The volume of a right circular cylinder of radius r and height h is defined to be

$$(\text{Area of the circular base}) \cdot (\text{Height}) = (\pi r^2) \cdot h = \pi r^2 h.$$

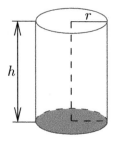   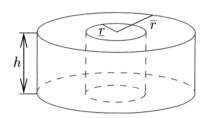

Cylinder volume $:= \pi r^2 h$ Washer volume $= \pi(\bar{r}^2 - \underline{r}^2)h$

Now consider a 'washer', namely, a cylinder with a hole, where the outer radius of the cylinder is $\bar{r}$, and the radius of the concentric hole is $\underline{r}$ ($\leq \bar{r}$). Suppose that the height is h. Then the volume of the washer is

$$\pi \bar{r}^2 h - \pi \underline{r}^2 h = \pi(\bar{r}^2 - \underline{r}^2)h.$$

Using these formulae and Riemann integration, we can calculate the volumes of solids of revolution of planar regions between the graphs of two functions using the 'washer method', where we imagine slicing the solid of revolution into infinitesimal slices, where each slice is a washer, and adding these tiny contributions to obtain the whole volume.

Definition 5.24 (Volume of a solid of revolution). Let $\overline{f}, \underline{f} : [a, b] \rightarrow \mathbb{R}$ be such that $0 \leq \underline{f} \leq \overline{f}$ pointwise:

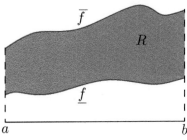

Let R be the planar region $R := \{(x, y) \in \mathbb{R}^2 : \text{for all } x \in [a, b], \underline{f}(x) \leq y \leq \overline{f}(x)\}$. The *volume of the solid of revolution obtained by revolving R about the x-axis* is defined to be

$$\int_a^b \pi((\overline{f}(x))^2 - (\underline{f}(x))^2) dx$$

(assuming that $\overline{f}^2 - \underline{f}^2 \in RI[a, b]$).

Here is the rationale behind the above definition. The volume of each elemental washer, assuming that it has been sliced vertically into thin washers corresponding to a partition $\{x_0 = a, x_1, \cdots, x_n = b\}$ of $[a, b]$, is approximated by

$$\pi\left((\overline{f}(x_i))^2 - (\underline{f}(x_i))^2\right) \cdot (x_i - x_{i-1}),$$

and the total volume is the sum of these. See the picture below. As n increases, this tends to the expression above.

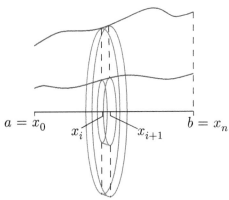

Example 5.43 (Volume of a cone with radius r and height h). See the following picture. We take $\underline{f} = 0$ and $\overline{f} = rx/h$ on the interval $[0, h]$. Then the volume of the cone is

$$\int_0^h \pi\left(\left(\frac{r}{h}x\right)^2 - 0^2\right) dx = \pi\frac{r^2}{h^2}\int_0^h x^2 dx = \pi\frac{r^2}{h^2} \cdot \frac{x^3}{3}\bigg|_0^h = \pi\frac{r^2}{h^2} \cdot \frac{h^3}{3} = \frac{\pi r^2 h}{3}.$$ ◇

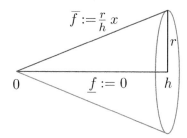

Example 5.44 (Volume of a sphere of radius r). On the interval $[-r, r]$, we consider the functions $\underline{f} = 0$ and $\overline{f} = \sqrt{r^2 - x^2}$. Then the volume of the sphere is

$$\int_{-r}^{r} \pi \left((\sqrt{r^2 - x^2})^2 - 0^2 \right) dx = \pi \int_{-r}^{r} (r^2 - x^2) dx = \pi \left(r^2 \cdot 2r - \frac{r^3 - (-r)^3}{3} \right)$$

$$= \pi \cdot \left(2r^3 - \frac{2r^3}{3} \right) = \pi \cdot 2r^3 \cdot \frac{2}{3} = \frac{4\pi r^3}{3}. \qquad \diamond$$

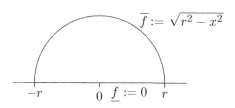

Exercise 5.88. Calculate the volume of a doughnut, with the radius of the greater circle equal to R (that is, of the central circle lying midway in the annular region obtained by taking a horizontal cross section of the doughnut), and that of the two little circles, obtained by taking a vertical cross section of the doughnut, equal to r.

Exercise 5.89. Calculate the volume of an ellipsoid, namely the solid of revolution obtained by revolving the region enclosed by the ellipse

$$\frac{x^2}{a^2} + \frac{y^2}{b^2} = 1,$$

where $a, b > 0$, in the upper half plane and the x-axis. What happens when $a = b$?

Exercise 5.90. A round hole of radius $\sqrt{3}$ is drilled through the center of a solid ball of radius 2 cm. Find the volume cut out.

Exercise 5.91 (Design of a clepsydra). A clepsydra (literally meaning 'water thief' in Greek) or a water clock is designed by revolving the graph of $x \mapsto C x^m : [0, r] \to \mathbb{R}$, where C, $m > 0$, as shown in the following figure.

Cx^m

r

What should m be if the level of water is to decrease linearly as time passes? You may use Toricelli's Law stating that the speed of water flowing out when the height of water is h is proportional to $\sqrt{h}$.

Exercise 5.92. Let $f : [0, \infty) \to [0, \infty)$ be a continuous function. If for each $a > 0$, the volume of the solid obtained by revolving the region under the graph of f over the interval $[0, a]$, about the x-axis is $a^2 + a$, then find f.

Arc length of a smooth curve

Recall that a *curve* γ is a map

$$t \mapsto \gamma(t) = (x(t), y(t)) : [a, b] \to \mathbb{R}^2$$

such that $t \mapsto x(t), t \mapsto y(t) : [a, b] \to \mathbb{R}$ are both continuous.

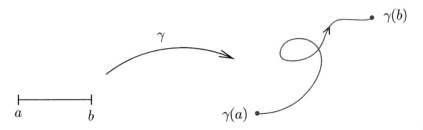

Definition 5.25 (Smooth curve). If $x, y : [a, b] \to \mathbb{R}$ are continuously differentiable (that is, $x, y \in C^1[a, b]$), then we will call the curve γ, given by

$$t \mapsto \gamma(t) = (x(t), y(t)) : [a, b] \to \mathbb{R}^2,$$

smooth.

Example 5.45. $t \mapsto (t, t^2) : [-1, 1] \to \mathbb{R}^2$ is a smooth curve. The graph is a segment of a parabola. See the leftmost graph in following picture.   ◇

Example 5.46. $t \mapsto (t^2, t^3) : [-1, 1] \to \mathbb{R}^2$ is a smooth curve. This is displayed in the middle in the following picture. Note that this smooth curve has a 'corner'[13].   ◇

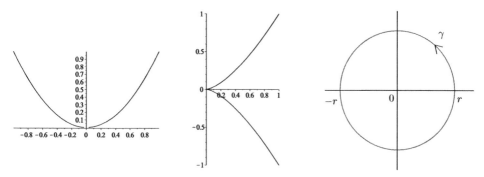

Example 5.47. Let $r > 0$ be fixed. Then $t \mapsto (r \cos t, r \sin t) : [0, 2\pi] \to \mathbb{R}^2$ is a smooth curve. See the rightmost picture in the above.   ◇

Definition 5.26 (Length of a smooth curve).

Let $t \mapsto \gamma(t) = (x(t), y(t)) : [a, b] \to \mathbb{R}^2$ be a smooth curve. We define the *arc length of* γ to be

$$\int_a^b \sqrt{(x'(t))^2 + (y'(t))^2} dt.$$

Rationale: Partition $[a, b]$ into $x_0 = a < x_1 < \cdots < x_{n-1} < x_n = b$.

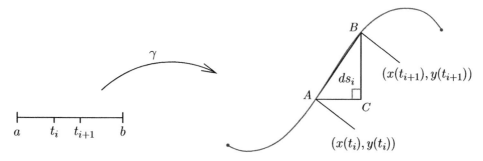

The length of the curve ds_i is approximated by the elemental chordal length of the line segment AB. Thus (by the Pythagoras Theorem in $\triangle ABC$),

$$ds_i \overset{(*)}{\approx} \ell(AB) = \sqrt{(x(t_{i+1}) - x(t_i))^2 + (y(t_{i+1}) - y(t_i))^2}$$

$$\overset{(**)}{\approx} \sqrt{(x'(t_i)(t_{i+1} - t_i))^2 + (y'(t_i)(t_{i+1} - t_i))^2}$$

$$= \sqrt{(x'(t_i))^2 + (y'(t_i))^2} \cdot (t_{i+1} - t_i).$$

---

[13] This example justifies the somewhat different definition of a smooth curve used at times; for example in considerations in the subject of differential geometry, where a curve is called smooth if at each point, one has a 'tangent vector', and this tangent vector varies smoothly with the point.

Thus the length of γ is $\sum_{i=1}^{n} \sqrt{(x'(t_i))^2 + (y'(t_i))^2} +$ error.

As n increases, the partitions get finer, and the sum on the right hand side approaches

$$\int_a^b \sqrt{(x'(t))^2 + (y'(t))^2} dt,$$

while the error in $(*)$ and $(**)$ goes to 0.

Example 5.48. Let $\gamma : [0, 1] \to \mathbb{R}^2$ be given by $\gamma(t) = (t, t^2), t \in [0, 1]$. Then the length of γ is

$$\int_0^1 \sqrt{1^2 + (2t)^2} dt = \int_0^1 \sqrt{1 + 4t^2} dt \quad (2t = u),$$

$$= \int_0^2 \sqrt{1 + u^2} \cdot \frac{1}{2} du$$

$$= \frac{1}{4} \left(u\sqrt{1 + u^2} + \log(u + \sqrt{1 + u^2}) \right) \Big|_0^2 \quad \text{(see Exercise 5.51)}$$

$$= \frac{1}{4}(2\sqrt{5} + \log(2 + \sqrt{5}))$$

$$= \frac{\sqrt{5}}{2} + \frac{1}{4} \log(2 + \sqrt{5}). \qquad \diamond$$

Example 5.49 (Circumference of a circle of radius r). Let $\gamma(t) := (r\cos t, r\sin t)$, $t \in [0, 2\pi]$. Then we have that the length of γ is

$$\int_0^{2\pi} \sqrt{(-r\sin t)^2 + (r\cos t)^2} dt = \int_0^{2\pi} \sqrt{r^2((\sin t)^2 + (\cos t)^2)} dt = \int_0^{2\pi} r dt = 2\pi r.$$

Thus the ratio

$$\frac{\text{circumference of a circle of radius } r}{\text{radius } r} = \frac{2\pi r}{r} = 2\pi = \text{constant},$$

a fact that we had accepted on faith in high school geometry, but which we have now proved. $\qquad \diamond$

Exercise 5.93. The position of a particle in the plane $\mathbb{R}^2$ at time $t \geq 0$ is given by

$$x(t) = \frac{1}{3}(2t + 3)^{3/2},$$

$$y(t) = \frac{t^2}{2} + t,$$

for $t \geq 0$.

(1) Find the distance it travels between $t = 0$ and $t = 3$.
(2) What is its average speed?

Exercise 5.94. Calculate the arclength of the cardioid given in polar coordinates by the equation $r = 2(1 + \cos\theta)$.

Exercise 5.95 (The elliptic integral). Show that the perimeter of an ellipse given by

$$\frac{x^2}{a^2} + \frac{y^2}{b^2} = 1,$$

where $b \geq a > 0$, is given by

$$b \int_0^{2\pi} \sqrt{1 - k^2(\sin\theta)^2}\,d\theta,$$

for a suitable constant k.

(This integral is called an *elliptic integral*, and cannot be expressed in terms of elementary functions when $b \neq a$.) What happens when $b = a$?

Exercise 5.96. Consider the following picture, where BC is a circular arc with radius 1 and center O, subtending an angle of $\theta \in (0, \pi/2)$ at O. Drop a perpendicular from C on OB, meeting OB in A. Note that $\ell(AC) = \sin\theta$. What is the arc length of circular arc BC in terms of θ? Observe that the hypotenuse BC in $\triangle ABC$ is bigger than $\ell(AC) = \sin\theta$ on the one hand, and being the straight line segment between the points B and C, it is visibly smaller than the circular arc length BC. Deduce from these considerations the inequality $\sin\theta \leq \theta$, which was also obtained analytically in Exercise 5.79(1).

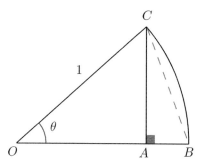

Area of a surface of revolution

Definition 5.27 (Surface of revolution). A *surface of revolution* is generated when a curve is revolved about a line.

Example 5.50 (Spherical bubble). If we revolve a semicircle with center at the origin about the x-axis, then the corresponding surface of revolution is the surface of a sphere.   ◇

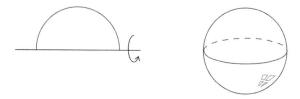

We will define the area of a surface of revolution by cutting it into strips or ribbons, and add up the results. It turns out that the area of each ribbon can be approximated by the area of the surface obtained by a line segment revolved about an axis or a 'frustum'. So a frustum is really just a strip cut out from a right circular cone. Thus we first need to find the surface area of a cone. But a cone can be cut out to form a sector, and *its* area we do know: it is $R^2\theta/2$, where the radius of the sector is θ and the angle subtended at the center is θ. However, in what is to follow, it will be convenient to express this area not using θ, but rather the *arc length* of the curved circular portion of the sector. All this might be confusing right now, so we will go step by step as follows.

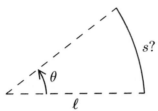

Step 1. Suppose that we have a sector of radius ℓ and angle θ subtended at the center. We ask: what is the length s of the circular arc? Well, we have just learnt a formula for the arclength! We have

$$s = \int_0^\theta \sqrt{(-\ell \sin t)^2 + (\ell \cos t)^2}\,dt = \int_0^\theta \ell\,dt = \ell \cdot \theta.$$

Step 2. Using the above formula for the arc length of a circular arc, and the area of a sector, we can find and express the curved surface area of a cone in a form that will be useful for the calculation of the surface area of revolution.

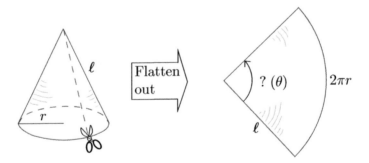

Consider a cone of slant height ℓ and radius r. What is its curved surface area? If we imagine the cone to be made of paper, cut along a straight line that passes through the apex, and flatten out the paper, then we obtain a sector whose area is the same as the curved surface area of the cone.

From Step 1, we know that $\theta \cdot \ell = 2\pi r$, and so

$$\theta = \frac{2\pi r}{\ell}.$$

Hence the curved surface area of the cone is equal to the area of the sector, which is given by

$$\frac{1}{2}\ell^2\theta = \frac{1}{2}\ell^2 \cdot \frac{2\pi r}{\ell} = \pi r \ell.$$

Step 3. Using Step 2, we can also calculate the area of a 'frustum'.

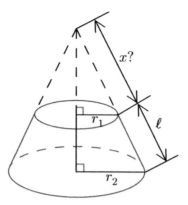

By considering the similar triangles shown in the picture, we obtain

$$\frac{r_1}{x} = \frac{r_2}{x+\ell},$$

and so

$$x = \frac{r_1 \ell}{r_2 - r_1}.$$

Thus the area of the frustum equals the difference of the curved surface areas of the bigger and the smaller cones, that is,

$$\pi r_2(\ell + x) - \pi r_1 x = \pi r_2 \ell + \pi(r_2 - r_1)x = \pi r_2 \ell + \pi r_1 \ell = \pi(r_1 + r_2)\ell.$$

Step 4. Now we look at the general case of the surface area of a surface of revolution. Suppose that $t \mapsto \gamma(t) = (x(t), y(t)) : [a, b] \to \mathbb{R}^2$ is a smooth curve.

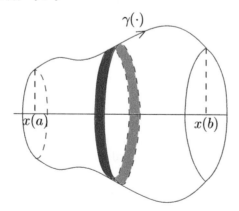

Partition $[a, b]$ into $a = t_0 < t_1 < \cdots < t_{n-1} < t_n = b$. Then the area of the elemental frustum is

$$\pi(y(t_i) + y(t_{i+1}))\sqrt{(x(t_{i+1}) - x(t_i))^2 + (y(t_{i+1}) - y(t_i))^2}$$

$$\approx 2\pi y(t_i)\sqrt{(x'(t_i))^2 + (y'(t_i))^2} \cdot (t_{i+1} - t_i).$$

We add such elemental contributions. This prompts the following definition.

Definition 5.28 (Surface area of a surface of revolution). Suppose that the curve γ, $t \mapsto \gamma(t) = (x(t), y(t)) : [a, b] \to \mathbb{R}^2$, is a smooth curve. Then the *surface area of the surface of revolution about the x-axis corresponding to* γ is defined to be

$$\int_a^b 2\pi y(t) \sqrt{(x'(t))^2 + (y'(t))^2} dt.$$

Example 5.51 (Surface area of a sphere of radius r). Consider the semicircular curve γ given by $\gamma(t) = (r\cos t, r\sin t)$, $t \in [0, \pi]$. Then the surface area of the corresponding surface of revolution is

$$\int_0^\pi 2\pi \cdot r\sin t \cdot \sqrt{(-r\sin t)^2 + (r\cos t)^2} dt$$

$$= \int_0^\pi 2\pi \cdot r\sin t \cdot r dt = 2\pi r^2 \int_0^\pi \sin t \, dt = 2\pi r^2 (-\cos t)\Big|_0^\pi = 2\pi r^2 \cdot 2$$

$$= 4\pi r^2.$$

We note that the ratio $\dfrac{\text{surface area of a sphere of radius } r}{r^2} = 4\pi = $ a constant.   ◇

Remark 5.3 (Solid angle and the steradian measure). If S is a surface lying on a sphere of radius r, then S is said to subtend a *solid angle* of measure

$$\frac{\text{surface area of } S}{r^2} \text{ steradian}$$

at the center of the sphere. (The motivation is that except for a multiplicative factor of 4π, this ratio measures the fraction of the sphere occupied by S.)

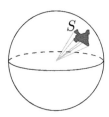

Thus the solid angle subtended at the center by the *entire* spherical surface is 4π steradian, while that of the hemisphere is 2π steradian, and so on. The solid angle subtended by Antarctica at the center of the Earth is

$$\frac{\text{area of antarctica}}{(\text{radius of Earth})^2} = \frac{14000000 \text{ km}^2}{(6371 \text{ km})^2} \approx 0.345 \text{ steradian}.$$

The solid angle is a measure of how large an object appears to an observer at the center. A *small* object *nearby* may subtend the same solid angle as a *large* object *far away*. For example, although the Moon is much smaller than the Sun, it is also much closer to the Earth, and in fact, the solid angles they subtend for an observer on the Earth are approximately the same. This is visually clear during a solar eclipse!

Remark 5.4 (A paradox? Toricelli's Trumpet or Gabriel's Horn). Consider the 'infinite trumpet' obtained by revolving the graph of $x \mapsto 1/x : [1, \infty) \to \mathbb{R}$ about the x-axis.

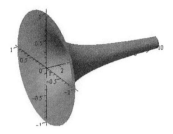

We will see[14] that the 'volume' of the solid of revolution is finite, but its 'surface area' is infinite. Indeed, we have

$$\text{volume:} = \int_1^\infty \pi \cdot \frac{1}{x^2} \, dx = \pi \left(-\frac{1}{x} \right) \Big|_1^\infty = \pi,$$

while

$$\text{surface area:} = \int_1^\infty 2\pi \cdot \frac{1}{x} \sqrt{1 + \left(-\frac{1}{x^2} \right)^2} \, dx \geq \int_1^\infty 2\pi \cdot \frac{1}{x} \cdot 1 \, dx = \lim_{x \to \infty} 2\pi \log x = \infty.$$

So this seems to be a paradox, since the above suggests that we can fill the trumpet with a *finite* amount of paint, but we would need an *infinite* amount of paint to cover its surface! But how can that be? After all, we can fill the trumpet with a finite amount of paint, and 'pour it out' while having automatically painted its surface—but painting its surface was supposed to be impossible, as it should have needed an infinite amount of paint.

The paradox arises from confusing our mental model of *real* paint with 'mathematical' paint. Real paint has a finite thickness, say the thickness of the paint molecule. At some point, the trumpet becomes thinner than this, and so with real paint we could neither fill the trumpet nor cover its surface. The only paint that could do this is 'mathematical' paint that has infinitely small thickness. A finite amount of infinitely thin paint could cover an infinite surface area.

Exercise 5.97. The aim of this exercise is to show that the reverse situation in Toricelli's Trumpet Paradox can't happen. That is, we want to establish the following result. If $f : [1, \infty) \to [0, \infty)$ is a continuously differentiable function, and if the surface area S of the surface of revolution obtained by revolving the graph of f about the x-axis, defined by

$$S := \int_1^\infty 2\pi f(x) \sqrt{1 + (f'(x))^2} \, dx < \infty$$

then so is the volume V of the solid of revolution, defined by

$$V := \int_1^\infty \pi (f(x))^2 \, dx.$$

---

[14] Strictly speaking, we have not defined the volume and surface area of solids of revolution when the interval is not compact by means of improper integrals, but in this example, we just use what seem to be the intuitively appropriate definitions.

296 THE HOW AND WHY OF ONE VARIABLE CALCULUS

Proceed as follows.

(1) Show that f is bounded by applying the Fundamental Theorem of Calculus to the integrand $(f^2)'$.

(2) Try getting an upper bound on V in terms of S.

Exercise 5.98. Calculate the surface area of a doughnut, with the radius of the greater circle equal to R (that is, of the central circle lying midway in the annular region obtained by taking a horizontal cross section of the doughnut), and that of the two little circles, obtained by taking a vertical cross section of the doughnut, equal to r.

Exercise 5.99 (Catenary). Let $a \in \mathbb{R}$ with $a > 0$. An arc of the 'catenary' given by

$$y = a \cosh \left(\frac{x}{a} \right),$$

whose end points have x-coordinates 0 and a, is revolved about the x-axis. Show that the surface area A and the volume V of the solid thus generated are related by the formula

$$A = \frac{2V}{a}.$$

Notes

Exercises 5.8, 5.9, 5.10, 5.15, 5.16, 5.21, 5.63, and 5.72 are based on [G]. Example 5.26 is based on [G, Exercise 39, page 223]. The proof of Theorem 5.33 is based on [N5]. Exercise 5.77 is based on [L, 1.6.3]. Exercise 5.91 is based on [B2]. Exercise 5.85 is based on [K2, Problem A-1]. Exercise 5.74 is based on [N3].

6

Series

In this chapter, we study 'series' of real numbers.

6.1 Series

Given a sequence $(a_n)_{n \in \mathbb{N}}$, one can form a new sequence $(s_n)_{n \in \mathbb{N}}$ of its *partial sums*:

$$s_1 := a_1,$$
$$s_2 := a_1 + a_2,$$
$$s_3 := a_1 + a_2 + a_3,$$
$$\cdots$$

The sequence $(s_n)_{n \in \mathbb{N}}$ is called the *sequence of partial sums* associated with $(a_n)_{n \in \mathbb{N}}$.

Example 6.1. Here are a couple of examples.

(1) The sequence of partial sums associated with the constant sequence $(1)_{n \in \mathbb{N}}$ is $(s_n)_{n \in \mathbb{N}}$, where

$$s_1 = 1,$$
$$s_2 = 1 + 1 = 2,$$
$$s_3 = 1 + 1 + 1 = 3,$$
$$\cdots$$

that is, $(s_n)_{n \in \mathbb{N}}$ is the sequence $(n)_{n \in \mathbb{N}}$ of natural numbers.

The How and Why of One Variable Calculus, First Edition. Amol Sasane.
© 2015 John Wiley & Sons, Ltd. Published 2015 by John Wiley & Sons, Ltd.

(2) The sequence of partial sums associated with the alternating sequence $((-1)^n)_{n\in\mathbb{N}}$ is $(s_n)_{n\in\mathbb{N}}$, where

$$s_1 = -1,$$

$$s_2 = -1 + 1 = 0,$$

$$s_3 = -1 + 1 - 1 = -1,$$

$$\cdots$$

that is, $(s_n)_{n\in\mathbb{N}}$ is the sequence $-1, 0, -1, 0, -1, 0, \cdots$.

(3) The sequence of partial sums associated with the geometric sequence $(1/2^n)_{n\in\mathbb{N}}$ is $(s_n)_{n\in\mathbb{N}}$, where

$$s_n = \frac{1}{2} + \frac{1}{2^2} + \frac{1}{2^3} + \cdots + \frac{1}{2^n} = \frac{\frac{1}{2}\left(1 - \frac{1}{2^n}\right)}{1 - \frac{1}{2}} = 1 - \frac{1}{2^n}.$$

Thus $(s_n)_{n\in\mathbb{N}}$ is the sequence $1/2, 3/4, 7/8, 15/16, \cdots$.

(4) The sequence of partial sums associated with the sequence $(1/n)_{n\in\mathbb{N}}$ is the sequence $1, 1 + \frac{1}{2}, 1 + \frac{1}{2} + \frac{1}{3}, \cdots$. ◇

The sequence $(s_n)_{n\in\mathbb{N}}$ of partial sums associated with a sequence $(a_n)_{n\in\mathbb{N}}$ may converge or it may diverge. We then describe these two possible situations as follows.

Definition 6.1 (Convergence/Divergence of $\displaystyle\sum_{n=1}^{\infty} a_n$).

Let $(a_n)_{n\in\mathbb{N}}$ be a sequence and let $(s_n)_{n\in\mathbb{N}}$ be the sequence of its partial sums.

(1) If $(s_n)_{n\in\mathbb{N}}$ converges, we say that 'the *series* $\displaystyle\sum_{n=1}^{\infty} a_n$ *converges*'.

Then we write $\displaystyle\sum_{n=1}^{\infty} a_n = \lim_{n\to\infty} s_n$ and call it 'the sum of the series'.

(2) If $(s_n)_{n\in\mathbb{N}}$ does not converge, we say that 'the series $\displaystyle\sum_{n=1}^{\infty} a_n$ *diverges*'.

Example 6.2.

(1) The series $\displaystyle\sum_{n=1}^{\infty} (-1)^n$ diverges.

Indeed, the sequence of partial sums is $-1, 0, -1, 0, \cdots$ which is divergent.

(2) The series $\displaystyle\sum_{n=1}^{\infty} \frac{1}{n(n+1)}$ converges. Its nth partial sum 'telescopes'[1]:

$$s_n = \sum_{k=1}^{n} \frac{1}{k(k+1)} = \sum_{k=1}^{n} \left(\frac{1}{k} - \frac{1}{k+1}\right)$$

$$= \left(1 - \frac{1}{2}\right) + \left(\frac{1}{2} - \frac{1}{3}\right) + \cdots + \left(\frac{1}{n} - \frac{1}{n+1}\right) = 1 - \frac{1}{n+1}.$$

---

[1] Loosely, this term describes the situation when in a sum/product the intermediate terms don't matter, and only the initial and final ones are the ones which do matter; so the final term is brought in contact with the initial term just like a telescope brings a far away object in close vision.

Since $\lim_{n\to\infty} s_n = 1 - 0 = 1$, we have $\displaystyle\sum_{n=1}^{\infty} \frac{1}{n(n+1)} = 1$.

(3) Let $(a_n)_{n\in\mathbb{N}}$ be the geometric sequence $\left(\dfrac{1}{2^n}\right)_{n\in\mathbb{N}}$.

Then $(s_n)_{n\in\mathbb{N}} = \left(1 - \dfrac{1}{2^n}\right)_{n\in\mathbb{N}}$ is convergent with limit 1. Thus $\displaystyle\sum_{n=1}^{\infty} \frac{1}{2^n} = 1$.

A pictorial proof is given below. ◇

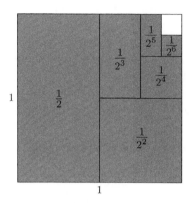

Exercise 6.1 (Tantalising $\tan^{-1}$). Show that $\displaystyle\sum_{n=1}^{\infty} \tan^{-1}\frac{1}{2n^2} = \frac{\pi}{4}$.

Hint: Write $\dfrac{1}{2n^2} = \dfrac{(2n+1) - (2n-1)}{1 + (2n+1)(2n-1)}$ and use $\tan(a-b) = \dfrac{\tan a - \tan b}{1 + \tan a \tan b}$.

Exercise 6.2. Show that for every real number $x > 1$, the series

$$\frac{1}{1+x} + \frac{2}{1+x^2} + \frac{4}{1+x^4} + \cdots + \frac{2^n}{1+x^{2^n}} + \cdots$$

converges. *Hint:* Add $\dfrac{1}{1-x}$.

Exercise 6.3. The Fibonnaci sequence $(F_n)_{n\in\mathbb{N}}$ is defined recursively by $F_0 = F_1 = 1$ and $F_{n+1} = F_n + F_{n-1}$ for $n \in \mathbb{N}$. Show that

$$\sum_{n=2}^{\infty} \frac{1}{F_{n-1}F_{n+1}} = 1.$$

Exercise 6.4. Show that for $0 < x < \pi/2$, $\tan\dfrac{x}{2} = \cot\dfrac{x}{2} - 2\cot x$.

Find the sum of the series $\displaystyle\sum_{n=1}^{\infty} \frac{1}{2^n} \tan\frac{\pi/4}{2^n}$.

What will we learn in this chapter? Just like we learnt ways of deducing the convergence of sequences and ways of finding their limits, we will learn about tests for checking convergence of series.

A natural next question is: Why bother learning about such things about series? It turns out that series play an important role in solutions to various problems that arise in Mathematics and applications of Mathematics in other disciplines. For example, in the theory of differential equations, in functional analysis, Fourier/harmonic analysis, and so on.

A necessary condition for the convergence of $\sum\limits_{n=1}^{\infty} a_n$

Note that in the above example of the divergent series

$$\sum_{n=1}^{\infty} (-1)^n,$$

the sequence $(a_n)_{n\in\mathbb{N}} = ((-1)^n)_{n\in\mathbb{N}}$ was not convergent. In fact, we have the following necessary condition for convergence of a series.

Proposition 6.1. *If the series* $\sum\limits_{n=1}^{\infty} a_n$ *converges, then* $\lim\limits_{n\to\infty} a_n = 0$.

Proof. Let $s_n := a_1 + \cdots + a_n$. Since the series converges, we have

$$\lim_{n\to\infty} s_n = L$$

for some $L \in \mathbb{R}$. But as $(s_{n+1})_{n\in\mathbb{N}}$ is a subsequence of $(s_n)_{n\in\mathbb{N}}$, it follows that

$$\lim_{n\to\infty} s_{n+1} = L.$$

By the Algebra of Limits,

$$\lim_{n\to\infty} a_{n+1} = \lim_{n\to\infty} (s_{n+1} - s_n) = \lim_{n\to\infty} s_{n+1} - \lim_{n\to\infty} s_n = L - L = 0.$$

Let $\epsilon > 0$. Then there exists an $N \in \mathbb{N}$ such that for all $n > N$, $|a_{n+1} - L| < \epsilon$. Then for all $n > N + 1$, $|a_n - L| < \epsilon$. Hence $(a_n)_{n\in\mathbb{N}}$ also converges with limit 0. $\square$

Being a necessary condition for the convergence of a series, this result helps us to conclude the divergence of a series.

Example 6.3.

(1) $\sum\limits_{n=1}^{\infty} (-1)^n$ diverges because $((-1)^n)_{n\in\mathbb{N}}$ does not converge, much less to 0.

(2) $\sum\limits_{n=1}^{\infty} \dfrac{n}{n+1}$ diverges because $\left(\dfrac{n}{n+1}\right)_{n\in\mathbb{N}}$ converges, but not to 0. $\diamond$

Exercise 6.5. Does the series $\sum\limits_{n=1}^{\infty} \cos \dfrac{1}{n}$ converge?

This condition $\lim\limits_{n\to\infty} a_n = 0$ is *not sufficient* for the convergence of the series

$$\sum_{n=1}^{\infty} a_n.$$

So just because $(a_n)_{n\in\mathbb{N}}$ converges to 0, we cannot conclude from this fact that the series above converges. Indeed, we had seen in Example 2.16 that the Harmonic Series[2]

$$\sum_{n=1}^{\infty} \frac{1}{n}$$

diverges[3], but clearly $\lim_{n\to\infty} \frac{1}{n} = 0$.

Theorem 6.2. *Let $s \in \mathbb{R}$. The series[4]* $\sum_{n=1}^{\infty} \frac{1}{n^s}$ *converges if and only if $s > 1$.*

Proof. Let $s > 1$. Clearly the sequence of partial sums is increasing as $\frac{1}{n^s} > 0$ for each n. We now show that it is bounded too. Then consider the interval $[1, n]$ and let σ_n be the 'step function' $\sigma_n(x) = (k+1)^{-s}$ if $x \in [k, k+1)$, $k = 1, \cdots, n-1$.

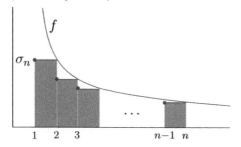

---

[2] The name arises from the concept of overtones/harmonics in music: if we imagine a string of unit length vibrating between its two fixed enpoints, then the wavelengths of the harmonics are $1, \frac{1}{2}, \frac{1}{3}, \cdots$.

[3] In connection with the divergence of the harmonic series, we also mention the **Erdös conjecture on arithmetic progressions** (APs): If the sums of the reciprocals of the numbers of a set A of natural numbers diverges, then A contains arbitrarily long APs. That is, if

$$\sum_{n\in A} \frac{1}{n}$$

diverges, then A contains APs of any given length. We know that

$$\sum_{n=1}^{\infty} \frac{1}{n}$$

diverges, and in this case, the claim is trivially true. It can be shown that

$$\sum_{p \text{ is prime}} \frac{1}{p}$$

diverges. So one may ask: Does the claim hold in this special case? The answer is 'Yes', and this is the Green-Tao Theorem from 2008. Terence Tao was awarded the Fields Medal in 2006, among other things, for this result of his.

[4] The function

$$s \mapsto \sum_{n=1}^{\infty} \frac{1}{n^s}$$

is called the *Riemann-zeta function*, which is an important function in number theory; the connection with number theory is brought out by Euler's identity, which says that

$$\zeta(s) := \sum_{n=1}^{\infty} \frac{1}{n^s} = \prod_{p \text{ prime}} \frac{1}{1-p^{-s}}.$$

Then we have

$$1 + \frac{1}{2^s} + \cdots + \frac{1}{n^s} = 1 + \int_1^n \sigma_n(x)dx \le 1 + \int_1^n \frac{1}{x^s}dx \le 1 + \int_1^\infty \frac{1}{x^s}dx,$$

where the rightmost improper integral is finite since $s > 1$. As the sequence of partial sums is monotone (increasing) and bounded, it is convergent. Thus

$$\sum_{n=1}^\infty \frac{1}{n^s}$$

converges for $s > 1$.

If on the other hand $s \le 1$, then the proof is similar to that of showing that the harmonic series diverges. Indeed, if the series was convergent, then

$$\lim_{n\to\infty} (S_{2n} - S_n) = 0,$$

while on the other hand,

$$S_{2n} - S_n = \frac{1}{(n+1)^s} + \frac{1}{(n+2)^s} + \cdots + \frac{1}{(2n)^s} \ge n\frac{1}{(2n)^s} \ge n\frac{1}{2n} = \frac{1}{2},$$

where we have used the fact that $s \le 1$ in order to obtain the last inequality. $\qquad\square$

Now let us see an important example of a convergent series. In fact, it lies at the core of most of the convergence results in Real Analysis.

Theorem 6.3. *Let $r \in \mathbb{R}$. The geometric series*

$$\sum_{n=0}^\infty r^n$$

converges if and only if $|r| < 1$. Moreover, if $|r| < 1$, then $\sum_{n=0}^\infty r^n = \frac{1}{1-r}$.

Proof. Let $|r| < 1$. Recall that in Example 2.11, we had shown that

$$\lim_{n\to\infty} r^n = 0.$$

Let $s_n := 1 + r + r^2 + \cdots + r^n$. Then $rs_n = r + r^2 + \cdots + r^n + r^{n+1}$, and so

$$(1 - r)s_n = s_n - rs_n = 1 - r^{n+1}.$$

As $\lim_{n\to\infty} r^{n+1} = 0$, it follows that $\lim_{n\to\infty} (1 - r)s_n = 1$. Hence

$$\sum_{n=1}^\infty r^n = \lim_{n\to\infty} s_n = \frac{1}{1-r}.$$

Now suppose that $|r| \ge 1$. If $r = 1$, then

$$\lim_{n\to\infty} r^n = 1 \ne 0,$$

and so by Proposition 6.1, the series diverges. Similarly if $r = -1$, then we have that $(r^n)_{n \in \mathbb{N}} = ((-1)^n)_{n \in \mathbb{N}}$ diverges, and so the series is divergent. Also if $|r| > 1$, then the sequence $(r^n)_{n \in \mathbb{N}}$ has the subsequence $(r^{2n})_{n \in \mathbb{N}}$ which is not bounded, and hence not convergent. Consequently, $(r^n)_{n \in \mathbb{N}}$ diverges, and hence the series diverges.   □

The name comes from the associated similarity in geometry.

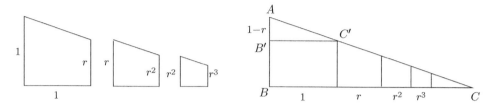

Since the triangles $AB'C'$ and ABC are similar,

$$\frac{BC}{AB} = \frac{1 + r + r^2 + r^3 + \cdots}{1} = \frac{B'C'}{AB'} = \frac{1}{1 - r}.$$

For a sequence $(a_n)_{n \in \mathbb{N}}$ with nonnegative terms, we sometimes write

$$\sum_{n=1}^{\infty} a_n < +\infty$$

to mean that the series converges. Similarly, we say that the series diverges to ∞ if the sequence of partial sums $(s_n)_{n \in \mathbb{N}}$ is such that for all $M \in \mathbb{R}$, there exists an index $N \in \mathbb{N}$ such that for every $n > N$, $s_n > M$. Analogously, we say that the series diverges to $-\infty$ if the sequence of partial sums $(s_n)_{n \in \mathbb{N}}$ is such that for all $M \in \mathbb{R}$, there exists an index $N \in \mathbb{N}$ such that for every $n > N$, $s_n < M$.

Exercise 6.6.

(1) (∗) Prove that if $a_1 \geq a_2 \geq a_3 \cdots$ is a sequence of nonnegative numbers, and if

$$\sum_{n=1}^{\infty} a_n < +\infty,$$

then a_n approaches 0 faster than $1/n$, that is, $\lim_{n \to \infty} n a_n = 0$.

Hint: Consider the inequalities $a_{n+1} + \cdots + a_{2n} \geq n \cdot a_{2n}$ and $a_{n+1} + \cdots + a_{2n+1} \geq n \cdot a_{2n+1}$.

(2) Show that the assumption $a_1 \geq a_2 \geq a_3 \cdots$ above cannot be dropped by considering the *lacunary series* whose n^2th term is $1/n^2$ and all other terms are zero.

Exercise 6.7. Consider the Arithmetic-Geometric Progression $1, 2r, 3r^2, 4r^3, \cdots$ where $r \in \mathbb{R}$. Note that $1, 2, 3, 4, \cdots$ form an AP, while $1, r, r^2, r^3, \cdots$ form a geometric progression. Show that if $|r| < 1$, then

$$1 + 2r + 3r^2 + 4r^3 + \cdots = \frac{1}{(1 - r)^2}.$$

Exercise 6.8 (Decimal representation of real numbers). Every nonnegative real number x can be written as

$$x = N + \frac{d_1}{10} + \frac{d_2}{10^2} + \frac{d_3}{10^3} + \cdots, \tag{6.1}$$

where N is a nonnegative integer and $0 \le d_1, d_2, d_3, \cdots \le 9$. One writes the left hand side above in short as

$$N.d_1 d_2 \cdots$$

and the '.' above is called the *decimal point*. The above claim can easily be justified based on our geometric picture of the real numbers as being points on the number line.

$$x = N.37\cdots$$

Indeed, if x is an integer N, then we simply take $d_1 = d_2 = d_3 = \cdots = 0$. If x is not an integer, then we take $N = \lfloor x \rfloor$. Then we divide the interval $[N, N+1)$ into 10 equal parts, and if x is not one of the subdivision points, then we take d_1 to be the number in $D := \{0, 1, 2, \cdots, 9\}$ such that

$$N + \frac{d_1}{10} < x < N + \frac{d_1 + 1}{10}.$$

Otherwise we take $d_2 = d_3 = \cdots = 0$ and d_1 such that $N + \frac{d_1}{10} = x$. We can then divide $[N + \frac{d_1}{10}, N + \frac{d_1+1}{10})$ into 10 equal parts (each of length 10^{-2}) and continue the process. Note that at the nth stage, x satisfies

$$N + \frac{d_1}{10} + \cdots + \frac{d_n}{10^n} \le x < N + \frac{d_1}{10} + \cdots + \frac{d_n + 1}{10^n},$$

and this shows that the nth partial sum s_n of the series

$$N + \frac{d_1}{10} + \frac{d_2}{10^2} + \frac{d_3}{10^3} + \cdots$$

satisfies $|s_n - x| < 10^{-n}$, showing the validity of (6.1).

Vice versa, every decimal expansion $N.d_1 d_2 d_3 \cdots$ converges to *some* nonnegative real number. The reason is that the sequence of partial sums is clearly increasing, and moreover bounded above by

$$N + \frac{9}{10} + \frac{9}{10^2} + \frac{9}{10^3} + \cdots = N + \frac{\frac{9}{10}}{1 - \frac{1}{10}} = N + 1.$$

(1) Show that $0.999 \cdots = 1.000 \cdots$.

(2) Show that every terminating decimal expansion (that is, one in which there is some K such that $d_k = 0$ for all $k > K$—the digits are all eventually zeros) is a nonnegative rational number.

(3) Show that every nonterminating but repeating decimal expansion, that is, a decimal expansion of the form

$$N.d_1 \cdots d_n \boxed{d_{n+1} \cdots d_{n+m}} \boxed{d_{n+1} \cdots d_{n+m}} \boxed{d_{n+1} \cdots d_{n+m}} \cdots$$

where a block of digits $\boxed{d_{n+1} \cdots d_{n+m}}$ keeps repeating, is a positive rational number.

Conversely, it can be shown that every nonnegative rational number has either a terminating or a repeating decimal expansion. See the appendix to this chapter on page 335.

(4) Find the rational number corresponding to $0.123123123\cdots$.

(5) Is $0.12345678910111213\cdots$ a rational number?

Exercise 6.9 (Integral Test). As in the proof of Theorem 6.2, it can sometimes be easy to determine whether or not the improper integral

$$\int_1^\infty f(x)dx$$

converges or diverges, and this can be used to deduce the convergence status of the series

$$\sum_{n=1}^\infty f(n).$$

This result is known as the *Integral Test*, and the first aim of this exercise is to prove this.

(1) Let $f : [1, \infty) \to [0, \infty)$ be a decreasing function, such that f is Riemann integrable on $[1, n]$ for all $n \in \mathbb{N}$. Show the inequalities

$$\sum_{k=2}^n f(k) \le \int_1^n f(x)dx \le \sum_{k=1}^{n-1} f(k),$$

for all $n \in \mathbb{N}$. Conclude the Integral Test, which says that

$$\sum_{n=1}^\infty f(n) \text{ converges if and only if } \int_1^\infty f(x)dx \text{ does.}$$

(2) Does $\displaystyle\sum_{n=2}^\infty \frac{1}{n \log n}$ converge? What about $\displaystyle\sum_{n=2}^\infty \frac{1}{n(\log n)^2}$? *Hint:* See Exercise 5.46.

6.2   Absolute convergence

Definition 6.2. The series $\displaystyle\sum_{n=1}^\infty a_n$ *converges absolutely if* $\displaystyle\sum_{n=1}^\infty |a_n|$ *converges.*

The name is justified, thanks to the following result.

Proposition 6.4. *If* $\displaystyle\sum_{n=1}^\infty |a_n|$ *converges, then* $\displaystyle\sum_{n=1}^\infty a_n$ *converges.*

Proof. Let $s_n := a_1 + \cdots + a_n$, $\sigma_n := |a_1| + \cdots + |a_n|$, for $n \in \mathbb{N}$. Since

$$\sum_{n=1}^{\infty} |a_n| < +\infty,$$

$(\sigma_n)_{n \in \mathbb{N}}$ is convergent, and in particular a Cauchy sequence. Let $\epsilon > 0$. Then there exists an $N \in \mathbb{N}$ such that for all $n, m > N$, $|\sigma_n - \sigma_m| < \epsilon$. So for $n > m > N$,

$$\epsilon > |\sigma_n - \sigma_m| = \sigma_n - \sigma_m$$
$$= (|a_1| + \cdots + |a_n|) - (|a_1| + \cdots + |a_m|) = |a_{m+1}| + \cdots + |a_n|$$
$$\geq |a_{m+1} + \cdots + a_n| = |(a_1 + \cdots + a_n) - (a_1 + \cdots + a_m)| = |s_n - s_m|.$$

Thus $(s_n)_{n \in \mathbb{N}}$ is Cauchy, and so it is convergent. Hence $\sum_{n=1}^{\infty} a_n$ converges. $\qquad\square$

Exercise 6.10. Does the series $\sum_{n=1}^{\infty} \dfrac{\sin n}{n^2}$ converge?

Example 6.4.

(1) The series $\sum_{n=1}^{\infty} \dfrac{(-1)^n}{n^2}$ converges absolutely, since $\sum_{n=1}^{\infty} \left| \dfrac{(-1)^n}{n^2} \right| = \sum_{n=1}^{\infty} \dfrac{1}{n^2} < \infty.$

In particular, $\sum_{n=1}^{\infty} \dfrac{(-1)^n}{n^2}$ converges.

(2) The series $\sum_{n=1}^{\infty} \dfrac{(-1)^n}{n}$ does not converge absolutely, since

$$\sum_{n=1}^{\infty} \left| \dfrac{(-1)^n}{n} \right| = \sum_{n=1}^{\infty} \dfrac{1}{n},$$

and we have seen that the harmonic series diverges. Does $\sum_{n=1}^{\infty} \dfrac{(-1)^n}{n}$ converge?

(See the following discussion to find out the answer.) $\qquad\diamond$

Definition 6.3 (Alternating Series). A series of the form

$$\sum_{n=1}^{\infty} (-1)^n a_n$$

with $a_n \geq 0$ for all $n \in \mathbb{N}$ is called an *alternating series*.

We note that $\sum_{n=1}^{\infty} \dfrac{(-1)^n}{n}$ is an alternating series $\sum_{n=1}^{\infty} (-1)^n a_n$ with $a_n := \dfrac{1}{n}$ $(n \in \mathbb{N})$.

We will now learn a result below, called the Leibniz Alternating Series Theorem, which will enable us to conclude that in fact this alternating series is convergent (since the sufficiency conditions for convergence in the Leibniz Alternating Series Theorem are satisfied:

$$a_1 = 1 \geq a_2 = \frac{1}{2} \geq a_3 = \frac{1}{3} \geq \cdots$$

and $\lim\limits_{n \to \infty} a_n = \lim\limits_{n \to \infty} \frac{1}{n} = 0$). Thus

$$\sum_{n=1}^{\infty} \frac{(-1)^n}{n}$$

converges, but does not converge absolutely, showing that:

$$\text{Absolute convergence} \quad \underset{\not\Leftarrow}{\Rightarrow} \quad \text{Convergence}.$$

Theorem 6.5 (Leibniz Alternating Series Theorem). *Let* $(a_n)_{n \in \mathbb{N}}$ *be a sequence such that*

(1) *it has nonnegative terms* ($a_n \geq 0$ *for all* n),

(2) *it is decreasing* ($a_1 \geq a_2 \geq a_3 \geq \cdots$), *and*

(3) $\lim\limits_{n \to \infty} a_n = 0$.

Then the series $\sum\limits_{n=1}^{\infty} (-1)^n a_n$ *converges.*

A pictorial 'proof without words' is shown below. The sum of the lengths of the disjoint dark intervals is at most the length of $(0, a_1)$.

Proof. We may just as well prove the convergence of

$$\sum_{n=1}^{\infty} (-1)^{n+1} a_n \left(= -\sum_{n=1}^{\infty} (-1)^n a_n \right).$$

Let $s_n = a_1 - a_2 + a_3 - + \cdots + (-1)^{n-1} a_n$. Clearly,

$$s_{2n+1} = s_{2n-1} - a_{2n} + a_{2n+1} \leq s_{2n-1},$$

$$s_{2n+2} = s_{2n} + a_{2n+1} - a_{2n+2} \geq s_{2n},$$

and so the sequence $s_2, s_4, s_6, \cdots$ is increasing, while the sequence $s_3, s_5, s_7, \cdots$ is decreasing. Also,

$$s_{2n} \leq s_{2n} + a_{2n+1} = s_{2n+1} \leq s_{2n-1} \leq \cdots \leq s_3.$$

So $(s_{2n})_{n\in\mathbb{N}}$ is a bounded $(s_2 \leq s_{2n} \leq s_3$ for all $n)$, increasing sequence, and hence it is convergent. But as $(a_{2n+1})_{n\in\mathbb{N}}$ is also convergent with limit 0, it follows that

$$\lim_{n\to\infty} s_{2n+1} = \lim_{n\to\infty} (s_{2n} + a_{2n+1}) = \lim_{n\to\infty} s_{2n}.$$

Hence $(s_n)_{n\in\mathbb{N}}$ is convergent, and so the series converges.   □

Exercise 6.11. Let $s > 0$. Show that $\displaystyle\sum_{n=1}^{\infty} \frac{(-1)^n}{n^s}$ converges.

Exercise 6.12. Prove that $\displaystyle\sum_{n=1}^{\infty} (-1)^n \frac{\sqrt{n}}{n+1}$ converges.

Exercise 6.13. Prove that $\displaystyle\sum_{n=1}^{\infty} (-1)^n \sin \frac{1}{n}$ converges.

6.2.1   Rearrangement of series

One might tend to think of a series as an 'infinite sum', and hence be tempted to attribute to it the usual properties associated with finite sums. For example, while adding a finite bunch of numbers, we know that we can do so in any order and grouping, as addition is associative and commutative. But with series, it turns out that rearrangements *can* give different answers!

Definition 6.4 (Permutation; Rearrangement). A bijective mapping $p : \mathbb{N} \to \mathbb{N}$ is called a *permutation* (of $\mathbb{N}$). The series

$$\sum_{n=1}^{\infty} a_{p(n)}$$

is called a *rearrangement of the series* $\displaystyle\sum_{n=1}^{\infty} a_n$.

Example 6.5 (Rearrangements of series may have different sums). Let S be the sum of the convergent series

$$1 - \frac{1}{2} + \frac{1}{3} - + \cdots = S.$$

(We will see later that $S = \log 2 \neq 0$.) Consider its rearrangement,

$$1 + \frac{1}{3} - \frac{1}{2} + \frac{1}{5} + \frac{1}{7} - \frac{1}{4} + \cdots .$$

Indeed, this series has the same terms as in the original series, but the pattern of signs is

$$+ + - \ + + - \ + + - \ \cdots$$

instead of the original pattern

$$+ - \ + - \ + - \ \cdots .$$

Note that since

$$S = 1 - \frac{1}{2} + \frac{1}{3} - \frac{1}{4} + \frac{1}{5} - \frac{1}{6} + \cdots ,$$

$$\frac{1}{2}S = \quad \frac{1}{2} \quad - \frac{1}{4} \quad + \frac{1}{6} - \cdots ,$$

and so we obtain

$$\frac{3}{2}S = 1 + \frac{1}{3} - \frac{1}{2} + \frac{1}{5} + \frac{1}{7} - \frac{1}{4} + \cdots .$$

This shows that the rearranged series has a sum which is one and a half times the sum of the original series!   ◇

Note that in the above example, the series was certainly convergent but it wasn't *absolutely* convergent. The next result shows that arbitrary rearrangements don't change the sum of an absolutely convergent series.

Theorem 6.6. *Let p be any permutation of* $\mathbb{N}$. *If the series*

$$\sum_{n=1}^{\infty} a_n$$

is absolutely convergent, then so is $\displaystyle\sum_{n=1}^{\infty} a_{p(n)}$, *and moreover, their sums coincide.*

Proof. Consider first the case when each a_n is nonnegative, and let s_n, s'_n denote their respective partial sums. Then $(s_n)_{n \in \mathbb{N}}$ and $(s'_n)_{n \in \mathbb{N}}$ are both increasing sequences and $(s_n)_{n \in \mathbb{N}}$ converges to $\ell := \sup\{s_n : n \in \mathbb{N}\}$. But for each n, there is some m such that the terms in s'_n all appear in the sum s_m, so that

$$s'_n \le s_m \le \ell.$$

Hence, for all $n \in \mathbb{N}$, $s'_n \le \ell$. So, the increasing sequence $(s'_n)_{n \in \mathbb{N}}$ is bounded above and hence convergent to $\ell' := \sup\{s'_n : n \in \mathbb{N}\}$. The above also shows that $\ell' \le \ell$. By considering

$$\sum_{n=1}^{\infty} a_n$$

as a rearrangement of the absolutely convergent series (having nonnegative terms)

$$\sum_{n=1}^{\infty} a_{p(n)},$$

we also get the reverse inequality $\ell \le \ell'$.

Now let us consider the general case. To this end, we first note that

$$S_1 := \sum_{n=1}^{\infty} (|a_n| - a_n)$$

is a convergent series of nonnegative terms. By the previous part,

$$\sum_{n=1}^{\infty} (|a_{p(n)}| - a_{p(n)})$$

is convergent too, with the same sum S_1. Also, from the convergence of the series (with nonnegative terms)

$$S_2 := \sum_{n=1}^{\infty} |a_n|,$$

it follows that $\displaystyle\sum_{n=1}^{\infty} |a_{p(n)}|$ is convergent too, with the same sum S_2.

Putting all of this together, it follows that $\displaystyle\sum_{n=1}^{\infty} a_{p(n)}$ is absolutely convergent, and

$$\sum_{n=1}^{\infty} a_{p(n)} = \sum_{n=1}^{\infty} |a_{p(n)}| - \sum_{n=1}^{\infty} (|a_{p(n)}| - a_{p(n)}) = S_2 - S_1$$

$$= \sum_{n=1}^{\infty} |a_n| - \sum_{n=1}^{\infty} (|a_n| - a_n) = \sum_{n=1}^{\infty} a_n.$$

This completes the proof.   □

In light of the above results and the previous example, one might wonder what happens with series that are convergent, but not absolutely convergent. (Such series are sometimes called *conditionally convergent*.) Well, the behaviour is radically different, as demonstrated by the following result. It is surprising enough that the naive expectation of 'commutativity' fails, but even more striking is the fact that the rearrangement can be done so as to get any limit whatsoever!

Theorem 6.7 (Riemann Rerrangement Theorem).

Let $\displaystyle\sum_{n=1}^{\infty} a_n$ *be a conditionally convergent series.*

(1) *If* $L \in \mathbb{R}$, *then there exists a permutation* $p_L : \mathbb{N} \to \mathbb{N}$ *such that*

$$\sum_{n=1}^{\infty} a_{p_L(n)} = L.$$

(2) *Similarly, there exist permutations* p_∞ *and* $p_{-\infty}$ *such that*

$$\sum_{n=1}^{\infty} a_{p_\infty(n)} \quad and \quad \sum_{n=1}^{\infty} a_{p_{-\infty}(n)}$$

diverge to ∞ *and* $-\infty$, *respectively.*

The proof of this interesting result is a bit long, although elementary, and so we will skip it[5].

Incidentally, the Riemann Rearrangement Theorem appears in the paper where Riemann also defines the Riemann integral: '*Ueber die Darstellbarkeit einer Function durch eine trigonometrische Reihe*' (On the representability of a function by a trigonometric series). This paper was submitted to the University of Göttingen in 1854 as Riemann's Habilitationsschrift (an academic qualification earned after obtaining a research doctorate as a qualification

---

[5] The gist is as follows: The series of nonnegative terms a_n^+ diverges to ∞, and the series of negative terms a_n^- diverges to $-\infty$ (Why? One can argue that otherwise the original series would be *absolutely* convergent.) Let $L \in \mathbb{R}$. As the series of positive terms diverges to ∞, there is a first index n so that $P = a_1^+ + a_2^+ + \cdots + a_n^+ > L$. As the series of negative terms diverges to $-\infty$, there's a first index m such that $PM = P + a_1^- + a_2^- + \cdots + a_m^- < L$. Similarly, let n' and m' be the smallest indices such that $PMP = PM + a_{n+1}^+ + a_{n+2}^+ + \cdots + a_{n'}^+ > L$ and $PMPM = PMP + a_{m+1}^- + a_{m+2}^- + \cdots + a_{m'}^- < L$. And so on. The sequence $P, PM, PMP, PMPM, PMPMP, \cdots$ so obtained converges to L since a_n^+ and a_n^- tend to 0, implying that, with every step in the construction, the difference between the resulting sum and L becomes smaller. An interested reader may wish to work out the details above or look them up for example in [A] or [R].

to become an instructor or supervise doctoral students). It was published in 1868 in the Proceedings of the Royal Philosophical Society at Göttingen. Riemann's definition of his integral appears in Section 4, while the rearrangement theorem appears in Section 3.

Exercise 6.14 (Inserting or removing parenthesis).

(1) Show that if a series converges, then the new series one obtains by 'inserting parentheses' in the original one (that is, adding up finite blocks of consecutive terms) converges to the same sum.

(2) Give a (simple) example to show that *removing* parentheses can change a convergent series into a divergent one!

(3) It is easy to show the inequality that for all $n \in \mathbb{N}$, $\frac{1}{3n-1} + \frac{1}{3n} + \frac{1}{3n+1} > 3 \cdot \frac{1}{3n}$.
Give another proof of the divergence of the Harmonic Series based on grouping terms.

6.2.2 Comparison, ratio, root

We will now learn three important tests for the convergence of a series:

(1) the comparison test (where we compare with a series whose convergence status is known)

(2) the ratio test (where we look at the behaviour of the ratio $\dfrac{a_{n+1}}{a_n}$ of terms)

(3) the root test (where we look at the behaviour of $\sqrt[n]{|a_n|}$)

We summarise them in the table below.

		Comparison	Ratio	Root
Absolute convergence	$\Leftarrow$	$\|a_n\| \leq c_n$ for all large n; $\displaystyle\sum_{n=1}^{\infty} c_n$ converges.	$\left\|\dfrac{a_{n+1}}{a_n}\right\| \leq r < 1$ for all large n.	$\sqrt[n]{\|a_n\|} \leq r < 1$ for all large n.
Divergence	$\Leftarrow$	$a_n \geq d_n \geq 0$ for all large n; $\displaystyle\sum_{n=1}^{\infty} d_n$ diverges.	$\left\|\dfrac{a_{n+1}}{a_n}\right\| \geq 1$ for all large n.	$\sqrt[n]{\|a_n\|} \geq 1$ infinitely often.

We prove these results in the following.

Theorem 6.8 (Comparison test).

(C): *If* $(a_n)_{n\in\mathbb{N}}$ *and* $(c_n)_{n\in\mathbb{N}}$ *are such that*

(1) *there exists an* $N \in \mathbb{N}$ *such that* $|a_n| \leq c_n$ *for all* $n \geq N$, *and*

(2) $\displaystyle\sum_{n=1}^{\infty} c_n$ *converges,*

then $\displaystyle\sum_{n=1}^{\infty} a_n$ *converges absolutely.*

(D): *If* $(a_n)_{n\in\mathbb{N}}$ *and* $(d_n)_{n\in\mathbb{N}}$ *are such that*

    (1) *there exists an* $N \in \mathbb{N}$ *such that* $a_n \geq d_n \geq 0$ *for all* $n \geq N$, *and*

    (2) $\displaystyle\sum_{n=1}^{\infty} d_n$ *diverges,*

then $\displaystyle\sum_{n=1}^{\infty} a_n$ *diverges.*

Proof.

(Convergence): Let $s_n := |a_1| + \cdots + |a_n|$ and $\sigma_n := c_1 + \cdots + c_n$. For $n > m$,

$$|s_n - s_m| = |a_{m+1}| + \cdots + |a_n| \leq c_{m+1} + \cdots + c_n = |\sigma_n - \sigma_m|,$$

and since $(\sigma_n)_{n\in\mathbb{N}}$ is Cauchy, it follows that $(s_n)_{n\in\mathbb{N}}$ is also Cauchy. Hence $(s_n)_{n\in\mathbb{N}}$ is absolutely convergent.

(Divergence): By the previous part, if $\displaystyle\sum_{n=1}^{\infty} a_n$ converges, then so must $\displaystyle\sum_{n=1}^{\infty} d_n$. $\qquad\square$

Example 6.6.

(1) $\displaystyle\sum_{n=1}^{\infty} \frac{\cos n}{n^2}$ converges.

We have $\left|\dfrac{\cos n}{n^2}\right| \leq \dfrac{1}{n^2} =: c_n$ for all $n \in \mathbb{N}$.

Since

$$\sum_{n=1}^{\infty} c_n = \sum_{n=1}^{\infty} \frac{1}{n^2} < \infty,$$

it follows by the Comparison Test that $\displaystyle\sum_{n=1}^{\infty} \frac{\cos n}{n^2}$ converges (absolutely).

(2) $\displaystyle\sum_{n=1}^{\infty} \sin \frac{1}{n}$ diverges.

Indeed, by Taylor's Formula, for each $n \in \mathbb{N}$, there exists a $c_n \in (0, n)$ such that

$$\sin \frac{1}{n} = \sin 0 + \frac{\cos 0}{1!} \cdot \frac{1}{n} - \frac{\sin c_n}{2!} \cdot \frac{1}{n^2}$$

$$= \frac{1}{n} - \frac{\sin c_n}{2n^2} \geq \frac{1}{n} - \frac{1}{2n^2}$$

$$\geq \frac{1}{n} - \frac{1}{2n} = \frac{1}{2n} =: d_n \geq 0.$$

Since

$$\sum_{n=1}^{\infty} d_n = \frac{1}{2} \sum_{n=1}^{\infty} \frac{1}{n}$$

diverges, it follows by the Comparison Test that $\displaystyle\sum_{n=1}^{\infty} \sin \frac{1}{n}$ diverges.

(3) $\displaystyle\sum_{n=1}^{\infty} \left(\sin \frac{1}{n} - \frac{1}{n} \right)$ converges.

By Taylor's Formula, for each $n \in \mathbb{N}$, there exists a $c_n \in (0, n)$ such that

$$\sin \frac{1}{n} = \sin 0 + \frac{\cos 0}{1!} \cdot \frac{1}{n} - \frac{\sin 0}{2!} \cdot \frac{1}{n^2} - \frac{\cos c_n}{3!} \cdot \frac{1}{n^3},$$

and so

$$\left| \sin \frac{1}{n} - \frac{1}{n} \right| = \left| -\frac{\cos c_n}{6n^3} \right| \leq \frac{1}{6n^3}.$$

As $\displaystyle\sum_{n=1}^{\infty} \frac{1}{n^3} < \infty$, by the Comparison Test, $\displaystyle\sum_{n=1}^{\infty} \left(\sin \frac{1}{n} - \frac{1}{n} \right)$ converges absolutely.

(4) $\displaystyle\sum_{n=2}^{\infty} \frac{1}{\log n}$ diverges.

For all $n \in \mathbb{N}$, $\log n \leq n$. (By the Mean Value Theorem, there exists a $c \in (1, n)$ such that

$$\frac{\log n - \log 1}{n - 1} = \frac{\log n}{n - 1} = \frac{1}{c} < 1,$$

and by rearranging, $\log n < n - 1 < n$ for $n > 1$.) For all $n \geq 2$,

$$\frac{1}{\log n} \geq \frac{1}{n} =: d_n,$$

and so by the Comparison Test, $\displaystyle\sum_{n=2}^{\infty} \frac{1}{\log n}$ diverges.   ◇

Theorem 6.9 (Ratio test). *Let* $(a_n)_{n \in \mathbb{N}}$ *be a sequence of nonzero terms.*

(C): *If there exists an* $r \in (0, 1)$ *and an* $N \in \mathbb{N}$ *such that for all* $n > N$, $\left| \dfrac{a_{n+1}}{a_n} \right| \leq r$,

then $\displaystyle\sum_{n=1}^{\infty} a_n$ *converges absolutely.*

(D): *If there exists an* $N \in \mathbb{N}$ *such that for all* $n > N$, $\left| \dfrac{a_{n+1}}{a_n} \right| \geq 1$,

then $\displaystyle\sum_{n=1}^{\infty} a_n$ *diverges.*

Proof.

(Convergence): We have

$$|a_{N+1}| \leq r|a_N|,$$

$$|a_{N+2}| \leq r|a_{N+1}| \leq r^2|a_N|,$$

$$|a_{N+3}| \leq r|a_{N+2}| \leq r^3|a_N|,$$

$$\cdots$$

Since $\displaystyle\sum_{n=1}^{\infty} r^n$ converges, $\displaystyle\sum_{n=N+1}^{\infty} |a_n| < +\infty$ by the Comparison Test.

Adding $|a_1| + \cdots + |a_N|$ to the partial sums of this last series, we see that

$$\sum_{n=1}^{\infty} |a_n|$$

converges too.

(Divergence): The given condition implies that

$$\cdots \geq |a_{N+3}| \geq |a_{N+2}| \geq |a_{N+1}|. \tag{6.2}$$

If the series $\displaystyle\sum_{n=1}^{\infty} a_n$ was convergent, then

$$0 = \lim_{n \to \infty} a_n = \lim_{k \to \infty} a_{N+k}.$$

Hence

$$\lim_{k \to \infty} |a_{N+k}| = 0$$

as well. But by (6.2), we see that $\displaystyle\lim_{k \to \infty} |a_{N+k}| \geq |a_{N+1}| > 0$, a contradiction. □

[6]It does *not* suffice for convergence of the series that

$$\text{for all sufficiently large } n, \quad \left| \frac{a_{n+1}}{a_n} \right| < 1.$$

Indeed, for the harmonic series, $\left| \dfrac{a_{n+1}}{a_n} \right| = \dfrac{\frac{1}{n+1}}{\frac{1}{n}} = \dfrac{n}{n+1} < 1$, but $\displaystyle\sum_{n=1}^{\infty} \frac{1}{n}$ diverges.

So the ratios have to uniformly separated from 1 (by a positive distance $1 - r$).

Note that in the case of the Harmonic Series, there is no $r \in (0, 1)$ such that

$$\left| \frac{a_{n+1}}{a_n} \right| = \frac{n}{n+1} \leq r < 1 \quad \text{for all large } n,$$

since if there were such an r, then

$$\lim_{n \to \infty} \frac{n}{n+1} = 1 \leq r < 1,$$

a contradiction.

---

[6] This 'dangerous bend' symbol was used by the 'Bourbaki group' of mathematicians, and appears in the margins of mathematics books written by the group to mark cautionary notes.

Corollary 6.10. *If* $\lim\sup\limits_{n\to\infty}\left|\dfrac{a_{n+1}}{a_n}\right| < 1$, *then* $\sum\limits_{n=1}^{\infty} a_n$ *converges absolutely.*

Proof. Let $L := \lim\sup\limits_{n\to\infty}\left|\dfrac{a_{n+1}}{a_n}\right| < 1$. Let r be such that $L < r < 1$. Set

$$b_n := \sup\left\{\left|\frac{a_{k+1}}{a_k}\right| : k \geq n\right\}.$$

Recalling the definition of $\lim\sup$ from Exercise 2.10, we see that the sequence $(b_n)_{n\in\mathbb{N}}$ is decreasing and that

$$L = \lim\sup\limits_{n\to\infty}\left|\frac{a_{n+1}}{a_n}\right| = \lim\limits_{n\to\infty} b_n = \inf\limits_{n\in\mathbb{N}} b_n.$$

Thus there exists an $N \in \mathbb{N}$ such that $L \leq b_N = \sup\left\{\left|\dfrac{a_{k+1}}{a_k}\right| : k \geq N\right\} < r < 1$.

So for all $n > N$, $\left|\dfrac{a_{n+1}}{a_n}\right| < r < 1$.

Using Part (C) from Theorem 6.9, we conclude that $\sum\limits_{n=1}^{\infty} a_n$ converges absolutely. $\qquad\square$

Remark 6.1. We know from Exercise 2.21 that if

$$\lim\limits_{n\to\infty}\left|\frac{a_{n+1}}{a_n}\right|$$

exists, then $\lim\sup\limits_{n\to\infty}\left|\dfrac{a_{n+1}}{a_n}\right|$ exists as well, and $\lim\sup\limits_{n\to\infty}\left|\dfrac{a_{n+1}}{a_n}\right| = \lim\limits_{n\to\infty}\left|\dfrac{a_{n+1}}{a_n}\right|$.

Example 6.7 (The exponential series). For all $x \in \mathbb{R}$, $\sum\limits_{n=0}^{\infty}\dfrac{x^n}{n!}$ converges.

If $x = 0$, then this is trivial. If $x \neq 0$, then convergence follows from the Ratio Test:

$$\left|\frac{\dfrac{x^{n+1}}{(n+1)!}}{\dfrac{x^n}{n!}}\right| = \frac{|x|}{n+1} \xrightarrow{n\to\infty} 0.$$

$\diamond$

Theorem 6.11 (Root test).

(C): *If there exists an $r \in (0,1)$ and an $N \in \mathbb{N}$ such that for all $n > N$, $\sqrt[n]{|a_n|} \leq r$, then* $\sum\limits_{n=1}^{\infty} a_n$ *converges absolutely.*

(D): *If for infinitely many n, $\sqrt[n]{|a_n|} \geq 1$, then* $\sum\limits_{n=1}^{\infty} a_n$ *diverges.*

Proof.

(Convergence): As $|a_n| \leq r^n$ for all $n > N$, by the Comparison Test,

$$\sum_{n=N+1}^{\infty} |a_n|$$

converges.

(Divergence): Suppose that for the subsequence $(a_{n_k})_{k \in \mathbb{N}}$, we have

$$\sqrt[n_k]{|a_{n_k}|} \geq 1.$$

Then $|a_{n_k}| \geq 1$. If the series was convergent, then

$$\lim_{n \to \infty} a_n = 0,$$

and so also

$$\lim_{n \to \infty} |a_{n_k}| = 0,$$

a contradiction. Thus the series $\displaystyle\sum_{n=1}^{\infty} a_n$ cannot converge. □

 It does not suffice for convergence of the series that

$$\text{for sufficiently large } n, \ \sqrt[n]{|a_n|} < 1.$$

For example, for the harmonic series $\sqrt[n]{|a_n|} = \dfrac{1}{\sqrt[n]{n}} < 1$, but $\displaystyle\sum_{n=1}^{\infty} \dfrac{1}{n}$ diverges.

So again, one needs the uniform separation from 1 (by a positive distance $1 - r$).

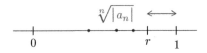

Corollary 6.12. *If* $\limsup\limits_{n \to \infty} \sqrt[n]{|a_n|} < 1$, *then* $\displaystyle\sum_{n=1}^{\infty} a_n$ *converges absolutely.*

Proof. Let $L := \limsup\limits_{n \to \infty} \sqrt[n]{|a_n|} < 1$.

Let r be such that $L < r < 1$. Set $b_n := \sup\{\sqrt[k]{|a_k|} : k \geq n\}$. The sequence $(b_n)_{n \in \mathbb{N}}$ is decreasing, and $L = \limsup\limits_{n \to \infty} \sqrt[n]{|a_n|} = \lim\limits_{n \to \infty} b_n = \inf\limits_{n \in \mathbb{N}} b_n$.

Thus there exists an $N \in \mathbb{N}$ such that $L \leq b_N = \sup\{\sqrt[k]{|a_k|} : k \geq N\} < r < 1$. So for all $n > N$, $\sqrt[n]{|a_n|} < r < 1$.

By Part (C) of Theorem 6.11, $\displaystyle\sum_{n=1}^{\infty} a_n$ converges absolutely. □

Example 6.8 (Ratio Test inconclusive; but Root Test decisive).

$\sum\limits_{n=1}^{\infty} \dfrac{1}{2^{n+(-1)^n}}$ converges. We have

$$n: \quad 1 \xrightarrow{1} 2 \xrightarrow{1} \quad 3 \xrightarrow{1} 4 \xrightarrow{1} \quad 5 \xrightarrow{1} 6 \cdots$$

$$(-1)^n: \ -1 \xrightarrow{2} 1 \xrightarrow{-2} -1 \xrightarrow{2} 1 \xrightarrow{-2} -1 \xrightarrow{2} 1 \cdots$$

$$n+(-1)^n: \quad \cdot \xrightarrow{3} \cdot \xrightarrow{-1} \quad \cdot \xrightarrow{3} \cdot \xrightarrow{-1} \quad \cdot \xrightarrow{3} \cdots$$

So,

$$\left| \frac{a_{n+1}}{a_n} \right|$$

alternates between $2^{-3} = 1/8$ and $2^1 = 2$, and the Ratio Test is inconclusive. But

$$\sqrt[n]{|a_n|} = \frac{1}{2^{\frac{n+(-1)^n}{n}}} = \frac{1}{2^{1+\frac{(-1)^n}{n}}} \xrightarrow{n\to\infty} \frac{1}{2^{1+0}} = \frac{1}{2} < 1,$$

and so, by the Root Test, $\sum\limits_{n=1}^{\infty} \dfrac{1}{2^{n+(-1)^n}}$ converges.   ◇

Exercise 6.15. Determine if the following series is convergent or not.

(1) $\sum\limits_{n=1}^{\infty} \dfrac{n^2}{2^n}.$

(2) $\sum\limits_{n=1}^{\infty} \dfrac{(n!)^2}{(2n)!}.$

(3) $\sum\limits_{n=1}^{\infty} \left(\dfrac{4}{5}\right)^n n^5.$

Exercise 6.16. ($*$) Prove that $\sum\limits_{n=1}^{\infty} \dfrac{n}{n^4 + n^2 + 1}$ converges. Also, find its value.

Hint: Write the denominator as $(n^2 + 1)^2 - n^2$.

Exercise 6.17. If $\sum\limits_{n=1}^{\infty} a_n^{2014}$ converges, then show that $\sum\limits_{n=1}^{\infty} a_n^{2015}$ converges too.

Hint: First conclude that for large n, $|a_n| < 1$.

Exercise 6.18. Determine if the following statements are true or false.

(1) If $\sum\limits_{n=1}^{\infty} |a_n|$ is convergent, then so is $\sum\limits_{n=1}^{\infty} a_n^2.$

(2) If $\sum_{n=1}^{\infty} a_n$ is convergent, then so is $\sum_{n=1}^{\infty} a_n^2$.

(3) If $\lim_{n \to \infty} a_n = 0$, then $\sum_{n=1}^{\infty} a_n$ converges.

(4) If $\lim_{n \to \infty} (a_1 + \cdots + a_n) = 0$, then $\sum_{n=1}^{\infty} a_n$ converges.

(5) $\sum_{n=1}^{\infty} \log \dfrac{n+1}{n}$ converges.

(6) If all $a_n > 0$ and the partial sums of $(a_n)_{n \in \mathbb{N}}$ are bounded, then $\sum_{n=1}^{\infty} a_n$ converges.

(7) If $a_n \neq 0$ $(n \in \mathbb{N})$ and $\sum_{n=1}^{\infty} a_n$ converges, then $\sum_{n=1}^{\infty} \dfrac{1}{a_n}$ diverges.

Exercise 6.19 (Fourier Series). In order to understand a complicated situation, it is natural to try to break it up into simpler things. For example, from Calculus we learn that an analytic function can be expanded into a Taylor series, where we break it down into the simplest possible analytic functions, namely monomials $1, x, x^2, \cdots$ as follows:

$$f(x) = f(0) + f'(0)x + \frac{f''(0)}{2!}x^2 + \cdots.$$

The idea behind the Fourier series is similar. In order to understand a complicated *periodic* function, we break it down into the simplest periodic functions, namely sines and cosines. Thus if $T \geq 0$ and $f : \mathbb{R} \to \mathbb{R}$ is T-*periodic*, that is, $f(x) = f(x + T)$ $(x \in \mathbb{R})$, then one tries to find coefficients $a_0, a_1, a_2, \cdots$ and $b_1, b_2, b_3, \cdots$ such that

$$f(x) = a_0 + \sum_{n=1}^{\infty} \left(a_n \cos\left(\frac{2\pi n}{T}x\right) + b_n \sin\left(\frac{2\pi n}{T}x\right) \right). \tag{6.3}$$

Suppose that the Fourier series given in (6.3) converges pointwise to f on $\mathbb{R}$. Show that if

$$\sum_{n=1}^{\infty} (|a_n| + |b_n|) < \infty,$$

then in fact the series converges uniformly.

The aim of this part of the exercise is to give experimental evidence for two things. Firstly, the plausibility of the Fourier expansion, and secondly, that the uniform convergence might fail if the condition in the previous part of this exercise does not hold. To this end, let us consider the *square wave* $f : \mathbb{R} \to \mathbb{R}$ given by

$$f(x) = \begin{cases} 1 & \text{if } x \in [n, n+1) \text{ for } n \text{ even,} \\ -1 & \text{if } x \in [n, n+1) \text{ for } n \text{ odd.} \end{cases}$$

Then f is a 2-periodic signal. From the theory of Fourier Series, which we will not discuss in this course, the coefficients can be calculated, and they happen to be

$$0 = a_0 = a_1 = a_2 = a_3 = \cdots$$

and

$$b_n = \begin{cases} \dfrac{4}{n\pi} & \text{if } n \text{ is odd,} \\ 0 & \text{if } n \text{ is even.} \end{cases}$$

Write a Maple program to plot the graphs of the partial sums of the series in (6.3) with, say, 3, 33, 333 terms. See Figure 6.1. Discuss your observations.

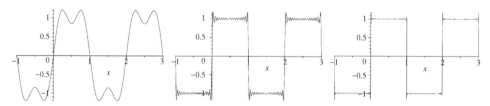

Figure 6.1   Partial sums of the Fourier series for the square wave.

Exercise 6.20. Show that if all $a_n \geq 0$ and $\displaystyle\sum_{n=1}^{\infty} a_n$ converges, then so does $\displaystyle\sum_{n=1}^{\infty} \sqrt{a_n a_{n+1}}$.

Exercise 6.21. If all $a_n \geq 0$, then show that $\displaystyle\sum_{n=1}^{\infty} a_n < \infty$ if and only if $\displaystyle\sum_{n=1}^{\infty} \frac{a_n}{1 + a_n} < \infty$.

Exercise 6.22. Let ℓ^1, ℓ^2 be the 'sequence spaces' defined by

$$\ell^1 := \left\{ (a_n)_{n\in\mathbb{N}} : \sum_{n=1}^{\infty} |a_n| < \infty \right\}, \qquad \ell^2 := \left\{ (a_n)_{n\in\mathbb{N}} : \sum_{n=1}^{\infty} |a_n|^2 < \infty \right\}.$$

Show that $\ell^1 \subset \ell^2$. Is $\ell^1 = \ell^2$?

Exercise 6.23. As $\displaystyle\sum_{n=1}^{\infty} \frac{1}{n}$ diverges, the reciprocal $1/s_n$ of the nth partial sum

$$s_n := 1 + \frac{1}{2} + \frac{1}{3} + \cdots + \frac{1}{n}$$

approaches 0 as $n \to \infty$. So the necessary condition for the convergence of the series

$$\sum_{n=1}^{\infty} \frac{1}{s_n}$$

is satisfied. But we don't know yet whether or not it actually converges. It is clear that the harmonic series diverges very slowly, which means that $1/s_n$ decreases very slowly, and this prompts the guess that this series diverges. Show that in fact our guess is correct.

Hint: $s_n < n$.

Exercise 6.24. Show that the series $\displaystyle\sum_{n=1}^{\infty} \frac{1}{n^n}$ converges.

Exercise 6.25. The Fibonnaci sequence $(F_n)_{n\in\mathbb{N}}$ is defined recursively by $F_0 = F_1 = 1$ and $F_{n+1} = F_n + F_{n-1}$ for $n \in \mathbb{N}$. Show that

$$\sum_{n=0}^{\infty} \frac{1}{F_n} < +\infty.$$

Hint: $F_{n+1} = F_n + F_{n-1} > F_{n-1} + F_{n-1} = 2F_{n-1}.$

Using this, show that both $\dfrac{1}{F_0} + \dfrac{1}{F_2} + \dfrac{1}{F_4} + \cdots$ and $\dfrac{1}{F_1} + \dfrac{1}{F_3} + \dfrac{1}{F_5} + \cdots$ converge.

Exercise 6.26. Determine if the series $\displaystyle\sum_{n=1}^{\infty} (\sqrt{1+n^2} - n)$ is convergent or not.

Exercise 6.27. (∗) Show that $\displaystyle\sum_{k=1}^{\infty} \sin\left(\pi\sqrt{k^4 + 1}\right)$ converges absolutely.

6.3   Power series

Let $(c_n)_{n\in\mathbb{N}}$ be a real sequence (thought of as a sequence of 'coefficients'). An expression of the type

$$\sum_{n=0}^{\infty} c_n x^n$$

is called a *power series* in the 'real variable'[7] $x \in \mathbb{R}$.

This is generalisation of the familiar polynomial function

$$c_0 + c_1 x + c_2 x^2 + c_3 x^3 + \cdots + c_n x^n.$$

Indeed, all polynomial expressions are (finite) power series, with the coefficients being eventually all zeros. For example,

$$1 + 399x - x^3 = \underset{c_0}{1} + \underset{c_1}{399x} + \underset{c_2}{0x^2} + \underset{c_3}{(-1)x^3} + \underset{c_4}{0x^4} + \underset{c_5}{0x^5} + \underset{c_6}{0x^6} + \cdots.$$

$\displaystyle\sum_{n=0}^{\infty} x^n, \sum_{n=0}^{\infty} \frac{1}{n!} x^n$ are examples of power series which are not polynomials.

Power series arise naturally in applications. For example, it can be shown that the following ordinary differential equation (ODE)

$$f''(x) + xf'(x) + x^2 f(x) = 0 \text{ with } f(0) = 1, f(1) = 0$$

---

[7] That is, there is a possibility of putting in various different values of x.

has the 'power series solution':

$$f(x) = 1 - \frac{1}{12}x^4 + \frac{1}{90}x^6 + \frac{1}{3360}x^8 + \cdots, \quad x \in [0, 1].$$

So questions about the convergence of power series are also natural.

Note that as yet we have not said anything about the set of $x \in \mathbb{R}$ where the power series converges. Of course, the power series always converges for $x = 0$. We ask:

$$\text{For which } x \in \mathbb{R} \text{ does } \sum_{n=0}^{\infty} c_n x^n \text{ converge?}$$

We will discover that the answer is: for all x in an interval like this:

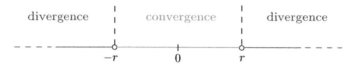

And the 'radius of convergence' r can be found from the coefficients $(c_n)_{n \geq 0}$. If the power series converges for all $x \in \mathbb{R}$, that is, if the above maximal interval is $(-\infty, \infty)$, we say that the power series has 'infinite radius of convergence'.

Example 6.9.

(1) The radius of convergence of $\sum_{n=0}^{\infty} x^n$ is 1.

Indeed, the geometric series converges for $x \in (-1, 1)$ and diverges if $|x| \geq 1$.

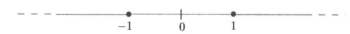

(2) The radius of convergence of $\sum_{n=0}^{\infty} \frac{1}{n!}x^n$ is infinite, as we had seen in Example 6.7.

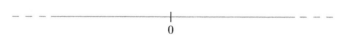

(3) The radius of convergence of $\sum_{n=0}^{\infty} n^n x^n$ is zero.

Indeed, if $x \neq 0$, then $\sqrt[n]{|n^n x^n|} = n|x| > 1$ for all n large enough[8]. By the Root test, it follows that the power series diverges for all nonzero real numbers. ◇

---

[8] $> 1/|x|$

Theorem 6.13. *Let* $(c_n)_{n\geq 0}$ *be a real sequence. Then*

either $\displaystyle\sum_{n=0}^{\infty} c_n x^n$ *is absolutely convergent for all* $x \in \mathbb{R}$

or *there exists a unique* $r \geq 0$ *such that*

(1) $\displaystyle\sum_{n=0}^{\infty} c_n x^n$ *is absolutely convergent for* $x \in (-r, r)$ *and*

(2) $\displaystyle\sum_{n=0}^{\infty} c_n x^n$ *diverges for* $x \notin [-r, r]$.

That is:

Proof. Let

$$S := \left\{ y \in [0, \infty) : \exists x \in \mathbb{R} \text{ such that } y = |x| \text{ and } \sum_{n=0}^{\infty} c_n x^n \text{ converges} \right\}.$$

Clearly $0 \in S$. Only two cases are possible:

$\underline{1°}$ S is not bounded above (in which case we'll show '$r = \infty$').

$\underline{2°}$ S is bounded above (in which case we'll show $r = \sup S$).

Here are the details.

$\underline{1°}$ Suppose that S is not bounded above. Let $x \in \mathbb{R}$. Then $|x|$ can't be an upper bound for S. So there must be an element $y \in S$ that prevents $|x|$ from being an upper bound, that is, we can find a $y = |x_0| \in S$ such that $|x| < |x_0|$ and

$$\sum_{n=0}^{\infty} c_n x_0^n$$

converges. It follows that its nth term goes to 0 as $n \to \infty$, and in particular, the sequence of terms is bounded: $|c_n x_0^n| \leq M$. Then noting that $|x_0| > 0$ (because $|x_0| = y > |x| \geq 0$), we have with $\rho := \frac{|x|}{|x_0|}$ (< 1), that

$$|c_n x^n| = |c_n x_0^n| \left(\frac{|x|}{|x_0|} \right)^n \leq M\rho^n \quad (n \in \mathbb{N}).$$

As the geometric series $\sum_{n=0}^{\infty} M r^n$ converges, by the Comparison Test,

$$\sum_{n=0}^{\infty} c_n x^n$$

is absolutely convergent. As $x \in \mathbb{R}$ was arbitrary, the claim follows.

$2°$ Now suppose that S is bounded above. We will show that

(1) If $|x| < \sup S$, then $\sum_{n=0}^{\infty} c_n x^n$ is absolutely convergent, and

(2) if $|x| > \sup S$, then $\sum_{n=0}^{\infty} c_n x^n$ diverges.

(1) If $x \in \mathbb{R}$ and $|x| < \sup S$, then by the definition of supremum, there exists a $y \in S$ such that $|x| < y$. Then we repeat the proof in $1°$ above as follows.

Since $y \in S$, there exists an $x_0 \in \mathbb{R}$ such that $y = |x_0|$ and $\sum_{n=0}^{\infty} c_n x^n$ converges.

Hence $|c_n x_0^n| \xrightarrow{n \to \infty} 0$, and in particular, there exists an $M > 0$ such that for all n, $|c_n x_0^n| \le M$. Then with $\rho := \frac{|x|}{|x_0|}$ (< 1), we have

$$|c_n x^n| = |c_n x_0^n| \left(\frac{|x|}{|x_0|} \right)^n \le M \rho^n \quad (n \in \mathbb{N}).$$

As $\rho < 1$, $\sum_{n=0}^{\infty} M \rho^n$ converges.

By the Comparison Test, $\sum_{n=0}^{\infty} c_n x^n$ is absolutely convergent.

(2) If $x \in \mathbb{R}$ and $|x| > \sup S$, then setting $y := |x|$, we see that $y \notin S$.

So by the definition of S, $\sum_{n=0}^{\infty} c_n x^n$ diverges (for otherwise $y \in S$).

The uniqueness of the radius of convergence is obvious, since if r, r' are distinct numbers having the property described in the theorem and $r < r'$, then

$$r < \rho := \frac{r + r'}{2} < r',$$

and as $0 < \rho < r'$,

$$\sum_{n=1}^{\infty} c_n \rho^n$$

ought to *converge*, while as $0 < \rho < r$, it ought to *diverge*, a contradiction.   □

If r is the radius of convergence of a power series, then $(-r, r)$ is called the *interval of convergence* of that power series. We note that the interval of convergence is the empty set if $r = 0$, and we set the interval of convergence to be $\mathbb{R}$ when the radius of convergence is infinite.

Note that whether or not the power series converges at $x = r$ and $x = -r$ is not answered by Theorem 6.13. In fact, this is a delicate issue, and either convergence or divergence can take place at these points, as demonstrated by the examples below.

Example 6.10. We have the following:

Power series	Radius of convergence	Set of x's for which the power series converges
$\displaystyle\sum_{n=1}^{\infty} x^n$	1	$(-1, 1)$
$\displaystyle\sum_{n=1}^{\infty} \frac{x^n}{n^2}$	1	$[-1, 1]$
$\displaystyle\sum_{n=1}^{\infty} \frac{x^n}{n}$	1	$[-1, 1)$
$\displaystyle\sum_{n=1}^{\infty} (-1)^n \frac{x^n}{n}$	1	$(-1, 1]$

We will check these claims in Exercise 6.28, after learning about a convenient way of calculating the radius of convergence below.   ◇

How to calculate the radius of convergence

Theorem 6.14. *Let the power series* $\displaystyle\sum_{n=0}^{\infty} c_n x^n$ *have radius of convergence r.*

(1) *If* $(\sqrt[n]{|c_n|})_{n \in \mathbb{N}}$ *is not bounded, then* $r = 0$.

(2) *If* $(\sqrt[n]{|c_n|})_{n \in \mathbb{N}}$ *is bounded, and we define*

$$M_n := \sup\{\sqrt[m]{|c_m|} : m \geq n\}, \quad n \in \mathbb{N},$$

then $(M_n)_{n \in \mathbb{N}}$ *is convergent. Set* $L := \lim_{n \to \infty} M_n = \limsup_{n \to \infty} \sqrt[n]{|c_n|}$.

  If $L = 0$, *then* $r = \infty$.

  If $L \neq 0$, *then* $r = \dfrac{1}{L}$.

Proof. We have

$$M_1 = \sup\{|c_1|, \sqrt{|c_2|}, \sqrt[3]{|c_3|}, \cdots\},$$
$$M_2 = \sup\{\quad \sqrt{|c_2|}, \sqrt[3]{|c_3|}, \cdots\},$$
$$M_3 = \sup\{\qquad\qquad \sqrt[3]{|c_3|}, \cdots\},$$
$$\cdots.$$

Clearly, $(M_n)_{n\in\mathbb{N}}$ is decreasing and bounded below (by 0), and so it is convergent to $L := \inf\limits_{n\in\mathbb{N}} M_n$.

If $x = 0$, then convergence of the power series is obvious.

Suppose that $x \in \mathbb{R}$ is such that $\frac{1}{|x|} > L$. As there is a gap between $\frac{1}{|x|}$ and L, we can find an α such that

$$L = \inf_{n\in\mathbb{N}} M_n < \alpha < \frac{1}{|x|}.$$

Then there exists an $N \in \mathbb{N}$ such that for all $n > N$, $M_n \le M_N < \alpha$. Thus for all $n > N$, $\sqrt[n]{|c_n|} < \alpha < \frac{1}{|x|}$, and upon rearranging, $\sqrt[n]{|c_n x^n|} < \alpha|x| =: r < 1$. By the Root Test, it follows that the series

$$\sum_{n=0}^{\infty} c_n x^n$$

is absolutely convergent.

Hence we have shown that

(1) if $L > 0$, then $\sum\limits_{n=0}^{\infty} c_n x^n$ is absolutely convergent for all $x \in (-L, L)$, and

(2) if $L = 0$, then $\sum\limits_{n=0}^{\infty} c_n x^n$ is absolutely convergent for all $x \in \mathbb{R}$!

It remains to show that in the case when $L > 0$, we have divergence for x such that $|x| > \frac{1}{L}$. For such x,

$$\frac{1}{|x|} < L = \inf_{n\in\mathbb{N}} M_n.$$

So for all n, $\frac{1}{|x|} < M_n = \sup\{\sqrt[n]{|c_n|}, \sqrt[n+1]{|c_{n+1}|}, \cdots\}$, and in particular, there exists an $m_n > n$ such that $|c_{m_n} x^{m_n}| > 1$. So:

For all $n \in \mathbb{N}$, there exists an $m_n > n$ such that $|c_{m_n} x^{m_n}| > 1$.

So it is *not* the case that $\lim\limits_{n\to\infty} c_n x^n = 0$, and hence $\sum\limits_{n=0}^{\infty} c_n x^n$ diverges. $\qquad\square$

Remark 6.2. If $L := \lim\limits_{n\to\infty} \sqrt[n]{|c_n|}$ exists, then the radius r of convergence of the power series is given by $r = 1/L$ if $L \ne 0$, and is infinite if $L = 0$.

The calculation of the radius of convergence is facilitated in some cases by the following result.

Theorem 6.15. *Consider the power series* $\displaystyle\sum_{n=0}^{\infty} c_n x^n$. *If*

$$L := \lim_{n \to \infty} \left| \frac{c_{n+1}}{c_n} \right|$$

exists, then the radius r of convergence of the power series is given by $r = 1/L$ if $L \neq 0$, and is infinite if $L = 0$.

Proof. We have that for all nonzero x such that $L < \frac{1}{|x|}$ that there exists a $q < 1$ and an N large enough such that

$$\frac{|c_{n+1} x^{n+1}|}{|c_n x^n|} = \left| \frac{c_{n+1}}{c_n} \right| |x| \leq q < 1$$

for all $n > N$. (This is because

$$\left| \frac{c_{n+1}}{c_n} x \right| \xrightarrow{n \to \infty} L|x| < 1.$$

So we may take for example $q = (L|x| + 1)/2 < 1$.) Thus by the Ratio Test, the power series converges absolutely for such x.

Hence if $L = 0$, then the above gives that $r = \infty$, while if $L \neq 0$, then we see that $r \geq \frac{1}{L}$. On the other hand, if $L \neq 0$ and $|x| > 1/L$, then

$$\left| \frac{c_{n+1}}{c_n} x \right| \xrightarrow{n \to \infty} L|x| > 1.$$

So there exists an N such that for all $n > N$, $\left| \left| \frac{c_{n+1}}{c_n} \right| - L \right| < \epsilon = L - \frac{1}{|x|}$, and so

$$L|x| - \left| \frac{c_{n+1}}{c_n} \right| |x| < L|x| - 1.$$

Thus for $n > N$,

$$\left| \frac{c_{n+1} x^{n+1}}{c_n x^n} \right| = \left| \frac{c_{n+1}}{c_n} \right| |x| > 1,$$

and by the Ratio Test, the power series diverges. $\qquad\square$

Exercise 6.28. Check all the claims in Example 6.10.

Power series are infinitely differentiable

We will now show that just like polynomials, power series are infinitely many times differentiable in their respective intervals of convergence, and moreover the derivative is again given by a power series, obtained by termwise differentiation of the original series, and this power series for the derivative has the same radius of convergence as the original series. This makes it possible to relate the coefficients of the power series with the successive derivatives at 0 of the function defined by the power series.

Let $\sum_{n=0}^{\infty} c_n x^n$ have radius of convergence $r > 0$ and let

$$f(x) := \sum_{n=0}^{\infty} c_n x^n = c_0 + c_1 x + c_2 x^2 + c_3 x^3 + \cdots, \quad x \in (-r, r).$$

If termwise differentiation were allowed, then

$$f'(x) = 0 + c_1 \cdot 1 + c_2 \cdot 2x + c_3 \cdot 3x^2 + \cdots = \sum_{n=1}^{\infty} n c_n x^{n-1}, \quad x \in (-r, r).$$

We justify this now.

Theorem 6.16. *Let $r > 0$, and let the power series*

$$f(x) := \sum_{n=0}^{\infty} c_n x^n$$

converge for $x \in (-r, r)$. Then f is differentiable in $(-r, r)$, and

$$f'(x) = \sum_{n=1}^{\infty} n c_n x^{n-1}, \quad x \in (-r, r).$$

Proof.
Step 1. First we show that the power series

$$g(x) := \sum_{n=1}^{\infty} n c_n x^{n-1} = c_1 + 2 c_2 x + \cdots + n c_n x^{n-1} + \cdots$$

is absolutely convergent in $(-r, r)$.

Fix $x \in (-r, r)$ and let ρ satisfy $|x| < \rho < r$. By hypothesis,

$$\sum_{n=0}^{\infty} c_n \rho^n$$

converges, and so

$$\lim_{n \to \infty} c_n \rho^n = 0.$$

In particular, $(c_n \rho^n)_{n \in \mathbb{N}}$ is bounded, and there is some positive number M such that $|c_n \rho^n| < M$ for all n. Now let $\alpha := |x|/\rho$. Then $0 \le \alpha < 1$, and

$$|n c_n x^{n-1}| = |c_n \rho^n| \cdot \frac{1}{\rho} \cdot n \left|\frac{x}{\rho}\right|^{n-1} \le \frac{M n \alpha^{n-1}}{\rho}.$$

But as $\alpha \in [0, 1)$, by Exercise 6.7, $\displaystyle\sum_{n=1}^{\infty} n\alpha^{n-1} = \frac{1}{(1-\alpha)^2}$.

Hence from the Comparison Test, it follows that $\displaystyle\sum_{n=1}^{\infty} nc_n x^{n-1}$ converges absolutely.

Step 2. Now we show that $f'(x_0) = g(x_0)$ for $|x_0| < R$, that is,

$$\lim_{x \to x_0} \left(\frac{f(x) - f(x_0)}{x - x_0} - g(x_0) \right) = 0.$$

As before, let ρ be such that $|x_0| < \rho < r$ and since $x \to x_0$, we may also restrict x so that $|x| < \rho$.

Let $\epsilon > 0$. As $\displaystyle\sum_{n=1}^{\infty} nc_n\rho^{n-1}$ converges absolutely, there is an N such that

$$\sum_{n=N}^{\infty} |nc_n\rho^{n-1}| < \frac{\epsilon}{4}. \tag{6.4}$$

Keep N fixed. We have $f(x) - f(x_0) = \displaystyle\sum_{n=1}^{\infty} c_n(x^n - x_0^n)$, and so for $x \neq x_0$,

$$\frac{f(x) - f(x_0)}{x - x_0} = \sum_{n=1}^{\infty} c_n \frac{x^n - x_0^n}{x - x_0} = \sum_{n=1}^{\infty} c_n(x^{n-1} + x^{n-2}x_0 + \cdots + xx_0^{n-2} + x_0^{n-1}).$$

Thus

$$\frac{f(x) - f(x_0)}{x - x_0} - g(x_0) = \sum_{n=1}^{\infty} c_n(x^{n-1} + x^{n-2}x_0 + \cdots + xx_0^{n-2} + x_0^{n-1} - nx_0^{n-1}).$$

We let S_1 be the sum of the first $N - 1$ terms of this series (that is, from $n = 1$ to $n = N - 1$) and S_2 be the sum of the remaining terms (from $n = N$ to ∞). Then since $|x|, |x_0| < \rho$, it follows that

$$|S_2| \leq \sum_{n=N}^{\infty} |c_n| \Big(\underbrace{\rho^{n-1} + \rho^{n-1} + \cdots + \rho^{n-1}}_{n \text{ terms}} + n\rho^{n-1} \Big) = \sum_{n=N}^{\infty} 2n|c_n|\rho^{n-1} < \frac{\epsilon}{2}.$$

The last inequality holds thanks to (6.4). Also,

$$S_1 = \sum_{n=1}^{N} c_n(x^{n-1} + x^{n-2}x_0 + \cdots + xx_0^{n-2} + x_0^{n-1} - nx_0^{n-1})$$

is a polynomial in x and by the algebra of limits,

$$\lim_{x \to x_0} S_1 = \sum_{n=1}^{N} c_n(x_0^{n-1} + x_0^{n-2}x_0 + \cdots + x_0x_0^{n-2} + x_0^{n-1} - nx_0^{n-1})$$

$$= \sum_{n=1}^{N} c_n(nx_0^{n-1} - nx_0^{n-1}) = 0.$$

So there is a $\delta > 0$ such that whenever $|x - x_0| < \delta$, we have $|S_1| < \epsilon/2$. Thus for $|x| < \rho$ and $0 < |x - x_0| < \delta$, we have

$$\left| \frac{f(x) - f(x_0)}{x - x_0} - g(x_0) \right| \le |S_1| + |S_2| < \frac{\epsilon}{2} + \frac{\epsilon}{2} = \epsilon.$$

This means that $f'(x_0) = g(x_0)$, as wanted. □

By a repeated application of the previous result, we have the following.

Corollary 6.17. *Let $r > 0$ and let $f(x) := \sum_{n=0}^{\infty} c_n x^n$ converge for $|x| < r$.*

Then for $k \ge 1$,

$$f^{(k)}(x) = \sum_{n=k}^{\infty} n(n-1)(n-2) \cdots (n-k+1)c_n x^{n-k} \quad \text{for } |x| < r. \tag{6.5}$$

In particular, for $n \ge 0$, $c_n = \frac{1}{n!}f^{(n)}(0)$.

Proof. This is straightforward, and the last claim follows by setting $x = 0$ in (6.5):

$$f^{(k)}(0) = k(k-1) \cdots 1 c_k + x \sum_{n=k+1}^{\infty} n(n-1) \cdots (n-k+1)c_n x^{n-k-1} \Big|_{x=0} = k! c_k.$$

Also, $f(0) = c_0$. □

There is nothing special about taking power series centered at 0. One can also consider

$$\sum_{n=0}^{\infty} c_n(x-a)^n,$$

where a is a fixed real number. Then we have the following result, analogous to the foregoing.

Corollary 6.18. *For a power series* $\sum_{n=0}^{\infty} c_n(x-a)^n$:

either *it is absolutely convergent for all $x \in \mathbb{R}$.*

or *there is a unique nonnegative real number r such that*

(1) $\sum_{n=0}^{\infty} c_n(x-a)^n$ *is absolutely convergent for $|x-a| < r$, and*

(2) $\sum_{n=0}^{\infty} c_n(x-a)^n$ *is divergent for $|x-a| > r$.*

If $f(x) := \sum_{n=0}^{\infty} c_n(x-a)^n$ *for $|x-a| < r$, then for all $k \geq 0$,*

$$f^{(k)}(x) = \sum_{n=k}^{\infty} n(n-1)\cdots(n-k+1)c_n(x-a)^{n-k} \text{ for } |x-a| < r.$$

In particular, for $n \geq 0$, $c_n = \dfrac{1}{n!} f^{(n)}(a)$.

Exercise 6.29. It can be shown that the real power series

$$f(x) := \sum_{n=0}^{\infty} \frac{x^{2n}}{(2n)!} \text{ and } g(x) := \sum_{n=0}^{\infty} \frac{x^{2n+1}}{(2n+1)!}$$

have infinite radius of convergence.
Show that for all $x \in \mathbb{R}$, $f'(x) = g(x)$ and $g'(x) = f(x)$.
Using this, show that for all $x \in \mathbb{R}$, $(f(x))^2 - (g(x))^2 = 1$. *Hint:* Differentiate!

Exercise 6.30. Find $1 + \dfrac{2^2}{1!} + \dfrac{3^2}{2!} + \dfrac{4^2}{3!} + \cdots$.

Exercise 6.31 (Power series method for solving differential equations). Assuming that the solution to the differential equation $f'(x) = 2xf(x)$ has a power series expansion

$$f(x) = \sum_{n=0}^{\infty} c_n x^n, \quad x \in \mathbb{R},$$

find f.

Taylor series

We have seen that power series define infinitely differentiable functions in the respective regions of convergence.

Now suppose that we *start* with an infinitely differentiable function f in an interval $\{x \in \mathbb{R} : |x - a| < r\}$. Then does it have a 'power series expansion'? Well, we can certainly form the power series

$$\sum_{n=0}^{\infty} \frac{f^{(n)}(a)}{n!}(x - a)^n. \tag{6.6}$$

Now we may ask:

(1) Does this power series converge for an $x \neq a$?

(2) If it does converge for an $x \neq a$, then is its sum equal to $f(x)$?

The answer to (1) is, rather surprisingly, '*Not always*'!

And the answer to (2), to even greater astonishment, is '*Not always*'!!

In other words,

(1) There exist infinitely differentiable functions f for which no matter which $a \in \mathbb{R}$ we take, the corresponding power series (6.6) has radius of convergence 0. We won't give an explicit example here.[9]

(2) There exist infinitely differentiable functions f for which the power series (6.6) converges for $x \neq a$, but the sum of the series is different from $f(x)$. Consider, for example, the function $f : \mathbb{R} \to \mathbb{R}$ given by

$$f(x) = \begin{cases} e^{-1/x^2} & \text{if } x \neq 0, \\ 0 & \text{if } x = 0. \end{cases}$$

Then in Exercise 6.32, we will show that $f^{(n)}(0) = 0$ for all $n \geq 0$. Hence, the power series

$$\sum_{n=0}^{\infty} \frac{f^{(n)}(0)}{n!} x^n \equiv 0,$$

which does not equal $f(x)$ for any nonzero x.

After all this talk about doom and gloom, we will now see that Taylor's Formula with remainder helps one to answer the above questions affirmatively for many nice functions. Recall that Taylor's Formula with Remainder gives a *finite* expansion

$$f(x) = \sum_{k=0}^{n} \frac{f^{(k)}(a)}{k!}(x - a)^k + \frac{f^{(n+1)}(c_{x,n})}{(n + 1)!}(x - a)^{n+1}$$

for some $c_{x,n}$ between a and x.

---

[9] The interested reader is referred to [**K**].

Theorem 6.19. *Let*

(1) $a \in \mathbb{R}$,

(2) $r > 0$,

(3) $f \in C^\infty(a - r, a + r)$.

Suppose that

$$\boxed{\exists M > 0 \text{ such that } \forall x \in (a - r, a + r), \ \forall n \geq 0, \ |f^{(n)}(x)| \leq M^n.}$$

Then

$$f(x) = \sum_{n=0}^{\infty} \frac{f^{(n)}(a)}{n!}(x - a)^n \text{ for all } x \in (a - r, a + r).$$

The power series $\displaystyle\sum_{n=0}^{\infty} \frac{f^{(n)}(a)}{n!}(x - a)^n$ is called the *Taylor series* of f centered at a.

Proof. Essentially, the boxed condition allows one to estimate the remainder in Taylor's Formula with Remainder, in order to show that the Taylor series does converge. Here are the details. From Taylor's Formula with Remainder, we know that for each fixed x,

$$f(x) = \sum_{k=0}^{n} \frac{f^{(k)}(a)}{k!}(x - a)^k + \frac{f^{(n+1)}(c_{x,n})}{(n + 1)!}(x - a)^{n+1}$$

for some $c_{x,n}$ between a and x. But

$$\left| \frac{f^{(n+1)}(c_{x,n})}{(n + 1)!}(x - a)^{n+1} \right| \leq \frac{M^{n+1}|x - a|^{n+1}}{(n + 1)!} \xrightarrow{n \to \infty} 0.$$

Indeed, the limit is zero since the left hand side of the first inequality is the $(n + 1)$st term of the convergent series

$$\sum_{k=0}^{\infty} \frac{(M|x - a|)^k}{k!} = \exp(M|x - a|).$$

Done!   □

Example 6.11 (Taylor Series for sin, cos).

$$\sin x = x - \frac{x^3}{3!} + \frac{x^5}{5!} - \frac{x^7}{7!} + - \cdots, x \in \mathbb{R}.$$

With $f := \sin$, we have

$$\sin' x = \cos x,$$
$$\sin'' x = -\sin x,$$
$$\sin''' x = -\cos x,$$
$$\sin'''' x = \sin x,$$

and so for all $n = 0, 1, 2, 3, \cdots$,

$$f^{(2n)}(0) = 0,$$
$$f^{(2n+1)}(0) = (-1)^n.$$

As $|\sin x|, |\cos x| \le 1$ for all $x \in \mathbb{R}$, by the previous result, the claim follows.

Similarly, $\cos x = 1 - \dfrac{x^2}{2!} + \dfrac{x^4}{4!} - \dfrac{x^6}{6!} + - \cdots$ for all $x \in \mathbb{R}$.   ◇

Example 6.12 (Taylor Series for exp). $\exp x = \displaystyle\sum_{n=0}^{\infty} \dfrac{x^n}{n!}$ for all $x \in \mathbb{R}$.

First consider a *fixed* $R > 0$. For *all* $x \in (-R, R)$ and *all* $n \ge 0$,

$$|f^{(n)}(x)| = |\exp^{(n)} x| = |\exp x| \le \exp R \le (\exp R)^n.$$

So

$$\exp x = \sum_{n=0}^{\infty} \frac{x^n}{n!}$$

for all $x \in (-R, R)$. But the choice of $R > 0$ was arbitrary. Hence the claim follows for all $x \in \mathbb{R}$.   ◇

Example 6.13 (Taylor Series for log). $\log(1 + x) = \displaystyle\sum_{n=1}^{\infty} \dfrac{(-1)^{n-1}}{n} x^n$ for all $x \in (-1, 1)$.

Let $a = 0$, $r = 1$, and let $f(x) = \log(1 + x)$ for $x \in (a - r, a + r) = (-1, 1)$. Then we have that $f(0) = \log 1 = 0$ and

$$f'(x) = \frac{1}{1 + x},$$
$$f''(x) = -\frac{1}{(1 + x)^2},$$
$$f'''(x) = \frac{2}{(1 + x)^3},$$
$$\cdots$$
$$f^{(n)}(x) = \frac{(-1)^{n-1}(n - 1)!}{(1 + x)^n}.$$

In particular, $f^{(n)}(0) = (-1)^{n-1} \cdot (n - 1)!$ for $n \in \mathbb{N}$. Hence

$$\sum_{n=0}^{\infty} \frac{f^{(n)}(0)}{n!} x^n = \sum_{n=1}^{\infty} \frac{(-1)^{n-1}}{n} x^n.$$

But

$$|f^{(n)}(x)| = \frac{(n-1)!}{(1+x)^n},$$

and if there existed an $M > 0$ such that for all n, $|f^{(n)}(x)| \le M^n$, then we would obtain

$$\frac{1}{M} \le \frac{|1+x|^n}{(n-1)!} \xrightarrow{n\to\infty} 0,$$

which is clearly impossible. So we can't use the result in Theorem 6.19.

In order to show that

$$\log(1+x) = \sum_{n=1}^{\infty} \frac{(-1)^{n-1}}{n} x^n = x - \frac{x^2}{2} + \frac{x^3}{3} - + \cdots \quad \text{for } x \in (-1,1),$$

we will first show the equality of their derivatives.

Note that the power series

$$\sum_{n=1}^{\infty} \frac{(-1)^{n-1}}{n} x^n = x - \frac{x^2}{2} + \frac{x^3}{3} - + \cdots$$

has radius of convergence $1/1 = 1$ since

$$\lim_{n\to\infty} \left| \frac{c_{n+1}}{c_n} \right| = \lim_{n\to\infty} \left| \frac{\frac{(-1)^n}{n+1}}{\frac{(-1)^{n-1}}{n}} \right| = \lim_{n\to\infty} \frac{n}{n+1} = 1.$$

So $g(x) := \sum_{n=1}^{\infty} \frac{(-1)^{n-1}}{n} x^n$ converges for $x \in (-1,1)$. Thus

$$g'(x) = \sum_{n=1}^{\infty} \frac{(-1)^{n-1}}{n} \cdot n x^{n-1} = \sum_{n=1}^{\infty} (-1)^{n-1} x^{n-1}$$

$$= 1 - x + x^2 - x^3 + x^4 - + \cdots$$

$$= \frac{1}{1+x}, \quad x \in (-1,1).$$

Hence

$$g'(x) = \frac{1}{1+x} = \log'(1+x) = f'(x), \quad x \in (-1,1).$$

By the Fundamental Theorem of Calculus,

$$g(x) = g(x) - 0 = g(x) - g(0) = \int_0^x g'(\xi)d\xi$$

$$= \int_0^x f'(\xi)d\xi = f(x) - f(0) = f(x) - 0 = f(x).$$

Consequently, $\log(1+x) = f(x) = g(x) = \sum_{n=1}^{\infty} \frac{(-1)^{n-1}}{n} x^n$, $x \in (-1,1)$.

Remark 6.3. Functions $f : (a, b) \to \mathbb{R}$ which have the property that

$$\forall \xi \in (a, b), \ \exists r_\xi > 0 \ \text{such that} \ \forall x \in (\xi - r_\xi, \xi + r_\xi), \ f(x) = \sum_{n=0}^{\infty} \frac{f^{(n)}(\xi)}{n!} (x - \xi)^n,$$

are called *real analytic*. The set of all real analytic functions on (a, b) is denoted by $C^\omega(a, b)$. We have $C^\omega(a, b) \subsetneq C^\infty(a, b)$; see the following exercise.

There is a link between real analytic functions and complex Analysis. 'Complex analysis' is all about doing Calculus in $\mathbb{C}$. It turns out that Complex Analysis is a very specialised branch of analysis that acquires a somewhat peculiar character owing to the special geometric meaning associated with the multiplication of complex numbers in the complex plane, and is rather different as a subject than what we have been doing in Calculus with real numbers, or 'Real Analysis'. For example, a theorem in Complex Analysis asserts that complex (once) differentiable functions are infinitely many times complex differentiable, which is certainly not true for the analogous claim in Calculus/Real Analysis (recall Exercise 4.29). Nevertheless, the two worlds are linked. For instance, apropos the remark that was made earlier, real analytic functions are precisely restrictions of complex differentiable functions defined on an 'open' set in $\mathbb{C}$ containing the real interval under consideration.

Exercise 6.32 (A C^∞ function which is not analytic). (*)

Let $f : \mathbb{R} \to \mathbb{R}$ be given by $f(x) = \begin{cases} e^{-1/x^2} & \text{if } x \neq 0, \\ 0 & \text{if } x = 0. \end{cases}$

(1) Sketch the graph of f.

(2) Prove that for every $n \in \mathbb{N}$, $\displaystyle\lim_{x \to 0} \frac{f(x)}{x^n} = 0$.

(3) Show that for each $n \in \mathbb{N}$, there is a polynomial p_n such that for all $x \neq 0$,

$$f^{(n)}(x) = e^{-1/x^2} p_n\left(\frac{1}{x}\right).$$

(4) Prove that $f^{(n)}(0) = 0$ for all $n \geq 1$ (showing that the Taylor series for f at 0 is identically zero, clearly not equal to f).

Appendix

Let us show that every nonnegative rational number has either a terminating or a repeating decimal expansion, which was mentioned in Exercise 6.8. Let

$$x = \frac{p}{q},$$

where p is a nonnegative integer and $q \in \mathbb{N}$. We can factorise $q = 2^i 5^j q'$, where $2, 5$ do not divide $q' \in \mathbb{N}$. Choose a natural number $n > i, j$. Then $10^n/(2^i 5^j)$ is a natural number, and so

$$10^n x = 10^n \frac{p}{q} = \frac{p'}{q'},$$

where p' is a nonnegative integer and q' is such that q' is coprime to 10. But if we look at the remainders we get when we divide $10, 10^2, 10^3, \cdots$ by q', then it must be the case that for some $K > k$, 10^K, 10^k leave the same remainder when divided by q', so that

$$10^K - 10^k = 10^k(10^{K-k} - 1)$$

is divisible by q'. But q' is coprime to 10, and hence also to 10^k, and thus q' must divide $10^{K-k} - 1$. So we have with $m := K - k$ that

$$10^n(10^m - 1)x = (10^{K-k} - 1)\frac{p'}{q'}$$

is an integer. Now let $x = N.d_1d_2d_3\cdots$. Then

$$10^{n+m}x - 10^n x$$

$$= Nd_1 \cdots d_{n+m} - Nd_1 \cdots d_n$$

$$+ \left(\frac{d_{n+m+1}}{10} + \frac{d_{n+m+2}}{10^2} + \frac{d_{n+m+3}}{10^3} + \cdots - \frac{d_{n+1}}{10} - \frac{d_{n+2}}{10^2} - \frac{d_{n+3}}{10^3} - \cdots \right),$$

is an integer, implying that the part in the brackets, say Δ, above is also an integer. But

$$\Delta = \frac{d_{n+m+1} - d_{n+1}}{10} + \frac{d_{n+m+2} - d_{n+2}}{10^2} + \frac{d_{n+m+3} - d_{n+3}}{10^3} + \cdots \in \mathbb{Z}.$$

We claim that this must imply that $d_{n+m+k} = d_{n+k}$ for all $k \in \mathbb{N}$, giving the desired conclusion.

1° It cannot be the case that the sequence $(d_{n+m+k} - d_k)_{n \in \mathbb{N}}$ is eventually the constant sequence $9, 9, 9, \cdots$. Indeed then the $d_{n+m+k} - d_{n+k} = 9$ for all $k > K$ for some K, and this means that $d_{n+m+k} = 9$ and $d_{n+k} = 0$ for all $k > K$, which is impossible.

2° Similar to 1°, it can't be the case that the sequence $(d_{n+m+k} - d_k)_{n \in \mathbb{N}}$ is eventually the constant sequence $-9, -9, -9, \cdots$ either.

3° Suppose that $k_* \in \mathbb{N}$ is the smallest number such that $d_{n+m+k_*} \neq d_{n+k_*}$.
   If $d_{n+m+k_*} > d_{n+k_*}$, then by 2°, there exists a number $K \in \mathbb{N}$ such that we have $d_{n+m+k_*+K} - d_{n+k_*+K} \neq -9$, and so

$$\Delta \geq \frac{1}{10^{k_*}} - \frac{9}{10^{k_*+1}} - \cdots - \frac{9}{10^{k_*+K-1}} - \frac{8}{10^{k_*+K}} - \frac{9}{10^{k_*+K+1}} \cdots$$

$$= \frac{1}{10^{k_*+K}},$$

while

$$\frac{1}{10^{k_*}} = \frac{9}{10^{k_*+1}} + \frac{9}{10^{k_*+2}} + \cdots \geq \Delta,$$

a contradiction to the fact that Δ is an integer.

On the other hand, if $d_{n+m+k_*} < d_{n+k_*}$, then by 1°, there exists a K such that $d_{n+m+k_*+K} - d_{n+k_*+K} \neq 9$, and so

$$\Delta \leq \frac{-1}{10^{k_*}} + \frac{9}{10^{k_*+1}} + \cdots + \frac{9}{10^{k_*+K-1}} + \frac{8}{10^{k_*+K}} + \frac{9}{10^{k_*+K+1}} \cdots$$

$$= -\frac{1}{10^{k_*+K}},$$

while

$$-\frac{1}{10^{k_*}} = -\frac{9}{10^{k_*+1}} - \frac{9}{10^{k_*+2}} + \cdots \leq \Delta,$$

again a contradiction to the fact that Δ is an integer.

Hence $d_{n+m+k} = d_{n+k}$ for all $k \in \mathbb{N}$, which means that the block of digits $d_{n+1} \cdots d_{n+m}$ keeps repeating.

Notes

Exercises 6.16 and 6.17 are based on [**R2**]. Exercise 6.30 is based on [**L**, Example 5.4.4].

Solutions

Solutions to the exercises from Chapter 1

Solution to Exercise 1.1

See the following picture for the construction of $-11/6$ on the real number line.

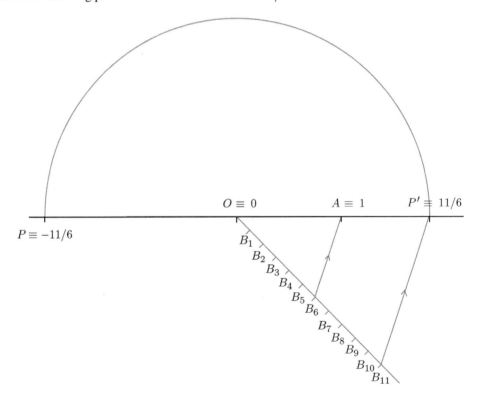

The How and Why of One Variable Calculus, First Edition. Amol Sasane.

The steps of the construction are as follows:

(1) Draw a ray OA and take $\ell(OA) = 1$.

(2) Draw a ray OB_1 not parallel to OA.

(3) Cut equal lengths $\ell(OB_1) = \ell(B_1B_2) = \ell(B_2B_3) = \cdots = \ell(B_{10}B_{11})$ along the ray OB_1.

(4) Join A to B_6 and construct $B_{11}P'$ parallel to B_6A, meeting OA (extended) at P'. Then $P' \equiv 11/6$.

(5) With O as center and radius $\ell(OP')$, draw a circle that meets the line passing through O and A at the point $P \neq P'$. Then $P \equiv -11/6$.

See the following picture for the construction of $\sqrt{3}$ on the real number line.

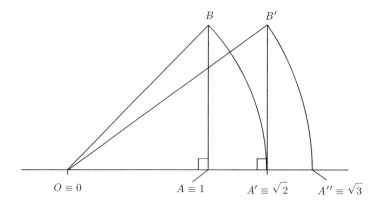

The steps of the construction are as follows:

(1) Draw a ray OA and take $\ell(OA) = 1$.

(2) Construct AB perpendicular to OA such that $\ell(AB) = 1$. Then

$$\ell(OB) = \sqrt{1^2 + 1^2} = \sqrt{2}.$$

(3) With center O and radius $\ell(OB)$, draw a circular arc meeting OA extended at A'. Then $A' \equiv \sqrt{2}$.

(4) Construct $A'B'$ perpendicular to OA' such that $\ell(A'B') = 1$. Then

$$\ell(OB') = \sqrt{(\sqrt{2})^2 + 1^2} = \sqrt{2+1} = \sqrt{3}.$$

(5) With center O and radius $\ell(OB')$, draw a circular arc meeting OA' extended at A''. Then $A'' \equiv \sqrt{3}$.

Remark 1. In fact, one can construct $\sqrt{n}$ for all $n \in \mathbb{N}$: indeed, given the length $\sqrt{n}$, the hypotenuse of the right angled triangle with side lengths $\sqrt{n}$ and 1 is equal to $\sqrt{n+1}$, and so knowing the unit length, we can construct $\sqrt{2}, \sqrt{3}, \sqrt{4} (= 2), \sqrt{5}, \cdots$.

Solution to Exercise 1.2

(1) We have $c_0 + c_1\dfrac{p}{q} + \cdots + c_d\dfrac{p^d}{q^d} = 0$. Multiplying throughout by q^d, we obtain

$$c_d p^d = -(c_0 q^d + c_1 p q^{d-1} + \cdots + c_{d-1}p^{d-1}q). \tag{1}$$

As q divides the right hand side, q divides $c_0 p^d$. But q has no common factors with p, and this implies that q divides c_d. (This is because if we decompose p, q as products of powers of primes, then there can be no common prime occurring in their respective decompositions.) Also by rearranging (1), we obtain

$$c_0 q^d = -(c_1 p q^{d-1} + \cdots + c_{d-1}p^{d-1}q + c_d p^d),$$

and since p divides the right hand side, p must divide $c_0 q^d$. But p and q have no common factor. So p must divide c_0.

(2) Suppose that $\sqrt{2}$ is rational, and let

$$\sqrt{2} = \frac{p}{q},$$

where $p, q \in \mathbb{Z}$, $q > 0$, and p, q have no common factor. Then p/q is a rational zero of the polynomial $x^2 - 2$. By the Rational Zeros Theorem, p divides -2 and q divides 1. So $p \in \{2, -2, 1, -1\}$ and $q = 1$. But then

$$\frac{p}{q} \in \{2, -2, 1, -1\}.$$

But clearly $\sqrt{2}$ is not equal to any of the values $2, -2, 1, -1$. This contradiction shows that $\sqrt{2} \notin \mathbb{Q}$.

(3) Suppose that $\sqrt[3]{6}$ is rational, and let

$$\sqrt[3]{6} = \frac{p}{q},$$

where $p, q \in \mathbb{Z}$, $q > 0$, and p, q have no common factor. Then p/q is a rational zero of the polynomial $x^3 - 6$. By the Rational Zeros Theorem, p divides -6 and q divides 1. So $p \in \{6, -6, 3, -3, 2, -2, 1, -1\}$ and $q = 1$. But then

$$\frac{p}{q} \in \{6, -6, 3, -3, 2, -2, 1, -1\}.$$

But none of these values satisfy $x^3 - 6 = 0$. This contradiction shows that $\sqrt[3]{6} \notin \mathbb{Q}$.

Solution to Exercise 1.3

(1) Let $a \in \mathbb{R}$ and $b_1, b_2 \in \mathbb{R}$ are such that $a + b_1 = 0 = b_1 + a$ and $a + b_2 = 0 = b_2 + a$. Then $b_1 = b_1 + 0 = b_1 + (a + b_2) = (b_1 + a) + b_2 = 0 + b_2 = b_2$. So $b_1 = b_2$.

(2) Let $a \in \mathbb{R}$. Then $a + (-1) \cdot a = 1 \cdot a + (-1) \cdot a = (1 + (-1)) \cdot a = 0 \cdot a = 0$. By the uniqueness of inverses, $-a = (-1) \cdot a$.

(3) As $1 + (-1) = 0 = -1 + 1$, by the uniqueness of additive inverses, it follows that $-(-1) = 1$. From part (2) above, we also know that $(-1) \cdot (-1) = -(-1)$. Hence $(-1) \cdot (-1) = 1$.

Solution to Exercise 1.4

(1) We have the following cases:

1° $a = 0$. Then $a^2 = 0 \cdot 0 = 0$. So $a^2 \geq 0$.

2° $a \in \mathbb{P}$. Then $a, a \in \mathbb{P}$ gives $a^2 = a \cdot a \in \mathbb{P}$. So $a^2 \geq 0$.

3° $-a \in \mathbb{P}$.

Then $-a, -a \in \mathbb{P}$ gives $a^2 = 1 \cdot a \cdot a = (-1) \cdot (-1) \cdot a \cdot a = (-a) \cdot (-a) \in \mathbb{P}$.
So $a^2 \geq 0$.

(2) Suppose that there exists x such that $x^2 + 1 = 0$.
Clearly with $x = 0$, $x^2 + 1 = 0^2 + 1 = 0 + 1 = 1 \neq 0$. So $x \neq 0$.
But then $x^2 = x \cdot x = (-x) \cdot (-x) \in \mathbb{P}$. Also $1 \in \mathbb{P}$. Hence we obtain $x^2 + 1 \in \mathbb{P}$. On the other hand, $x^2 + 1 = 0 \notin \mathbb{P}$, a contradiction.

Solution to Exercise 1.5

(1) $S = (0, 1]$.

An upper bound of S. 1 is an upper bound, since for all $x \in S = (0, 1]$, we have $x \leq 1$. In fact, any real number $u \geq 1$ is an upper bound.

A lower bound of S. 0 is a lower bound, since for all $x \in S = (0, 1]$, we have $0 < x$. In fact, any real number $\ell \leq 0$ is a lower bound.

Is S bounded? Yes. S is bounded above, since 1 is an upper bound of S. S is also bounded below, since 0 is a lower bound. Since S is bounded above as well as bounded below, it is bounded.

Supremum of S. $\sup S = 1$. Indeed, 1 is an upper bound, and moreover, if u is also an upper bound, then $1 \leq u$ (since $1 \in S$).

Infimum of S. $\inf S = 0$. First of all, 0 is a lower bound. Let ℓ be a lower bound of S. We prove that $\ell \leq 0$. (We do this by supposing that $\ell > 0$, and arriving at a contradiction. The contradiction is obtained as follows: if $\ell > 0$, then we will see that the average of 0 and ℓ, namely $\frac{\ell}{2}$, is an element in S that is less than the lower bound ℓ, which is a contradiction to the definition of a lower bound!) If $\ell > 0$, then $0 < \frac{\ell}{2}$. Moreover, since $\ell \leq 1$ (ℓ is a lower bound of S and $1 \in S$), it follows that $\frac{\ell}{2} \leq \frac{1}{2} \leq 1$. Thus we have $\frac{\ell}{2} \in S$. But since $\ell > 0$, it follows that $\frac{\ell}{2} < \ell$, a contradiction. Hence $\ell \leq 0$.

Maximum of S. $\max S = 1$, since $\sup S = 1 \in S$.

Minimum of S. $\min S$ does not exist since $\inf S = 0 \notin S$.

(2) $S = [0, 1]$.

An upper bound of S. 1 is an upper bound, since for all $x \in S = [0, 1]$, we have $x \leq 1$.

A lower bound of S. 0 is a lower bound, since for all $x \in S = (0, 1]$, we have $0 \leq x$.

Is S bounded? Yes, since S is bounded above (1 is an upper bound) and it is bounded below (0 is a lower bound).

Supremum of S. $\sup S = 1$. Indeed, 1 is an upper bound, and moreover, if u is also an upper bound, then $1 \leq u$ (since $1 \in S$).

Infimum of S. $\inf S = 0$. Indeed, 0 is a lower bound, and moreover, if ℓ is also a lower bound, then $\ell \leq 0$ (since $0 \in S$).

Maximum of S. $\max S = 1$, since $\sup S = 1 \in S$.

Minimum of S. $\min S = 0$, since $\inf S = 0 \in S$.

(3) $S = (0, 1)$.

An upper bound of S. 1 is an upper bound, since for all $x \in S = (0, 1)$, $x < 1$.

A lower bound of S. 0 is a lower bound, since for all $x \in S = (0, 1)$, $0 < x$.

Is S bounded? Yes, since S is bounded above (1 is an upper bound) and it is bounded below (0 is a lower bound).

Supremum of S. $\sup S = 1$. First of all, 1 is an upper bound. Let u be an upper bound of S. We prove that $1 \leq u$. (We do this by supposing that $u < 1$ and arriving at a contradiction. The contradiction is obtained as follows: if $u < 1$, then we will see that the average of u and 1, namely $\frac{u+1}{2}$, is an element in S that is larger than the upper bound u, which is a contradiction to the definition of an upper bound!) Since u is an upper bound and since $\frac{1}{2} \in S$, it follows that $0 < u$ (since $0 < \frac{1}{2} \leq u$). So if $u < 1$, then $0 < u = \frac{u+u}{2} < \frac{u+1}{2} < \frac{1+1}{2} = 1$. Hence $\frac{u+1}{2} \in S$. But $u < \frac{u+1}{2}$ contradicts the fact that u is an upper bound of S.

Infimum of S. $\inf S = 0$. First of all, 0 is a lower bound. Let ℓ be a lower bound of S. We prove that $\ell \leq 0$. If $\ell > 0$, then $0 < \frac{\ell}{2}$. Moreover, since $\frac{1}{2} \in S$ and ℓ is a lower bound of S, it follows that $\ell \leq \frac{1}{2}$. Thus we have $0 < \frac{\ell}{2} < l \leq \frac{1}{2} < 1$, and so $\frac{\ell}{2} \in S$. But $\frac{\ell}{2} < l$ contradicts the fact that ℓ is a lower bound of S.

Maximum of S. $\max S$ does not exist, since $\sup S = 1 \notin S$.

Minimum of S. $\min S$ does not exist, since $\inf S = 0 \notin S$.

Solution to Exercise 1.6

S is a subset of $\mathbb{R}$, $S \neq \emptyset$ (for example $a_1 \in S$) and S is bounded above. Thus by the Least Upper Bound Property of $\mathbb{R}$, $u_* := \sup S \in \mathbb{R}$ exists. We claim that this is the smallest number bigger than each of the terms of the sequence. Indeed, as u_* is an upper bound of S, $a_n \leq u_*$ for all $n \in \mathbb{N}$. So first of all, we see that u_* is bigger than each of the terms of the sequence. Is it the smallest such number? Let u be any number such that $a_n \leq u$ for all $n \in \mathbb{N}$. Then u is an upper bound of S, and as u_* is the least upper bound, we have $u_* \leq u$. This proves the claim.

Solution to Exercise 1.7

(1) Let ℓ be a lower bound of S: for all $x \in S$, $\ell \leq x$. So for all $x \in S$, $-x \leq -\ell$, that is, for all $y \in -S$, $y \leq -\ell$. Thus $-S$ is bounded above because $-\ell$ is an upper bound of $-S$. Since S is nonempty, it follows that there exists an element $x \in S$, and so we obtain that $-x \in -S$. Hence $-S$ is nonempty.

As $-S$ is nonempty and bounded above, it follows that $\sup(-S)$ exists, by the Least Upper Bound Property of $\mathbb{R}$.

Since $\sup(-S)$ is an upper bound of $-S$, we have that for all $y \in -S$, $y \leq \sup(-S)$, that is, for all $x \in S$, $-x \leq \sup(-S)$. Hence for all $x \in S$, $-\sup(-S) \leq x$. So $-\sup(-S)$ is a lower bound of S.

Next we prove that $-\sup(-S)$ is the greatest lower bound of S. Suppose that ℓ' is a lower bound of S such that $-\sup(-S) < \ell'$. Then for all $x \in S$, $-\sup(-S) < \ell' \leq x$, that is, for all $x \in S$, $-x \leq -\ell' < \sup(-S)$. Hence, for all $y \in -S$, $y \leq -\ell' < \sup(-S)$. So $-\ell'$ is an upper bound of $-S$, and $-\ell' < \sup(-S)$, which contradicts the fact that $\sup(-S)$ is the least upper bound of $-S$. Hence $\ell' \leq -\sup(-S)$.

Consequently, inf S exists and inf $S = -\sup(-S)$.

(2) Suppose that S is a nonempty subset of $\mathbb{R}$, which is bounded below. Then by the above, inf S exists.

Solution to Exercise 1.8

Suppose that S is a nonempty subset of $\mathbb{R}$, which is bounded above and $\alpha > 0$. By the Least Upper Bound Property of $\mathbb{R}$, $\sup S$ exists. Also, for every $y \in \alpha \cdot S$, we have $y = \alpha x$ for some $x \in X$, and so $y = \alpha x \leq \alpha \sup S$. So $\alpha \sup S$ is an upper bound for $\alpha \cdot S$. Since S is nonempty, it follows that $\alpha \cdot S$ is also nonempty. Hence, by the Least Upper Bound Property of $\mathbb{R}$, $\sup(\alpha \cdot S)$ exists. Also, since $\alpha \sup S$ is an upper bound for $\alpha \cdot S$, we obtain $\sup(\alpha \cdot S) \leq \alpha \cdot \sup S$.

It would be great if we also obtained the reverse of this last inequality. We can do this by just replacing α by its reciprocal and replacing S by $\alpha \cdot S$, and noticing that

$$\frac{1}{\alpha} \cdot (\alpha \cdot S) = \left\{ \frac{1}{\alpha}(\alpha x) = x : x \in S \right\} = S.$$

Thus $\sup S = \sup \left(\frac{1}{\alpha} \cdot (\alpha \cdot S) \right) \leq \frac{1}{\alpha} \cdot \sup(\alpha \cdot S)$, and upon rearranging,

$$\alpha \cdot \sup S \leq \sup(\alpha \cdot S).$$

This completes the proof of $\sup(\alpha \cdot S) = \alpha \cdot \sup S$.

Next, suppose now that S is a nonempty subset of $\mathbb{R}$, which is bounded below and $\alpha > 0$. Using Exercise 1.7 and the above, we have that

$$\inf(\alpha \cdot S) = -\sup(-(\alpha \cdot S)) = -\sup(\alpha \cdot (-S)) = -\alpha \sup(-S)$$
$$= -\alpha(-\inf S) = \alpha \cdot \inf S.$$

Solution to Exercise 1.9

Since $\sup B$ is an upper bound of B, we have $x \leq \sup B$ for all $x \in B$. Since $A \subset B$, in particular, we obtain $x \leq \sup B$ for all $x \in A$. Thus $\sup B$ is an upper bound of A, and so by the definition of the least upper bound of A, we obtain $\sup A \leq \sup B$.

Solution to Exercise 1.10

Since S is bounded, in particular, it is bounded above, and furthermore, since it is nonempty, $\sup S$ exists, by the Least Upper Bound Property of $\mathbb{R}$.

Since S is bounded, in particular, it is bounded below, and furthermore, since it is nonempty, inf S exists, by the Greatest Lower Bound Property of $\mathbb{R}$.

Let $x \in S$. Since inf S is a lower bound of S,

$$\inf S \leq x. \tag{2}$$

Moreover, since $\sup S$ is an upper bound of S,

$$x \leq \sup S. \tag{3}$$

From (2) and (3), we obtain inf $S \leq \sup S$.

Let $\inf S = \sup S$. If $x \in S$, then we have

$$\inf S \leq x \leq \sup S, \tag{4}$$

and so $\inf S = x(= \sup S)$ (for if $\inf S < x$, then from (4), $\inf S < \sup S$, a contradiction). Thus S is a singleton set. Conversely, if $S = \{x\}$, then clearly x is an upper bound. If $u < x$ is an upper bound, then $x \leq u < x$ gives $x < x$, a contradiction. So $\sup S = x$. Clearly x is also a lower bound. If $\ell > x$ is also a lower bound, then $x > \ell \geq x$ gives $x > x$, a contradiction. So $\inf S = x = \sup S$.

Solution to Exercise 1.11

If $a \in A$, then $a \leq \sup A \leq \max\{\sup A, \sup B\}$. Similarly we have that if $b \in B$, then $b \leq \sup B \leq \max\{\sup A, \sup B\}$. So if $x \in A \cup B$, then either $x \in A$ or $x \in B$, and from the above, $x \leq \max\{\sup A, \sup B\}$. Hence $A \cup B$ is bounded above by $\max\{\sup A, \sup B\}$. Also, as A is nonempty, $A \cup B$ is nonempty. Hence, by the Least Upper Bound Property of $\mathbb{R}$, $\sup(A \cup B)$ exists. Also, it follows from the above that

$$\sup(A \cup B) \leq \max\{\sup A, \sup B\}. \tag{5}$$

Since $A \subset A \cup B$, we also have that $\sup A \leq \sup(A \cup B)$. Similarly, as $B \subset A \cup B$, we have $\sup B \leq \sup(A \cup B)$. From here, we obtain

$$\max\{\sup A, \sup B\} \leq \sup(A \cup B). \tag{6}$$

From (5) and (6), it follows that $\sup(A \cup B) = \max\{\sup A, \sup B\}$.

Solution to Exercise 1.12

(1) False (if $S = \{1\}$, then we have that $u = 3$ is an upper bound of S, and although we have $u' = 2 < 3 = u$, $u'(= 2)$ is an upper bound of $\{1\} = S$).

(2) True (if $\epsilon > 0$, then $u_* - \epsilon < u_*$, and so $u_* - \epsilon$ cannot be an upper bound of S).

(3) False ($\mathbb{N}$ has no maximum).

(4) False ($\mathbb{N}$ has no supremum).

(5) False ($[0, 1)$ is bounded, but it does not have a maximum).

(6) False ($\emptyset$ is bounded, but it does not have a supremum).

(7) True (Least Upper Bound Property of $\mathbb{R}$).

(8) True (the supremum itself is an upper bound).

(9) True (definition of maximum).

(10) False (the set $[0, 1)$ has supremum 1, but $1 \notin [0, 1)$).

(11) False ($-\mathbb{N}$ is bounded above since 0 is an upper bound, but $|-\mathbb{N}| = \mathbb{N}$ is not bounded).

(12) True ($\ell \leq x \leq u$ implies $x \leq u$ and $-x \leq -\ell$, and so we have

$$x \leq u \leq \max\{-\ell, u\} \text{ and } -x \leq -\ell \leq \max\{-\ell, u\}.$$

Thus $|x| \leq \max\{-\ell, u\}$ and so $\max\{-\ell, u\}$ is an upper bound of $|S|$. Moreover, for every $y \in |S|$, we have $y = |x|$ for some $x \in S$, and so $y = |x| \geq 0$. Hence 0 is a lower bound of $|S|$.)

(13) False (if $S = \{0, 1\}$, then $\inf S = 0 < \frac{1}{2} < 1 = \sup S$, but $\frac{1}{2} \notin S$).

Solution to Exercise 1.13

Clearly $A + B$ is nonempty. Indeed, since A is nonempty, there exists some element $a \in A$, and as B is nonempty, there exists an element $b \in B$, and so $a + b \in A + B$, that is, $A + B$ is not empty.

Since $\sup A$ is an upper bound of A, we have that for all $a \in A$, $a \leq \sup A$. Also, as $\sup B$ is an upper bound of B, we have that for all $b \in B$, $b \leq \sup B$. So for all $a \in A$ and $b \in B$, $a + b \leq \sup A + \sup B$. Thus $A + B$ is bounded above.

Since $A + B$ is bounded above and it is not empty, by the Least Upper Bound Property of $\mathbb{R}$, it follows that $\sup(A + B)$ exists. Moreover, since we have shown above that $\sup A + \sup B$ is an upper bound for $A + B$, we also obtain

$$\sup(A + B) \leq \sup A + \sup B. \tag{7}$$

Now let $a \in A$ and $b \in B$. Then we have $a + b \in A + B$, and so $a + b \leq \sup(A + B)$, that is, $a \leq \sup(A + B) - b$. This inequality holds for *all* $a \in A$, that is, $\sup(A + B) - b$ is an upper bound for A, giving $\sup A \leq \sup(A + B) - b$. So $b \leq \sup(A + B) - \sup A$. But again, this inequality holds for *all* $b \in B$. Hence $\sup B \leq \sup(A + B) - \sup A$, that is,

$$\sup A + \sup B \leq \sup(A + B). \tag{8}$$

From (7) and (8), it follows that $\sup(A + B) = \sup A + \sup B$.

Solution to Exercise 1.14

(S is nonempty and bounded below (0 is a lower bound), and so by the Greatest Lower Bound Property of $\mathbb{R}$, $\inf S$ exists.)

'If' part: Let $\inf S > 0$. Since $\inf S$ is a lower bound, it follows that for all $x \in S$, $\inf S \leq x$, that is, for all $x \in S$, $\frac{1}{x} \leq \frac{1}{\inf S}$. Hence for all $y \in S^{-1}$, $y \leq \frac{1}{\inf S}$. So $\frac{1}{\inf S}$ is an upper bound of S^{-1}. Thus S^{-1} is bounded above.

'Only if' part: Suppose that S^{-1} is bounded above (with an upper bound u, say). Then for all $y \in S^{-1}$, $y \leq u$, that is,

$$\text{for all } x \in S, \ \frac{1}{x} \leq u. \tag{9}$$

Since S is not empty, $\exists x_* \in S$ and so $\frac{1}{x_*} \leq u$. But $x_* \in S$ implies that $x_* > 0$, and so $\frac{1}{x_*} > 0$. Consequently $u > 0$. Hence from (9), we have that for all $x \in S$, $\frac{1}{u} \leq x$. Thus $\frac{1}{u}$ is a lower bound of S, and so $\frac{1}{u} \leq \inf S$. But $u > 0$ implies $\frac{1}{u} > 0$, and consequently $\inf S (\geq \frac{1}{u}) > 0$.

If $\inf S > 0$, then as in the 'If' part, $\frac{1}{\inf S}$ is an upper bound of S^{-1}, and so we obtain

$$\sup S^{-1} \leq \frac{1}{\inf S}. \tag{10}$$

Furthermore, since $u := \sup S^{-1}$ is an upper bound of S^{-1}, as in the 'Only if' part, $\frac{1}{u} = \frac{1}{\sup S^{-1}}$ is a lower bound of S, and so $\frac{1}{\sup S^{-1}} \leq \inf S$, that is,

$$\frac{1}{\inf S} \leq \sup S^{-1}. \tag{11}$$

From (10) and (11), it follows that $\sup S^{-1} = \frac{1}{\inf S}$.

Solution to Exercise 1.15

(1) $S = \{\frac{1}{n} : n \in \mathbb{Z}\setminus\{0\}\} = \{\frac{1}{n} : n \in \mathbb{N}\} \cup \{-\frac{1}{n} : n \in \mathbb{N}\}$.

An upper bound of S. 1 is an upper bound, since for all $n \in \mathbb{N}$, $\frac{1}{n} \leq 1$ and $-\frac{1}{n} \leq 0 \leq 1$.

A lower bound of S. -1 is a lower bound, since for all $n \in \mathbb{N}$, $-1 \leq 0 \leq \frac{1}{n}$ and $-1 \leq -\frac{1}{n}$.

Is S bounded? Yes, since S is bounded above (1 is an upper bound) and it is bounded below (-1 is a lower bound).

Supremum of S. $\sup S = 1$. 1 is an upper bound. Moreover, if u is also an upper bound, then since $1 = \frac{1}{1} \in S$, $1 \leq u$.

Infimum of S. $\inf S = -1$. -1 is a lower bound. Moreover, if ℓ is also a lower bound, then as $-1 = \frac{1}{-1} \in S$, it follows that $\ell \leq -1$.

Maximum of S. $\max S = 1$, since $\sup S = 1 \in S$.

Minimum of S. $\min S = -1$, since $\inf S = -1 \in S$.

(2) $S = \{\frac{n}{n+1} : n \in \mathbb{N}\}$.

An upper bound of S. 1 is an upper bound, since for all $n \in \mathbb{N}$, $\frac{n}{n+1} < 1$.

A lower bound of S. $\frac{1}{2}$ is a lower bound because for all $n \in \mathbb{N}$, $\frac{1}{2} \leq \frac{n}{n+1}$ (since $n+1 \leq 2n$, that is, $1 \leq n$).

Is S bounded? Yes, since S is bounded above (1 is an upper bound) and it is bounded below ($\frac{1}{2}$ is a lower bound).

Supremum of S. $\sup S = 1$. 1 is an upper bound of S. If $u < 1$ is an upper bound, then let $N \in \mathbb{N}$ be such that $\frac{u}{u-1} < N$. Then $u < \frac{N}{N+1}$, contradicting the fact that u is an upper bound.

Infimum of S. $\inf S = \frac{1}{2}$. $\frac{1}{2}$ is a lower bound, and if ℓ is a lower bound, then since $\frac{1}{2} = \frac{1}{1+1} \in S$, it follows that $\ell \leq \frac{1}{2}$.

Maximum of S. $\max S$ does not exist since $\sup S = 1 \notin S$.

Minimum of S. $\min S$ exists since $\inf S = \frac{1}{2} \in S$.

(3) $S = \{(-1)^n \left(1 + \frac{1}{n}\right) : n \in \mathbb{N}\}$.

   (This set has the elements $-\frac{2}{1}, \frac{3}{2}, -\frac{4}{3}, \frac{5}{4}, \cdots$.)

An upper bound of S. $\frac{3}{2}$ is an upper bound.

   If $n \in \mathbb{N}$ and n is even, then $(-1)^n \left(1 + \frac{1}{n}\right) = 1 + \frac{1}{n} \leq 1 + \frac{1}{2} = \frac{3}{2}$.

   If $n \in \mathbb{N}$ and n is odd, then $(-1)^n \left(1 + \frac{1}{n}\right) = -1 - \frac{1}{n} < 0 < \frac{3}{2}$.

A lower bound of S. -2 is a lower bound.

   If $n \in \mathbb{N}$ and n is even, then $(-1)^n \left(1 + \frac{1}{n}\right) = 1 + \frac{1}{n} > 0 > -2$.

   If $n \in \mathbb{N}$ and n is odd, then $(-1)^n \left(1 + \frac{1}{n}\right) = -1 - \frac{1}{n} \geq -1 - 1 = -2$.

Is S bounded? Yes, since S is bounded above ($\frac{3}{2}$ is an upper bound) and it is bounded below (-2 is a lower bound).

Supremum of S. $\sup S = \frac{3}{2}$. $\frac{3}{2}$ is an upper bound, and if u is also an upper bound, then since $\frac{3}{2} = (-1)^2 \left(1 + \frac{1}{2}\right) \in S$, it follows that $\frac{3}{2} \leq u$.

Infimum of S. $\inf S = -2$. -2 is a lower bound, and if ℓ is also a lower bound, then since $-2 = (-1)^1 \left(1 + \frac{1}{1}\right) \in S$, it follows that $\ell \leq -2$.

Maximum/minimum of S. $\max S = \sup S = \frac{3}{2} \in S$ and $\min S = \inf S = -2 \in S$.

Solution to Exercise 1.16

(a) As $x^2 \geq 0$ for all $x \in \mathbb{R}$, we have $(xy - 1)^2 + x^2 \geq 0 + 0 = 0$ for all $(x, y) \in \mathbb{R}^2$. Hence 0 is a lower bound for S, and so S is bounded below.

(b) We will show that $\inf S = 0$. We already know that 0 is *a* lower bound. Suppose that ℓ is any lower bound. We claim that $\ell \leq 0$. Suppose not, that is, $\ell > 0$. Then choose an $n \in \mathbb{N}$ such that $n > 1/\sqrt{\ell}$ (possible by the Archimedean Property). Then with $(x, y) := (1/n, n)$, we have

$$(xy - 1)^2 + x^2 = \left(\frac{1}{n} \cdot n - 1\right)^2 + \left(\frac{1}{n}\right)^2 = (1 - 1)^2 + \frac{1}{n^2} = 0 + \frac{1}{n^2} = \frac{1}{n^2} < \ell,$$

a contradiction to ℓ being a lower bound of S. Hence $\ell \leq 0$. Consequently, $\inf S = 0$.

(c) No, since if $\inf S = 0 \in S$, we would have $0 = (xy - 1)^2 + x^2$ for some $(x, y) \in \mathbb{R}^2$, and so $x = 0$ as well as $xy - 1 = 0$, but this is impossible. (Indeed, $x = 0$ implies that we have $xy - 1 = 0 \cdot y - 1 = 0 - 1 = -1 \neq 0$.)

Solution to Exercise 1.17

By the density of $\mathbb{Q}$ in $\mathbb{R}$, there exists an $r \in \mathbb{Q}$ such that

$$a + \sqrt{2} < r < b + \sqrt{2},$$

that is, $a < r - \sqrt{2} < b$, and clearly $r - \sqrt{2} \in \mathbb{R} \backslash \mathbb{Q}$.

Solution to Exercise 1.18

Given any $x \in (a, b)$, we have $a < x < b$. Motivated by the following picture, let us take $\delta = \min\{x - a, b - x\}$. Then $\delta > 0$, and if $|y - x| < \delta$, we have $-\delta < y - x < \delta$. So

$$a = x - (x - a) \leq x - \delta < y < x + \delta \leq x + (b - x) = b,$$

that is, $y \in (a, b)$. Hence $(x - \delta, x + \delta) \subset (a, b)$.

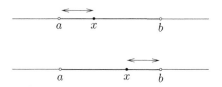

On the other hand, the interval $[a, b]$ does not have the stipulated property. Indeed, if we take $x := a \in [a, b]$, then no matter how small a $\delta > 0$ we take, the set $(a - \delta, a + \delta)$ contains points that do not belong to $[a, b]$: for example,

$$a - \frac{\delta}{2} \in (a - \delta, a + \delta) \text{ but } a - \frac{\delta}{2} \notin [a, b].$$

The following picture illustrates this.

Solution to Exercise 1.19

(1) If $x \in \bigcap_{n \in \mathbb{N}} \left(0, \dfrac{1}{n} \right)$, then

$$\text{for all } n \in \mathbb{N}, \ 0 < x < \frac{1}{n}. \tag{12}$$

Let $N \in \mathbb{N}$ be such that $\frac{1}{x} < N$ (Archimedean Property). Thus $\frac{1}{N} < x$, which contradicts (12).

So $\neg \left(\exists x \in \bigcap_{n \in \mathbb{N}} \left(0, \dfrac{1}{n} \right) \right)$ that is, $\bigcap_{n \in \mathbb{N}} \left(0, \dfrac{1}{n} \right) = \emptyset$.

(2) Clearly $\{0\} \subset \left[0, \dfrac{1}{n} \right]$ for all $n \in \mathbb{N}$ and so

$$\{0\} \subset \bigcap_{n \in \mathbb{N}} \left[0, \frac{1}{n} \right]. \tag{13}$$

Let $x \in \bigcap_{n \in \mathbb{N}} \left[0, \dfrac{1}{n} \right]$.

Then $x \in [0, 1]$ and so $x \geq 0$. If $x > 0$, then let $N \in \mathbb{N}$ be such that $\frac{1}{x} < N$ (Archimedean Property), that is, $\frac{1}{N} < x$. So $x \notin [0, \frac{1}{N}]$, and hence

$$x \notin \bigcap_{n \in \mathbb{N}} \left[0, \frac{1}{n} \right].$$

Consequently, if $x \in \bigcap_{n \in \mathbb{N}} \left[0, \dfrac{1}{n} \right]$, then $x = 0$, that is,

$$\bigcap_{n \in \mathbb{N}} \left[0, \frac{1}{n} \right] \subset \{0\}. \tag{14}$$

From (13) and (14), $\bigcap_{n \in \mathbb{N}} \left[0, \dfrac{1}{n} \right] = \{0\}$.

(3) Let $n \in \mathbb{N}$.

If $x \in \left[\dfrac{1}{n+2}, 1 - \dfrac{1}{n+2} \right]$, then $0 < \dfrac{1}{n+2} \leq x \leq 1 - \dfrac{1}{n+2} < 1$, and so $x \in (0, 1)$.
Hence

$$\bigcup_{n \in \mathbb{N}} \left[\frac{1}{n+2}, 1 - \frac{1}{n+2} \right] \subset (0, 1). \tag{15}$$

If $x \in (0, 1)$, then $0 < x < 1$. Let $N_1 \in \mathbb{N}$ be such that $\frac{1}{x} - 2 < N_1$ (Archimedean Property), that is, $\frac{1}{N_1 + 2} < x$. Let $N_2 \in \mathbb{N}$ be such that $\frac{1}{1-x} - 2 < N_2$ (Archimedean Property), that is, $x < 1 - \frac{1}{N_2 + 2}$. Thus with $N := \max\{N_1, N_2\}$, we have

$$\frac{1}{N+2} \leq \frac{1}{N_1 + 2} < x < 1 - \frac{1}{N_2 + 2} \leq 1 - \frac{1}{N+2},$$

that is, $x \in \left[\dfrac{1}{N+2}, 1 - \dfrac{1}{N+2} \right] \subset \bigcup_{n \in \mathbb{N}} \left[\dfrac{1}{n+2}, 1 - \dfrac{1}{n+2} \right]$. So we have

$$(0, 1) \subset \bigcup_{n \in \mathbb{N}} \left[\frac{1}{n+2}, 1 - \frac{1}{n+2} \right]. \tag{16}$$

From (15) and (16), we obtain $(0, 1) = \bigcup_{n \in \mathbb{N}} \left[\frac{1}{n+2}, 1 - \frac{1}{n+2} \right]$.

(4) If $x \in [0, 1]$, then for any $n \in \mathbb{N}$, $-\frac{1}{n} < 0 \le x \le 1 < 1 + \frac{1}{n}$, and so $x \in \left(-\frac{1}{n}, 1 + \frac{1}{n} \right)$. Hence

$$[0, 1] \subset \bigcap_{n \in \mathbb{N}} \left(-\frac{1}{n}, 1 + \frac{1}{n} \right). \tag{17}$$

Let $x \in \bigcap_{n \in \mathbb{N}} \left(-\frac{1}{n}, 1 + \frac{1}{n} \right)$. Then

$$-\frac{1}{n} \le x \le 1 + \frac{1}{n} \quad \text{for all } n \in \mathbb{N}. \tag{18}$$

We prove that this gives $0 \le x \le 1$. For if $x < 0$, then let $N_1 \in \mathbb{N}$ be such that $-\frac{1}{x} < N_1$, that is, $x < -\frac{1}{N_1}$, a contradiction to (18). Similarly, if $x > 1$, then let $N_2 \in \mathbb{N}$ be such that $\frac{1}{x-1} < N_2$, that is, $x > 1 + \frac{1}{N_2}$, a contradiction to (17). Hence we see that neither $x < 0$ nor $x > 1$ are possible, and so $x \in [0, 1]$. Thus

$$\bigcap_{n \in \mathbb{N}} \left(-\frac{1}{n}, 1 + \frac{1}{n} \right) \subset [0, 1]. \tag{19}$$

(17) and (19) imply $\bigcap_{n \in \mathbb{N}} \left(-\frac{1}{n}, 1 + \frac{1}{n} \right) = [0, 1]$.

Solution to Exercise 1.20

If S is bounded, then it is bounded above and it is bounded below. Thus S has an upper bound, say u, and a lower bound, say ℓ. So for all $x \in S$, $\ell \le x \le u$, that is, $x \le u$ and $-x \le -\ell$, and so we have

$$x \le u \le \max\{-\ell, u\} \text{ and } -x \le -\ell \le \max\{-\ell, u\}.$$

Thus $|x| \le \max\{-\ell, u\} =: M$.

Conversely, if there exists an M such that for all $x \in S$, $|x| \le M$, we have $-M \le x \le M$. So $-M$ is a lower bound of S and M is an upper bound of S. Thus S is bounded.

Solution to Exercise 1.21

From the inequality $|a + b| \le |a| + |b|$ for all $a, b \in \mathbb{R}$, we have that if $x, y \in \mathbb{R}$, then by taking $a := x - y$ and $b := y$ in the previous inequality, we obtain $|x| = |x - y + y| \le |x - y| + |y|$, that is,

$$|x| - |y| \le |x - y|. \tag{20}$$

Interchanging x and y in (20), we obtain

$$|y| - |x| \le |y - x| = |-(x - y)| = |-1| \cdot |x - y| = 1 \cdot |x - y| = |x - y|,$$

and so,

$$-(|x| - |y|) \le |x - y| \tag{21}$$

for all $x, y \in \mathbb{R}$. From (20) and (21), we obtain $||x| - |y|| \le |x - y|$ for all $x, y \in \mathbb{R}$.

Solution to Exercise 1.22

(1) This follows easily from induction on n.

If $n = 1$, then clearly equality holds ($|a_1| = |a_1|$).

If the claim is true for some $n \in \mathbb{N}$, and if $a_1, \cdots, a_n, a_{n+1} \in \mathbb{R}$, then we have using the triangle inequality and the induction hypothesis that

$$|a_1 + \cdots + a_n + a_{n+1}| \leq |a_1 + \cdots + a_n| + |a_{n+1}| \leq |a_1| + \cdots + |a_n| + |a_{n+1}|,$$

and so the result follows by induction.

(2) Suppose that $a, b \in \mathbb{R}$ are such that $|a| + |b| = |a + b|$. Suppose that a, b don't have the same sign. Then one must be positive and the other must be negative. Without loss of generality, we may assume that $a < 0 < b$. Then we have two cases:

$1°$ $a + b \geq 0$: Then $-a + b = |a| + |b| = |a + b| = a + b$ gives $a = 0$, a contradiction.
$2°$ $a + b < 0$: Then $-a + b = |a| + |b| = |a + b| = -(a + b)$ gives $b = 0$, a contradiction.

So a, b must have the same sign. Conversely, if a, b have the same sign, then clearly we have equality.

For the general equality case, we use induction on n again. Suppose that the result holds for some n and that

$$|a_1| + \cdots + |a_n| + |a_{n+1}| = |a_1 + \cdots + a_n + a_{n+1}|.$$

By the triangle inequality, we have

$$|a_1| + \cdots + |a_n| + |a_{n+1}| = |a_1 + \cdots + a_n + a_{n+1}| \leq |a_1 + \cdots + a_n| + |a_{n+1}|$$
$$\leq |a_1| + \cdots + |a_n| + |a_{n+1}|,$$

and so all the inequalities above must be equalities. In particular,

$$|a_1 + \cdots + a_n| + |a_{n+1}| = |a_1| + \cdots + |a_n| + |a_{n+1}|$$

gives $|a_1 + \cdots + a_n| = |a_1| + \cdots + |a_n|$. By the induction hypothesis, the numbers $a_1, \cdots, a_n$ must have the same sign. Also, $|a_1 + \cdots + a_n + a_{n+1}| = |a_1 + \cdots + a_n| + |a_{n+1}|$ shows that a_{n+1} must have the same sign as $a_1 + \cdots + a_n$. But, as $a_1, \cdots, a_n$ all have the same sign, $a_1 + \cdots + a_n$ has the same sign as $a_1, \cdots, a_n$. Consequently $a_1, \cdots, a_n, a_{n+1}$ have the same sign.

Solution to Exercise 1.23

We have the following two cases:

$1°$ If $a \geq b$, then $\max\{a, b\} = a = \dfrac{a + b + a - b}{2} = \dfrac{a + b + |a - b|}{2}$.

$2°$ If $a < b$, then $\max\{a, b\} = b = \dfrac{a + b + b - a}{2} = \dfrac{a + b + |a - b|}{2}$.

Also, $\min\{a, b\} = a + b - \max\{a, b\}$, and so from the above it follows that

$$\min\{a, b\} = \frac{a + b - |a - b|}{2}.$$

Solution to Exercise 1.24

We have $f(3) = 1 + 3^2 = 10$ and $g(3) = 1 - 3^2 = -8$, and so

(1) $f(3) + g(3) = 10 + (-8) = 2$,

(2) $f(3) - 3 \cdot g(3) = 10 - 3 \cdot (-8) = 10 + 24 = 34$,

(3) $f(3) \cdot g(3) = 10 \cdot (-8) = -80$,

(4) $(f(3))/(g(3)) = 10/(-8) = -5/4$,

(5) $f(g(3)) = f(-8) = 1 + (-8)^2 = 65$,

(6) for $a \in \mathbb{R}$, $f(a) + g(-a) = 1 + a^2 + (1 - (-a)^2) = 2$,

(7) for $t \in \mathbb{R}$, $f(t) \cdot g(-t) = (1 + t^2)(1 - (-t)^2) = 1 - t^4$.

Solution to Exercise 1.25

Let $x_1, x_2 \in \mathbb{R}$ be such that $f(x_1) = f(x_2)$. Then we have

$$x_1 |x_1| = x_2 |x_2|. \tag{22}$$

We have the following three possible cases:

1° $x_1 > 0$. Then $|x_1| = x_1$, and so the above implies first of all that x_2 can't be zero, and also that

$$x_2 = \frac{x_1^2}{|x_2|} > 0.$$

So (22) gives $x_1^2 = x_2^2$, and by taking square roots, $x_1 = x_2$.

2° $x_2 < 0$. By multiplying (22) by -1, we have

$$(-x_1)|x_1| = (-x_1)| - x_1| = (-x_2)| - x_2| = (-x_2)|x_2|.$$

By 1° and the innermost equality above, it follows that $-x_1 = -x_2$ and so $x_1 = x_2$.

3° $x_1 = 0$. If $x_2 \neq 0$, then the right hand side of (22) is nonzero, while the left hand side is 0, a contradiction. So $x_2 = 0 = x_1$.

Hence f is injective.

Let $y \in \mathbb{R}$. Then we have the following two cases:

1° If $y \geq 0$, then $f(\sqrt{y}) = \sqrt{y}|\sqrt{y}| = (\sqrt{y})^2 = y$.

2° If $y < 0$, then $f(-\sqrt{y}) = -\sqrt{y}| - \sqrt{y}| = -(\sqrt{y})^2 = -|y| = y$.

So f is also surjective.

Solution to Exercise 1.26

See Figures 1 and 2.

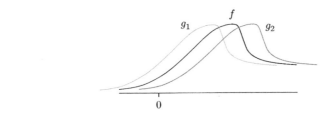

Figure 1.   The graphs of f, g_1, g_2.

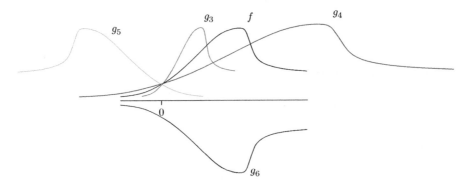

Figure 2.   The graphs of f, g_3, g_4, g_5, and g_6.

Solution to Exercise 1.27

The Maple plot is shown in the following picture. We note that the graph is symmetric about the y-axis. This is because the polynomial is an *even function*, that is, it satisfies

$$p(x) = p(-x), \quad x \in \mathbb{R}.$$

As a consequence, a point (x, y) belongs to graph of p if and only if $(-x, y)$ belongs to the graph of p. See Figure 3.

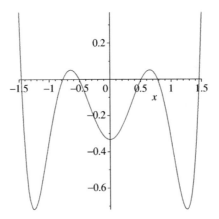

Figure 3.   The graph of $x \mapsto x^6 - 3 \cdot x^4 + 2 \cdot x^2 - \frac{1}{3}$, for $x \in (-\frac{3}{2}, \frac{3}{2})$, plotted using Maple.

Solutions to the exercises from Chapter 2

Solution to Exercise 2.1

(1) Yes. For instance, the constant sequence $(1)_{n \in \mathbb{N}}$ converges to 1.

(2) Suppose that the terms of the convergent sequence $(a_n)_{n \in \mathbb{N}}$ (with limit, say, L) lie in the finite set $\{v_1, \cdots, v_m\}$. If $L \notin \{v_1, \cdots, v_m\}$, then with

$$\epsilon := \min\{|v_1 - L|, \cdots, |v_m - L|\} > 0,$$

let $N \in \mathbb{N}$ be such that for all $n > N$, $|a_n - L| < \epsilon$. In particular, with $n = N + 1 > N$, we have $|a_{N+1} - L| < \epsilon$. But, $a_{N+1} \in \{v_1, \cdots, v_m\}$. Let $a_{N+1} = v_k$ for some $k \in \{1, \cdots, m\}$. Then we have

$$|v_k - L| = |a_{N+1} - L| < \epsilon = \min\{|v_1 - L|, \cdots, |v_m - L|\} \le |v_k - L|,$$

a contradiction. So $L \in \{v_1, \cdots, v_m\}$, that is, L must be one of the terms. Thus we have shown that

$$\boxed{\begin{array}{c} \text{terms of the sequence} \\ \text{take finitely many values} \end{array}} \Rightarrow \boxed{\begin{array}{c} L \text{ must be one} \\ \text{of the terms} \end{array}},$$

that is,

$$\boxed{\begin{array}{c} L \text{ is not equal to} \\ \text{any of the terms} \end{array}} \Rightarrow \boxed{\begin{array}{c} \text{terms of the sequence cannot} \\ \text{consist of finitely many values} \end{array}}.$$

(3) Suppose that $((-1)^n)_{n \in \mathbb{N}}$ is a convergent sequence with limit L. Then from part (2) above, it follows that $L = 1$ or $L = -1$: indeed, the terms of the sequence take finitely many values, namely 1 and -1, and so L must be one of these terms. So we have the following two cases:

$1°$ If the limit is 1, then given $\epsilon = 1 > 0$, let $N \in \mathbb{N}$ be such that for all $n > N$, we have $|(-1)^n - 1| < \epsilon = 1$. Let n be any odd number $> N$. Then we have that for such n, $2 = |-2| = |-1 - 1| = |(-1)^n - 1| < \epsilon = 1$, a contradiction.

$2°$ If the limit is -1, then given $\epsilon = 1 > 0$, let $N \in \mathbb{N}$ be such that for all $n > N$, we have $|(-1)^n - (-1)| < \epsilon = 1$. Let n be any even number $> N$. Then we have that for such n, $2 = |2| = |1 + 1| = |(-1)^n - (-1)| < \epsilon = 1$, a contradiction.

So $((-1)^n)_{n \in \mathbb{N}}$ is divergent.

Solution to Exercise 2.2

(1) We have seen that the sequence $((-1)^n)_{n \in \mathbb{N}}$ is divergent. Let $\epsilon > 0$, and let $N \in \mathbb{N}$. Then for all even $n > N$ (there are obviously infinitely many such n), we have $|a_n - L| = |(-1)^n - 1| = |1 - 1| = 0 < \epsilon$.

(2) Again, for the divergent sequence $((-1)^n)_{n \in \mathbb{N}}$, with $\epsilon = 3 > 0$, for all $N \in \mathbb{N}$ and all $n > N$, we have $|a_n - L| \le |a_n| + |L| \le 1 + 1 = 2 < 3 = \epsilon$.

Solution to Exercise 2.3

S is nonempty and bounded above, and so by the Least Upper Bound Property of $\mathbb{R}$, it follows that $\sup S$ exists.

Given $n \in \mathbb{N}$, we have $\frac{1}{n} > 0$, and so $\sup S - \frac{1}{n} < \sup S$. Thus $S - \frac{1}{n}$ is not an upper bound of S. Hence there must exist an element in S, which we denote by a_n, such that

$$\neg \left(a_n \leq \sup S - \frac{1}{n} \right),$$

that is, $a_n > \sup S - \frac{1}{n}$. In this way we construct the sequence $(a_n)_{n \in \mathbb{N}}$.

As $\sup S$ is an upper bound of S, we also have $a_n \leq \sup S$ for all $n \in \mathbb{N}$. Consequently,

$$\text{for all } n \in \mathbb{N}, \quad \sup S - \frac{1}{n} < a_n \leq \sup S < \sup S + \frac{1}{n},$$

that is, for all $n \in \mathbb{N}$, $-\frac{1}{n} < a_n - \sup S < \frac{1}{n}$, and so for all $n \in \mathbb{N}$, $|a_n - \sup S| < \frac{1}{n}$.

Given $\epsilon > 0$, let the number $N \in \mathbb{N}$ be such that $N > \frac{1}{\epsilon}$. Then for all $n > N$, we have

$$|a_n - \sup S| < \frac{1}{n} < \frac{1}{N} < \epsilon.$$

Hence $(a_n)_{n \in \mathbb{N}}$ is convergent with limit equal to $\sup S$.

Solution to Exercise 2.4

Suppose $L < 0$. Then $\epsilon := -L/2 > 0$, and so there exists an $N \in \mathbb{N}$ such that for all $n > N$, $|a_n - L| < \epsilon = -L/2$. Hence with $n = N + 1 \, (> N)$,

$$a_{N+1} - L \leq |a_{N+1} - L| < -\frac{L}{2},$$

that is, $a_{N+1} < \frac{L}{2} < 0$, a contradiction.

Solution to Exercise 2.5

(C).

Solution to Exercise 2.6

(1) Let $M > 0$ be such that

$$\text{for all } n \in \mathbb{N}, \quad |b_n| \leq M.$$

Given $\epsilon > 0$, let $N \in \mathbb{N}$ be such that $M/\epsilon < N$, that is, $1/N < \epsilon/M$. Then for all $n > N$,

$$\left| \frac{b_n}{n} - 0 \right| = \frac{|b_n|}{n} \leq \frac{M}{n} < \frac{M}{N} < M \cdot \frac{\epsilon}{M} = \epsilon.$$

Hence, $\left(\dfrac{b_n}{n} \right)_{n \in \mathbb{N}}$ is convergent with limit 0.

(2) For all real $x \in \mathbb{R}$, $\sin x \in [-1, 1]$, and so by the above, $\displaystyle\lim_{n \to \infty} \frac{\sin n}{n} = 0$.

Solution to Exercise 2.7

Given $\epsilon > 0$, there exists an $N_1 \in \mathbb{N}$ such that

$$\text{for all } n > N_1, \quad |a_n - L| < \epsilon/2.$$

Since $(a_n)_{n\in\mathbb{N}}$ is convergent, it is bounded: there exists an $M > 0$ such that for all $n \in \mathbb{N}$, $|a_n| \leq M$. Choose $^1N \in \mathbb{N}$ such that

$$\max\left\{N_1, \frac{N_1(M + |L|)}{\epsilon/2}\right\} < N,$$

and so, $N > N_1$ and $\dfrac{N_1(M + |L|)}{N} < \dfrac{\epsilon}{2}$. Then for $n > N$, we have

$$\left|\frac{a_1 + \cdots + a_{N_1} + a_{N_1+1} + \cdots + a_n}{n} - L\right|$$

$$= \left|\frac{a_1 + \cdots + a_{N_1} + a_{N_1+1} + \cdots + a_n - nL}{n}\right|$$

$$= \frac{|a_1 + \cdots + a_{N_1} + a_{N_1+1} + \cdots + a_n - nL|}{n}$$

$$\leq \frac{|a_1 - L| + \cdots + |a_{N_1} - L| + |a_{N_1+1} - L| + \cdots + |a_n - L|}{n}$$

$$\leq \frac{(|a_1| + |L| + \cdots + |a_{N_1}| + |L|) + \epsilon/2 + \cdots + \epsilon/2}{n}$$

$$\leq \frac{N_1(M + |L|) + (n - N_1)(\epsilon/2)}{n}$$

$$\leq \frac{N_1(M + |L|)}{n} + \left(1 - \frac{N_1}{n}\right) \cdot \frac{\epsilon}{2} < \frac{N_1(M + |L|)}{N} + 1 \cdot \frac{\epsilon}{2} < \frac{\epsilon}{2} + \frac{\epsilon}{2} = \epsilon.$$

So $\left(\dfrac{a_1 + \cdots + a_n}{n}\right)_{n\in\mathbb{N}}$ is a convergent sequence with limit L.

If $a_n = (-1)^n$, $n \in \mathbb{N}$, then $(a_n)_{n\in\mathbb{N}}$ is divergent, but the sequence with nth term

$$\frac{a_1 + \cdots + a_n}{n} = \frac{(-1)^1 + (-1)^2 + \cdots + (-1)^n}{n} = \begin{cases} 0 & \text{if } n \text{ is even} \\ -\dfrac{1}{n} & \text{if } n \text{ is odd} \end{cases}$$

is convergent with limit equal to 0. Indeed, given $\epsilon > 0$, let $N \in \mathbb{N}$ be such that $\frac{1}{\epsilon} < N$. Then for $n > N$, we have

$$|a_n - 0| = |a_n| = \begin{cases} 0 & \text{if } n \text{ is even} \\ \dfrac{1}{n} & \text{if } n \text{ is odd} \end{cases} \leq \frac{1}{n} < \frac{1}{N} < \epsilon.$$

So $\left(\dfrac{a_1 + \cdots + a_n}{n}\right)_{n\in\mathbb{N}}$ is convergent with limit 0.

Solution to Exercise 2.8

Let $(a_n)_{n\in\mathbb{N}}$ be a bounded decreasing sequence of real numbers. Let ℓ_* be the *greatest lower bound* of $\{a_n : n \in \mathbb{N}\}$. The existence of ℓ_* is guaranteed by the *Greatest Lower Bound Property* of the set of real numbers. We show that ℓ^* is the *limit* of

---

^{1} This is arrived at by working backwards; we wish to make $|\frac{a_1+\cdots+a_n}{n} - L|$ less than ϵ for all $n > N$, so we manipulate this (as shown in the chain of inequalities that follow) to see if we can indeed achieve this by choosing the N large enough.

$(a_n)_{n\in\mathbb{N}}$. Taking $\epsilon > 0$, we must show that there exists a positive integer N such that $|a_n - \ell_*| < \epsilon$ for all $n > N$. Since $\ell_* + \epsilon > \ell_*$, $\ell_* + \epsilon$ is not *a lower bound* of $\{a_n : n \in \mathbb{N}\}$. Therefore there exists N with $\ell_* \leq a_N < \ell_* + \epsilon$. Since $(a_n)_{n\in\mathbb{N}}$ is *decreasing*, we have for all $n \geq N$ that $\ell_* - \epsilon < \ell_* \leq a_n \leq a_N < \ell_* + \epsilon$, and so $|a_n - \ell_*| < \epsilon$.

Solution to Exercise 2.9

(1) We prove that $|a_n| \leq 1$ for all $n \in \mathbb{N}$. We prove this using induction. We have $|a_1| = |1| = 1$. If $k \in \mathbb{N}$ is such that $|a_k| \leq 1$, then

$$|a_{k+1}| = \left|\frac{2k+3}{3k+3}a_k\right| = \left|\frac{2k+3}{3k+3}\right||a_k| = \left(\frac{2k+3}{3k+3}\right)|a_k| \leq 1 \cdot 1 = 1,$$

and so the claim follows from induction. So the sequence is bounded.

(2) Since $n \geq 1$, it follows that $2n + 1 \leq 3n$, and so

$$\frac{2n+1}{3n} \leq 1$$

for all $n \in \mathbb{N}$. Furthermore, note that for all $n \in \mathbb{N}$, $a_n \geq 0$ (induction!). So for all $n \geq 2$,

$$a_n = \frac{2n+1}{3n}a_{n-1} \leq 1 \cdot a_{n-1} = a_{n-1}.$$

So $(a_n)_{n\in\mathbb{N}}$ is decreasing.

(3) As the sequence is bounded and monotone, it is convergent.

Solution to Exercise 2.10

Since $(a_n)_{n\in\mathbb{N}}$ is bounded, it follows that there exists an M such that for all $n \in \mathbb{N}$, $|a_n| \leq M$, that is, $-M \leq a_n \leq M$. If $k \in \mathbb{N}$, then in particular, for all $n \geq k$, $-M \leq a_n \leq M$, and so the set $\{a_n : n \geq k\}$ is bounded. By the Least Upper Bound Property of $\mathbb{R}$, it then follows that $\inf\{a_n : n \geq k\}$ and $\sup\{a_n : n \geq k\}$ exist, that is, ℓ_k and u_k are well-defined. Furthermore, for each k,

$$-M \leq \inf\{a_n : n \geq k\} \leq \sup\{a_n : n \geq k\} \leq M,$$

and so we see that the sequences $(\ell_k)_{k\in\mathbb{N}}$ and $(u_k)_{k\in\mathbb{N}}$ are bounded.

Clearly $\{a_n : n \geq k+1\} \subset \{a_n : n \geq k\}$, and so

$$u_{k+1} = \sup\{a_n : n \geq k+1\} \leq \sup\{a_n : n \geq k\} = u_k,$$

and so $(u_k)_{k\in\mathbb{N}}$ is a decreasing sequence.

Similarly, again since $\{a_n : n \geq k+1\} \subset \{a_n : n \geq k\}$, we have[2]

$$\inf\{a_n : n \geq k+1\} \geq \inf\{a_n : n \geq k\},$$

that is, $\ell_{k+1} \geq \ell_k$. Consequently, $(\ell_k)_{k\in\mathbb{N}}$ is a increasing sequence.

As the sequences $(u_k)_{k\in\mathbb{N}}$, $(\ell_k)_{k\in\mathbb{N}}$ are both bounded and monotone, it follows that they are convergent.

---

[2] If $\emptyset \neq A \subset B \subset \mathbb{R}$, and A, B are bounded below, then we have $\inf A \geq \inf B$. This fact follows from the Exercises 1.9 and 1.7. Indeed, we have the two sets $-A$, $-B$ are nonempty, bounded above, and also $-A \subset -B$, giving $\inf A = -\sup(-A) \geq -\sup(-B) = \inf B$.

Solution to Exercise 2.11

If $\lim\limits_{n\to\infty} a_n = L$, then for a *fixed* $k \in \mathbb{N}$, $\lim\limits_{n\to\infty} a_n^k = L^k$.

So the first equality is obtained by an incorrect application of the above.

(In fact, it will be shown later on that $\lim\limits_{n\to\infty}\left(1 + \dfrac{1}{n}\right)^n = e > 1$.)

Solution to Exercise 2.12

We have $c_n = \dfrac{a_n b_n + 5n}{a_n^2 + n} = \dfrac{a_n \cdot \frac{b_n}{n} + 5}{a_n \cdot a_n \cdot \frac{1}{n} + 1}$ for all $n \in \mathbb{N}$.

(N) The sequence $\left(a_n \cdot \dfrac{b_n}{n} + 5\right)_{n\in\mathbb{N}}$ is convergent.

The sequence $(a_n)_{n\in\mathbb{N}}$ is convergent with limit, say L. Since $(b_n)_{n\in\mathbb{N}}$ is bounded, the sequence $(b_n/n)_{n\in\mathbb{N}}$ is convergent with limit 0. Hence $((a_n b_n)/n)_{n\in\mathbb{N}}$ is convergent with limit $L \cdot 0 = 0$. The sequence $(5)_{n\in\mathbb{N}}$ is convergent with limit 5.

So the sequence $\left(a_n \cdot \dfrac{b_n}{n} + 5\right)_{n\in\mathbb{N}}$ is convergent with limit $0 + 5 = 5$.

(D) The sequence

$$\left(\frac{a_n^2}{n} + 1\right)_{n\in\mathbb{N}}$$

has nonzero terms for all $n \in \mathbb{N}$ and it is convergent with the nonzero limit 1.

We have $\dfrac{a_n^2}{n} + 1 \geq 1$, and so $\dfrac{a_n^2}{n} + 1 \neq 0$ for all $n \in \mathbb{N}$.

Since the sequence $(a_n)_{n\in\mathbb{N}}$ is convergent with limit L, it follows that the sequence $(a_n^2)_{n\in\mathbb{N}}$ is convergent with limit L^2. Since $(1/n)_{n\in\mathbb{N}}$ is convergent with limit 0, it follows that

$$\left(a_n^2 \cdot \frac{1}{n}\right)_{n\in\mathbb{N}}$$

is convergent with limit $L^2 \cdot 0 = 0$. Finally, as $(1)_{n\in\mathbb{N}}$ is convergent with limit 1, it follows that the sequence

$$\left(\frac{a_n^2}{n} + 1\right)_{n\in\mathbb{N}}$$

is convergent with limit $0 + 1 = 1 (\neq 0)$.

From (N) and (D) and the Algebra of Limits, $(c_n)_{n\in\mathbb{N}}$ is convergent to $\dfrac{5}{1} = 5$.

Solution to Exercise 2.13

We begin by showing that $L \geq 0$. Suppose on the contrary that $L < 0$. Set $\epsilon := -L/2 > 0$. Let $N \in \mathbb{N}$ be such that for all $n > N$, $|a_n - L| < \epsilon = -L/2$. Then we have

$$a_n - L \leq |a_n - L| < -\frac{L}{2},$$

and so $a_n < L/2 < 0$ for all $n > N$, a contradiction to the fact that

$$a_n \geq 0 \text{ for all } n \in \mathbb{N}.$$

So $L \geq 0$.

Now we show that $(\sqrt{a_n})_{n \in \mathbb{N}}$ is convergent with limit $\sqrt{L}$. Let $\epsilon > 0$. We consider the only two possible cases, namely $L = 0$ or $L > 0$:

$\underline{1°}$ If $L = 0$, then let $N \in \mathbb{N}$ be such that for all $n > N$,

$$|a_n - L| = |a_n - 0| = |a_n| = a_n < \epsilon^2.$$

Then for $n > N$, we have $\sqrt{a_n} < \epsilon$, that is,

$$|\sqrt{a_n} - \sqrt{L}| = |\sqrt{a_n} - \sqrt{0}| = |\sqrt{a_n}| = \sqrt{a_n} < \epsilon.$$

So $(\sqrt{a_n})_{n \in \mathbb{N}}$ is convergent with limit $\sqrt{L}$.

$\underline{2°}$ If $L > 0$, then let $N \in \mathbb{N}$ be such that for $n > N$, $|a_n - L| < \epsilon\sqrt{L}$. Then for all $n > N$, we obtain

$$\epsilon\sqrt{L} > |a_n - L| = |(\sqrt{a_n} - \sqrt{L})(\sqrt{a_n} + \sqrt{L})|$$
$$= |\sqrt{a_n} - \sqrt{L}||\sqrt{a_n} + \sqrt{L})| = |\sqrt{a_n} - \sqrt{L}|(\sqrt{a_n} + \sqrt{L})$$

and so $|\sqrt{a_n} - \sqrt{L}| < \dfrac{\epsilon\sqrt{L}}{\sqrt{a_n} + \sqrt{L}} \leq \dfrac{\epsilon\sqrt{L}}{\sqrt{L}} = \epsilon.$

Hence $(\sqrt{a_n})_{n \in \mathbb{N}}$ is convergent with limit $\sqrt{L}$.

Solution to Exercise 2.14

For all $n \in \mathbb{N}$, we have

$$\sqrt{n^2 + n} - n = (\sqrt{n^2 + n} - n) \cdot \frac{\sqrt{n^2 + n} + n}{\sqrt{n^2 + n} + n} = \frac{n^2 + n - n^2}{\sqrt{n^2 + n} + n} = \frac{n}{\sqrt{n^2 + n} + n}$$

$$= \frac{n(1)}{n\left(\dfrac{1}{n}\sqrt{n^2 + n} + 1\right)} = \frac{1}{\sqrt{\dfrac{n^2 + n}{n^2}} + 1} = \frac{1}{\sqrt{1 + \dfrac{1}{n}} + 1}.$$

By the Algebra of Limits, $\lim\limits_{n \to \infty}\left(1 + \dfrac{1}{n}\right) = 1 + 0 = 1$, and as

$$1 + \frac{1}{n} \geq 0 \text{ for all } n \in \mathbb{N},$$

by the previous part, it follows that $\lim\limits_{n \to \infty}\sqrt{1 + \dfrac{1}{n}} = \sqrt{1} = 1$. Hence

$$\lim_{n \to \infty}\left(\sqrt{1 + \frac{1}{n}} + 1\right) = 1 + 1 = 2(\neq 0).$$

Also $\sqrt{1 + \dfrac{1}{n}} + 1 > 1 > 0$. Thus by the Algebra of Limits,

$$\lim_{n \to \infty}\frac{1}{\sqrt{1 + \dfrac{1}{n}} + 1} = \frac{1}{2},$$

that is, $(\sqrt{n^2 + n} - n)_{n \in \mathbb{N}}$ is convergent with limit $\dfrac{1}{2}$.

Solution to Exercise 2.15

(1) Consider the sequence $(b_n - a_n)_{n \in \mathbb{N}}$. Since $a_n \leq b_n$, it follows that $b_n - a_n \geq 0$ for all $n \in \mathbb{N}$. From the Algebra of Limits, it follows that the sequence $(b_n - a_n)_{n \in \mathbb{N}}$ is convergent (being the sum of the convergent sequence $(b_n)_{n \in \mathbb{N}}$ and the convergent sequence $(-a_n)_{n \in \mathbb{N}}$). Moreover, its limit is

$$\lim_{n \to \infty} b_n - \lim_{n \to \infty} a_n.$$

From Exercise 2.4 on page (54), $\lim_{n \to \infty} b_n - \lim_{n \to \infty} a_n \geq 0$, that is, $\lim_{n \to \infty} b_n \geq \lim_{n \to \infty} a_n$.

(2) The inequality

$$\liminf_{n \to \infty} a_n \leq \limsup_{n \to \infty} a_n$$

for a bounded sequence $(a_n)_{n \in \mathbb{N}}$ follows immediately by applying the first part of this exercise to the two convergent sequences $(\ell_n)_{n \in \mathbb{N}}$ and $(u_n)_{n \in \mathbb{N}}$ and observing that for all $n \in \mathbb{N}$, we have that $\ell_n := \inf\{a_k : k \geq n\} \leq u_n := \sup\{a_k : k \geq n\}$.

Consider the bounded sequence $((-1)^n)_{n \in \mathbb{N}}$. For any n, we have

$$\ell_n = \inf\{a_k : k \geq n\} = \inf\{-1, 1\} = -1,$$
$$u_n = \sup\{a_k : k \geq n\} = \sup\{-1, 1\} = 1,$$

and so $\liminf_{n \to \infty} a_n = -1 < \limsup_{n \to \infty} a_n$.

Solution to Exercise 2.16

For all $n \in \mathbb{N}$, we have

$$0 \leq \frac{n!}{n^n} = \frac{1}{n} \cdot \frac{2}{n} \cdots \frac{n-1}{n} \cdot \frac{n}{n} \leq \frac{1}{n} \cdot 1 \cdots 1 \cdot 1 = \frac{1}{n}.$$

Since $(0)_{n \in \mathbb{N}}$ and $(\frac{1}{n})_{n \in \mathbb{N}}$ are both convergent with the same limit 0, from the Sandwich Theorem, it follows that $(\frac{n!}{n^n})_{n \in \mathbb{N}}$ is convergent with the limit 0.

Solution to Exercise 2.17

Let $k \in \mathbb{N}$. For all $n \in \mathbb{N}$, we have

$$0 \leq \frac{1^k + 2^k + 3^k + \cdots + n^k}{n^{k+2}} \leq \frac{n^k + n^k + n^k + \cdots + n^k}{n^{k+2}} \leq \frac{n \cdot n^k}{n^{k+2}} = \frac{1}{n}.$$

Thus

$$0 \leq \frac{1^k + 2^k + 3^k + \cdots + n^k}{n^{k+2}} \leq \frac{1}{n}$$

for all $n \in \mathbb{N}$. As $(0)_{n \in \mathbb{N}}$ and $(\frac{1}{n})_{n \in \mathbb{N}}$ are convergent with limit 0, from the Sandwich Theorem,

$$\lim_{n \to \infty} \frac{1^k + 2^k + 3^k + \cdots + n^k}{n^{k+2}} = 0.$$

Solution to Exercise 2.18

(1) We prove the claim using induction. Let $x \geq -1$. Clearly

$$(1+x)^1 = 1 + x = 1 + 1 \cdot x.$$

If for some $k \in \mathbb{N}$, $(1+x)^k \geq 1 + kx$, then we have

$$
\begin{aligned}
(1+x)^{k+1} &= (1+x)^k(1+x) \\
&\geq (1+kx)(1+x) \quad \text{(induction hypothesis and as } 1+x \geq 0) \\
&= 1 + kx + x + x^2 = 1 + (k+1)x + x^2 \\
&\geq 1 + (k+1)x.
\end{aligned}
$$

Hence by induction, the result follows.

(2) For all $n \in \mathbb{N}$,

$$
n^{\frac{1}{n}} \geq 1 \tag{23}
$$

(for if $n^{\frac{1}{n}} < 1$, then $n = (n^{\frac{1}{n}})^n < 1^n = 1$, a contradiction!). Clearly for all $n \in \mathbb{N}$,

$$
n^{\frac{1}{n}} = (\sqrt{n^2})^{\frac{1}{n}} = \sqrt{n^{\frac{2}{n}}} < (1+\sqrt{n})^{\frac{2}{n}}. \tag{24}
$$

Finally, $\left(1 + \dfrac{1}{\sqrt{n}}\right)^n \geq 1 + n \cdot \dfrac{1}{\sqrt{n}} = 1 + \sqrt{n}$, and so

$$
\left(1 + \frac{1}{\sqrt{n}}\right)^2 = \left(\left(1 + \frac{1}{\sqrt{n}}\right)^n\right)^{\frac{2}{n}} \geq (1+\sqrt{n})^{\frac{2}{n}}. \tag{25}
$$

Combining (23), (24), and (25), we obtain that for all $n \in \mathbb{N}$

$$
1 \leq n^{\frac{1}{n}} < (1+\sqrt{n})^{\frac{2}{n}} \leq \left(1 + \frac{1}{\sqrt{n}}\right)^2. \tag{26}
$$

(3) As $\lim\limits_{n \to \infty} \dfrac{1}{n} = 0$, it follows that $\lim\limits_{n \to \infty} \dfrac{1}{\sqrt{n}} = 0$. Hence

$$
\lim_{n \to \infty} \left(1 + \frac{1}{\sqrt{n}}\right)^2 = (1+0)^2 = 1 = \lim_{n \to \infty} 1.
$$

Using (26), and the Sandwich Theorem, $(n^{\frac{1}{n}})_{n \in \mathbb{N}}$ is convergent and $\lim\limits_{n \to \infty} n^{\frac{1}{n}} = 1$.

Remark 2. Here are two alternative proofs.

(A) For $n \geq 2$, we have $n^{\frac{1}{n}} > 1^{\frac{1}{n}} = 1$. Set $b_n := n^{\frac{1}{n}} - 1 > 0$ for $n > 2$. Then $n^{\frac{1}{n}} = 1 + b_n$, and so

$$
n = (1+b_n)^n = 1 + nb_n + \frac{n(n-1)}{2}b_n^2 + \cdots + nb_n^{n-1} + b_n^n > \frac{n(n-1)}{2}b_n^2.
$$

Thus

$$
\frac{2}{n-1} > b_n^2 > 0
$$

for $n > 2$, and so by the Sandwich Theorem, $\lim\limits_{n \to \infty} b_n^2 = 0$. Hence also

$$
\lim_{n \to \infty} b_n = 0,
$$

and so $\lim\limits_{n \to \infty} n^{\frac{1}{n}} = \lim\limits_{n \to \infty} (1 + b_n) = 1 + 0 = 1$.

(B) Yet another proof can be given using the Arithmetic Mean-Geometric Mean Inequality, which says that for nonnegative real numbers $a_1, \cdots, a_n$, there holds that

$$
\text{their arithmetic mean:} = \frac{a_1 + \cdots + a_n}{n} \geq \sqrt[n]{a_1 \cdots a_n} =: \text{their geometric mean.}
$$

Applying this to the n numbers $\underbrace{1, \cdots, 1}_{(n-2) \text{ times}}, \sqrt{n}, \sqrt{n}$ gives

$$\frac{(n-2) \cdot 1 + \sqrt{n} + \sqrt{n}}{n} \geq \sqrt[n]{\sqrt{n} \cdot \sqrt{n}} = n^{\frac{1}{n}},$$

and so,

$$1 - \frac{2}{n} + \frac{2}{\sqrt{n}} \geq n^{\frac{1}{n}} (\geq 1).$$

By the Sandwich Theorem, it follows that $\lim_{n \to \infty} n^{\frac{1}{n}} = 1$.

Solution to Exercise 2.19

Consider the sequence $(a_n - a)_{n \in \mathbb{N}}$. As $a_n \in (a, b)$ for all $n \in \mathbb{N}$, we have $a_n - a \geq 0$. From the Algebra of Limits, it follows that the sequence $(a_n - a)_{n \in \mathbb{N}}$ is convergent (being the sum of the convergent sequence $(a_n)_{n \in \mathbb{N}}$ and the convergent sequence $(-a)_{n \in \mathbb{N}}$). Moreover, its limit is $L - a$. From Exercise 2.4 on page 54, we obtain $L - a \geq 0$, that is $a \leq L$.

Next consider the sequence $(b - a_n)_{n \in \mathbb{N}}$. As $a_n \in (a, b)$ for all $n \in \mathbb{N}$, we have that $b - a_n \geq 0$. From the Algebra of Limits, it follows that the sequence $(b - a_n)_{n \in \mathbb{N}}$ is convergent (being the sum of the convergent sequence $(-a_n)_{n \in \mathbb{N}}$ and the convergent sequence $(b)_{n \in \mathbb{N}}$). Moreover, its limit is $-L + b$. From Exercise 2.4 on page 54, we obtain $-L + b \geq 0$, that is $L \leq b$.

Consequently $a \leq L \leq b$, that is, $L \in [a, b]$.

Consider $\left(\dfrac{1}{n+1} \right)_{n \in \mathbb{N}}$ contained in $(0, 1)$. It is convergent with limit $0 \notin (0, 1)$.

Solution to Exercise 2.20

For all $n \in \mathbb{N}$, we have

$$-\frac{1}{n} < b_n - a_n < \frac{1}{n},$$

and so by adding a_n, we have

$$-\frac{1}{n} + a_n < b_n < \frac{1}{n} + a_n.$$

By the Algebra of Limits, we know that

$$\lim_{n \to \infty} \left(-\frac{1}{n} + a_n \right) = \lim_{n \to \infty} -\frac{1}{n} + \lim_{n \to \infty} a_n = 0 + \lim_{n \to \infty} a_n = \lim_{n \to \infty} a_n$$

$$\lim_{n \to \infty} \left(\frac{1}{n} + a_n \right) = \lim_{n \to \infty} \frac{1}{n} + \lim_{n \to \infty} a_n = 0 + \lim_{n \to \infty} a_n = \lim_{n \to \infty} a_n.$$

So by the Sandwich Theorem, it follows that $(b_n)_{n \in \mathbb{N}}$ is convergent with the limit $\lim_{n \to \infty} a_n$.

Solution to Exercise 2.21

('If' part): Suppose that $\lim_{n \to \infty} \inf a_n = L = \lim_{n \to \infty} \sup a_n$. Then for all n,

$$\ell_n = \inf\{a_n, a_{n+1}, \cdots\} \leq a_n \leq \sup\{a_n, a_{n+1}, \cdots\} = u_n,$$

and so by the Sandwich Theorem, it follows that $(a_n)_{n \in \mathbb{N}}$ converges to L too.

('Only if' part): Now suppose that $(a_n)_{n\in\mathbb{N}}$ is convergent with limit L. Let $\epsilon > 0$. Then there exists an $N \in \mathbb{N}$ such that for all $n > N$, $|a_n - L| < \epsilon$, that is, $L - \epsilon < a_n < L + \epsilon$. Hence for all $n > N$, we have $u_n = \sup\{a_n, a_{n+1}, \cdots\} \leq L + \epsilon$, and so

$$\limsup_{n\to\infty} a_n := \lim_{n\to\infty} u_n \leq L + \epsilon. \tag{27}$$

Similarly, $\ell_n = \inf\{a_n, a_{n+1}, \cdots\} \geq L - \epsilon$, and so

$$\liminf_{n\to\infty} a_n := \lim_{n\to\infty} \ell_n \geq L - \epsilon. \tag{28}$$

From Exercise 2.15 and the inequalities (27) and (28) above, we obtain

$$L - \epsilon \leq \liminf_{n\to\infty} a_n \leq \limsup_{n\to\infty} a_n \leq L + \epsilon.$$

As the choice of $\epsilon > 0$ was arbitrary, this implies that $\liminf_{n\to\infty} a_n = L = \limsup_{n\to\infty} a_n$.

Solution to Exercise 2.22

$(a_{n^2})_{n\in\mathbb{N}} = (\frac{1}{n^4})$ is a subsequence of $(a_n)_{n\in\mathbb{N}} := (\frac{1}{n^2})_{n\in\mathbb{N}}$. But the sequence $(\frac{1}{n^3})$ has the terms

$$1, \frac{1}{8}, \frac{1}{27}, \cdots,$$

and so it is not a subsequence of $(\frac{1}{n^2})_{n\in\mathbb{N}}$: for example, the term $\frac{1}{8}$ does not appear in

$$1, \frac{1}{4}, \frac{1}{9}, \frac{1}{16}, \cdots.$$

Solution to Exercise 2.23

Observe that

> the terms $2, 8, 2, 8$ appear adjacently
> and so the terms $1, 6, 1, 6$ appear adjacently
> and so the terms $6, 6, 6$ appear adjacently
> and so the terms $3, 6, 3, 6$ appear adjacently
> and so the terms $1, 8, 1, 8$ appear adjacently

> and so the terms $8, 8, 8$ appear adjacently
> and so the terms $6, 4, 6, 4$ appear adjacently
> and so the terms $2, 4, 2, 4$ appear adjacently
> and so the terms $8, 8, 8$ appear adjacently

$$\cdots$$

Hence we get the loop

$$\cdots, 8, 8, 8, \cdots \to \cdots, 6, 4, 6, 4, \cdots \to \cdots, 2, 4, 2, 4, \cdots \to \cdots, 8, 8, 8, \cdots \to,$$

which contains 6, and so 6 appears infinite number of times. Thus we can choose indices

$$n_1 < n_2 < n_3 < \cdots$$

such that for all $k \in \mathbb{N}$, $a_{n_k} = 6$. So $(6)_{k\in\mathbb{N}}$ is a subsequence of the given sequence.

Solution to Exercise 2.24

The sequence $(a_n)_{n\in\mathbb{N}}$ satisfies

$$a_{n+1} = \frac{2(n+1)+1}{3(n+1)}a_n = \frac{2+\frac{3}{n}}{3+\frac{3}{n}}a_n,$$

for all $n \in \mathbb{N}$. Since $(\frac{1}{n})_{n\in\mathbb{N}}$ is convergent with limit 0, by the Algebra of Limits,

$$\lim_{n\to\infty} \frac{2+\frac{3}{n}}{3+\frac{3}{n}} = \frac{2+3\cdot 0}{3+3\cdot 0} = \frac{2}{3}.$$

Again applying the Algebra of Limits, we obtain

$$L = \lim_{n\to\infty} a_{n+1} = \lim_{n\to\infty} \left(\frac{2+\frac{3}{n}}{3+\frac{3}{n}}a_n\right) = \lim_{n\to\infty}\left(\frac{2+\frac{3}{n}}{3+\frac{3}{n}}\right)\lim_{n\to\infty} a_n = \frac{2}{3}L.$$

Hence $\frac{1}{3}L = 0$, that is, $L = 0$. So $\lim_{n\to\infty} a_n = 0$.

Solution to Exercise 2.25

(1) True.

Let $(a_n)_{n\in\mathbb{N}}$ be convergent with limit L, and let $(a_{n_k})_{k\in\mathbb{N}}$ be a subsequence. If $\epsilon > 0$, then there exists an $N \in \mathbb{N}$ such that for all $n > N$, $|a_n - L| < \epsilon$. Choose $K \in \mathbb{N}$ such that for all $k > K$, $n_k > N$. Then for all $k > K$, $n_k > N$ and so $|a_{n_k} - L| < \epsilon$. Thus $(a_{n_k})_{k\in\mathbb{N}}$ is convergent with limit L.

(2) False.

The sequence $((-1)^n)_{n\in\mathbb{N}}$ is divergent, but it possesses the convergent subsequence $((-1)^{2n})_{n\in\mathbb{N}} = (1)_{n\in\mathbb{N}}$ with limit 1.

(3) True.

Let $(a_n)_{n\in\mathbb{N}}$ be bounded, and let $M > 0$ be a number such that for all $n \in \mathbb{N}$, $|a_n| \leq M$. If $(a_{n_k})_{k\in\mathbb{N}}$ is a subsequence, then also $|a_{n_k}| \leq M$ for all $k \in \mathbb{N}$, and so it is bounded as well.

(4) False.

The sequence $1, 0, 2, 0, 3, 0, \cdots$ is unbounded, but has as a subsequence $0, 0, 0, \cdots$, which is bounded.

(5) True.

Let $(a_n)_{n\in\mathbb{N}}$ be an increasing sequence. (The argument is similar for a decreasing sequence.) Suppose that $(a_{n_k})_{k\in\mathbb{N}}$ is a subsequence. For each $k \in \mathbb{N}$, $n_{k+1} > n_k$ and so $a_{n_{k+1}} \geq a_{n_{k+1}-1} \geq a_{n_{k+1}-2} \geq \cdots \geq a_{n_k}$. Thus $(a_{n_k})_{k\in\mathbb{N}}$ is increasing as well.

(6) False.

The sequence $1, 0, 2, 0, 3, 0, \cdots$ is not monotone, but it has the monotone subsequence $1, 2, 3, \cdots$.

(7) True.

The sequence itself is a subsequence of itself, and so it must be convergent too.

(8) False.

The sequence $(a_n)_{n\in\mathbb{N}} = ((-1)^n)_{n\in\mathbb{N}}$ is divergent, but we have that both the sequences $(a_{2n})_{n\in\mathbb{N}} = (1)_{n\in\mathbb{N}}$ and $(a_{2n+1})_{n\in\mathbb{N}} = (-1)_{n\in\mathbb{N}}$ are convergent (with limits 1 and -1, respectively).

(9) True.

Let L be the common limit, and let $\epsilon > 0$. There exists an $N_1 \in \mathbb{N}$ such that for all $n > N_1$, $|a_{2n} - L| < \epsilon$. Also, there exists an $N_2 \in \mathbb{N}$ such that for all $n > N_2$, $|a_{2n+1} - L| < \epsilon$. Choose $N \in \mathbb{N}$ such that $N > \max\{2N_1, 2N_2 + 1\}$. Let $n > N$. If n is even, say $n = 2k$ for some $k \in \mathbb{N}$, then $2k > N \geq 2N_1$ gives $k > N_1$, and so we have $|a_{2k} - L| = |a_n - L| < \epsilon$. If, on the other hand, n is odd, say $n = 2k + 1$ for some $k \in \mathbb{N}$, then $2k + 1 = n > N \geq 2N_2 + 1$ gives $k > N_2$, and so $|a_{2k+1} - L| = |a_n - L| < \epsilon$. Hence for all $n > N$, we have $|a_n - L| < \epsilon$. Consequently, $(a_n)_{n \in \mathbb{N}}$ is convergent with limit L.

Solution to Exercise 2.26

We know that

$$\neg \left(\forall \epsilon > 0, \ \exists N \in \mathbb{N} \text{ such that } \forall n > N, \ |a_n - L| < \epsilon \right),$$

that is,

$$\exists \epsilon > 0 \text{ such that } \forall N \in \mathbb{N}, \ \exists n > N \text{ such that } |a_n - L| \geq \epsilon. \qquad (29)$$

Take $N = 1$ in (29). Then there exists an $n_1 > N = 1$ such that $|a_{n_1} - L| \geq \epsilon$. We construct the subsequence terms inductively as follows.

$$N = n_1 : \quad \exists n_2 > n_1 \text{ such that } |a_{n_2} - L| \geq \epsilon,$$

$$N = n_2 : \quad \exists n_3 > n_2 \text{ such that } |a_{n_3} - L| \geq \epsilon,$$

$$N = n_3 : \quad \exists n_4 > n_3 \text{ such that } |a_{n_4} - L| \geq \epsilon.$$

$$\cdots$$

Suppose that $a_{n_1}, \cdots, a_{n_k}$ have been constructed. Take $N = n_k$ in (29). Then there exists an $n_{k+1} > N = n_k$ such that $|a_{n_{k+1}} - L| \geq \epsilon$. Thus we obtain the subsequence $(a_{n_k})_{k \in \mathbb{N}}$ which satisfies $|a_{n_k} - L| \geq \epsilon$ for all $k \in \mathbb{N}$.

Solution to Exercise 2.27

(a) Clearly $\sqrt{2} < 2$ because $2 < 4$. Suppose that for some $n \in \mathbb{N}$, $a_n \leq 2$. Then

$$a_{n+1} = \sqrt{2 + a_n} \leq \sqrt{2 + 2} = \sqrt{4} = 2.$$

It follows by induction that $a_n \leq 2$ for all $n \in \mathbb{N}$.

(b) We have

$$a_{n+1}^2 - a_n^2 = (\sqrt{2 + a_n})^2 - a_n^2 = 2 + a_n - a_n^2 = 2 + 2a_n - a_n - a_n^2$$

$$= 2(1 + a_n) - a_n(1 + a_n) = (2 - a_n)(1 + a_n).$$

From part (a), we know that $2 - a_n \geq 0$, and clearly $1 + a_n \geq 0$. Thus $a_{n+1}^2 \geq a_n^2$, and as a_n is nonnegative for each n, we conclude that $a_{n+1} \geq a_n$ for all $n \in \mathbb{N}$. Consequently, the sequence $(a_n)_{n \in \mathbb{N}}$ is increasing.

(c) Yes, since the sequence $(a_n)_{n \in \mathbb{N}}$ is monotone (increasing) and bounded (above).

Suppose that $(a_n)_{n \in \mathbb{N}}$ converges to $L \in \mathbb{R}$. Then the subsequence $(a_{n+1})_{n \in \mathbb{N}}$ converges to L too. On the other hand, as $(2 + a_n)_{n \in \mathbb{N}}$ is nonnegative and convergent to $2 + L$, it follows that $(a_{n+1})_{n \in \mathbb{N}} = (\sqrt{2 + a_n})_{n \in \mathbb{N}}$ converges to $\sqrt{2 + L}$. Thus

$$L = \sqrt{2 + L},$$

and so $L^2 = 2 + L$, that is, $0 = L^2 - L - 2 = (L-2)(L+1)$. Hence $L = 2$ or $L = -1$. As $a_n > 0$ for all $n \in \mathbb{N}$, $L \geq 0$, and so L can't be -1. Consequently $L = 2$.

Solution to Exercise 2.28

For all $n \in \mathbb{N}$, $-1 \leq \sin n \leq 1$, and so the sequence $(\sin n)_{n \in \mathbb{N}}$ is bounded. So by the Bolzano–Weierstrass Theorem, it has a convergent subsequence.

Suppose that $(n)_{n \in \mathbb{N}}$ has a convergent subsequence $(n_k)_{k \in \mathbb{N}}$ with limit L. Then also the sequence $(n_{k+1})_{k \in \mathbb{N}}$ is convergent with limit L. So $(n_{k+1} - n_k)_{k \in \mathbb{N}}$ must be convergent with limit $L - L = 0$. But for all $k \in \mathbb{N}$, we have

$$|(n_{k+1} - n_k) - 0| = n_{k+1} - n_k \geq 1,$$

a contradiction. So no subsequence of $(n)_{n \in \mathbb{N}}$ can be convergent.

Solution to Exercise 2.29

Since $(a_n)_{n \in \mathbb{N}}$ is bounded, it follows from the Bolzano–Weierstrass Theorem that $(a_n)_{n \in \mathbb{N}}$ has a convergent subsequence $(a_{n_k})_{k \in \mathbb{N}}$ with some limit, say L_1.

But since the given sequence $(a_n)_{n \in \mathbb{N}}$ is divergent, in particular, it can't converge to L_1. So by Exercise 2.26, there exists an $\epsilon > 0$ and a subsequence, say $(a_{m_k})_{k \in \mathbb{N}}$ such that for all $k \in \mathbb{N}$, $|a_{m_k} - L_1| \geq \epsilon$.

As $(a_{n_k})_{k \in \mathbb{N}}$ converges to L_1, we know that there exists a K large enough so that for all $k \geq K$, $|a_{n_k} - L_1| < \epsilon$.

Then clearly the subsequence $a_{m_1}, a_{m_2}, a_{m_3}, \cdots$ has terms which are all distinct from the subsequence $a_{n_K}, a_{n_{K+1}}, a_{n_{K+2}}, \cdots$ (since for $k \geq K$, $|a_{n_k} - L_1| < \epsilon$, while for all k, $|a_{m_k} - L_1| \geq \epsilon$).

Now by the Bolzano–Weierstrass Theorem, there exists a subsequence of the bounded sequence $a_{m_1}, a_{m_2}, a_{m_3}, \cdots$ which converges, say, to L_2.

But from the inequalities $|a_{m_k} - L_1| \geq \epsilon$, $k \geq 1$, we see that $|L_2 - L_1| \geq \epsilon > 0$, and so $L_1 \neq L_2$.

Solution to Exercise 2.30

Let $\epsilon > 0$. Since $(a_n)_{n \in \mathbb{N}}$ is a Cauchy sequence, there exists an $N \in \mathbb{N}$ such that for all n, $m > N$, $|a_n - a_m| < \epsilon$. In particular for $n > N$, we have $m := n + 1 > N$ and so it follows that $|a_n - a_{n+1}| < \epsilon$, that is, $|(a_{n+1} - a_n) - 0| < \epsilon$. So $(a_{n+1} - a_n)_{n \in \mathbb{N}}$ is convergent with limit 0.

Solution to Exercise 2.31

(A), (B), (C), (D).

Solution to Exercise 2.32

(1) False. The sequence $(a_n)_{n \in \mathbb{N}} := ((-1)^n)_{n \in \mathbb{N}}$ is not Cauchy as it is divergent, but $(a_n^2)_{n \in \mathbb{N}} = (1)_{n \in \mathbb{N}}$ is convergent, and hence Cauchy.

(2) True. If $(a_n)_{n \in \mathbb{N}}$ is Cauchy, then it is convergent. By the Algebra of Limits, it follows that $(a_n^2)_{n \in \mathbb{N}}$ is convergent too, and in particular, it is Cauchy.

Solution to Exercise 2.33

('If' part) Suppose that $(a_n)_{n \in \mathbb{N}}$ converges to 0. Let $\epsilon > 0$. Then there exists an $N \in \mathbb{N}$ such that whenever $n > N$, $|a_n - 0| < \epsilon$. As $a_n \geq 0$, we have that $a_n < \epsilon$ for all $n > N$. But by the definition of a_n, this means that for any $x \in I$,

$$|f_n(x) - f(x)| \leq a_n < \epsilon.$$

Hence $(f_n)_{n \in \mathbb{N}}$ is uniformly convergent to f.

('Only if' part) Now suppose that $(f_n)_{n \in \mathbb{N}}$ is uniformly convergent to f. Let $\epsilon > 0$. Then there exists an $N \in \mathbb{N}$ such that whenever $n > N$, we have for all $x \in I$, $|f_n(x) - f(x)| < \epsilon$. But this says that ϵ is an upper bound for the set $\{|f_n(x) - f(x)| : x \in I\}$. By the definition of the least upper bound, it now follows that $a_n \leq \epsilon$. Since a_n is nonnegative, it follows that $|a_n - 0| = |a_n| = a_n \leq \epsilon$ for all $n > N$. In other words, the sequence $(a_n)_{n \in \mathbb{N}}$ converges to 0.

For a fixed $x \in (0, \infty)$, we have

$$\lim_{n \to \infty} f_n(x) = \lim_{n \to \infty} x e^{-nx} = x \cdot \lim_{n \to \infty} e^{-nx} = x \cdot 0 = 0.$$

So the pointwise limit of $(f_n)_{n \in \mathbb{N}}$ is the function f, which is identically zero on $(0, \infty)$. We now show that the convergence is uniform using the result proved earlier. For $x \in (0, \infty)$ and $n \in \mathbb{N}$, we have $nx > 0$, and so

$$e^{nx} = \sum_{k=0}^{\infty} \frac{1}{k!}(nx)^k \geq \frac{1}{1!}nx.$$

Hence $0 < x e^{-nx} \leq \frac{1}{n}$. Consequently

$$0 \leq a_n := \sup\{|f_n(x) - f(x)| : x \in (0, \infty)\} = \sup_{x \in (0, \infty)} x e^{-nx} \leq \frac{1}{n},$$

and so by the Sandwich Theorem,

$$\lim_{n \to \infty} a_n = 0.$$

Hence $(f_n)_{n \in \mathbb{N}}$ converges uniformly to the function which is identically zero on $(0, \infty)$.

Solution to Exercise 2.34

We will show that $(f_n)_{n \in \mathbb{N}}$ converges uniformly to the function, which is identically 0 on $[0, 1]$. We have for all $x \in [0, 1]$ that

$$|f_n(x) - 0| = \frac{x}{1 + nx} = \frac{1}{n} \cdot \frac{nx}{1 + nx} \leq \frac{1}{n} \cdot 1.$$

Let $\epsilon > 0$. Choose an $N \in \mathbb{N}$ such that $N > 1/\epsilon$. Then whenever $n > N$, we have for all $x \in [0, 1]$ that

$$|f_n(x) - 0| \leq \frac{1}{n} < \frac{1}{N} < \epsilon.$$

Consequently, $(f_n)_{n \in \mathbb{N}}$ converges uniformly to the function which is identically zero on $[0, 1]$.

Solution to Exercise 2.35

$(f_n)_{n \in \mathbb{N}}$ converges pointwise to the zero function f, defined by $f(x) = 0$ $(x \in (0, 1))$. Indeed, for each $x \in (0, 1)$,

$$\lim_{n \to \infty} x^n = 0,$$

and so the following statement is true:

$$\forall \epsilon > 0, \ \forall x \in (0,1), \ \exists N \in \mathbb{N} \text{ such that } \forall n > N, \ |f_n(x) - f(x)| < \epsilon.$$

If fact, we can choose $N > \dfrac{\log \epsilon}{\log x}$ so that $x^N < \epsilon$, and if $n > N$, we are guaranteed that

$$|f_n(x) - f(x)| = |x^n - 0| = x^n < x^N < \epsilon.$$

It is clear that our choice of N in the above depends not only on ϵ but also on the point $x \in (0,1)$. The closer x is to 1, the larger N is. The question arises: Is there an N so that for $n > N$, $|f_n(x) - f(x)| < \epsilon$ for *all* $x \in (0,1)$? We will show below that the answer is 'no'. In other words, $(f_n)_{n \in \mathbb{N}}$ does not converge uniformly to the zero function f.

For example, let $\epsilon = 1/2 > 0$. Let us suppose that there does exist an $N \in \mathbb{N}$ such that for all $n > N$,

$$\forall x \in (0,1), \ |f_n(x) - f(x)| = x^n < \frac{1}{2} = \epsilon.$$

In particular, we would have $\forall x \in (0,1)$, $x^{N+1} < \dfrac{1}{2}$. Take $x = 1 - \dfrac{1}{m}$, $m \in \mathbb{N}$, $m \geq 2$. Then

$$\left(1 - \frac{1}{m}\right)^{N+1} < \frac{1}{2}.$$

Letting $m \to \infty$, we obtain $1 \leq \dfrac{1}{2}$, a contradiction.

The following picture explains this visually. If $(f_n)_{n \in \mathbb{N}}$ were to converge to the zero function uniformly, then in particular, for all n large enough, the graph of f_n would lie in a strip of width $\epsilon = 1/2$ around the graph of the zero function. But no matter how large an n we take, some part of the graph of f_n always falls outside the strip.

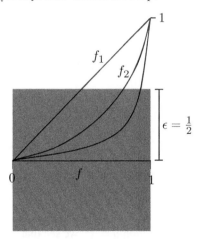

Solution to Exercise 2.36

If $x \neq 0$, then $0 < \dfrac{1}{1 + x^2} < 1$, and so

$$\lim_{n \to \infty} f_n(x) = 1 - \lim_{n \to \infty} \left(\frac{1}{1 + x^2}\right)^n = 1 - 0 = 1.$$

On the other hand,

$$\lim_{n \to \infty} f_n(0) = \lim_{n \to \infty} \left(1 - \frac{1}{(1 + 0^2)^n}\right) = 1 - 1 = 0.$$

So the sequence $(f_n)_{n \in \mathbb{N}}$ converges pointwise to f. Clearly f is discontinuous at 0: for example, the sequence $(\frac{1}{n})_{n \in \mathbb{N}}$ converges to 0, but the sequence $(f(\frac{1}{n}))_{n \in \mathbb{N}} = (1)_{n \in \mathbb{N}}$ converges to $1 \neq 0 = f(0)$.

Solution to Exercise 2.37

(1) We have $\lim\limits_{m \to \infty} a_{m,n} = \lim\limits_{m \to \infty} \dfrac{m}{m + n} = \lim\limits_{m \to \infty} \dfrac{1}{1 + \frac{n}{m}} = \dfrac{1}{1 + 0} = 1$, and

$$\lim_{n \to \infty} a_{m,n} = \lim_{n \to \infty} \frac{m}{m + n} = \lim_{n \to \infty} \frac{\frac{m}{n}}{\frac{m}{n} + 1} = \frac{0}{0 + 1} = 0.$$

Hence $\lim\limits_{m \to \infty} \lim\limits_{n \to \infty} a_{m,n} = \lim\limits_{m \to \infty} 0 = 0 \neq 1 = \lim\limits_{n \to \infty} 1 = \lim\limits_{n \to \infty} \lim\limits_{m \to \infty} a_{m,n}$.

(2) We have for $x \in \mathbb{R}$ that

$$|f_n(x) - f(x)| = |f_n(x)| = \frac{|\sin(nx)|}{\sqrt{n}} \leq \frac{1}{\sqrt{n}}.$$

Let $\epsilon > 0$. Choose an $N \in \mathbb{N}$ such that $N > 1/\epsilon^2$. Then whenever $n > N$, we have for all $x \in \mathbb{R}$ that

$$|f_n(x) - f(x)| \leq \frac{1}{\sqrt{n}} < \frac{1}{\sqrt{N}} < \epsilon.$$

Consequently, $(f_n)_{n \in \mathbb{N}}$ converges pointwise (in fact even uniformly!) to the function f, which is identically zero on $\mathbb{R}$.

We have

$$f_n'(x) := \frac{d}{dx} f_n(x) = \frac{n \cos(nx)}{\sqrt{n}} = \sqrt{n} \cos(nx).$$

f' on the other hand is identically 0 on $\mathbb{R}$. If $x = 0$, then $(f_n'(x))_{n \in \mathbb{N}} = (\sqrt{n})_{n \in \mathbb{N}}$ is clearly unbounded, and thus not convergent. Hence $(f_n')_{n \in \mathbb{N}}$ is not pointwise convergent (to f').

(3) If $x = 0$ or $x = 1$, $f_n(x) = 0$, and so $(f_n(x))_{n \in \mathbb{N}} = (0)_{n \in \mathbb{N}}$ is convergent with limit 0 $(= f(x))$. Now suppose that $x \in (0, 1)$. Then $0 < 1 - x^2 < 1$. So we can find a positive number h such that $1 - x^2 = 1/(1 + h)$. By the binomial expansion for $(1 + h)^n$, we have the inequality $(1 + h)^n > n(n - 1)h^2$, which yields

$$0 < nx(1 - x^2)^n < \frac{nx}{n(n - 1)h^2}.$$

Thus by the Sandwich Theorem, we obtain that $(f_n(x))_{n \in \mathbb{N}}$ converges to 0. Consequently, $(f_n)_{n \in \mathbb{N}}$ converges pointwise to the zero function f on $[0, 1]$.

Using the variable substitution $u = x^2$, we see that

$$\int_0^1 f_n(x)dx = \int_0^1 nx(1 - x^2)^n dx = \frac{1}{2} \int_0^1 nu^n du = \frac{1}{2} \frac{n}{n + 1}.$$

Thus

$$\lim_{n \to \infty} \int_0^1 f_n(x)dx = \lim_{n \to \infty} \frac{1}{2} \frac{n}{n + 1} = \frac{1}{2} \neq 0 = \int_0^1 f(x)dx = \int_0^1 \lim_{n \to \infty} f_n(x)dx.$$

Solutions to the exercises from Chapter 3

Solution to Exercise 3.1

(1) Let $\epsilon > 0$. If $\delta := \sqrt{\epsilon}$, then we note that $\delta > 0$. Moreover, if $x \in \mathbb{R}$ and this x satisfies $|x - 0| = |x| < \delta = \sqrt{\epsilon}$, then $|x^2 - 0^2| = |x^2| = |x| \cdot |x| < \delta \cdot \delta = \sqrt{\epsilon} \cdot \sqrt{\epsilon} = \epsilon$. So f is continuous at 0.

(2) Let $c \in \mathbb{R}$ and suppose that $c \neq 0$. Let $\epsilon > 0$. If $x \in \mathbb{R}$, then

$$|x^2 - c^2| = |(x - c)(x + c)| = |x - c| \cdot |x + c|.$$

If $x \in \mathbb{R}$ is such that $|x - c| < \delta$, then

$$x < c + \delta \leq |c + \delta| \leq |c| + |\delta| = |c| + \delta, \text{ and}$$

$$-x < \delta - c \leq |\delta - c| \leq |\delta| + |-c| = \delta + |c|.$$

Thus if $x \in \mathbb{R}$ satisfies $|x - c| < \delta$, then $|x| < \delta + |c|$, and so

$$|x + c| \leq |x| + |c| < \delta + |c| + |c| = \delta + 2|c|.$$

So if $|x - c| < \delta$, we have $|x^2 - c^2| = |x - c| \cdot |x + c| < \delta \cdot (\delta + 2|c|)$. Thus in order to make $|x^2 - c^2|$ less than ϵ, we choose δ such that $\delta(\delta + 2|c|) < \epsilon$: indeed, let

$$\delta := \min \left\{ \frac{\epsilon}{2|c| + 1}, 1 \right\}.$$

Since $\epsilon > 0$, it follows that δ is positive. Furthermore, if $x \in \mathbb{R}$ satisfies $|x - c| < \delta$, then we obtain

$$|x^2 - c^2| < \delta(\delta + 2|c|) \leq \frac{\epsilon}{2|c| + 1}(1 + 2|c|) = \epsilon.$$

Solution to Exercise 3.2

(1) Let $c' \in \mathbb{R}$ and $\epsilon > 0$. Since f is continuous at c, there exists a $\delta > 0$ such that for all $x \in \mathbb{R}$ satisfying $|x - c| < \delta$, $|f(x) - f(c)| < \epsilon$. Then for all $x \in \mathbb{R}$ satisfying $|x - c'| < \delta$, we have[3]

$$|f(x) - f(c')| = |f(x) - f(c') + f(c) - f(c)|$$

$$= |f(x - c' + c) - f(c)| < \epsilon,$$

since $|(x - c' + c) - c| = |x - c'| < \delta$. So f is continuous at c'. Since the choice of $c' \in \mathbb{R}$ was arbitrary, it follows that f is continuous on $\mathbb{R}$.

(2) Let $\alpha \in \mathbb{R}$, and let $f : \mathbb{R} \to \mathbb{R}$ be given by $f(x) = \alpha x$, for all $x \in \mathbb{R}$. Then

$$f(x + y) = \alpha(x + y) = \alpha x + \alpha y = f(x) + f(y).$$

Remark 3. $(*)$ The 'Axiom of Choice' is an axiom of set theory equivalent to the statement that the Cartesian product of any collection of nonempty sets is nonempty. Using this Axiom of Choice, one can show that there *are* functions $f : \mathbb{R} \to \mathbb{R}$ that are additive, but not continuous.

---

[3] Here we use the fact that f is 'additive'. First of all, $f(0) = f(0 + 0) = f(0) + f(0)$, and so $f(0) = 0$. Hence it follows that $-f(c') = f(-c')$, since $0 = f(0) = f(c' - c') = f(c') + f(-c')$. Finally, using this, and the additivity of f, $f(x) - f(c') + f(c) = f(x) + f(-c') + f(c) = f(x - c') + f(c) = f(x - c' + c)$.

So in light of the above exercise, such a function must necessarily fail to be continuous at every real number! Hence such a function obviously can't be linear. In fact, although it can be shown that such a function exists, it is not a *constructive* proof, in that an explicit example of such an f is not given.

Solution to Exercise 3.3

We observe that as $|f(0)| \le M|0| = M \cdot 0 = 0$, it follows that $|f(0)| = 0$, that is, $f(0) = 0$. Given $\epsilon > 0$, we define $\delta = \epsilon/M$. Then for all $x \in \mathbb{R}$ satisfying $|x| = |x - 0| < \delta$, we have

$$|f(x) - f(0)| = |f(x) - 0| = |f(x)| \le M|x| = M|x - 0| < M\delta = M\frac{\epsilon}{M} = \epsilon.$$

Hence f is continuous at 0.

Solution to Exercise 3.4

Let $c \in \mathbb{R}$. Suppose that f is continuous at c. Consider $\epsilon = \frac{1}{2} > 0$. Then there exists a $\delta > 0$ such that for all $x \in \mathbb{R}$ satisfying $|x - c| < \delta$, $|f(x) - f(c)| < \epsilon = \frac{1}{2}$. We have the following two cases:

   $1°$ $c \in \mathbb{Q}$. Then there exists $x \in \mathbb{R}\backslash\mathbb{Q}$ such that $|x - c| < \delta$.
        But $|f(x) - f(c)| = |1 - 0| = |1| = 1 > \frac{1}{2}$, a contradiction.
   $2°$ $c \in \mathbb{R}\backslash\mathbb{Q}$. Then there exists $x \in \mathbb{Q}$ such that $|x - c| < \delta$.
        But $|f(x) - f(c)| = |0 - 1| = |-1| = 1 > \frac{1}{2}$, a contradiction.

Hence f is not continuous at c.

Solution to Exercise 3.5

Since $\epsilon := \frac{f(c)}{2} > 0$, there exists a $\delta > 0$ such that for all $x \in (a, b)$ satisfying $|x - c| < \delta$,

$$|f(x) - f(c)| < \frac{f(c)}{2}.$$

Thus for all $x \in (a, b)$ satisfying $|x - c| < \delta$ (that is, $c - \delta < x < c + \delta$), we have

$$f(c) - f(x) \le |f(c) - f(x)| = |f(x) - f(c)| < \frac{f(c)}{2},$$

and so $f(x) > \frac{f(c)}{2} > 0$.

Remark 4. Thus if a continuous function $f : (a, b) \to \mathbb{R}$ is positive (respectively, negative) at a point $c \in (a, b)$, then it stays positive (respectively, negative) in a small open interval $(c - \delta, c + \delta)$ containing the point c.

Solution to Exercise 3.6

We will show the four implications (1)$\Rightarrow$(2), (2)$\Rightarrow$(3), (3)$\Rightarrow$(4), and (4)$\Rightarrow$(1), which are enough to get all the four equivalences (and eight implications) given in the statement. In other words, we show the chain:

$$
\begin{array}{ccc}
(1) & \Rightarrow & (2) \\
\Uparrow & & \Downarrow \\
(4) & \Leftarrow & (3)
\end{array}
$$

One may ask for instance: Have we really shown that (3)⇒(2)? Well, if (3) is true, then:

    since we have shown (3)⇒(4), (4) is true,

    since we have shown (4)⇒(1), (1) is true, and finally

    since we have shown (1)⇒(2), (2) is true!

Thus (3)⇒(2) is true too.

(1)⇒(2): Let $\epsilon > 0$. Then by (1), there exists a $\delta > 0$ such that whenever $x \in I$ satisfies $|x - c| < \delta$, we have $|f(x) - f(c)| < \epsilon$, and so $|f(x) - f(c)| \leq \epsilon$. So (2) holds.

(2)⇒(3): Let $\epsilon > 0$. Then by (2), there exists a $\delta' > 0$ such that whenever $x \in I$ satisfies $|x - c| < \delta'$, we have $|f(x) - f(c)| \leq \epsilon$. Thus with $\delta := \frac{\delta'}{2} > 0$, we have that whenever $x \in I$ satisfies $|x - c| \leq \delta = \frac{\delta'}{2} < \delta'$, we have $|f(x) - f(c)| \leq \epsilon$. So (3) holds.

(3)⇒(4): Let $\epsilon > 0$. Then $\epsilon/2 > 0$, and so by (3), there exists a $\delta > 0$ such that whenever $x \in I$ satisfies $|x - c| < \delta$, we have $|f(x) - f(c)| \leq \frac{\epsilon}{2} < \epsilon$. So (4) holds.

(4)⇒(1): Let $\epsilon > 0$. Then by (4), there exists a $\delta > 0$ such that whenever $x \in I$ satisfies $|x - c| \leq \delta$, we have $|f(x) - f(c)| < \epsilon$. In particular, whenever $x \in I$ satisfies $|x - c| < \delta$, we have $|f(x) - f(c)| < \epsilon$. So (1) holds.

Solution to Exercise 3.7

Let $(x_n)_{n \in \mathbb{N}}$ be a convergent sequence with limit c. Then $(x_n^2)_{n \in \mathbb{N}}$ is also convergent with limit c^2. Thus $(f(x_n))_{n \in \mathbb{N}}$ is convergent with limit $f(c)$. So by Theorem 3.1, f is continuous.

Solution to Exercise 3.8

We only have to show that $f(x) < f(c)$ for all $x \in (c - \delta, c)$. Let $(x_n)_{n \in \mathbb{N}}$ be a sequence such that $x < x_1 < x_2 < \cdots$ and $\lim_{n \to \infty} x_n = c$.

Then $f(x) < f(x_1) \leq f(x_n)$ for all $n \in \mathbb{N}$. So

$$f(x) < f(x_1) \leq \lim_{n \to \infty} f(x_n) = f\left(\lim_{n \to \infty} x_n\right) = f(c),$$

where we use the continuity of f in order to get the last but one equality. Thus $f(x) < f(c)$.

Remark 5. From here it follows that a continuous function $f : (c - \delta, c + \delta) \to \mathbb{R}$, which is strictly increasing on $(c - \delta, c)$ and $(c, c + \delta)$, is strictly increasing on $(c - \delta, c + \delta)$.

Solution to Exercise 3.9

Let $x \in \mathbb{R}$. For each $n \in \mathbb{N}$, pick a $q_n \in \mathbb{Q}$ such that

$$x < q_n < x + 1/n.$$

Then the sequence $(q_n)_{n \in \mathbb{N}}$ is convergent with limit x. (Let $\epsilon > 0$. By the Archimedean Property, there exists an $N \in \mathbb{N}$ such that $1/\epsilon < N$. Consequently, for all $n > N$, we have that $|q_n - x| = q_n - x < 1/n < 1/N < \epsilon$.)

Since f is continuous at c, we have $f(c) = \lim_{n \to \infty} f(q_n) = \lim_{n \to \infty} 0 = 0$.

Solution to Exercise 3.10

If $x_1 \neq x_2$, then the sequence $x_1, x_2, x_1, x_2, \cdots$ is divergent. (Indeed, the subsequence $x_1, x_1, x_1, \cdots$ converges to x_1, while the subsequence $x_2, x_2, x_2, \cdots$ converges to x_2, and so it follows that the sequence $x_1, x_2, x_1, x_2, \cdots$ is divergent.)

Thus $f(x_1), f(x_2), f(x_1), f(x_2), \cdots$ is divergent. Consequently $f(x_1) \neq f(x_2)$: for otherwise if $f(x_1) = f(x_2)$, then the sequence

$$f(x_1), \quad f(x_2), \quad f(x_1), \quad f(x_2) \quad \cdots$$
$$\| \qquad\qquad \| \qquad\qquad \| \qquad\qquad \|$$
$$f(x_1), \quad f(x_1), \quad f(x_1), \quad f(x_1) \quad \cdots$$

is a constant sequence, and so it is convergent with limit $f(x_1)$ $(= f(x_2))$.

So we have shown that if $x_1 \neq x_2$, then $f(x_1) \neq f(x_2)$, that is, the function f is one-to-one.

Solution to Exercise 3.11

That (1) implies (2) is immediate from Theorem 3.1.

Now, suppose that (2) holds. Let $(x_n)_{n \in \mathbb{N}}$ be a sequence contained in I that converges to $c \in I$. Then the new sequence $x_1, c, x_2, c, x_3, c, \cdots$ also converges to c. By the hypothesis (2), it follows that the image of this sequence under f, namely the sequence $f(x_1), f(c), f(x_2), f(c), f(x_3), f(c), \cdots$ must be convergent, say with limit L. But then each of its subsequences must also be convergent with the same limit L. By looking at the even indexed terms of this sequence, we obtain that the subsequence $f(c), f(c), f(c), \cdots$ converges to L, and so $L = f(c)$. On the other hand, by looking at the odd indexed terms, we conclude that the sequence $f(x_1), f(x_2), f(x_3), \cdots$ must also be convergent with limit L $(= f(c))$. Consequently, f is continuous at c.

Solution to Exercise 3.12

Let $x \neq 0$. We consider two possible cases.

$\underline{1^\circ}$ Let x be rational. Then we can find a sequence $(y_n)_{n \in \mathbb{N}}$ of irrational numbers that converges to x. If f was continuous at x, then $(f(y_n))_{n \in \mathbb{N}} = (-y_n)_{n \in \mathbb{N}}$ would have to be convergent with limit $f(x) = x$. But then we obtain $x = -x$ and so $x = 0$, a contradiction.

$\underline{2^\circ}$ Let x be irrational. Then we can find a sequence $(r_n)_{n \in \mathbb{N}}$ of rational numbers that converges to x. If f was continuous at x, then $(f(r_n))_{n \in \mathbb{N}} = (r_n)_{n \in \mathbb{N}}$ would have to be convergent with limit $f(x) = -x$. But then we obtain $x = -x$ and so $x = 0$, a contradiction.

So f is not continuous at x whenever $x \neq 0$. We now show that f is continuous at 0. Let $\epsilon > 0$. Let $\delta := \epsilon > 0$. Since $f(x) = x$ if x is rational, and $-x$ if x is irrational, it follows that in either case, $|f(x)| = |x|$. Thus whenever $x \in \mathbb{R}$ and it satisfies $|x - 0| = |x| < \delta = \epsilon$, we have that $|f(x) - f(0)| = |f(x) - 0| = |f(x)| = |x| < \delta = \epsilon$. So f is continuous at 0.

Solution to Exercise 3.13

Let x be a rational number. Then $f(x) > 0$. We can find a sequence $(y_n)_{n\in\mathbb{N}}$ of irrational numbers that converges to x. If f was continuous at x, then $(f(y_n))_{n\in\mathbb{N}} = (0)_{n\in\mathbb{N}}$ would have to be convergent with limit $f(x) > 0$, which is clearly not true. Thus f is not continuous at x.

Now suppose that x is irrational. Then there exists an integer N such that x belongs to the interval $(N, N+1)$. Let $\epsilon > 0$. Suppose that r is a rational number in $(N, N+1)$ for which $f(r) \geq \epsilon$. Let $r = n/d$, where n, d are integers without any common divisors and $d > 0$. Then we have $f(r) = 1/d \geq \epsilon$ and so $d \leq 1/\epsilon$. Thus there are just finitely many possibilities for d. Moreover, from $N < n/d < N+1$, it follows that n satisfies $Nd < n < (N+1)d$, and so there are only finitely many possibilities for n as well. Hence there are just finitely many rational numbers r in $(N, N+1)$ for which $f(r) \geq \epsilon$. Now among these finitely many rational numbers, let r_* be one which is the closest to x, and set $\delta := \min\{|x - r_*|/2, |x - N|, |x - (N+1)|\} > 0$. Then whenever $z \in \mathbb{R}$ and $|z - x| < \delta$, we are guaranteed that $z \in (N, N+1)$, and moreover if z is rational, then it can't be one of the rational r in $(N, N+1)$ for which $f(r) \geq \epsilon$, and so it must be the case that $f(z) < \epsilon$. If z is irrational, then $f(z) = 0$ by definition of f, and so then too we have $f(z) = 0 < \epsilon$. So we have that for all $z \in \mathbb{R}$ such that $|z - x| < \delta$, $|f(z) - f(x)| = |f(z) - 0| = |f(z)| = f(z) < \epsilon$. This completes the proof that f is continuous at every irrational number.

Solution to Exercise 3.14

We show by induction that $f(n) = nf(1)$ for all natural numbers n. For $n = 1$, we have $f(n) = f(1) = 1 \cdot f(1) = nf(1)$, and so the claim is true when $n = 1$. If for some $n \in \mathbb{N}$ there holds $f(n) = nf(1)$, then we have

$$f(n+1) = f(n) + f(1) = nf(1) + f(1) = (n+1)f(1).$$

This completes the induction step, and so the result holds for all natural numbers.

Also, $f(0) = f(0+0) = f(0) + f(0)$ shows that $f(0) = 0 = 0 \cdot f(1)$. So $f(n) = nf(1)$ for all nonnegative integers. Let m be a negative integer. Then $-m$ is a positive integer, and so

$$0 = f(0) = f(m + (-m)) = f(m) + f(-m) = f(m) + (-m)f(1),$$

that is, $f(m) = -(-m)f(1) = mf(1)$. So we have that $f(n) = nf(1)$ for all integers n.

Now every rational number r can be expressed as $r = n/d$ for some integers n, d with $d > 0$. Then

$$nf(1) = f(n) = f(d \cdot (n/d)) = \underbrace{f(n/d) + \cdots + f(n/d)}_{d \text{ times}} = d \cdot f(n/d).$$

Consequently, $f(n/d) = (nf(1))/d$, that is, $f(r) = rf(1)$.

Let $x \in \mathbb{R}$. Then we can find a sequence $(r_n)_{n\in\mathbb{N}}$ of rational numbers which converges to x. Using the continuity of f, we obtain

$$f(x) = f\left(\lim_{n\to\infty} r_n\right) = \lim_{n\to\infty} f(r_n) = \lim_{n\to\infty} r_n f(1) = xf(1).$$

Consequently, $f(x) = xf(1)$ for all $x \in \mathbb{R}$.

Solution to Exercise 3.15

Let $x \in \mathbb{R}$. Clearly $f(2x) = -f(x)$, and so

$$f(x) = -f(x/2) = (-1)^2 f(x/4) = \cdots = (-1)^n f(x/2^n)$$

for all $x \in \mathbb{R}$. Since the sequence $(x/2^n)_{n \in \mathbb{N}}$ converges to 0, it follows that

$$f(0) = f\left(\lim_{n \to \infty} x/2^n\right) = \lim_{n \to \infty} f(x/2^n) = (-1)^n f(x).$$

But $f(0) = f(2 \cdot 0) = -f(0)$, and so $f(0) = 0$. It follows from the above that

$$f(x) = (-1)^n f(0) = (-1)^n 0 = 0.$$

So if f is continuous and it satisfies the given identity, then it must be the constant function 0.

Conversely, the constant function 0 is continuous and also $f(2x) + f(x) = 0 + 0 = 0$ for all $x \in \mathbb{R}$.

Solution to Exercise 3.16

Define $f : (0, \infty) \to \mathbb{R}$ by $f(x) = 1/x$, $x > 0$. Then f is continuous. Take $(x_n)_{n \in \mathbb{N}}$ to be the Cauchy sequence $\left(\frac{1}{n}\right)_{n \in \mathbb{N}}$. Then $f(x_n) = n$, and $|f(x_n) - f(x_m)| = |n - m| \geq 1$ for all $n \neq m$, showing that the sequence $(f(x_n))_{n \in \mathbb{N}}$ is not Cauchy.

Solution to Exercise 3.17

(1) True. Indeed, the sequences

$$\left(\frac{1}{2n + 7}\right)_{n \in \mathbb{N}} \quad \text{and} \quad \left(\frac{n}{n^2 + 1}\right)_{n \in \mathbb{N}}$$

both converge with limit 0, and as f is continuous at 0, it follows that

$$f(0) = f\left(\lim_{n \to \infty} \frac{1}{2n + 7}\right) = \lim_{n \to \infty} f\left(\frac{1}{2n + 7}\right) = \lim_{n \to \infty} g\left(\frac{n}{n^2 + 1}\right) = g\left(\lim_{n \to \infty} \frac{n}{n^2 + 1}\right) = g(0).$$

(2) True. This has nothing to do with the continuity of f on $\mathbb{R}$. We just note that since $(g(n))_{n \in \mathbb{N}}$ converges to L, its subsequence $(g(n^2))_{n \in \mathbb{N}}$ also converges to L. But since $f(n) = g(n^2)$ for all n, we obtain that $(f(n))_{n \in \mathbb{N}}$ converges to L.

Solution to Exercise 3.18

We apply Theorem 3.2 several times in order to prove this.

Since the function $x \mapsto x$ is continuous on $\mathbb{R}$, it follows that the function $x \mapsto x^2$ is continuous on $\mathbb{R}$ as well. Moreover, the function $x \mapsto 1$ is continuous on $\mathbb{R}$, and so we obtain that the function $x \mapsto 1 + x^2$ is continuous on $\mathbb{R}$. As $1 + x^2 \geq 1 > 0$ for all real x, we conclude that the function $x \mapsto \frac{1}{1+x^2}$ is continuous on $\mathbb{R}$. Hence the function $x \mapsto x^2 \cdot \frac{1}{1+x^2} = \frac{x^2}{1+x^2}$ is continuous on $\mathbb{R}$, that is, f is continuous on $\mathbb{R}$.

Solution to Exercise 3.19

Since $x \mapsto |x|, x + 1$ are continuous, so is f. Also the composition $f \circ f$ is then continuous. So

$$\lim_{x \to -2} (f \circ f)(x) = (f \circ f)(-2).$$

But $f(-2) = |-2+1| - |-2| = |-1| - 2 = 1 - 2 = -1$, and so

$$\lim_{x \to -2}(f \circ f)(x) = (f \circ f)(-2) = f(f(-2))$$

$$= f(-1) = |-1+1| - |-1| = |0| - 1 = 0 - 1 = -1.$$

Solution to Exercise 3.20

(1), (2): False. We will just show that (2) is false, from which it follows immediately that (1) is false too. Let $f = 1_{\mathbb{Q}}$ be the indicator function of the rational numbers, that is, it is 1 if the argument is rational, and zero otherwise. Let $g := 1 - 1_{\mathbb{Q}}$. If $x \in \mathbb{Q}$, then $(g \circ f)(x) = g(1) = 0$, and if $x \notin \mathbb{Q}$, then $(g \circ f)(x) = g(0) = 0$. So $g \circ f \equiv 0$ is clearly continuous. But neither f nor g is continuous at any point of $\mathbb{R}$.

(3) False. Let $f \equiv x$, and g be the same as above. Then $g \circ f = g$ isn't continuous at any point of $\mathbb{R}$, but f is continuous on $\mathbb{R}$.

(4) True. This is the contrapositive of the statement of Theorem 3.3.

Solution to Exercise 3.21

We have

$$x \mapsto \frac{1}{x} : \mathbb{R} \setminus \{0\} \to \mathbb{R} \ \text{ is continuous,}$$

$$\sin : \mathbb{R} \to \mathbb{R} \ \text{ is continuous,}$$

and so $x \mapsto \sin \frac{1}{x} : \mathbb{R} \setminus \{0\} \to \mathbb{R}$ is continuous. Hence for every $\epsilon > 0$, there exists a $\delta > 0$ such that whenever $|x - c| < \delta$,

$$\left| x \sin \frac{1}{x} - c \sin \frac{1}{c} \right| = |f(x) - f(c)| < \epsilon.$$

So f is continuous on $\mathbb{R} \setminus \{0\}$. The continuity at 0 can be checked directly as follows. Let $\epsilon > 0$. Set $\delta = \epsilon > 0$. Then if $0 < |x - 0| < \delta$, we have

$$|f(x) - f(0)| = \left| x \sin \frac{1}{x} - 0 \right| = |x| \cdot \left| \sin \frac{1}{x} \right| \le |x| \cdot 1 = |x - 0| < \delta = \epsilon.$$

So f is also continuous at 0.

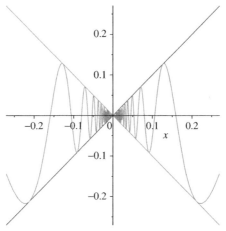

Solution to Exercise 3.22

Yes. The maps $x \mapsto f(x) + g(x)$, $f(x) - g(x)$ are continuous, and thus also

$$x \mapsto \frac{f(x) + g(x) + |f(x) - g(x)|}{2} = \max\{f(x), g(x)\} : I \to \mathbb{R}.$$

Hence $\max\{f, g\}$ is continuous on I.

Solution to Exercise 3.23

Since the function $f : [0, 1] \to \mathbb{R}$ and the function $x \mapsto x$ from $[0, 1]$ to $\mathbb{R}$ are both continuous on the interval $[0, 1]$, it follows that also the function $g : [0, 1] \to \mathbb{R}$ defined by

$$g(x) = f(x) - x, \quad \text{for all } x \in [0, 1]$$

is continuous on $[0, 1]$. Since $0 \leq f(x) \leq 1$ for all $x \in [0, 1]$, we have

$$g(0) = f(0) - 0 = f(0) \geq 0, \quad \text{and}$$
$$g(1) = f(1) - 1 \leq 0.$$

So by the Intermediate Value Theorem, there exists a $c \in [0, 1]$ such that $g(c) = 0$, that is, $f(c) = c$.

Solution to Exercise 3.24

Consider the function $g : [0, 1] \to \mathbb{R}$ given by

$$g(x) = f(x) - f(1) - (f(0) - f(1))x, \quad x \in \mathbb{R}.$$

As f and the map $x \mapsto x$ are continuous, it follows that g is continuous too. We have

$$g(0) = f(0) - f(1) - (f(0) - f(1)) \cdot 0 = f(0) - f(1),$$
$$g(1) = f(1) - f(1) - (f(0) - f(1)) \cdot 1 = -(f(0) - f(1)).$$

Thus $y = 0$ lies between $g(0)$ and $g(1)$, and so by the Intermediate Value Theorem, there exists a $c \in [0, 1]$ such that $g(c) = 0$, that is, $f(c) - f(1) = (f(0) - f(1))c$.

Solution to Exercise 3.25

Imagine a directed line ℓ such that the pancake lies entirely to one side of the line ℓ (say if we look along the direction of the line, the pancake appears to our right). Now translate the line to the right parallel to itself till the whole pancake appears to the left of the line. Suppose that the total distance by which the line is translated is d. At each intermediate distance $x \in [0, d]$, let $A(x)$ denote the area of the part of the pancake that lies to the right of our line. Thus if S is the total area of the pancake, then $A(0) = A$ and $A(d) = 0$. As the map $A : [0, d] \to \mathbb{R}$ is continuous and $f(0) = S \geq S/2 \geq 0 = f(d)$, it follows by the Intermediate Value Theorem, that there is a $y \in [0, d]$ such that $A(y) = S/2$, or in other words, at some position of our line, the pancake is divided into two parts with equal areas.

As the direction of the line in the above process was inconsequential, given any direction, we can choose a straight line cut having that direction that divides the pancake into two equal parts.

Solution to Exercise 3.26

True. Consider $f : \mathbb{R} \to \mathbb{R}$ given by

$$f(x) = x^{399} + \frac{1976}{1 + x^2(\cos x)^2} - 28, \quad x \in \mathbb{R}.$$

We have with $a := \sqrt[399]{28}$ that

$$f(a) = 28 + \frac{1976}{1 + \text{(something positive)}} - 28 > 0,$$

and with $b := -\sqrt[399]{1976}$ that

$$f(a) = -1976 + \frac{1976}{1 + \text{(something positive)}} - 28 < 0.$$

Thus $f(a) > 0 =: y > f(b)$. As f is continuous[4], it follows by the Intermediate Value Theorem that there must be a c between a and b such that $f(c) = 0$, that is,

$$c^{399} + \frac{1976}{1 + c^2(\cos c)^2} = 28.$$

Solution to Exercise 3.27

Let the weekend campsite be at altitude H. Let $u : [0, 1] \to \mathbb{R}$ be the position function for the walk up, and $d : \left[0, \frac{1}{2}\right] \to \mathbb{R}$ be the position function for the walk down. (We assume that these are continuous functions.) Consider the function $f : [0, 1/2] \to \mathbb{R}$ given by $f(t) = u(t) - d(t)$, $t \in [0, 1/2]$. Then f is also continuous, and moreover

$$f(0) = u(0) - d(0) = 0 - H = -H < 0, \quad \text{while}$$

$$f\left(\frac{1}{2}\right) = u\left(\frac{1}{2}\right) - d\left(\frac{1}{2}\right) = u\left(\frac{1}{2}\right) - 0 = u\left(\frac{1}{2}\right) \geq 0.$$

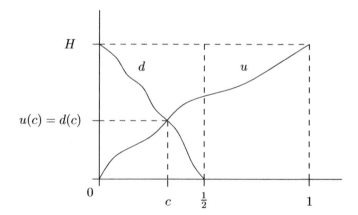

Hence by the Intermediate Value Theorem, it follows that there exists a $c \in \left[0, \frac{1}{2}\right]$ such that $f(c) = 0$, that is, $u(c) = d(c)$. So at time c past 8:00, the hiker was exactly at the same spot on Saturday and Sunday.

[4] the continuity of cos will be proved in Chapter 5

Solution to Exercise 3.28

The polynomial function $p : [-1, 2] \to \mathbb{R}$ is continuous on the interval $[-1, 2]$. Moreover,

$$p(-1) = 2 \cdot (-1)^3 - 5 \cdot (-1)^2 - 10 \cdot (-1) + 5 = -2 - 5 + 10 + 5 = 8 > 0, \text{ and}$$

$$p(2) = 2 \cdot (2)^3 - 5 \cdot (2)^2 - 10 \cdot (2) + 5 = 16 - 20 - 20 + 5 = -19 < 0.$$

Since $p(-1) > 0 > p(2)$, from the Intermediate Value Theorem applied to the continuous function p on the interval $[-1, 2]$, we conclude that there must exist a $c \in [-1, 2]$ such that $p(c) = 0$. So p has a real root in the interval $[-1, 2]$.

Solution to Exercise 3.29

Suppose that $c, d \in [a, b]$ such that c, d are such that the nonzero numbers $f(c)$ and $f(d)$ have opposite signs. Then by the Intermediate Value Theorem applied to the continuous function $f : [c, d] \to \mathbb{R}$, with $y := 0$ (which lies between $f(c)$ and $f(d)$), it follows that there must be an $x_* \in [c, d] \subset [a, b]$ such that $f(x_*) = y = 0$, a contradiction to the fact that f is never zero on $[a, b]$.

Solution to Exercise 3.30

Obviously $S \subset \mathbb{R}$. We now show the reverse inclusion. Let $y \in \mathbb{R}$.

As S is not bounded above, y is not an upper bound of S, that is, there exists a $x_0 \in \mathbb{R}$ such that $f(x_0) < y$.

Similarly, since S is not bounded below, y is not a lower bound of S, and so there exists a $x_1 \in \mathbb{R}$ such that $f(x_1) > y$.

Now consider the restriction of f to the interval with endpoints x_0 and x_1 with the endpoints included in the interval. Applying the Intermediate Value Theorem to this continuous function, it follows that there exists a real number c such that $f(c) = y$.

Since y was arbitrary, this shows that $S = \mathbb{R}$.

Solution to Exercise 3.31

The following three cases are possible:

$\underline{1°}$ $f(0) = 0$. Let $x_0 = 0$ and let $m \in \mathbb{Z} \backslash \{0\}$. Clearly $f(x_0) = f(0) = 0 = m0 = mx_0$.

$\underline{2°}$ $f(0) > 0$. Choose $N \in \mathbb{N}$ satisfying $N > f(1)$ (that such an N exists follows from the Archimedean Property). Consider the function $g : [0, 1] \to \mathbb{R}$ defined by $g(x) = f(x) - Nx$, $x \in [0, 1]$. As f and $x \mapsto Nx$ are continuous, so is g. Note that $g(0) = f(0) - N \cdot 0 = f(0) > 0$, while $g(1) = f(1) - N < 0$. Applying the Intermediate Value Theorem to g (with $y = 0$), it follows that there exists an $x_0 \in [0, 1]$ such that $g(x_0) = 0$, that is, $f(x_0) = Nx_0$.

$\underline{3°}$ $f(0) < 0$. Choose an $N \in \mathbb{N}$ such that $N > -f(1)$ (again the Archimedean Property guarantees the existence of such an N), and consider the continuous function $g : [0, 1] \to \mathbb{R}$ defined by $g(x) = f(x) + Nx$. We have $g(0) = f(0) < 0$, and $g(1) = f(1) + N > 0$, and so by the Intermediate Value Theorem, it follows that there exists an $x_0 \in [0, 1]$ such that $g(x_0) = 0$, that is, $f(x_0) = -Nx_0$.

This completes the proof.

Solution to Exercise 3.32

Suppose that such a continuous function exists. From Exercise 3.31, it follows that there exists an $x_0 \in \mathbb{R}$ and an $m \in \mathbb{Z}\backslash\{0\}$ such that $f(x_0) = mx_0$. We have the following two possible cases:

   $\underline{1°}$ $x_0 \in \mathbb{Q}$. But then $f(x_0)$ is irrational, while mx_0 is rational, a contradiction.

   $\underline{2°}$ $x_0 \notin \mathbb{Q}$. But then $f(x_0)$ is rational, whereas mx_0 is irrational, a contradiction.

So f cannot be continuous.

(Alternately, one could observe that $f(\mathbb{R}\backslash\mathbb{Q})$, being a subset of $\mathbb{Q}$, is countable. Also $f(\mathbb{Q})$ is clearly countable. Hence $f(\mathbb{R}) = f(\mathbb{R}\backslash\mathbb{Q}) \cup f(\mathbb{Q})$ is countable. But then $f(\mathbb{R})$ must be a singleton, since if it contained two distinct points $a < b$, then it would also contain the whole interval $[a, b]$ by the Intermediate Value Theorem, and $f(\mathbb{R})$ would then be uncountable. Suppose that $f(\mathbb{R}) = \{x_*\}$. But $f(0) = x_* = f(\sqrt{2})$, and so x_* belongs to $\mathbb{Q}$ and to $\mathbb{R}\backslash\mathbb{Q}$, a contradiction.)

Solution to Exercise 3.33

That f is strictly decreasing: If $x_1 > x_2 \geq 0$, then

$$x_1^2 = x_1 \cdot x_1 > x_1 \cdot x_2 \geq x_2 \cdot x_2 = x_2^2 \geq 0,$$

and so $1 + x_1^2 > 1 + x_2^2$, giving

$$f(x_1) = \frac{1}{1 + x_1^2} < \frac{1}{1 + x_2^2} = f(x_2).$$

So f is strictly decreasing.

That $f([0, \infty)) = (0, 1]$: Let $\epsilon > 0$ and $n \in \mathbb{N}$ be such that $n > 1/\epsilon - 1$. Then we have $n^2 > n > 1/\epsilon - 1$, and so

$$f(n) = \frac{1}{1 + n^2} < \frac{1}{1 + 1/\epsilon - 1} = \epsilon.$$

On the other hand, $f(0) = 1$. So by the Intermediate Value Theorem, $(\epsilon, 1] \subset f([0, \infty))$. As $\epsilon > 0$ was arbitrary, it follows that $(0, 1] \subset f([0, \infty))$. The reverse inclusion follows by observing that

$$0 \leq \frac{1}{1 + x^2} \leq \frac{1}{1 + 0} = 1,$$

for $x \in [0, \infty)$ so that $f([0, \infty)) \subset [0, 1]$, and clearly $f(x) = 1/(1 + x^2)$ is never zero for $x \in [0, \infty)$.

Expression for f^{-1}: Let $y \in (0, 1]$ and $x \in [0, \infty)$. Then

$$\frac{1}{1 + x^2} = y \Leftrightarrow \frac{1}{y} = 1 + x^2 \Leftrightarrow x^2 = \frac{1}{y} - 1 \Leftrightarrow x = \sqrt{\frac{1}{y} - 1}.$$

So $f^{-1} : (0, 1] \to [0, \infty)$ is given by $f^{-1}(y) = \sqrt{\frac{1}{y} - 1}$, $y \in (0, 1]$.

$y \mapsto \frac{1}{y}$ is a continuous function on $(0, 1]$, and thus so is $y \mapsto \frac{1}{y} - 1$. Also, the range of the mapping $y \mapsto \frac{1}{y} - 1 : (0, 1] \to \mathbb{R}$ is contained in $[0, \infty)$. Composing with the continuous square root function $\sqrt{\cdot} : [0, \infty) \to [0, \infty)$, we obtain the desired continuity of f^{-1}.

(Alternately, we can also establish the continuity as follows. For any $n > 0$, the restriction $f|_{[0,n]} : [0,n] \to \mathbb{R}$ is strictly decreasing and continuous. Clearly we have

$$f|_{[0,n]}([0,n]) = f([0,n]) = [f(n), f(0)] = \left[\frac{1}{1+n^2}, 1\right].$$

So its inverse $(f|_{[0,n]})^{-1} : \left[\frac{1}{1+n^2}, 1\right] \to [0,n]$ is continuous. But if

$$y \in \left[\frac{1}{1+n^2}, 1\right] = f|_{[0,n]}([0,n]),$$

then $y = f(x)$ for some $x \in [0,n]$, and so $(f|_{[0,n]})^{-1}(y) = x = f^{-1}(y)$. This shows that

$$(f|_{[0,n]})^{-1} = (f^{-1})|_{[\frac{1}{1+n^2}, 1]}.$$

But as n was arbitrary, it follows that f^{-1} is continuous on $(0, 1]$.) See the picture below for the graphs of f, f^{-1}.

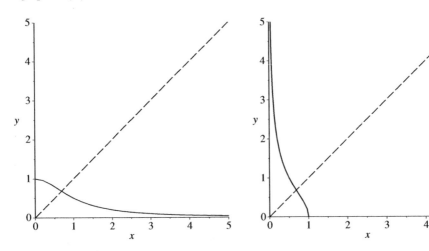

Solution to Exercise 3.34

(1) Let

$$f(x) = \begin{cases} 2x & \text{if } x \in (0, \tfrac{1}{2}), \\ 2 - 2x & \text{if } x \in [\tfrac{1}{2}, 1). \end{cases}$$

Then f is continuous on $(0, 1)$ and $f(0, 1) = (0, 1]$.

(2) If there existed such a continuous f, then since $0, 1 \in T$ and since $f(S) = T$, there would exist $a, b \in S = (0, 1)$ such that $f(a) = 0$ and $f(b) = 1$. But then since we have that $f(a) = 0 < 1/2 < 1 = f(b)$, it follows by the Intermediate Value Theorem that there is a c between a and b such that $f(c) = 1/2$. Thus $1/2 \in f(S) = T = \{0, 1\}$, a contradiction.

(Or note that $S = (0, 1)$ is an interval while $T = \{0, 1\}$ is not, and the existence of such an f would contradict Corollary 3.7.)

Solution to Exercise 3.35

Consider the function $g : [0, T] \to \mathbb{R}$, given by $g(x) = f(x)$ for all $x \in [0, T]$. Then g is continuous on $[0, T]$. Applying the Extreme Value Theorem to g, we conclude that there exist $c, d \in [0, T]$ such that $g(c) = \max\{g(x) : x \in [0, T]\}$ and $g(d) = \min\{g(x) : x \in [0, T]\}$. So for all $x \in [0, T]$, $g(d) \leq g(x) \leq g(c)$, that is, $f(d) \leq f(x) \leq f(c)$.

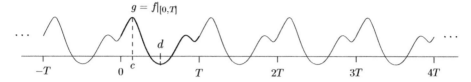

So far we have proved the fact that f is bounded on $[0, T]$. We now prove that f is bounded on $\mathbb{R}$ using the periodicity of f.

Now if x is any real number, there exists a $n \in \mathbb{Z}$ such that $x = nT + r$, where $r \in \mathbb{R}$ is such that $r \in [0, T)$. (Indeed, we have

$$\frac{x}{T} = \left\lfloor \frac{x}{T} \right\rfloor + \Theta$$

where $\Theta \in [0, 1)$. Consequently, $x = nT + r$, where $n := \left\lfloor \frac{x}{T} \right\rfloor \in \mathbb{Z}$ and $r := T \cdot \Theta \in [0, T)$.) Thus $f(x) = f(nT + r) = f(r)$. As $f(d) \leq f(r) \leq f(c)$, it follows that $f(d) \leq f(x) \leq f(c)$. Since the choice of $x \in \mathbb{R}$ was arbitrary, it follows that $f(d) \leq f(x) \leq f(c)$ for all $x \in \mathbb{R}$. So $f(c)$ and $f(d)$ are upper and lower bounds, respectively, of the set $\{f(x) : x \in \mathbb{R}\}$, and so it is bounded.

Solution to Exercise 3.36

(1) If $x \in (a, b]$, then let $f|_{[a,x]}$ denote the restriction of the function f to the interval $[a, x]$, defined by $f|_{[a,x]}(y) = f(y)$ for all $y \in [a, x]$. We note that $f|_{[a,x]}$ is a continuous function. Applying the Extreme Value Theorem to $f|_{[a,x]}$, we see that

$$\max\{f|_{[a,x]}(y) : y \in [a,x]\} = \max\{f(y) : y \in [a,x]\}$$

exists, and so f_* is well-defined.

(2) f_* is given by $f_*(x) = \begin{cases} x - x^2 & \text{if } 0 \leq x \leq \frac{1}{2}, \\ \frac{1}{4} & \text{if } \frac{1}{2} < x \leq 1. \end{cases}$

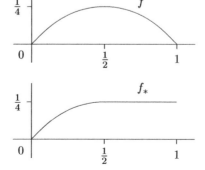

Solution to Exercise 3.37

True. By the Extreme Value Theorem, there is a $c \in [a, b]$ such that $f(c) = \max f([a, b]) =: \delta$. As $f(c) > 0$, we have $\delta > 0$, and moreover, for all $x \in [a, b]$, $f(x) \geq \max f([a, b]) = \delta$.

Solution to Exercise 3.38

(B).

Solution to Exercise 3.39

By the Extreme Value Theorem, there exist $c, d \in [a, b]$ such that $f(c) \leq f(x)$ and $f(x) \leq f(d)$ for all $x \in [a, b]$. Hence

$$f(c) = \frac{f(c) + \cdots + f(c)}{n} \leq \frac{f(c_1) + \cdots + f(c_n)}{n} \leq \frac{f(d) + \cdots + f(d)}{n} = f(d).$$

So applying the Intermediate Value Theorem to f on the interval with end points c and d, we obtain the existence of a c_* in this interval such that

$$f(c_*) = \frac{f(c_1) + \cdots + f(c_n)}{n}.$$

But then clearly this c_* belongs to $[a, b]$ (since c, d do and because c_* lies in the interval with the end points c and d).

Alternative solution: The use of the Extreme Value Theorem can be avoided as follows. We note that if k_*, k_{**} are such that

$$f(c_{k_*}) = \min\{f(c_1), \cdots, f(c_n)\}$$
$$f(c_{k_{**}}) = \max\{f(c_1), \cdots, f(c_n)\},$$

then we have again

$$f(c_{k_*}) = \frac{f(c_{k_*}) + \cdots + f(c_{k_*})}{n} \leq \frac{f(c_1) + \cdots + f(c_n)}{n} \leq \frac{f(c_{k_{**}}) + \cdots + f(c_{k_{**}})}{n} = f(c_{k_{**}}).$$

So applying the Intermediate Value Theorem to f on the interval with end points c_{k_*} and $c_{k_{**}}$, we obtain the existence of a c_* in this interval such that

$$f(c_*) = \frac{f(c_1) + \cdots + f(c_n)}{n}.$$

Solution to Exercise 3.40

Let f be uniformly continuous on $(0, 1)$ and $\epsilon = \frac{1}{2} > 0$. Then there exists a $\delta > 0$ such that whenever $x, y \in (0, 1)$ satisfy $|x - y| < \delta$, we have $|\frac{1}{x} - \frac{1}{y}| < \epsilon = \frac{1}{2}$. Take any natural number $n > 1$ such that $n > \frac{1}{2\delta}$. Set $x := \frac{1}{n}$ and $y := \frac{1}{2n}$. Then $x, y \in (0, 1)$, and $|x - y| = \frac{1}{2n} < \delta$. Thus

$$\frac{1}{2} > \left| \frac{1}{x} - \frac{1}{y} \right| = |n - 2n| = |n| = n,$$

a contradiction.

Solution to Exercise 3.41

We have for $x, y \in \mathbb{R}$ that $||x| - |y|| \leq |x - y|$. If $\epsilon > 0$, then with $\delta := \epsilon > 0$, we have for all $x, y \in \mathbb{R}$ satisfying $|x - y| < \delta = \epsilon$ that

$$||x| - |y|| \leq |x - y| < \delta = \epsilon.$$

So the absolute value function is uniformly continuous.

Solution to Exercise 3.42

Let $\epsilon > 0$. Set $\delta = \epsilon/L > 0$. If $x, y \in \mathbb{R}$ and $|x - y| < \delta$, then $|f(x) - f(y)| \leq L|x - y| < L\delta = \epsilon$. Thus f is uniformly continuous.

The converse is not true. In other words, there *are* uniformly continuous functions that are *not* Lipschitz. For example, it follows from Example 3.17 that $x \mapsto \sqrt{|x|} : \mathbb{R} \to \mathbb{R}$ is uniformly continuous. It is, however, not Lipschitz. Indeed, if it were Lipschitz, then there would exist an $L > 0$ such that for all $x, y \in \mathbb{R}$, $|\sqrt{|x|} - \sqrt{|y|}| \leq L|x - y|$, and in particular with $x = 1/n^2$ ($n \in \mathbb{N}$) and $y = 0$, we would obtain $n \leq L$ for all $n \in \mathbb{N}$, a contradiction to the Archimedean Property of $\mathbb{R}$.

Solution to Exercise 3.43

Suppose that $(x_n)_{n \in \mathbb{N}}$ is a Cauchy sequence in I. We want to show that $(f(x_n))_{n \in \mathbb{N}}$ is a Cauchy sequence in $\mathbb{R}$. Let $\epsilon > 0$. Since f is uniformly continuous, there exists a $\delta > 0$ such that whenever $|x - y| < \delta$, we have $|f(x) - f(y)| < \epsilon$. As $(x_n)_{n \in \mathbb{N}}$ is Cauchy, there exists an index $N \in \mathbb{N}$ such that for all $n, m > N$, $|x_n - x_m| < \delta$. For all $n, m > N$, $|f(x_n) - f(y_n)| < \epsilon$. Hence $(f(x_n))_{n \in \mathbb{N}}$ is a Cauchy sequence in $\mathbb{R}$.

Exercise 3.16 shows that continuous functions don't necessarily preserve Cauchyness.

Solution to Exercise 3.44

(1) Let $\epsilon > 0$. Since the function f is uniformly continuous, there exists a $\delta_f > 0$ such that whenever $x, y \in I$ satisfy $|x - y| < \delta_f$, we have $|f(x) - f(y)| < \epsilon/2$. Similarly, since g is uniformly continuous, there exists a $\delta_g > 0$ such that whenever $x, y \in I$ satisfy $|x - y| < \delta_g$, we have $|g(x) - g(y)| < \epsilon/2$. So with $\delta := \min\{\delta_f, \delta_g\} > 0$, we have that whenever $x, y \in I$ satisfy $|x - y| < \delta$,

$$|(f + g)(x) - (f + g)(y)| = |f(x) + g(x) - f(y) - g(y)|$$

$$\leq |f(x) - f(y)| + |g(x) - g(y)| < \frac{\epsilon}{2} + \frac{\epsilon}{2} = \epsilon.$$

(2) With $I := \mathbb{R}$ and $f(x) = g(x) = x$ (so that f, g are uniformly continuous), we have $(fg)(x) = x^2$, making fg not uniformly continuous.

(3) Let $M_f, M_g > 0$ be such that for all $x, y \in \mathbb{R}$, $|f(x)| < M_f$ and $|g(y)| < M_g$. Given $\epsilon > 0$, there exists a $\delta_f > 0$ such that whenever $x, y \in I$ satisfy $|x - y| < \delta_f$, we have

$$|f(x) - f(y)| < \frac{\epsilon}{2M_g}.$$

Similarly, since g is uniformly continuous, there exists a $\delta_g > 0$ such that whenever $x, y \in I$ satisfy $|x - y| < \delta_g$, we have

$$|g(x) - g(y)| < \frac{\epsilon}{2M_f}.$$

So with $\delta := \min\{\delta_f, \delta_g\} > 0$, we have that whenever $x, y \in I$ satisfy $|x - y| < \delta$,

$$|(fg)(x) - (fg)(y)| = |f(x)g(x) - f(x)g(y) + f(x)g(y) - f(y)g(y)|$$
$$\leq |f(x)||g(x) - g(y)| + |g(y)||f(x) - f(y)|$$
$$< M_f \cdot \frac{\epsilon}{2M_f} + M_g \cdot \frac{\epsilon}{2M_g} = \epsilon.$$

Solution to Exercise 3.45

(B), (C), (D).

Solution to Exercise 3.46

(1) The function is a ratio of two polynomials, and so it is continuous at all points where the denominator polynomial is nonzero. So we know that f is continuous on $\mathbb{R}\setminus\{2\}$. Moreover,

$$\lim_{x \to 2} f(x) = \lim_{x \to 2} \frac{x^3 - 3x - 2}{x - 2} = \lim_{x \to 2} \frac{3x^2 - 3}{1} = 3 \cdot 4 - 3 = 9.$$

(We used the $\frac{0}{0}$ form of l'Hôpital's Rule above.) Hence by defining $f(2) = 9$, f will be continuous on $\mathbb{R}$. So there does exist such a continuous function.

(2) This statement is false. Indeed,

$$\lim_{x \nearrow 2} f(x) = \lim_{x \nearrow 2} x = 2,$$

while

$$\lim_{x \searrow 2} f(x) = \lim_{x \searrow 2} 2x = 4,$$

and so $\lim_{x \to 2} f(x)$ does not exist.

Solution to Exercise 3.47

(If part): Suppose that $L := \lim_{x \to c+} f(x) = \lim_{x \to c-} f(x)$.

Let $\epsilon > 0$. Then there exists a $\delta_+ > 0$ such that for all x satisfying $c < x < c + \delta_+$, we have $|f(x) - L| < \epsilon$. Also, there exists a $\delta_- > 0$ such that for all x satisfying $c - \delta_- < x < c$, we have $|f(x) - L| < \epsilon$. Let $\delta := \min\{\delta_+, \delta_-\}$. Then for all x such that $0 < |x - c| < \delta$, we have either $c < x < c + \delta \leq c + \delta_+$ or $c - \delta_- \leq c - \delta < x < c$, and in either case there holds $|f(x) - L| < \epsilon$. Hence

$$\lim_{x \to c} f(x) = L \quad \left(= \lim_{x \to c+} f(x) = \lim_{x \to c-} f(x) \right).$$

(Only if part): Let $L := \lim_{x \to c} f(x)$.

Let $\epsilon > 0$. Then there exists a $\delta > 0$ such that for all x satisfying $0 < |x - c| < \delta$, we have $|f(x) - L| < \epsilon$. In particular, for all x satisfying $c < x < c + \delta$, we have $|f(x) - L| < \epsilon$, and so $\lim_{x \to c+} f(x) = L$.

Also, for all x satisfying $c - \delta < x < c$, we have $|f(x) - L| < \epsilon$, and so $\lim_{x \to c-} f(x) = L$. This implies that $\lim_{x \to c+} f(x) = L = \lim_{x \to c-} f(x)$.

Solution to Exercise 3.48

(1) As $x \mapsto |x|, x + 1$ are continuous, we have
$$\lim_{x \to 0} |x| = |0| = 0, \text{ and } \lim_{x \to 0}(x + 1) = 0 + 1 = 1 \neq 0.$$

Thus by the Algebra of Limits, $\lim_{x \to 0} \dfrac{|x|}{x + 1} = \dfrac{\lim_{x \to 0} |x|}{\lim_{x \to 0}(x + 1)} = \dfrac{0}{1} = 0$.

(2) We have $\lim_{x \to 1+} (\lfloor x \rfloor - x) = 1 - 1 = 0$, while $\lim_{x \to 1-} (\lfloor x \rfloor - x) = 0 - 1 = -1$. So $\lim_{x \to 1}(\lfloor x \rfloor - x)$ does not exist.

(3) We have $\lim_{x \to 0+} \lfloor x \rfloor = 0$, $\lim_{x \to 0-} \lfloor x \rfloor = -1$, and so
$$\lim_{x \to 0+} x \lfloor x \rfloor = 0 \cdot 0 = 0,$$
$$\lim_{x \to 0-} x \lfloor x \rfloor = 0 \cdot (-1) = 0.$$

Hence $\lim_{x \to 0} x \lfloor x \rfloor = 0$.

(4) Suppose that the limit exists and is some number L. Take any $\theta \in [-\pi/2, \pi/2]$ such that $\sin \theta \neq L$. Set
$$x_n := \frac{1}{\theta + 2n\pi}, \quad n \in \mathbb{N}.$$

Then $(x_n)_{n \in \mathbb{N}}$ converges to 0. But
$$\sin \frac{1}{x_n} = \sin(\theta + 2n\pi) = \sin \theta,$$

and so $\left(\sin \dfrac{1}{n}\right)_{n \in \mathbb{N}}$ converges to $\sin \theta \neq L$, a contradiction. So $\lim_{x \to 0} \sin \dfrac{1}{x}$ does not exist.

Solution to Exercise 3.49

We have that B is zero for only finitely many real values, and so for all sufficiently large x, $B(x) \neq 0$. We have
$$\frac{A(x)}{B(x)} = \frac{a_0 + \cdots + a_{\alpha-1}x^{\alpha-1} + a_\alpha x^\alpha}{b_0 + b_1 x + \cdots + b_{\beta-1}x^{\beta-1} + b_\beta x^\beta}$$
$$= \frac{a_\alpha}{b_\beta} \cdot x^{\alpha-\beta} \cdot \frac{\dfrac{a_0}{a_\alpha x^\alpha} + \cdots + \dfrac{a_{\alpha-1}}{a_\alpha x} + 1}{\dfrac{b_0}{b_\beta x^\beta} + \cdots + \dfrac{b_{\beta-1}}{b_\beta x} + 1} = \frac{a_\alpha}{b_\beta} \cdot x^{\alpha-\beta} \cdot \varphi(x),$$

where $\varphi(x) := \dfrac{\dfrac{a_0}{a_\alpha x^\alpha} + \cdots + \dfrac{a_{\alpha-1}}{a_\alpha x} + 1}{\dfrac{b_0}{b_\beta x^\beta} + \cdots + \dfrac{b_{\beta-1}}{b_\beta x} + 1}$. Clearly

$$\lim_{x\to\infty} \varphi(x) = \lim_{x\to\infty} \frac{\dfrac{a_0}{a_\alpha x^\alpha} + \cdots + \dfrac{a_{\alpha-1}}{a_\alpha x} + 1}{\dfrac{b_0}{b_\beta x^\beta} + \cdots + \dfrac{b_{\beta-1}}{b_\beta x} + 1} = \frac{0 + \cdots + 0 + 1}{0 + \cdots + 0 + 1} = 1.$$

Suppose that $\beta > \alpha$. Then $\lim\limits_{x\to\infty} x^{\alpha-\beta} = 0$, and so

$$\lim_{x\to\infty} \frac{A(x)}{B(x)} = \lim_{x\to\infty} \frac{a_\alpha}{b_\beta} \cdot x^{\alpha-\beta} \cdot \varphi(x) = \frac{a_\alpha}{b_\beta} \cdot 0 \cdot 1 = 0.$$

Next suppose that $\alpha = \beta$. Then

$$\lim_{x\to\infty} \frac{A(x)}{B(x)} = \lim_{x\to\infty} \frac{a_\alpha}{b_\beta} \cdot \varphi(x) = \frac{a_\alpha}{b_\beta} \cdot 1 = \frac{a_\alpha}{b_\beta}.$$

Finally suppose that $\alpha > \beta$. Consider first the case when $\dfrac{a_\alpha}{b_\beta} > 0$.

Then there exists an $R_1 > 0$ such that for all $x > R_1$, $1 - \varphi(x) \le |\varphi(x) - 1| < 1/2$, and so $\varphi(x) > 1/2$. Let $M > 0$. Then taking any R_2 such that

$$R_2 > \left(2M \cdot \frac{b_\beta}{a_\alpha}\right)^{1/(\alpha-\beta)},$$

we have for $x > R_2$ that $\dfrac{a_\alpha}{b_\beta} \cdot x^{\alpha-\beta} > 2M$. Hence for all $x > \max\{R_1, R_2\}$, we obtain

$$\frac{A(x)}{B(x)} = \frac{a_\alpha}{b_\beta} \cdot x^{\alpha-\beta} \cdot \varphi(x) > 2M \cdot \frac{1}{2} = M.$$

Thus $\lim\limits_{x\to\infty} \dfrac{A(x)}{B(x)} = +\infty$.

Now consider the case when $\dfrac{a_\alpha}{b_\beta} < 0$.

Then there exists an $R_1 > 0$ such that for all $x > R_1$, $\pm(\varphi(x) - 1) \le |\varphi(x) - 1| < 1/2$, and so $0 < 1/2 < \varphi(x) < 3/2$. Let $M > 0$. Then taking any R_2' such that

$$R_2' > \left(-\frac{2}{3}M \cdot \frac{b_\beta}{a_\alpha}\right)^{1/(\alpha-\beta)},$$

we have for $x > R_2'$ that $\dfrac{a_\alpha}{b_\beta} \cdot x^{\alpha-\beta} < -\dfrac{2}{3}M$. Hence for all $x > \max\{R_1, R_2'\}$, we obtain

$$\frac{A(x)}{B(x)} = \frac{a_\alpha}{b_\beta} \cdot x^{\alpha-\beta} \cdot \varphi(x) < -\frac{2}{3}M \cdot \frac{3}{2} = -M.$$

Thus $\lim\limits_{x\to\infty} \dfrac{A(x)}{B(x)} = -\infty$.

Solution to Exercise 3.50

(A), (B), (C).

Solution to Exercise 3.51

(1) Let $R(x) := \dfrac{2x+1}{x^2 - 2x - 3} = \dfrac{2x+1}{(x+1)(x-3)} = \dfrac{A_1}{x+1} + \dfrac{A_2}{x-3}$. We have

$$A_1 = \lim_{x \to -1} (x+1)R(x) = \lim_{x \to -1} \frac{2x+1}{x-3} = \frac{2 \cdot (-1) + 1}{-1 - 3} = \frac{1}{4}, \text{ and}$$

$$A_2 = \lim_{x \to 3} (x-3)R(x) = \lim_{x \to 3} \frac{2x+1}{x+1} = \frac{2 \cdot 3 + 1}{3 + 1} = \frac{7}{4}.$$

(2) Let $R(x) := \dfrac{x^2 + 3x + 9}{(x+1)(x-2)^2} = \dfrac{A}{x+1} + \dfrac{B_1}{x-2} + \dfrac{B_2}{(x-2)^2}$. We have

$$A = \lim_{x \to -1} (x+1)R(x) = \lim_{x \to -1} \frac{x^2 + 3x + 9}{(x-2)^2} = \frac{(-1)^2 + 3 \cdot (-1) + 9}{(-1-2)^2} = \frac{7}{9}.$$

Also, $B_2 = \lim\limits_{x \to 2} (x-2)^2 R(x) = \lim\limits_{x \to 2} \dfrac{x^2 + 3x + 9}{x+1} = \dfrac{2^2 + 3 \cdot 2 + 9}{2+1} = \dfrac{19}{3}$. Thus

$$R(x) - \frac{B_2}{(x-2)^2} = \frac{A}{x+1} + \frac{B_1}{x-2},$$

and so

$$B_1 = \lim_{x \to 2} (x-2) \left(R(x) - \frac{B_2}{(x-2)^2} \right) = \lim_{x \to 2} \left(\frac{x^2 + 3x + 9}{(x+1)(x-2)} - \frac{19}{3(x-2)} \right)$$

$$= \lim_{x \to 2} \frac{3(x^2 + 3x + 9) - 19(x+1)}{3(x+1)(x-2)} = \lim_{x \to 2} \frac{3x^2 - 10x + 8}{3(x+1)(x-2)}$$

$$= \lim_{x \to 2} \frac{3x^2 - 6x - 4x + 8}{3(x+1)(x-2)} = \lim_{x \to 2} \frac{3x(x-2) - 4(x-2)}{3(x+1)(x-2)}$$

$$= \lim_{x \to 2} \frac{3x - 4}{3(x+1)} = \frac{3 \cdot 2 - 4}{3(2+1)} = \frac{2}{9}.$$

Solutions to the exercises from Chapter 4

Solution to Exercise 4.1

Let $x_0 \in \mathbb{R}$ be arbitrary. For $x \neq x_0$, we have

$$\frac{f(x) - f(x_0)}{x - x_0} = \frac{\sqrt{1+x^2} - \sqrt{1+x_0^2}}{x - x_0} = \frac{x^2 + 1 - (x_0^2 + 1)}{(x - x_0)(\sqrt{1+x^2} + \sqrt{1+x_0^2})}$$

$$= \frac{x^2 - x_0^2}{(x - x_0)(\sqrt{1+x^2} + \sqrt{1+x_0^2})} = \frac{x + x_0}{\sqrt{1+x^2} + \sqrt{1+x_0^2}}.$$

So $\displaystyle\lim_{x \to x_0} \frac{f(x) - f(x_0)}{x - x_0} = \lim_{x \to x_0} \frac{x + x_0}{\sqrt{1+x^2} + \sqrt{1+x_0^2}} = \frac{x_0 + x_0}{\sqrt{1+x_0^2} + \sqrt{1+x_0^2}} = \frac{x_0}{\sqrt{1+x_0^2}}.$

Hence f is differentiable everywhere, and $f'(x) = \dfrac{x}{\sqrt{1+x^2}}, x \in \mathbb{R}.$

Solution to Exercise 4.2

(1)$\Rightarrow$(2): Let $\epsilon > 0$. Let $\delta > 0$ be such that whenever $0 < |x - c| < \delta$, we have

$$\left| \frac{f(x) - f(c)}{x - c} - f'(c) \right| < \epsilon.$$

Now take $\tilde{\delta} := \delta/c$. Thus $\tilde{\delta} > 0$ and if $0 < |k - 1| < \tilde{\delta}$, then $0 < |kc - c| < \delta$, so that

$$\left| \frac{f(kc) - f(c)}{kc - c} - f'(c) \right| < \epsilon,$$

that is, $\left| \dfrac{f(kc) - f(c)}{k - 1} - cf'(c) \right| < \epsilon c.$ So $\displaystyle\lim_{k \to 1} \frac{f(kc) - f(c)}{k - 1} = cf'(c).$

(2)$\Rightarrow$(1): Suppose that $\displaystyle\lim_{k \to 1} \frac{f(kc) - f(c)}{k - 1} = L.$

Let $\epsilon > 0$. Let $\delta > 0$ be such that whenever $0 < |k - 1| < \delta$, we have

$$\left| \frac{f(kc) - f(c)}{k - 1} - L \right| < \epsilon.$$

Now let $\tilde{\delta} := \delta c$. Then $\tilde{\delta} > 0$ and whenever $0 < |x - c| < \tilde{\delta}$, we have with $k := x/c$ that

$$0 < c \left| \frac{x}{c} - 1 \right| < \delta c,$$

that is, $0 < |k - 1| < \delta$, so that $\left| \dfrac{f(\frac{x}{c}c) - f(c)}{\frac{x}{c} - 1} - L \right| < \epsilon.$

So if $0 < |x - c| < \tilde{\delta}$, then we have $\left| \dfrac{f(x) - f(c)}{x - c} - \dfrac{L}{c} \right| < \dfrac{\epsilon}{c}$. Hence

$$\lim_{x \to c} \frac{f(x) - f(c)}{x - c}$$

exists and equals L/c. So f is differentiable at c and $f'(c) = \dfrac{L}{c} = \dfrac{1}{c} \cdot \lim_{k \to 1} \dfrac{f(kc) - f(c)}{k - 1}$.

Solution to Exercise 4.3

Let $c \in (-a, a)$. Let $\epsilon > 0$. Then there exists a $\delta > 0$ such that for all $x \in (-a, a)$ satisfying $0 < |x - c| < \delta$, we have

$$\left| \frac{f(x) - f(c)}{x - c} - f'(c) \right| < \epsilon.$$

Now let $x \in (-a, a)$ be such that $0 < |x - (-c)| = |x + c| = |(-x) - c| < \delta$. Then we have

$$\epsilon > \left| \frac{f(-x) - f(c)}{-x - c} - f'(c) \right|$$

$$= \left| \frac{f(x) - f(-c)}{-x - c} - f'(c) \right| = \left| -\frac{f(x) - f(-c)}{x + c} - f'(c) \right|$$

$$= \left| \frac{f(x) - f(-c)}{x + c} + f'(c) \right| = \left| \frac{f(x) - f(-c)}{x - (-c)} - (-f'(c)) \right|.$$

Thus $f'(-c) = -f'(c)$. As $c \in (-a, a)$ was arbitrary, it follows that f' is odd. $f'(0) = -f'(-0) = -f'(0)$, and so $2f'(0) = 0$, which implies that $f'(0) = 0$.

Solution to Exercise 4.4

We have

$$\frac{101}{100} = \frac{m_v}{m_0} = \frac{1}{m_0} \cdot m_v = \frac{1}{m_0} \cdot \frac{m_0}{\sqrt{1 - v^2/c^2}} = \frac{1}{\sqrt{1 - v^2/c^2}}.$$

As left hand side is close to 1, we expect that v should be small compared to c. We have for $x \approx 0$ that

$$f(x) := \frac{1}{\sqrt{1 - x}} \approx f(0) + f'(0)x = \frac{1}{\sqrt{1 - 0}} - \frac{1}{2(1 - x)^{3/2}} \cdot (-1)\Big|_{x=0} \cdot x = 1 + \frac{1}{2} \cdot x.$$

(The above calculation of the derivative may be accepted on faith now, but we will soon learn how to calculate the derivatives of fractional powers.) Thus we have

$$\frac{101}{100} \approx 1 + \frac{1}{2} \cdot \frac{v^2}{c^2},$$

and so $\dfrac{v^2}{c^2} = \dfrac{2}{100}$. Consequently, $v \approx \dfrac{\sqrt{2}}{10} c \approx 0.1414c$.

Solution to Exercise 4.5

Let $\epsilon > 0$. As f is differentiable at c, there is a $\delta_1 > 0$ such that whenever $0 < |x - c| < \delta_1$,

$$\left| \frac{f(x) - f(c)}{x - c} - f'(c) \right| < \epsilon.$$

Also, since g is differentiable at c, there is a $\delta_2 > 0$ such that whenever $0 < |x - c| < \delta_2$,

$$\left| \frac{g(x) - g(c)}{x - c} - g'(c) \right| < \epsilon.$$

Let $0 < x - c < \min\{\delta, \delta_1, \delta_2, b - c\}$. Then $0 < |x - c| < \delta_1, 0 < |x - c| < \delta_2$, and for such $x, f(x) = g(x)$, giving

$$|f'(c) - g'(c)| \leq \left| f'(c) - \frac{f(x) - f(c)}{x - c} + \frac{g(x) - g(c)}{x - c} - g'(c) \right|$$

$$\leq \left| f'(c) - \frac{f(x) - f(c)}{x - c} \right| + \left| \frac{g(x) - g(c)}{x - c} - g'(c) \right| < \epsilon + \epsilon = 2\epsilon.$$

As the choice of $\epsilon > 0$ was arbitrary, we obtain $|f'(c) - g'(c)| \leq 0$, that is, $f'(c) = g'(c)$.

Solution to Exercise 4.6

For $x \neq 0$, we have

$$\frac{f(x) - f(0)}{x - 0} = \begin{cases} \dfrac{x^2 - 0^2}{x - 0} & \text{for } x \in \mathbb{Q}, \\ \dfrac{0 - 0^2}{x - 0} & \text{for } x \notin \mathbb{Q} \end{cases} = \begin{cases} x & \text{if } x \in \mathbb{Q}, \\ 0 & \text{if } x \notin \mathbb{Q}. \end{cases}$$

Let $\epsilon > 0$. Set $\delta = \epsilon > 0$. Whenever $x \in \mathbb{R}$ satisfies $0 < |x - 0| < \delta$, we have

$$\left| \frac{f(x) - f(0)}{x - 0} - 0 \right| = \begin{cases} |x - 0| & \text{if } x \in \mathbb{Q}, \\ |0 - 0| = 0 & \text{if } x \notin \mathbb{Q}. \end{cases} < \delta = \epsilon.$$

So f is differentiable at 0. But f can't be differentiable at any nonzero x, since f is not continuous at any nonzero x. The lack of continuity of f at each nonzero x can be seen easily as follows using the sequential characterisation of continuity:

1° If $x \in \mathbb{Q}$, then consider a sequence $(r_n)_{n \in \mathbb{N}}$ of irrational numbers converging to x. If f were continuous at x, then

$$0 = \lim_{n \to \infty} 0 = \lim_{n \to \infty} f(r_n) = f(x) = x^2 \neq 0,$$

a contradiction.

2° If $x \notin \mathbb{Q}$, then consider a sequence $(q_n)_{n \in \mathbb{N}}$ of rational numbers converging to x. If f were continuous at x, then

$$x^2 = \lim_{n \to \infty} q_n^2 = \lim_{n \to \infty} f(q_n) = f(x) = 0,$$

again a contradiction.

Solution to Exercise 4.7

For $h > 0$, small enough, we have

$$\frac{f(c + h) - f(c - h)}{2h} = \frac{f(c + h) - f(c) + f(c) - f(c - h)}{2h}$$

$$= \frac{1}{2} \frac{f(c + h) - f(c)}{h} + \frac{1}{2} \frac{f(c) - f(c - h)}{h}.$$

Since f is differentiable at c, we have

$$\lim_{\substack{h \to 0 \\ h > 0}} \frac{f(c+h) - f(c)}{h} = f'(c) \text{ and } \lim_{\substack{h \to 0 \\ h > 0}} \frac{f(c) - f(c-h)}{h} = \lim_{\substack{h \to 0 \\ h > 0}} \frac{f(c + (-h)) - f(c)}{-h} = f'(c).$$

Consequently, $\displaystyle \lim_{\substack{h \to 0 \\ h > 0}} \frac{f(c+h) - f(c-h)}{2h} = \frac{1}{2} \cdot f'(c) + \frac{1}{2} \cdot f'(c) = f'(c).$

The converse is not true. For example, take f given by $f(x) = |x|$, $x \in \mathbb{R}$, and $c = 0$. Then for $h > 0$, we have

$$\frac{f(c+h) - f(c-h)}{2h} = \frac{|0+h| - |0-h|}{2h} = \frac{h-h}{2h} = 0.$$

Hence $\displaystyle \lim_{\substack{h \to 0 \\ h > 0}} \frac{f(c+h) - f(c-h)}{2h}$ exists (and is $= 0$). But f is not differentiable at $c = 0$.

Solution to Exercise 4.8

If $x \neq 0$, then f is differentiable at x and we have

$$f'(x) = 2x \sin \frac{1}{x} + x^2 \left(\cos \frac{1}{x} \right) \left(-\frac{1}{x^2} \right) = 2x \sin \frac{1}{x} - \cos \frac{1}{x}.$$

On the other hand, we have for $x \neq 0$ that

$$\frac{f(x) - f(0)}{x - 0} = x \sin \frac{1}{x},$$

and so, given $\epsilon > 0$, if we set $\delta = \epsilon > 0$, then for all $x \in \mathbb{R}$ satisfying $0 < |x - 0| = |x| < \delta = \epsilon$,

$$\left| \frac{f(x) - f(0)}{x - 0} - 0 \right| = \left| x \sin \frac{1}{x} \right| = |x| \cdot \left| \sin \frac{1}{x} \right| \leq |x| \cdot 1 < \delta = \epsilon.$$

Consequently, f is differentiable at 0, and its derivative is $f'(0) = 0$.

If f' were continuous, then since the sequence $\left(\frac{1}{2\pi n} \right)_{n \in \mathbb{N}}$ converges to 0, it would follow that also $\left(f'\left(\frac{1}{2\pi n} \right) \right)_{n \in \mathbb{N}}$ converges to $f'(0) = 0$. However, for all $n \in \mathbb{N}$,

$$f'\left(\frac{1}{2\pi n} \right) = \frac{2}{2\pi n} \sin(2\pi n) - \cos(2\pi n) = 0 - 1 = -1.$$

Hence f' is not continuous at 0. (So f is differentiable, but not 'continuously differentiable'.)

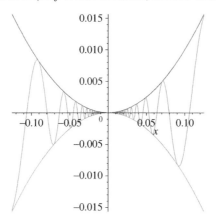

Solution to Exercise 4.9

(B), (C).

Solution to Exercise 4.10

We use the Product Rule repeatedly:

$$(fgh)'(c) = f'(c)(gh)(c) + f(c)(gh)'(c) = f'(c)g(c)h(c) + f(c)(g'(c)h(c) + g(c)h'(c))$$

$$= f'(c)g(c)h(c) + f(c)g'(c)h(c) + f(c)g(c)h'(c).$$

(For example, $(x^3)' = (x \cdot x \cdot x)' = 3 \cdot x' \cdot x \cdot x = 3 \cdot 1 \cdot x^2 = 3x^2$.)

Solution to Exercise 4.11

We have $W(x) = a(x)d(x) - b(x)c(x)$ $(x \in \mathbb{R})$, and so

$$W'(x) = a'(x)d(x) + a(x)d'(x) - b'(x)c(x) - b(x)c'(x)$$

$$= \det \begin{bmatrix} a'(x) & b(x) \\ c'(x) & d(x) \end{bmatrix} + \det \begin{bmatrix} a(x) & b'(x) \\ c(x) & d'(x) \end{bmatrix}.$$

Solution to Exercise 4.12

We have

$$f'(x) = \frac{3x^2}{3} - \frac{2x}{2} - 2 = x^2 - x - 2 = (x+1)(x-2).$$

Hence $\{x \in \mathbb{R} : f'(x) = 0\} = \{-1, 2\}$, and

$$\{x \in \mathbb{R} : (x+1)(x-2) > 0\} = (-\infty, -1) \bigcup (2, \infty).$$

The plots are shown below. We notice that f (solid line) has a maximum and a minimum in $[-3, 3]$ where f' (dashed line) is 0, and f is increasing where $f' > 0$.

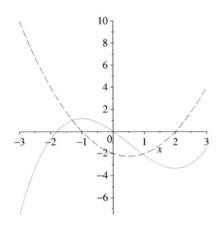

Solution to Exercise 4.13

Let $\alpha \in \mathbb{R}$ be such that $p(x) = (x - \alpha)^2 q(x)$ for some polynomial q. Then $p(\alpha) = 0$, that is,

$$1 + \frac{\alpha}{1!} + \frac{\alpha^2}{2!} + \cdots + \frac{\alpha^n}{n!} = 0. \tag{30}$$

Also, $p'(x) = 2(x - \alpha)q(x) + (x - \alpha)^2 q'(x)$, and so $p'(\alpha) = 0$. But

$$0 = p'(\alpha) = \frac{d}{dx}\left(1 + \frac{x}{1!} + \frac{x^2}{2!} + \cdots + \frac{x^n}{n!}\right)\bigg|_{x=\alpha}$$

$$= 0 + \frac{1}{1!} + \frac{2x}{2!} + \cdots + \frac{nx^{n-1}}{n!}\bigg|_{x=\alpha}$$

$$= 1 + \frac{x}{1!} + \frac{x^2}{2!} + \cdots + \frac{x^{n-1}}{(n-1)!}\bigg|_{x=\alpha}$$

$$= 1 + \frac{\alpha}{1!} + \frac{\alpha^2}{2!} + \cdots + \frac{\alpha^{n-1}}{(n-1)!}. \tag{31}$$

Subtracting (30) and (31), we obtain

$$\frac{\alpha^n}{n!} = 0,$$

and so $\alpha = 0$. But then $p(\alpha) = 1 \neq 0$, a contradiction!

Solution to Exercise 4.14

By the Quotient Rule, we have

$$\cot' x = \frac{d}{dx}\frac{\cos x}{\sin x} = \frac{(-\sin x)(\sin x) - (\cos x)(\cos x)}{(\sin x)^2} = -\frac{1}{(\sin x)^2} = -(\text{cosec } x)^2,$$

$$\sec' x = \frac{d}{dx}\frac{1}{\cos x} = \frac{-(-\sin x)}{(\cos x)^2} = (\tan x)(\sec x),$$

$$\text{cosec}' x = \frac{d}{dx}\frac{1}{\sin x} = \frac{-\cos x}{(\sin x)^2} = -(\cot x)(\text{cosec } x).$$

Solution to Exercise 4.15

We have by the Quotient Rule that

$$\frac{d}{dx}\frac{4\sin x}{2x + \cos x} = \frac{(4\cos x)(2x + \cos x) - 4(\sin x)(2 - \sin x)}{(2x + \cos x)^2}$$

$$= \frac{8x\cos x + 4(\cos x)^2 - 8\sin x + 4(\sin x)^2}{(2x + \cos x)^2}$$

$$= \frac{8x\cos x - 8\sin x + 4}{(2x + \cos x)^2}.$$

Solution to Exercise 4.16

For $x \in (0, \infty)$, we have

$$f'(x) = -\frac{1}{(1+x^2)^2} \cdot 2x = -\frac{2x}{(1+x^2)^2} \neq 0.$$

Hence by the Differentiable Inverse Theorem, f^{-1} is differentiable and for

$$y = f(x) = \frac{1}{1+x^2} \in (0,1) \quad \text{(so that } x = \sqrt{\frac{1}{y} - 1}\text{),}$$

we have

$$(f^{-1})'(y) = \frac{1}{f'(x)} = \frac{1}{-\dfrac{2x}{(1+x^2)^2}} = -\frac{(1+x^2)^2}{2x} = -\frac{(1+1/y-1)^2}{2\sqrt{1/y-1}}$$

$$= -\frac{1/y^2}{2\sqrt{1/y-1}} = -\frac{1}{2y\sqrt{y(1-y)}}.$$

Solution to Exercise 4.17

That f is strictly increasing follows from Corollary 4.10 that we will learn about later: indeed, $f'(x) = 1 + \cos x > 1 - 1 = 0$, where the last inequality follows from the fact that $\cos x > -1$ for all $x \in (-\frac{\pi}{2}, \frac{\pi}{2})$.

For $x \in (-\frac{\pi}{2}, \frac{\pi}{2})$, we have

$$f'(x) = 1 + \cos x \neq 0.$$

Hence by the Differentiable Inverse Theorem, f^{-1} is differentiable and for

$$y = f(x) = x + \sin x,$$

we have

$$(f^{-1})'(y) = \frac{1}{f'(x)} = \frac{1}{1 + \cos x} = \frac{1}{1 + \cos(f^{-1}(y))}$$

As $f(0) = 0 + \sin 0 = 0$, it follows that $f^{-1}(0) = 0$, and so

$$(f^{-1})'(0) = \frac{1}{1 + \cos(f^{-1}(0))} = \frac{1}{1 + \cos 0} = \frac{1}{1 + 1} = \frac{1}{2}.$$

Solution to Exercise 4.18

We have for $x \in \mathbb{R}$ that

$$f'(x) = \Big(\cos(\cos(1 + x^2))\Big) \cdot \Big(-\sin(1 + x^2)\Big) \cdot (2x) = -2x \cdot \sin(1 + x^2) \cdot \cos(\cos(1 + x^2)).$$

Solution to Exercise 4.19

For $x \neq 1$, we have

$$S(x) := \sum_{k=1}^{n} x^k = \frac{(1-x)(x + x^2 + \cdots + x^n)}{1 - x} = \frac{x - x^{n+1}}{1 - x}.$$

Differentiating with respect to x, we obtain

$$\sum_{k=1}^{n} kx^{k-1} = \frac{(1-(n+1)x^n)(1-x) - (x - x^{n+1})(-1)}{(1-x)^2}$$

$$= \frac{1 - (n+1)x^n - x + (n+1)x^{n+1} + x - x^{n+1}}{(1-x)^2}$$

$$= \frac{1 - (n+1)x^n + nx^{n+1}}{(1-x)^2}.$$

Multiplying by x, and differentiating again, we obtain

$$\sum_{k=1}^{n} k^2 x^{k-1} = \frac{2}{(1-x)^3}(x - (n+1)x^{n+1} + nx^{n+2}) + \frac{1}{(1-x)^2}(1 - (n+1)^2 x^n + n(n+2)x^{n+1})$$

$$= \frac{1 + x - (n+1)^2 x^n + (2n^2 + 2n - 1)x^{n+1} - n^2 x^{n+2}}{(1-x)^3}.$$

Multiplying by x again and then substituting $x = 1/2$ yields $\displaystyle\sum_{k=1}^{n} \frac{k^2}{2^k} = 6 - \frac{n^2 + 4n + 6}{2^n}$.

Solution to Exercise 4.20

(1) Putting $x = 3$, we get $4^n = (1+3)^n = \displaystyle\sum_{k=0}^{n} \binom{n}{k} 3^k$, and so $\displaystyle\sum_{k=1}^{n} \binom{n}{k} 3^k = 4^n - 1$.

(2) Differentiating $(1 + x)^n = \displaystyle\sum_{k=0}^{n} \binom{n}{k} x^k$, we get $n(1 + x)^{n-1} = \displaystyle\sum_{k=1}^{n} \binom{n}{k} kx^{k-1}$.

Multiplying by x, we get $nx(1 + x)^{n-1} = \displaystyle\sum_{k=1}^{n} \binom{n}{k} kx^k$, and differentiating we get,

$$n(1 + x)^{n-1} + n(n-1)(1 + x)^{n-2} = \sum_{k=1}^{n} \binom{n}{k} k^2 x^{k-1}.$$

Finally, by setting $x = 1$, we obtain

$$\sum_{k=1}^{n} \binom{n}{k} k^2 = n2^{n-1} + n(n-1)2^{n-2} = n2^{n-2}(2 + n - 1) = n(n+1)2^{n-2}.$$

(3) As $(1 + x^2)^n = \displaystyle\sum_{k=0}^{n} \binom{n}{k} x^{2k}$, $x(1 + x^2)^n = \displaystyle\sum_{k=0}^{n} \binom{n}{k} x^{2k+1}$, and differentiating,

$$1 \cdot (1 + x^2)^n + x \cdot n(1 + x^2)^{n-1} \cdot 2x = \sum_{k=0}^{n} \binom{n}{k} (2k+1)x^{2k}.$$

Setting $x = 1$, we get $\displaystyle\sum_{k=0}^{n} (2k+1) \binom{n}{k} = 2^n + n2^{n-1} \cdot 2 = (n+1)2^n$, and so

$$\sum_{k=1}^{n} (2k+1) \binom{n}{k} = (n+1)2^n - 1.$$

Solution to Exercise 4.21

We have $f(\cdot) + f(-\cdot) \equiv 0$ since f is odd. So its derivative is the zero function too. But on the other hand, using the Chain Rule, we also have that

$$0 = f'(x) \cdot 1 + f'(-x) \cdot (-1) = f'(x) - f'(-x),$$

and so $f'(x) = f'(-x)$ for all $x \in (-a, a)$. Consequently, f' is even.

Solution to Exercise 4.22

Using the Chain Rule, we obtain the following values.

x	$(f \circ g)(x)$	$(f \circ g)'(x)$	$(g \circ f)(x)$	$(g \circ f)'(x)$
0	3	$6 \cdot 3 = 18$	2	$9 \cdot 1 = 9$
1	0	$7 \cdot 9 = 63$	1	$9 \cdot (-9) = -81$
2	2	$(-9) \cdot 9 = -81$	3	$3 \cdot 7 = 21$
3	1	$1 \cdot (-3) = -3$	0	$(-3) \cdot 6 = -18$

Solution to Exercise 4.23

f	f'	$f' \circ f$	$f \circ f'$	$(f \circ f)'$
$1/x^3$	$-3/x^4$	$-3x^{12}$	$-x^{12}/27$	$3x^{12} \cdot (-3/x^4) = 9x^8$
$\cos x$	$-\sin x$	$-\sin(\cos x)$	$\cos(-\sin x)$	$(-\sin(\cos x)) \cdot (-\sin x)$
x^3	$3x^2$	$3x^6$	$27x^6$	$3x^6 \cdot 3x^2 = 9x^8$
3	0	0	3	$0 \cdot 0 = 0$
$3x$	3	3	9	$3 \cdot 3 = 9$

Solution to Exercise 4.24

(1) $g'(x) = f'(x^2) \cdot 2x = 2xf'(x^2)$.

(2) $g'(x) = 2f(x) \cdot f'(x)$.

(3) $g'(x) = f'(f(x)) \cdot f'(x)$.

(4) $g'(x^2) \cdot 2x = f'(x)$, and so for all $x > 0$, we have $g'(x^2) = \dfrac{f'(x)}{2x}$.

With $y := x^2$, we obtain $g'(y) = \dfrac{f'(\sqrt{y})}{2\sqrt{y}}$ for all $y > 0$.

Solution to Exercise 4.25

First we note that $f(1) = 0$ because

$$f(1) = f(1 \cdot 1) = f(1) + f(1).$$

We have

$$f'(c) = \lim_{k \to 1} \frac{f(kc) - f(c)}{kc - c} = \lim_{k \to 1} \frac{1}{c} \cdot \frac{f(k) + f(c) - f(c)}{k - 1}$$

$$= \lim_{k \to 1} \frac{1}{c} \cdot \frac{f(k)}{k - 1} = \lim_{k \to 1} \frac{1}{c} \cdot \frac{f(k) - 0}{k - 1}$$

$$= \frac{1}{c} \lim_{k \to 1} \frac{f(k) - f(1)}{k - 1} = \frac{f'(1)}{c}.$$

So $f'(x) = f'(1)/x$, $x \in (0, \infty)$. So f' is infinitely differentiable, and

$$f^{(n)}(x) = \frac{(-1)^{n-1}(n-1)! f'(1)}{x^n}, \quad x \in (0, \infty).$$

In particular, if $f'(1) = 2$, then $f^{(n)}(3) = \dfrac{(-1)^{n-1}(n-1)! 2}{3^n}$.

Solution to Exercise 4.26

(1) We have

$$(fg)'' = ((fg)')' = (f'g + fg')' = (f'g)' + (fg')'$$
$$= (f')'g + f'g' + f'g' + f(g')'$$
$$= f''g + 2f'g' + fg'',$$

and so the claim follows.

(2) When $n = 1$, we have for $x \in I$ that

$$(fg)'(x) = f'(x)g(x) + f(x)g'(x) = \binom{1}{0} f^{(0)}(x)g^{(1-0)}(x) + \binom{1}{1} f^{(1)}(x)g^{(1-1)}(x),$$

and so the claim is true when $n = 1$.

Suppose that the claim is true for some $n \in \mathbb{N}$. Then for $x \in I$,

$$(fg)^{(n+1)}(x) = \left((fg)^{(n)}\right)'(x) = \left(\sum_{k=0}^{n} \binom{n}{k} f^{(k)} g^{(n-k)}\right)'(x)$$

$$= \sum_{k=0}^{n} \binom{n}{k} \left(f^{(k)} g^{(n-k)}\right)'(x)$$

$$= \sum_{k=0}^{n} \binom{n}{k} \left(f^{(k+1)}(x)g^{(n-k)}(x) + f^{(k)}(x)g^{(n-k+1)}(x)\right).$$

Using the facts that

$$\binom{n}{0} = 1 = \binom{n+1}{0},$$

$$\binom{n}{n} = 1 = \binom{n+1}{n+1},$$

$$\binom{n}{k-1} + \binom{n}{k} = \binom{n+1}{k},$$

$$\sum_{k=0}^{n-1}\binom{n}{k}f^{(k+1)}(x)g^{(n-k)}(x) = \sum_{k=1}^{n}\binom{n}{k-1}f^{(k)}(x)g^{(n-k+1)}(x),$$

we obtain

$$(fg)^{(n+1)}(x) = \sum_{k=0}^{n}\binom{n}{k}f^{(k+1)}(x)g^{(n-k)}(x) + \sum_{k=0}^{n}f^{(k)}(x)g^{(n-k+1)}(x)$$

$$= \sum_{k=0}^{n-1}\binom{n}{k}f^{(k+1)}(x)g^{(n-k)}(x) + \binom{n}{n}f^{(n+1)}(x)g^{(0)}(x)$$

$$+ \binom{n}{0}f^{(0)}(x)g^{(n+1)}(x) + \sum_{k=1}^{n}\binom{n}{k}f^{(k)}(x)g^{(n-k+1)}(x)$$

$$= \binom{n+1}{0}f^{(0)}(x)g^{(n+1)}(x)$$

$$+ \sum_{k=1}^{n-1}\left(\binom{n}{k-1}+\binom{n}{k}\right)f^{(k)}(x)g^{(n-k+1)}(x)$$

$$+ \binom{n+1}{n+1}f^{(n+1)}(x)g^{(0)}(x)$$

$$= \sum_{k=0}^{n+1}\binom{n+1}{k}f^{(k+1)}(x)g^{(n+1-k)}(x).$$

Consequently, the claim follows by induction.

(3) Consider the map $\Theta_x : (0,\infty) \to \mathbb{R}$ given by $\Theta_x(t) = t^x, t > 0$. Then

$$\Theta_x^{(n)}(t) = x(x-1)\cdots(x-n+1)t^{x-n} = x^{[n]}t^{x-n},$$

and so by the first part of this exercise,

$$\Theta_{x+y}^{(n)}(t) = (t^{x+y})^{(n)} = (t^x t^y)^{(n)} = (\Theta_x\Theta_y)^{(n)}(t) = \sum_{k=0}^{n}\binom{n}{k}\Theta_x^{(k)}(t)\Theta_y^{(n-k)}(t),$$

that is, $(x+y)^{[n]}t^{x+y-n} = \sum_{k=0}^{n}\binom{n}{k}x^{[k]}t^{x-k}y^{[n-k]}t^{y-(n-k)}$. Setting $t=1$ gives

$$(x+y)^{[n]} = \sum_{k=0}^{n}\binom{n}{k}x^{[k]}y^{[n-k]}.$$

Solution to Exercise 4.27

(D). Indeed,

$$(f\circ g)'' = ((f\circ g)')' = ((f'\circ g)\cdot g')' = (f'\circ g)'\cdot g' + (f'\circ g)\cdot g''$$

$$= ((f''\circ g)\cdot g')\cdot g' + (f'\circ g)\cdot g''$$

$$= (f''\circ g)\cdot (g')^2 + (f'\circ g)\cdot g''.$$

Solution to Exercise 4.28

(1) We have

$$f'(x) = \frac{-n}{(x-a)^{n+1}} \text{ and } f''(x) = \frac{(-n)(-(n+1))}{(x-a)^{n+2}}.$$

If $f^{(k)}(x) = \dfrac{(-1)^k n(n+1)\cdots(n+(k-1))}{(x-a)^{n+k}}$, then

$$f^{(k+1)}(x) = \frac{(-1)^k n(n+1)\cdots(n+(k-1))(-(n+k))}{(x-a)^{n+k+1}}$$

$$= \frac{(-1)^{k+1} n(n+1)\cdots(n+(k-1))(n+(k+1-1))}{(x-a)^{n+k+1}}.$$

Thus it follows by induction that for all $k \in \mathbb{N}, f^{(k)}(x) = \dfrac{(-1)^k n(n+1)\cdots(n+(k-1))}{(x-a)^{n+k}}$.

(2) We have $f(x) = \dfrac{1}{2}\left(\dfrac{1}{x-1} - \dfrac{1}{x+1}\right)$, and so by the previous part,

$$f^{(k)}x = \frac{1}{2}\left(\frac{(-1)^k 1 \cdot 2 \cdots k}{(x-1)^{k+1}} - \frac{(-1)^k 1 \cdot 2 \cdots k}{(x+1)^{k+1}}\right) = \frac{(-1)^k k!}{2} \cdot \left(\frac{(x+1)^{k+1} - (x-1)^{k+1}}{(x^2-1)^{k+1}}\right).$$

Solution to Exercise 4.29

$f(x) = x^2$ if $x > 0$, and so $f'(x) = 2x = 2|x|$ if $x > 0$. Similarly, $f(x) = -x^2$ if $x < 0$, and so $f'(x) = -2x = 2|x|$ if $x < 0$. Moreover, we have

$$\lim_{x\to 0} \frac{f(x) - f(0)}{x - 0} = \lim_{x\to 0} \frac{x|x| - 0}{x} = \lim_{x\to 0}|x| = |0| = 0.$$

Hence f is differentiable at 0 as well, and $f'(0) = 0 = 2 \cdot 0 = 2|0|$. Consequently, f is differentiable on $\mathbb{R}$, and $f'(x) = 2|x|, x \in \mathbb{R}$.

f' is not differentiable at 0.

$f_n(x) = x^n|x|$ is n times continuously differentiable, but $f^{(n)}(x) = n!|x|, x \in \mathbb{R}$, and so $f^{(n)}$ is not differentiable at 0.

Solution to Exercise 4.30

We must have $f(1) = g(1) = 2$, since the graphs meet at $(1, 2)$. So we have $1 + a + b = 2$ and $1 - c = 2$, that is, $c = -1$ and $a + b = 1$.

The tangent line being the same at $(1, 2)$ implies that $f'(1) = g'(1)$, and so

$$(2x + a)\Big|_{x=1} = (3x^2)\Big|_{x=1},$$

that is, $2 + a = 3$, and so $a = 1$. Using $a + b = 1$ (derived earlier), we obtain $b = 0$. So we have $a = 1, b = 0, c = -1$. Then

$$f(x) = x^2 + x \text{ and}$$

$$g(x) = x^3 + 1.$$

So $f(1) = 2 = g(1)$. Also, $f'(x) = 2x + 1$ and $g'(x) = 3x^2$, so that $f'(1) = 3$ and $g'(1) = 3$. Hence the tangent lines will be the same.

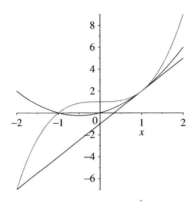

Solution to Exercise 4.31

The point $(1/9, 5)$ corresponds to $t = 3$. (Indeed, $1/t^2 = 1/9$ gives $t = \pm 3$, and as $t > 0$, $t = 3$. We check that then $y(3) = \sqrt{3^2 + 16} = \sqrt{25} = 5$.) We have

$$x'(t) = -\frac{2}{t^3}, \quad x'(3) = -\frac{2}{3^3} = -\frac{2}{27}, \quad \text{and}$$

$$y'(t) = \frac{t}{\sqrt{t^2 + 16}}, \quad y'(3) = \frac{3}{\sqrt{3^2 + 16}} = \frac{3}{5}.$$

So the equation of the tangent at the point $(1/9, 5)$ to the curve is given by

$$\frac{y - 5}{x - \frac{1}{9}} = \frac{y'(3)}{x'(3)} = \frac{3/5}{-2/27} = -\frac{81}{10},$$

that is, $y - 5 = -\frac{81}{10}\left(x - \frac{1}{9}\right)$.

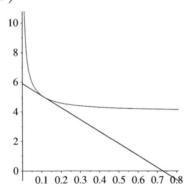

Solution to Exercise 4.32

(C) is true, since the tangent line is given by

$$y = f(c) + f'(c)(x - c),$$

and when $x = 0$, we obtain $y = f(c) - f'(c)c$, and so the tangent line intersects the y-axis at the point $(0, f(c) - f'(c)c)$. That (A),(B),(D) are false can be seen by considering the constant function $f \equiv 1$.

Solution to Exercise 4.33

(1) We want (x_0, y_0) satisfying $x_0^2 + x_0 y_0 + y_0^2 = 9$ such that $\left.\dfrac{dy}{dx}\right|_{x_0} = 0$.

We have $2x + 1 \cdot y + x\dfrac{dy}{dx} + 2y\dfrac{dy}{dx} = 0$, and so $\dfrac{dy}{dx} = \dfrac{-2x - y}{x + 2y}$.

So $2x_0 + y_0 = 0$, that is, $y_0 = -2x_0$ and upon substituting this in $x_0^2 + x_0 y_0 + y_0^2 = 9$ we obtain $x_0^2 + x_0(-2x_0) + (-2x_0)^2 = 9$, that is, $x_0 \in \{\sqrt{3}, -\sqrt{3}\}$. So the points where the tangent is horizontal are $(\sqrt{3}, -2\sqrt{3})$ and $(-\sqrt{3}, 2\sqrt{3})$.

(2) A similar calculation[5] gives $(-2\sqrt{3}, \sqrt{3})$ and $(2\sqrt{3}, -\sqrt{3})$.

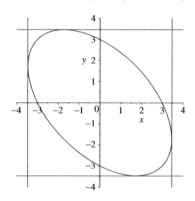

Solution to Exercise 4.34

Using implicit differentiation, we obtain

$$\sin(xy - y^2) + x\cos(xy - y^2) \cdot \left(y + x\frac{dy}{dx} - 2y \cdot \frac{dy}{dx}\right) = 2x. \qquad (32)$$

When $x = y = 1$, this gives $\dfrac{dy}{dx} = -1$.

So the tangent to the curve at $(1, 1)$ is given by $\dfrac{y - 1}{x - 1} = -1$, that is, $x + y = 2$.

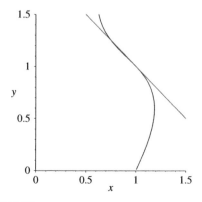

---

[5] or alternately, a symmetry argument

Solution to Exercise 4.35

(1) By Implicit Differentiation,

$$\frac{2}{3}x^{-1/3} + \frac{2}{3}y^{-1/3}\frac{dy}{dx} = 0, \tag{33}$$

and so $\dfrac{dy}{dx} = -\dfrac{y^{1/3}}{x^{1/3}}$. Thus $\dfrac{dy}{dx}\Big|_{x=1} = -\dfrac{y^{1/3}}{x^{1/3}}\Big|_{x=1} = -1.$

From $\dfrac{dy}{dx} = -\dfrac{y^{1/3}}{x^{1/3}}$, it follows using the Quotient Rule that

$$\frac{d^2y}{dx^2} = -\frac{\frac{1}{3}y^{-2/3}\frac{dy}{dx}\cdot x^{1/3} - y^{1/3}\cdot\frac{1}{3}x^{-2/3}}{x^{2/3}}$$

$$= -\frac{\frac{1}{3}y^{-2/3}\left(-\dfrac{y^{1/3}}{x^{1/3}}\right)\cdot x^{1/3} - y^{1/3}\cdot\frac{1}{3}x^{-2/3}.}{x^{2/3}} = \frac{y^{-1/3} + y^{1/3}x^{-2/3}}{3x^{2/3}}.$$

Consequently, $\dfrac{d^2y}{dx^2}\Big|_{x=1} = \dfrac{y^{-1/3} + y^{1/3}x^{-2/3}}{3x^{2/3}}\Big|_{x=1} = \dfrac{2}{3}.$

Alternately, we could have used implicit differentiation one more time in (33) to obtain:

$$\frac{2}{3}\cdot\left(-\frac{1}{3}\right)\cdot x^{-4/3} + \frac{2}{3}\cdot\left(-\frac{1}{3}\right)\cdot y^{-4/3}\frac{dy}{dx}\cdot\frac{dy}{dx} + \frac{2}{3}y^{-1/3}\cdot\frac{d^2y}{dx^2} = 0,$$

that is,

$$x^{-4/3} + y^{-4/3}\left(\frac{dy}{dx}\right)^2 - 3y^{-1/3}\frac{d^2y}{dx^2} = 0.$$

Substituting $x = 1 = y$ and $\dfrac{dy}{dx}\Big|_{x=1} = -1$, we can solve for $\dfrac{d^2y}{dx^2}\Big|_{x=1}$ to obtain $\dfrac{d^2y}{dx^2}\Big|_{x=1} = \dfrac{2}{3}.$

(2) The equation of the tangent line at the point $(1, 1)$ is given by

$$\frac{y - 1}{x - 1} = -1,$$

that is, $x + y = 2$. The plots are displayed in the following picture, showing the expected tangency.

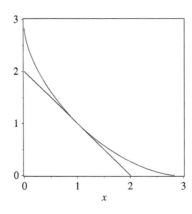

(3) From the picture, we see that as x increases from 1, the slopes of the tangent lines become more horizontal, that is, they *increase* from -1, while if x reduces from the value 1, then the slopes of the tangent lines become steeper, that is they *decrease* from -1. This means that

$$\frac{d^2y}{dx^2} = \frac{d}{dx}\frac{dy}{dx}$$

at $x = 1$ should be positive, and indeed we have found that $\left.\dfrac{d^2y}{dx^2}\right|_{x=1} = \dfrac{2}{3} > 0.$

Solution to Exercise 4.36

By implicit differentiation, we obtain

$$1 \cdot y^3 + x \cdot 3y^2\frac{dy}{dx} + 3x^2y + x^3\frac{dy}{dx} = 0.$$

If there was a horizontal tangent at a point, then

$$\frac{dy}{dx} = 0$$

there, and so $y^3 + 3x^2y = 0$, that is $y(y^2 + 3x^2) = 0$. Consequently, we have either $y = 0$ or $y^2 + 3x^2 = 0$. The latter equation implies that $y = x = 0$, which is impossible since $xy^3 + x^3y = 0 \neq 4$, showing that the point doesn't lie on the curve, a contradiction. Hence $y = 0$, but then again we get the contradiction that $xy^3 + x^3y = 0 \neq 4$. So the tangent line to the curve at any point can't be horizontal.

Solution to Exercise 4.37

With $f(x) = x^2 - 2$, we have $f'(x) = 2x$, and so the update equation in the Newton–Raphson method is

$$x_{n+1} = x_n - \frac{f(x_n)}{f'(x_n)} = x_n - \frac{x_n^2 - 2}{2x_n} = x_n - \frac{x_n}{2} + \frac{1}{x_n} = \frac{x_n}{2} + \frac{1}{x_n}.$$

Taking $x_0 = 1$, we get

$$x_1 = \frac{1}{2} + 1 = \frac{3}{2} = 1.5,$$

$$x_2 = \frac{3}{4} + \frac{2}{3} = \frac{17}{12} = 1.41666\cdots,$$

$$x_3 = \frac{17}{24} + \frac{12}{17} = \frac{577}{408} \approx 1.4142157.$$

Solution to Exercise 4.38

We have with $p(x) := x^4 - x^3 - 75$ that

$$p(3) = 3^4 - 3^3 - 75 = 81 - 27 - 75 = 6 - 27 < 0,$$

$$p(4) = 4^4 - 4^3 - 75 = 256 - 64 - 75 = 256 - 139 > 0.$$

By the intermediate value property, there exists an $x_* \in [3, 4]$ such that $p(x_*) = 0$.

The update equation in the Newton–Raphson method is

$$x_{n+1} = x_n - \frac{f(x_n)}{f'(x_n)}$$

$$= x_n - \frac{x_n^4 - x_n^3 - 75}{4x_n^3 - 3x_n^2} = \frac{4x_n^4 - 3x_n^3 - x_n^4 + x_n^3 + 75}{4x_n^3 - 3x_n^2} = \frac{3x_n^4 - 2x_n^3 + 75}{4x_n^3 - 3x_n^2}.$$

The following table gives the values of x_n obtained for successive values of n:

n	Approximate value of x_n
0	3.5
1	3.2611
2	3.2291
3	3.2286
4	3.2286

(Using MATLAB, one can check that $x_* = 3.228577 \cdots$.)

Solution to Exercise 4.39

We have

$$f'(x) = \begin{cases} \dfrac{1}{2\sqrt{x}} & \text{if } x \geq 0, \\[2mm] \dfrac{1}{2\sqrt{-x}} & \text{if } x < 0. \end{cases}$$

Hence the update equation becomes (assuming $x_n \neq 0$)

$$x_{n+1} = x_n - \frac{f(x_n)}{f'(x_n)} = \begin{cases} x_n - \dfrac{\sqrt{x_n}}{\frac{1}{2\sqrt{x_n}}} & \text{if } x_n > 0, \\[4mm] x_n - \dfrac{-\sqrt{-x_n}}{\frac{1}{2\sqrt{-x_n}}} & \text{if } x_n < 0. \end{cases} = -x_n.$$

This means that if we start with $x_0 \neq 0$, then the sequence of iterates generated by the Newton–Raphson method is $x_0, -x_0, x_0, -x_0, x_0, -x_0, \cdots$, which diverges.

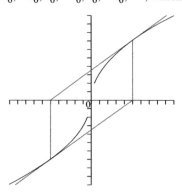

Solution to Exercise 4.40

The distance between $(0, b)$ and a point (x, x^2) on the parabola is given, using, Pythagoras's Theorem, by $d(x) = \sqrt{(x - 0)^2 + (x^2 - b)^2}$, $x \in \mathbb{R}$. We have

$$\lim_{x \to \pm\infty} d(x) = +\infty,$$

and so there exists an $R > 0$ such that whenever $|x| > R$, $d(x) > d(0) = b$. On $[-R, R]$, it follows from the Extreme Value Theorem that the continuous function d has a minimiser, say at $x_* \in [-R, R]$, and in particular $d(x_*) \le d(0) = b$. But for $x \in \mathbb{R}\backslash[-R, R]$, $d(x) > b \ge d(x_*)$. So x_* is a global minimiser. Hence $d'(x_*) = 0$. But

$$d'(x_*) = \frac{1}{2\sqrt{(x_* - 0)^2 + (x_*^2 - b)^2}}(2x_* + 2(x_*^2 - b)2x_*).$$

We have that if $b \in (0, 1/2]$, then $d'(x_*) = 0$ if and only if $x_* = 0$, while if $b > 1/2$, then $d'(x_*) = 0$ if and only if $x_* \in \{0, \sqrt{b - \frac{1}{2}}, -\sqrt{b - \frac{1}{2}}\}$. We have if $b > 1/2$ that

$$d\left(\pm\sqrt{b - \frac{1}{2}}\right) = \sqrt{b - \frac{1}{4}} < b,$$

where the last inequality follows from that fact that $(b - 1/2)^2 > 0$, that is, $b^2 > b - 1/4$ and so $b > \sqrt{b - 1/4}$. Thus the minimum value of the distance is

$$\begin{cases} \sqrt{b - \frac{1}{4}} & \text{if } b \ge \frac{1}{2}, \\ b & \text{if } 0 < b \le \frac{1}{2}. \end{cases}$$

Solution to Exercise 4.41

We have

$$f(x + \pi) = (\sin(x + \pi) - \cos(x + \pi))^2$$
$$= (-\sin x - (-\cos x))^2 = (\sin x - \cos x)^2 = f(x), \quad x \in \mathbb{R},$$

and so f is π-periodic. It is also continuous, and so by the Extreme Value Theorem, it possesses a maximum value on $[0, \pi]$, which serves as the maximum value on $\mathbb{R}$. We have

$$f'(x) = 2(\sin x - \cos x) \cdot (\cos x - (-\sin x)) = 2((\sin x)^2 - (\cos x)^2) = -2\cos(2x), \quad x \in \mathbb{R}.$$

So $f'(x) = 0$ if and only if

$$x \in \left\{(2n + 1)\frac{\pi}{4} : n \in \mathbb{Z}\right\}.$$

Hence these x are candidates for the maximisers. We look for maximisers in $[0, \pi]$. Then the candidates for maximisers are $\pi/4$ and $3\pi/4$. But $f(\pi/4) = 0$, while

$$f(3\pi/4) = \left(\frac{1}{\sqrt{2}} - \left(-\frac{1}{\sqrt{2}}\right)\right)^2 = 2.$$

Thus the maximum value of f is 2.

Solution to Exercise 4.42

Let $O \equiv (0,0)$, $P \equiv (x, y)$ and $Q \equiv (X, Y)$. By Pythagoras's Theorem, the length of OP is $\sqrt{x^2 + y^2}$, and the length of PQ is $\sqrt{(X - x)^2 + (Y - y)^2}$. Hence the total time of travel from $(0,0)$ to (X, Y) is

$$T(y) = \frac{\sqrt{x^2 + y^2}}{1} + \frac{\sqrt{(X - x)^2 + (Y - y)^2}}{\frac{1}{\mu}}.$$

Thus

$$T'(y) = \frac{y}{\sqrt{x^2 + y^2}} - \mu \cdot \frac{Y - y}{\sqrt{(X - x)^2 + (Y - y)^2}} = \sin(\theta_a(y)) - \mu \cdot \sin(\theta_g(y)).$$

Consequently, for the minimising y, we have $T'(y) = 0$, that is,

$$\frac{\sin \theta_a}{\sin \theta_g} = \mu.$$

Solution to Exercise 4.43

Let p be given by $p(x) = Ax^2 + Bx + C$, $x \in \mathbb{R}$, where A, B, C are constants, with $A \neq 0$. Then the slope of the tangent at the midpoint is

$$p'\left(\frac{a + b}{2}\right) = 2Ax + B\Big|_{x = \frac{a+b}{2}} = A \cdot (a + b) + B.$$

On the other hand, the slope of the chord is

$$\frac{p(b) - p(a)}{b - a} = \frac{Ab^2 + Bb + \mathcal{C} - Aa^2 - Ba - \mathcal{C}}{b - a} = A \cdot \frac{b^2 - a^2}{b - a} + B \cdot \frac{b - a}{b - a} = A \cdot (a + b) + B.$$

So the claim follows.

Solution to Exercise 4.44

(B).

Solution to Exercise 4.45

If $a, b \in \mathbb{R}$ and $a < b$, then applying the Mean Value Theorem to the function $\cos \cdot$, we obtain

$$\frac{\cos a - \cos b}{a - b} = -\sin c$$

for some $c \in (a, b)$. But as $|\sin \theta| \leq 1$ for all real θ, it follows that $|\cos a - \cos b| \leq |a - b|$.

Solution to Exercise 4.46

Let us suppose $f(0) \neq 0$. Consider first the case that $f(0) > 0$. Then by the Mean Value Theorem applied to $[-a, 0]$, we have

$$\frac{f(0) - f(-a)}{0 - (-a)} = f'(c)$$

for some c between $-a$ and 0. Hence

$$1 = 0 + 1 < \frac{f(0)}{a} + 1 = \frac{f(0) - (-a)}{a} = \frac{f(0) - f(-a)}{0 - (-a)} = f'(c) \le 1,$$

a contradiction.

Next suppose that $f(0) < 0$. Then by the Mean Value Theorem applied to $[0, a]$, we have

$$\frac{f(a) - f(0)}{a - 0} = f'(d)$$

for some d between 0 and a. Hence

$$1 = 1 - 0 < 1 - \frac{f(0)}{a} = \frac{a - f(0)}{a} = \frac{f(a) - f(0)}{a - 0} = f'(d) \le 1,$$

a contradiction.

Solution to Exercise 4.47

Given $\epsilon > 0$, let $R' > 0$ be such that for all $x > R'$, $|f'(x) - L'| < \epsilon$. Take any $R > R'$ such that for all $x \in \mathbb{R}$ with $x > R$, $|f(x) - L| < \epsilon/2$. Then by the Mean Value Theorem applied to f in $[R + 1, R + 2]$,

$$\frac{f(R + 2) - f(R + 1)}{1} = f'(\xi)$$

for some $\xi \in (R + 1, R + 2)$. So

$$|f'(\xi)| = |f(R + 2) - f(R + 1)| = |f(R + 2) - L + L - f(R + 1)| \le \frac{\epsilon}{2} + \frac{\epsilon}{2} = \epsilon.$$

Hence $|L'| = |L' - f'(\xi) + f'(\xi)| \le |L' - f'(\xi)| + |f'(\xi)| \le \epsilon + \epsilon = 2\epsilon$. As $\epsilon > 0$ was arbitrary, it follows that $L' = 0$.

Solution to Exercise 4.48

Let $x \in (a, b) \setminus \{c\}$. By the Mean Value Theorem applied to the compact interval with endpoints x and c, we have

$$\frac{f(x) - f(c)}{x - c} = f'(c_x),$$

for some point c_x lying between x and c. Note that if the distance of x to c is less than a certain amount δ, then the distance of c_x to c will also be less than the amount δ.

Let $L := \lim_{x \to c} f'(x)$.

Let $\epsilon > 0$. Then there exists a $\delta > 0$ such that for all $x \in (a, b)$ satisfying $0 < |x - c| < \delta$, we have that $|f'(x) - L| < \epsilon$. Hence for $x \in (a, b)$ satisfying $0 < |x - c| < \delta$, we have

$$\left| \frac{f(x) - f(c)}{x - c} - L \right| = |f'(c_x) - L| < \epsilon.$$

So $f'(c)$ exists and equals L.

Solution to Exercise 4.49

If $x, y \in (a, b)$ and $x < y$, then we have by the Mean Value Theorem applied to the interval $[x, y]$ that

$$\frac{f(x) - f(y)}{x - y} = f'(c)$$

for some $c \in (a, b)$. Since $|f'(c)| \leq M$, we obtain that $|f(x) - f(y)| \leq M|x - y|$. Clearly, this is also true if $x = y$, and (by interchanging the roles of x and y) if $x > y$. Hence for all x, y in (a, b), there holds $|f(x) - f(y)| \leq M|x - y|$. Let $\epsilon > 0$. Set $\delta = \epsilon/M$. Then if $x, y \in (a, b)$ and $|x - y| < \delta$, we have $|f(x) - f(y)| \leq M|x - y| < M\delta = \epsilon$. Hence f is uniformly continuous on (a, b).

Solution to Exercise 4.50

Consider f given by

$$f(x) = \frac{c_0}{1}x + \frac{c_1}{2}x^2 + \cdots + \frac{c_d}{d+1}x^{d+1}, \quad x \in \mathbb{R}.$$

Then $f(0) = 0$ and

$$f(1) = \frac{c_0}{1} + \frac{c_1}{2} + \cdots + \frac{c_d}{d+1} = 0.$$

By Rolle's Theorem, $f'(c) = 0$ for some $c \in (0, 1)$. But

$$f'(x) = \frac{c_0}{1} \cdot 1 + \frac{c_1}{2} \cdot 2x + \cdots + \frac{c_d}{d+1} \cdot (d+1)x^d = c_0 + c_1 x + \cdots + c_d x^d,$$

and so the claim follows.

Solution to Exercise 4.51

Let $x_1^{(0)} < \cdots < x_{n+1}^{(0)}$ be such that $f(x_k^{(0)}) = 0$ for all $k = 1, \cdots, n+1$. By Rolle's Theorem applied to each $[x_k^{(0)}, x_{k+1}^{(0)}]$, $k = 1, \cdots, n$, we get the existence of $x_k^{(1)} \in (x_k^{(0)}, x_{k+1}^{(0)})$, $k = 1, \cdots, n$ such that $f'(x_k^{(1)}) = 0$, $k = 1, \cdots, n$. Clearly, $x_1^{(1)} < \cdots < x_n^{(1)}$. By Rolle's Theorem applied to f' on each $[x_k^{(1)}, x_{k+1}^{(1)}]$, $k = 1, \cdots, n-1$, we get the existence of $x_k^{(2)} \in (x_k^{(1)}, x_{k+1}^{(1)})$, $k = 1, \cdots, n-1$ such that $f''(x_k^{(2)}) = 0$, $k = 1, \cdots, n-1$. Proceeding in this manner, we eventually get the existence of an $x_n^{(n)}$ such that $f^{(n)}(x_n^{(n)}) = 0$. See the following picture.

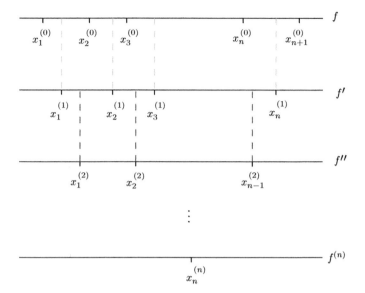

Solution to Exercise 4.52

Consider $f : \mathbb{R} \to \mathbb{R}$ given by $f(x) = x^2 - x\sin x - \cos x$, $x \in \mathbb{R}$. Then

$$f\left(-\frac{\pi}{2}\right) = \frac{\pi^2}{4} + \frac{\pi}{2} \cdot 1 - 0 = \frac{\pi^2}{4} + \frac{\pi}{2} > 0,$$

$$f(0) = 0 - 0 - 1 = -1 < 0,$$

$$f\left(\frac{\pi}{2}\right) = \frac{\pi^2}{4} - \frac{\pi}{2} \cdot 1 - 0 = \frac{\pi(\pi - 2)}{4} > 0.$$

So by the Intermediate Value Theorem, there exists a $c_1 \in (-\pi/2, 0)$ and also there exists a $c_2 \in (0, \pi/2)$ such that $f(c_1) = f(c_2) = 0$.

Suppose that f has more than two distinct roots. Then by Rolle's Theorem, f' would be zero for at least two distinct real numbers. But

$$f'(x) = 2x - 1 \cdot \sin x - x \cdot \cos x - (-\sin x) = x\underbrace{(2 - \cos x)}_{\substack{\neq 0 \\ \text{as } |\cos x| \leq 1}},$$

and so $f'(x) = 0$ if and only if $x = 0$. Consequently, f has exactly two zeros.

Solution to Exercise 4.53

Applying the Mean Value Theorem to $f' : [-1, 0] \to \mathbb{R}$, there exists a $c \in (-1, 0)$ such that

$$\frac{f'(0) - f'(-1)}{0 - (-1)} = \frac{0 - 1/2}{1} = -\frac{1}{2} = f''(c) \geq 0,$$

a contradiction. So a function with the stated properties does not exist.

Solution to Exercise 4.54

We have $|f(x) - f(y)| \leq (x - y)^2 = |x - y|^2$, and so for $x \neq y$ we obtain

$$\left| \frac{f(x) - f(y)}{x - y} - 0 \right| \leq |x - y|.$$

Let $\epsilon > 0$. Then with $\delta := \epsilon > 0$, we have that whenever y satisfies $0 < |x - y| < \delta = \epsilon$, we have

$$\left| \frac{f(x) - f(y)}{x - y} - 0 \right| \leq |x - y| < \delta = \epsilon.$$

Thus $f'(x) = 0$ for all $x \in \mathbb{R}$.

Now suppose that $a, b \in \mathbb{R}$ and that $a < b$. Then by the Mean Value Theorem applied to f on the interval $[a, b]$, we obtain

$$\frac{f(a) - f(b)}{a - b} = f'(c)$$

for some $c \in (a, b)$. But $f'(c) = 0$, and so $f(a) = f(b)$. Thus f is constant.

Solution to Exercise 4.55

Since $x \mapsto f(x + n)$ is differentiable too, it follows that

$$x \mapsto \frac{f(x + n) - f(x)}{n} = f'(x)$$

is differentiable. Also, for all $m \in \mathbb{N}$, we have

$$f'(x + m) = \frac{f(x + m + n) - f(x + m)}{n} = \frac{f(x + m + n) - f(x) + f(x) - f(x + m)}{n}$$

$$= \frac{n + m}{n} \cdot \frac{f(x + m + n) - f(x)}{n + m} - \frac{m}{n} \frac{f(x + m) - f(x)}{m}$$

$$= \frac{n + m}{n} f'(x) - \frac{m}{n} f'(x) = f'(x).$$

Thus $f''(x) = \dfrac{f'(x + n) - f'(x)}{n} = \dfrac{f'(x) - f'(x)}{n} = 0$.

By the Mean Value Theorem applied to f', this gives that f' is a constant, that is, there is a $c \in \mathbb{R}$ such that for all $x \in \mathbb{R}, f'(x) = c$. Applying the Mean Value Theorem to f, we obtain for every nonzero real x that

$$\frac{f(x) - f(0)}{x - 0} = f'(z)$$

for some z between 0 and x. But since f' is constant, it follows that $f(x) = f(0) + cx$ for all $x \in \mathbb{R}$.

Solution to Exercise 4.56

(1) Let $g := y^2 + (y')^2$. Then $g' = 2y \cdot y' + 2y' \cdot y'' = 2y'\underbrace{(y + y'')}_{=0} = 0$.

(2) We have

$$(A \cos x + B \sin x)' = -A \sin x + B \cos x, \text{ and}$$

$$(A \cos x + B \sin x)'' = -A \cos x - B \sin x.$$

Thus $(A \cos x + B \sin x)'' + (A \cos x + B \sin x) = 0$.

If $y = A \cos x + B \sin x$, then

$$y(0) = A \cdot 1 + B \cdot 0 = A, \text{ and}$$

$$y'(0) = -A \sin x + B \cos x \Big|_{x=0} = -A \cdot 0 + B \cdot 1 = B.$$

So if $y = A \cos x + B \sin x$, then $A = y(0)$ and $B = y'(0)$.

Now let y be any solution to $y'' + y = 0$. Set $f(x) := y(x) - y(0) \cos x - y'(0) \sin x$. Then

$$f(0) = y(0) - y(0) \cdot 1 - 0 = 0 \text{ and}$$

$$f'(0) = y'(0) + y(0) \cdot 0 - y'(0) \cdot 1 = 0.$$

Also, $f'' + f = y'' + y + 0 = 0 + 0 = 0$. So f is also a solution. Hence by the previous part,

$$f^2 + (f')^2 = (f(0))^2 + (f'(0))^2 = 0.$$

Consequently, $f = f' \equiv 0$. So $y(x) = y(0) \cos x + y'(0) \sin x$. Done!

(3) Let $y(x) = \sin(\alpha + x)$. Then $y'(x) = (\cos(\alpha + x)) \cdot 1$ and $y''(x) = (-\sin(\alpha + x)) \cdot 1$. So $y''(x) + y(x) = -\sin(\alpha + x) + \sin(\alpha + x) = 0$. Hence by the previous part,

$$\sin(\alpha + x) = y(0) \cos x + y'(0) \sin x = (\sin \alpha)(\cos x) + (\cos \alpha)(\sin x).$$

Setting $x = \beta$, we get $\sin(\alpha + \beta) = (\sin \alpha)(\cos \beta) + (\cos \alpha)(\sin \beta)$.

Next consider $y(x) := \cos(\alpha + x)$. Then $y'(x) = -\sin(\alpha + x)$ and $y''(x) = -\cos(\alpha + x)$. So $y''(x) + y(x) = -\cos(\alpha + x) + \cos(\alpha + x) = 0$. Hence by the previous part,

$$\cos(\alpha + x) = y(0) \cos x + y'(0) \sin x = (\cos \alpha)(\cos x) + (-\sin \alpha)(\sin x).$$

Setting $x = \beta$, we get $\cos(\alpha + \beta) = (\cos \alpha)(\cos \beta) - (\sin \alpha)(\sin \beta)$.

Solution to Exercise 4.57

(1) Suppose that there exist $a, b \in \mathbb{R}$ such that $a < b$ and $f(a) = a, f(b) = b$. Applying the Mean Value Theorem to f on $[a, b]$, we obtain

$$\frac{f(b) - f(a)}{b - a} = \frac{b - a}{b - a} = 1 = f'(c)$$

for some $c \in (a, b) \subset (0, 1)$. But as $f'(c) \neq 1$, we have arrived at a contradiction. Hence if f has a fixed point, then it is unique.

(2) Let $x_1 \in \mathbb{R}$, and set $x_{n+1} := f(x_n)$ for $n \in \mathbb{N}$. We claim that

$$|f(x_{n+1}) - f(x_n)| \le M|x_{n+1} - x_n| \quad (n \in \mathbb{N}). \tag{34}$$

This is clearly true if $x_{n+1} = x_n$. If they are not equal, then we apply the Mean Value Theorem to the interval with endpoints x_{n+1} and x_n to obtain

$$\frac{f(x_{n+1}) - f(x_n)}{x_{n+1} - x_n} = f'(c),$$

for some c between x_{n+1} and x_n. As $|f'(c)| \le M$, this yields the inequality (34).

Thus we have by a repeated application of (34) that

$$|x_{n+1} - x_n| = |f(x_n) - f(x_{n-1})| \le M|x_n - x_{n-1}|$$
$$\le M \cdot M|x_{n-1} - x_{n-2}|$$
$$\cdots$$
$$\le M^{n-1}|x_2 - x_1|.$$

Now consider the series $x_1 + \sum_{n=1}^{\infty}(x_{n+1} - x_n)$.

The estimate $|x_{n+1} - x_n| \le M^{n-1}|x_2 - x_1|$, with $M < 1$, shows that this series converges absolutely, by using the Comparison Test. Since the partial sums of the above series telescope, that is, $x_1 + (x_2 - x_1) + (x_3 - x_2) + \cdots + (x_{n+1} - x_n) = x_{n+1}$, it follows that

$$x_* := \lim_{n \to \infty} x_n$$

exists. Since f is continuous, we also have that

$$f(x_*) = f\left(\lim_{n \to \infty} x_n\right) = \lim_{n \to \infty} f(x_n) = \lim_{n \to \infty} x_{n+1} = x_*.$$

So x_* is a fixed point.

(3) See Figure 4.

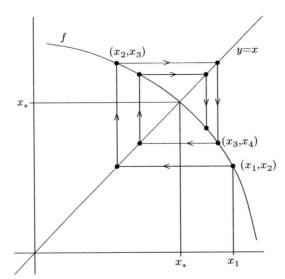

Figure 4.   The zig-zag path $(x_1, x_2) \to (x_2, x_2) \to (x_2, x_3) \to \cdots$.

(4) As $f'(x) = 1 - \dfrac{e^x}{(1 + e^x)^2} = \dfrac{1 + e^{2x} + e^x}{1 + e^{2x} + 2e^x}$, and so $0 < f'(x) < 1$ for all $x \in \mathbb{R}$.

If f had a fixed point x_*, then $f(x_*) = x_*$ would imply that

$$f(x_*) = x_* + \frac{1}{1 + e^{x_*}} = x_*,$$

and so we would have

$$\frac{1}{1 + e^{x_*}} = 0,$$

which is clearly impossible, since $e^{x_*} > 0$.

We observe that although $|f'(x)| < 1$ for all $x \in \mathbb{R}$, it is not uniformly bounded away from 1, that is, there does not exist an $M < 1$ such that for all $x \in \mathbb{R}$, $|f(x)| < M$. Suppose, on the contrary, that such an M exists. Then we would have for all $x \in \mathbb{R}$ that

$$1 - \frac{e^x}{(1 + e^x)^2} < M.$$

In particular, setting $x = -n$ ($n \in \mathbb{N}$), we obtain for all $n \in \mathbb{N}$ that $1 - \dfrac{e^{-n}}{(1 + e^{-n})^2} < M$. Passing the limit as $n \to \infty$ yields

$$1 - \frac{0}{(1 + 0)^2} = 1 \leq M,$$

a contradiction to the fact that $M < 1$. Thus there is no contradiction to the result from part (2).

Solution to Exercise 4.58

Taylor's Formula gives for $x \neq 0$ that there exists a c_x between 0 and x, such that

$$\sin x = \sin 0 + \frac{\cos 0}{1!}x - \frac{\sin 0}{2!}x^2 - \frac{\cos 0}{3!}x^3 + \frac{\sin 0}{4!}x^4 + \frac{\cos c_x}{5!}x^5 = x - \frac{x^3}{3!} + \frac{\cos c_x}{5!}x^5.$$

So for $x \neq 0$, $\left| \dfrac{\sin x - x}{x^3} + \dfrac{1}{6} \right| = \left| \dfrac{\cos c_x}{5!}x^2 \right| \leq \dfrac{1}{5!}|x|^2.$

Since $\lim\limits_{x \to 0} |x|^2 = 0$, it follows that $\lim\limits_{x \to 0} \dfrac{\sin x - x}{x^3} = -\dfrac{1}{6}$.

Solution to Exercise 4.59

(1) From Taylor's Formula, for each x such that $|x - a| \leq \epsilon$, there exists a $c_x \in (a - \epsilon, a + \epsilon)$ such that

$$f(x) = \sum_{k=0}^{n} \frac{f^{(k)}(a)}{k!}(x - a)^k + \frac{f^{(n+1)}(c_x)}{(n + 1)!}(x - a)^{n+1}.$$

For $|x - a| \leq \epsilon$, set $E_n(x) := \dfrac{f^{(n+1)}(c_x)}{(n + 1)!}(x - a)^{n+1}$. Then for $0 < |x - a| < \epsilon$,

$$0 \leq \left| \frac{E_n(x)}{(x - a)^n} \right| = \frac{|f^{(n+1)}(c_x)|}{(n + 1)!} \left| \frac{(x - a)^{n+1}}{(x - a)^n} \right| \leq \frac{M}{(n + 1)!}|x - a|,$$

and so by the Sandwich Theorem, $\lim\limits_{x \to a} \dfrac{E_n(x)}{(x - a)^n} = 0$. Hence the claim follows.

(2) tan is infinitely differentiable on $(-\pi/2, \pi/2)$, and in particular, $\tan^{(4)}$ is continuous. We have for $x \in (-\pi/2, \pi/2)$ that

$$\tan' x = 1 + (\tan x)^2,$$

$$\tan'' x = 2(\tan x)(1 + (\tan x)^2),$$

$$\tan''' x = 2(1 + (\tan x)^2) + 6(\tan x)^2(1 + (\tan x)^2).$$

Thus

$$\tan x = \tan 0 + \frac{\tan' 0}{1!}x + \frac{\tan'' 0}{2!}x^2 + \frac{\tan''' 0}{3!}x^3 + o(x^3)$$

$$= 0 + x + 0 + \frac{x^3}{3} + o(x^3) = x + \frac{x^3}{3} + o(x^3).$$

(3) We have

$$\tan x - x = \frac{x^3}{3} + o(x^3), \quad \sin x = x - \frac{x^3}{6} + o(x^3), \quad \cos x = 1 - \frac{x^2}{2} + o(x^2),$$

as $x \to 0$. Thus

$$\tan x - x = \frac{x^3}{3} + h_1(x), \quad \sin x = x - \frac{x^3}{6} + h_2(x), \quad \cos x = 1 - \frac{x^2}{2} + h_3(x),$$

where $\lim\limits_{x \to 0} \dfrac{h_1(x)}{x^3} = \lim\limits_{x \to 0} \dfrac{h_2(x)}{x^3} = \lim\limits_{x \to 0} \dfrac{h_3(x)}{x^2} = 0$. So

$$\lim_{x \to 0} \frac{\tan x - x}{\sin x - x \cos x} = \lim_{x \to 0} \frac{\frac{x^3}{3} + h_1(x)}{x - \frac{x^3}{6} + h_2(x) - x(1 - \frac{x^2}{2} + h_3(x))}$$

$$= \lim_{x \to 0} \frac{\frac{1}{3} + \frac{h_1(x)}{x^3}}{-\frac{1}{6} + \frac{h_2(x)}{x^3} + \frac{1}{2} + \frac{h_3(x)}{x^2}} = \frac{\frac{1}{3} + 0}{-\frac{1}{6} + 0 + \frac{1}{2} + 0} = 1.$$

Solution to Exercise 4.60

(It is clear visually that all chords lie above the graph of the function $|\cdot|$.) For $x_1, x_2 \in \mathbb{R}$ and $\alpha \in (0, 1)$, we have by the triangle inequality that

$$|(1 - \alpha)x_1 + \alpha x_2| \le |(1 - \alpha)x_1| + |\alpha x_2| = |1 - \alpha||x_1| + |\alpha||x_2| = (1 - \alpha)|x_1| + \alpha|x_2|.$$

Thus $|\cdot|$ is convex.

Solution to Exercise 4.61

('If' part) Suppose that $x_1, x_2 \in I$ and $\alpha \in (0, 1)$. Then we have $(x_1, f(x_1)) \in U(f)$ and $(x_2, f(x_2)) \in U(f)$. Since $U(f)$ is convex, we have that

$$(1 - \alpha) \cdot (x_1, f(x_1)) + \alpha \cdot (x_2, f(x_2)) = (\underbrace{(1 - \alpha) \cdot x_1 + \alpha \cdot x_2}_{=:x \in I}, \underbrace{(1 - \alpha)f(x_1) + \alpha f(x_2)}_{=:y}) \in U(f).$$

Consequently, $(1 - \alpha)f(x_1) + \alpha f(x_2) = y \ge f(x) = f((1 - \alpha) \cdot x_1 + \alpha \cdot x_2)$. Hence f is convex.

('Only if' part) Let $(x_1, y_1), (x_2, y_2) \in U(f)$ and $\alpha \in (0, 1)$. Then we know that $y_1 \ge f(x_1)$ and $y_2 \ge f(x_2)$ and so we also have that

$$(1 - \alpha)y_1 + \alpha y_2 \ge (1 - \alpha)f(x_1) + \alpha f(x_2) \ge f((1 - \alpha) \cdot x_1 + \alpha \cdot x_2),$$

where the last inequality follows from the convexity of f. Consequently,

$$((1 - \alpha) \cdot x_1 + \alpha \cdot x_2, (1 - \alpha)y_1 + \alpha y_2) \in U(f),$$

that is, $(1 - \alpha) \cdot (x_1, f(x_1)) + \alpha \cdot (x_2, f(x_2)) \in U(f)$. So $U(f)$ is convex.

Solution to Exercise 4.62

(A), (B), (D). The reasons are as follows:

(A) $\dfrac{d^2}{dx^2}\dfrac{1}{x} = \dfrac{d}{dx}\left(-\dfrac{1}{x^2}\right) = -\dfrac{-2}{x^3} = \dfrac{2}{x^3} > 0$ for $x > 0$.

(B) $\dfrac{d^2}{dx^2}(-\sin x) = \dfrac{d}{dx}(-\cos x) = -(-\sin x) = \sin x > 0$ for $x \in (0, \pi/2)$.

(C) $f := |x|$ is convex, but $\sqrt{f} = \sqrt{|x|}$ isn't: with $x_1 = 0$, $x_2 = 1$, $\alpha = \dfrac{1}{2}$, we have

$$\sqrt{\frac{0+1}{2}} = \frac{1}{\sqrt{2}} > \frac{1}{2} = \frac{\sqrt{0}+\sqrt{1}}{2}.$$

(D) First, since f is convex, we have for $x_1, x_2 \in \mathbb{R}$ and $\alpha \in (0, 1)$ that

$$f((1-\alpha)x_1 + \alpha x_2) \leq (1-\alpha)f(x_1) + \alpha f(x_2).$$

Furthermore, since f is nonnegative, it follows from the above inequality that

$$(f((1-\alpha)x_1 + \alpha x_2))^2 \leq ((1-\alpha)f(x_1) + \alpha f(x_2))^2. \tag{35}$$

Finally, using the convexity of $t \mapsto t^2$, we have with $t_1 := f(x_1)$ and $t_2 := f(x_2)$ that

$$((1-\alpha)f(x_1) + \alpha f(x_2))^2 = ((1-\alpha)t_1 + \alpha t_2)^2$$
$$\leq (1-\alpha)t_1^2 + \alpha t_2^2$$
$$= (1-\alpha)(f(x_1))^2 + \alpha(f(x_2))^2. \tag{36}$$

From (35) and (36), it follows that f^2 is convex.

Solution to Exercise 4.63

The total length of the path from A to B, $L(x)$, is given (using Pythagoras's Theorem) by $L(x) = \sqrt{a^2 + x^2} + \sqrt{(x-1)^2 + b^2}$, $x \in \mathbb{R}$. We have

$$L'(x) = \frac{x}{\sqrt{a^2 + x^2}} + \frac{x-1}{\sqrt{(x-1)^2 + b^2}}, \quad \text{and}$$

$$L''(x) = \frac{a^2}{(a^2 + x^2)^{3/2}} + \frac{b^2}{((x-1)^2 + b^2)^{3/2}} > 0.$$

Since $L''(x) > 0$ for all x, L is convex. Hence any x_* such that $L'(x_*) = 0$ will be a minimiser of L. $L'(x_*) = 0$ gives

$$\frac{x_*}{\sqrt{a^2 + x_*^2}} = \frac{1 - x_*}{\sqrt{(x_* - 1)^2 + b^2}}.$$

But if $x_* \leq 0$, then the left hand side is ≤ 0, while the right hand side is > 0, and this is impossible. Also, if $x_* \geq 1$, then the left hand side is > 0, while the right hand side is ≤ 0, and this is also not possible. So we must have $x_* \in (0, 1)$. But then it is clear that $\alpha, \beta \in (0, \pi/2)$ and

$$\cos\alpha = \frac{x_*}{\sqrt{a^2 + x_*^2}} = \frac{1 - x_*}{\sqrt{(x_* - 1)^2 + b^2}} = \cos\beta.$$

As $\cos : (0, \pi/2) \to \mathbb{R}$ is strictly decreasing, it follows that it is injective, and so $\alpha = \beta$ for the x_* where L is minimised.

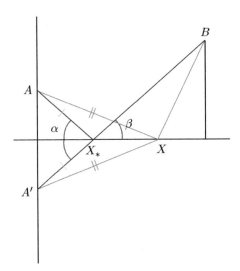

A geometric proof is obtained by reflecting the point $A = (0, a)$ in the x-axis to obtain the point $A' = (0, -a)$, and noticing that if $X = (x, 0)$ and $B = (1, b)$, then

$$L(x) = \ell(AX) + \ell(XB) = \ell(A'X) + \ell(XB),$$

and clearly this is minimised when A, X, B are collinear. Let X_* be the intersection of $A'B$ with the x-axis and $O = (0, 0)$. We have $\alpha = \angle AX_*O = \angle OX_*A' = \angle BX_*X = \beta$.

Solution to Exercise 4.64

We have

$$f'(x) = \sum_{k=1}^{n} 2(x - a_k) = 2nx - 2\sum_{k=1}^{n} a_k,$$

and so $f'(x) = 0$ if and only if $x = \dfrac{1}{n}\sum_{k=1}^{n} a_k$. Also, $f''(x) = 2n > 0$, $x \in \mathbb{R}$.

So f is convex. Hence $x_* := \dfrac{1}{n}\sum_{k=1}^{n} a_k$ is a minimiser of f and it is the only one.

The minimum value of f is

$$f(x_*) = \sum_{k=1}^{n} (x_* - a_k)^2 = nx_*^2 - 2x_* \sum_{k=1}^{n} a_k + \sum_{k=1}^{n} a_k^2$$

$$= n \cdot \frac{1}{n^2}\left(\sum_{k=1}^{n} a_k\right)^2 - 2 \cdot \frac{1}{n}\left(\sum_{k=1}^{n} a_k\right)\left(\sum_{k=1}^{n} a_k\right) + \sum_{k=1}^{n} a_k^2$$

$$= \sum_{k=1}^{n} a_k^2 - \frac{1}{n}\left(\sum_{k=1}^{n} a_k\right)^2.$$

Remark 6. Since $f(x) \geq 0$ for all $x \in \mathbb{R}$, in particular, $f(x_*) \geq 0$, that is,

$$\text{Quadratic mean:} = \sqrt{\frac{1}{n}\sum_{k=1}^{n} a_k^2} \geq \text{Arithmetic mean:} = \frac{1}{n}\sum_{k=1}^{n} a_k.$$

Solution to Exercise 4.65

(1) We have $f'(t) = 2at + b$ and $f''(t) = 2a > 0$. So f is convex. We have $f'(t) = 0$ if and only if $2at + b = 0$, that is, $t = -b/(2a)$. So it follows that $-b/(2a)$ is a minimiser of f, and it is the only one. We have

$$f\left(-\frac{b}{2a}\right) = a\left(-\frac{b}{2a}\right)^2 + b\left(-\frac{b}{2a}\right) + c = \frac{b^2}{4a} - \frac{b^2}{2a} + c = -\frac{b^2}{4a} + c = -\frac{b^2 - 4ac}{4a}.$$

(2) Let $f(t) := \sum_{k=1}^{n} (ta_k - b_k)^2$, $t \in \mathbb{R}$. Then $f(t) \geq 0$. Also,

$$f(t) = \sum_{k=1}^{n}(t^2 a_k^2 - 2a_k b_k t + b_k^2) = \left(\sum_{k=1}^{n} a_k^2\right)t^2 - 2\left(\sum_{k=1}^{n} a_k b_k\right)t + \sum_{k=1}^{n} b_k^2 = at^2 + bt + c,$$

where $a := \left(\sum_{k=1}^{n} a_k^2\right)$, $b := -2\left(\sum_{k=1}^{n} a_k b_k\right)$ and $c := \sum_{k=1}^{n} b_k^2$.

If $a = 0$, then each $a_k = 0$ for $k = 1, \cdots, n$, and so both sides of the inequality are 0, and so the claim holds trivially.

So we may assume that $a > 0$. Then f has the minimum value $-(b^2 - 4ac)/(4a)$. As $f(t) \geq 0$ for all $t \in \mathbb{R}$, we must have that in particular $-(b^2 - 4ac)/(4a) \geq 0$, that is, $b^2 \leq 4ac$. Hence

$$\left(-2\sum_{k=1}^{n} a_k b_k\right)^2 \leq 4\left(\sum_{k=1}^{n} a_k^2\right)\left(\sum_{k=1}^{n} b_k^2\right),$$

and so $\left(\sum_{k=1}^{n} a_k b_k\right)^2 \leq \left(\sum_{k=1}^{n} a_k^2\right)\left(\sum_{k=1}^{n} b_k^2\right)$.

Solution to Exercise 4.66

(1) We prove this using induction on n. The result is trivially true when $n = 1$, and in fact, we have equality in this case. Suppose that the inequality has been established for some $n \in \mathbb{N}$. Now if $x_1, \cdots, x_n, x_{n+1} \in I$, then we have with $t := 1/(n+1) \in (0, 1)$ that

$$f\left(\frac{1}{n+1}(x_1 + \cdots + x_n + x_{n+1})\right) = f\left(\frac{n}{n+1} \cdot \frac{1}{n}(x_1 + \cdots + x_n) + \frac{1}{n+1} \cdot x_{n+1}\right)$$

$$= f\left(\left(1 - \frac{1}{n+1}\right) \cdot \frac{1}{n}(x_1 + \cdots + x_n) + \frac{1}{n+1}x_{n+1}\right)$$

$$= f\left((1-t) \cdot \frac{1}{n}(x_1 + \cdots + x_n) + t \cdot x_{n+1}\right)$$

$$\leq (1-t) \cdot f\left(\frac{1}{n}(x_1 + \cdots + x_n)\right) + t \cdot f(x_{n+1})$$

$$\leq (1-t) \cdot \frac{f(x_1) + \cdots + f(x_n)}{n} + t \cdot f(x_{n+1})$$

$$= \frac{n}{n+1} \cdot \frac{f(x_1) + \cdots + f(x_n)}{n} + \frac{1}{n+1} \cdot f(x_{n+1})$$

$$= \frac{f(x_1) + \cdots + f(x_n) + f(x_{n+1})}{n+1},$$

and so the claim follows for all n.

(2) Let $f : (0, \infty) \to \mathbb{R}$ be defined by $f(x) = -\log x$, $x \in (0, \infty)$. Then

$$f'(x) = -\frac{1}{x} \text{ and } f''(x) = \frac{1}{x^2} \text{ for } x \in (0, \infty).$$

As $f''(x) = 1/x^2 > 0$ for all $x \in (0, \infty)$, it follows that f is convex.

(3) By the first two parts, it follows for $a_1, \cdots, a_n \in (0, \infty)$ that

$$-\log\left(\frac{a_1 + \cdots + a_n}{n}\right) \leq \frac{-\log a_1 - \cdots - \log a_n}{n} = -\frac{1}{n}\log(a_1 \cdots a_n),$$

that is, $\log\left(\dfrac{a_1 + \cdots + a_n}{n}\right) \geq \dfrac{1}{n}\log(a_1 \cdots a_n) = \log \sqrt[n]{a_1 \cdots a_n}$. As $\exp$ is increasing,

$$\frac{a_1 + \cdots + a_n}{n} = \exp \log\left(\frac{a_1 + \cdots + a_n}{n}\right) \geq \exp \log \sqrt[n]{a_1 \cdots a_n} = \sqrt[n]{a_1 \cdots a_n}.$$

Solution to Exercise 4.67

(B). If x is one of the sides, then the other side is $\frac{p}{2} - x$. Thus the area is

$$A(x) = x\left(\frac{p}{2} - x\right) = \frac{px}{2} - x^2,$$

which is a concave function (because $A''(x) = -2 < 0$), and hence A maximised at the x_* satisfying $A'(x_*) = 0$, that is, when the rectangle becomes a square with side length $x_* = p/4$.

The diagonal $d(x)$ satisfies $D(x) := (d(x))^2 = x^2 + \left(\dfrac{p}{2} - x\right)^2$.

Clearly, $D'(x) = 2x - 2(p/2 - x) = 4x - p = 0$ if and only if $x = p/4$ and $D''(x) = 4 > 0$. Consequently, D is convex, and so D is minimised when the rectangle becomes a square, that is, $D(x) = (d(x))^2 \geq D(x_*) = (d(x_*))^2$, and as d is always nonnegative, it follows that $d(x) \geq d(x_*)$.

Solution to Exercise 4.68

We have $y(x) = 2x^3 + 2x^2 - 2x - 1$, and so

$$y'(x) = 6x^2 + 4x - 2 = 6\left(x - \frac{1}{3}\right)(x + 1).$$

Thus $y'(x) > 0$ if and only if $[x > 1/3$ or $x < -1]$, and here y is increasing.

Also, $y'(x) < 0$ if and only if $-1 < x < 1/3$, and here y is decreasing. As

$$y''(x) = 12x + 4 = 12\left(x + \frac{1}{3}\right),$$

we see that $y''(x) > 0$ if and only if $x > -1/3$, where y is convex.

Similarly, $y''(x) < 0$ if and only if $x < -1/3$, where y is concave.

$y''(x) = 0$ if and only if $x = -1/3$, which is a point of inflection.

$y'(x) = 0$ if and only if $[x = 1/3$ or $x = -1]$.

Around $x = -1$, y is concave, and so $x = -1$ is a local maximiser.

Around $x = 1/3$, y is convex, and so $x = 1/3$ is a local minimiser.

$y(0) = -1, y(1) = 2 + 2 - 2 - 1 = 1, y(-1) = -2 + 2 + 2 - 1 = 1,$

$$y\left(-\frac{1}{3}\right) = -\frac{2}{27} + \frac{2}{9} + \frac{2}{3} - 1 = \frac{-2 + 4 + 18 - 27}{27} = -\frac{7}{27},$$

and $y(-2) = -16 + 8 + 4 - 1 < 0$. By the Intermediate Value Theorem, the graph of y crosses the x-axis between -2 and -1, between -1 and 0, and between 0 and 1. So there are at least three distinct zeros.

There can't be any more than 3 distinct zeros, because otherwise by Rolle's Theorem y' would be 0 at least 3 times, but we have seen earlier that y' has only two zeros.

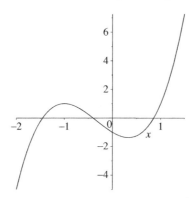

Solution to Exercise 4.69

Let x be the length of the piece from which the square is constructed. Then the length of the side of the square is $x/4$, and that of the equilateral triangle is $(\ell - x)/3$. Thus the total area is

$$A(x) := \left(\frac{x}{4}\right)^2 + \frac{\sqrt{3}}{4}\left(\frac{\ell - x}{3}\right)^2.$$

So $A'(x) = 2 \cdot \frac{x}{4} \cdot \frac{1}{4} + \frac{\sqrt{3}}{4} \cdot 2 \cdot \frac{\ell - x}{3} \cdot \frac{(-1)}{3}$, and $A'(x) = 0$ if and only if $x = \frac{4\ell}{4 + 3\sqrt{3}} =: x_*$.

Also, $A''(x) = \frac{1}{8} + \frac{1}{6\sqrt{3}} > 0$, and so A is convex. Thus x_* is a global minimiser.

Solution to Exercise 4.70

(1) True, since $f' > 0$ on I.

(2) False, since f' is not increasing on I.

(3) True, since $f' > 0$ on I, which implies that f is strictly increasing on I.

(4) True, since f' is strictly concave in a neighbourhood of 0, which implies that $f'''(0) \leq 0$. (Or because f'' is strictly increasing in a neighbourhood of 0 and so $f''' \leq 0$ near 0, and in particular, $f'''(0) \leq 0$.)

(5) There is not enough data to conclude this.

(6) False. It is clear that f' has local maximas, and f'' is zero at these points. If f'' were convex, this would mean that $f'' \leq 0$ between these points (Why? Since if ξ_1, ξ_2 are such that $f''(\xi_1) = f''(\xi_2) = 0$, then $f''((1-t)\xi_1 + t\xi_2) \leq (1-t)f''(\xi_1) + tf''(\xi_2) = 0$), so that f' would have to be decreasing, and this is visibly seen to be false from the given graph of f'.

(7) False. Indeed, $f' > 0$ on I implies that f is strictly increasing, and so f can't have a local maximum anywhere on I.

Solution to Exercise 4.71

We have

$$(x^3 + x - 2)\Big|_{x=1} = 1 + 1 - 2 = 0 \text{ and } (x^2 - 3x + 2)\Big|_{x=1} = 1 - 3 + 2 = 0.$$

So, by l'Hôpital's Rule, we do have

$$\lim_{x \to 1} \frac{x^3 + x - 2}{x^2 - 3x + 2} = \lim_{x \to 1} \frac{3x^2 + 1}{2x - 3},$$

provided that the latter exists. However,

$$(3x^2 + 1)\Big|_{x=1} = 3 \cdot 1 + 1 = 4 \neq 0,$$

and so l'Hôpital's Rule is not applicable a second time. This is the error. In fact,

$$\lim_{x \to 1} \frac{x^3 + x - 2}{x^2 - 3x + 2} = \lim_{x \to 1} \frac{3x^2 + 1}{2x - 3} = \frac{3 \cdot 1^2 + 1}{2 \cdot 1 - 3} = \frac{4}{-1} = -4.$$

Solution to Exercise 4.72

Using l'Hôpital's Rule, we have

$$\lim_{x \to 0} \frac{\sin x - x}{x^3} = \lim_{x \to 0} \frac{\cos x - 1}{3x^2} = \lim_{x \to 0} \frac{-\sin x}{6x} = \lim_{x \to 0} \frac{-\cos x}{6} = \frac{-1}{6}.$$

Solution to Exercise 4.73

If $f(x) := \tan^{-1}x - \pi/3$, $g(x) := x - \sqrt{3}$ then $f(\sqrt{3}) = \tan^{-1}\sqrt{3} - \pi/3 = \pi/3 - \pi/3 = 0$ and $g(\sqrt{3}) = \sqrt{3} - \sqrt{3} = 0$. g is nonzero in $\mathbb{R} \backslash \sqrt{3}$ and $g'(x) = 1 \neq 0$ for all $x \in \mathbb{R}$. Finally,

$$\lim_{x \to \sqrt{3}} \frac{f'(x)}{g'(x)} = \lim_{x \to \sqrt{3}} \frac{1/(x^2 + 1)}{1} = \frac{1}{(\sqrt{3})^2 + 1} = \frac{1}{3 + 1} = \frac{1}{4}.$$

By l'Hôpital's Rule, it follows that $\lim_{x \to \sqrt{3}} \frac{f(x)}{g(x)} = \lim_{x \to \sqrt{3}} \frac{\tan^{-1}x - \pi/3}{x - \sqrt{3}} = \frac{1}{4}$.

Alternately, one could just observe that

$$\lim_{x \to \sqrt{3}} \frac{\tan^{-1}x - \pi/3}{x - \sqrt{3}} = \lim_{x \to \sqrt{3}} \frac{\tan^{-1}x - \tan^{-1}\sqrt{3}}{x - \sqrt{3}} = \frac{d}{dx}\tan^{-1}x\Big|_{x=\sqrt{3}} = \frac{1}{1 + x^2}\Big|_{x=\sqrt{3}} = \frac{1}{4}.$$

Solution to Exercise 4.74

For $x \in [0, \pi^2/4)$, let $f(x) := \tan \sqrt{x}$, $g(x) := \sqrt{x}$. Then $f(0) = \tan 0 = 0$, $g(0) = 0$. Also, f, g are differentiable in $(0, \pi^2/4)$. Moreover,

$$f'(x) := (\sec \sqrt{x})^2 \cdot \frac{1}{2\sqrt{x}},$$

$$g'(x) := \frac{1}{2\sqrt{x}}.$$

Thus $g'(x) \neq 0$ in $(0, \pi^2/4)$, and

$$\lim_{x \to 0+} \frac{f'(x)}{g'(x)} = \lim_{x \to 0+} \frac{(\sec \sqrt{x})^2 \cdot \dfrac{1}{2\sqrt{x}}}{\dfrac{1}{2\sqrt{x}}}$$

$$= \lim_{x \to 0+} (\sec \sqrt{x})^2 = (\sec \sqrt{0})^2 = (\sec 0)^2 = \left(\frac{1}{\cos 0}\right)^2 = 1.$$

Hence by l'Hôpital's Rule, also $\displaystyle \lim_{x \to 0+} \frac{\tan \sqrt{x}}{\sqrt{x}} = \lim_{x \to 0+} \frac{f(x)}{g(x)} = \lim_{x \to 0+} \frac{f'(x)}{g'(x)} = 1.$

Solution to Exercise 4.75

Let

$$g(x) := 1 - x^{3/4},$$

$$f(x) := (2x - x^4)^{1/2} - x^{1/3},$$

$f(1) = 0$ and $g(1) = 0$. Also, f, g are differentiable in a neighbourhood of 1, and

$$g'(x) := -\frac{3}{4}x^{-1/4},$$

$$f'(x) := \frac{1}{2}(2x - x^4)^{-1/2} \cdot (2 - 4x^3) - \frac{1}{3}x^{-2/3}.$$

Thus $g'(1) = -\frac{3}{4}$, and by continuity, $g'(x) \neq 0$ in a neighbourhood of 1. We have

$$\lim_{x \to 1} \frac{f'(x)}{g'(x)} = \lim_{x \to 1} \frac{\frac{1}{2}(2x - x^4)^{-1/2} \cdot (2 - 4x^3) - \frac{1}{3}x^{-2/3}}{-\frac{3}{4}x^{-1/4}} = \frac{-\frac{4}{3}}{-\frac{3}{4}} = \frac{16}{9}.$$

Hence by l'Hôpital's Rule, also

$$\lim_{x \to 1} \frac{(2x - x^4)^{1/2} - x^{1/3}}{1 - x^{3/4}} = \lim_{x \to 1} \frac{f(x)}{g(x)} = \lim_{x \to 1} \frac{f'(x)}{g'(x)} = \frac{16}{9}.$$

Solutions to the exercises from Chapter 5

Solution to Exercise 5.1

(1) False. Partitions are always *finite* sets, while the given set is infinite.

(2) True. For example, for each $n \in \mathbb{N}$,

$$P_n := \left\{ a, a + \frac{b-a}{n}, a + 2\frac{b-a}{n}, \cdots, a + (n-1)\frac{b-a}{n}, b \right\}$$

is a partition of $[a, b]$.

(3) False. We have learnt the definition of partition for a *compact* interval in $\mathbb{R}$.

(4) False. Any partition of $[0, 1]$ must contain the endpoints 0 and 1.

Solution to Exercise 5.2

Let $P = \{x_0 = 0 < x_1 < \cdots < x_{n-1} < x_n = 1\}$ be any partition of $[0, 1]$. Since every interval $[x_k, x_{k+1}]$ contains an irrational number, it follows that the lower sum

$$\underline{S}(f, P) = \sum_{k=0}^{n-1} 0 \cdot (x_{k+1} - x_k) = 0.$$

On the other hand, for every interval $[x_k, x_{k+1}]$, we have an increasing sequence of rational numbers $(q_n)_{n \in \mathbb{N}}$ in $[x_k, x_{k+1}]$ converging to x_{k+1}, and so

$$x_{k+1} = \sup_{n \in \mathbb{N}} q_n = \sup_{n \in \mathbb{N}} f(q_n) \leq \sup_{x \in [x_k, x_{k+1}]} f(x) \leq x_{k+1}.$$

Hence, the upper sum is

$$\overline{S}(f, P) = \sum_{k=0}^{n-1} x_{k+1} \cdot (x_{k+1} - x_k) \geq \sum_{k=0}^{n-1} \frac{x_{k+1} + x_k}{2} \cdot (x_{k+1} - x_k)$$

$$= \frac{1}{2} \sum_{k=0}^{n-1} (x_{k+1}^2 - x_k^2) = \frac{1}{2}(1^2 - 0^2) = \frac{1}{2}.$$

So we have $\overline{S}(f) = \inf_{P \in \mathcal{P}} \overline{S}(f, P) \geq \frac{1}{2} > 0 = \sup_{P \in \mathcal{P}} \underline{S}(f, P) = \underline{S}(f)$. Thus $f \notin RI[0, 1]$.

Solution to Exercise 5.3

We have for $n \geq 2$

$$\overline{S}(f, P_n) = \left(\sup_{x \in [0,1]} f(x) \right) \cdot 1 + \left(\sup_{x \in [1, 1+\frac{1}{n}]} f(x) \right) \cdot \frac{1}{n} + \left(\sup_{x \in [1+\frac{1}{n}, 2]} f(x) \right) \cdot \left(1 - \frac{1}{n} \right)$$

$$= 1 \cdot 1 + 1 \cdot \frac{1}{n} + (-1) \cdot \left(1 - \frac{1}{n} \right) = \frac{2}{n}.$$

Also,

$$\underline{S}(f,P_n) = \left(\inf_{x\in[0,1]} f(x) \right) \cdot 1 + \left(\inf_{x\in[1,1+\frac{1}{n}]} f(x) \right) \cdot \frac{1}{n} + \left(\inf_{x\in[1+\frac{1}{n},2]} f(x) \right) \cdot \left(1 - \frac{1}{n}\right)$$

$$= 1\cdot 1 + (-1)\cdot\frac{1}{n} + (-1)\cdot\left(1 - \frac{1}{n}\right) = 0.$$

Thus

$$0 = \underline{S}(f,P_n) \le \underline{S}(f) \le \overline{S}(f) \le \overline{S}(f,P_n) = \frac{2}{n}$$

for all $n \ge 2$, and so, $\underline{S}(f) = \overline{S}(f) = 0$. Hence $f \in RI[0,2]$ and $\displaystyle\int_0^2 f(x)dx = 0$.

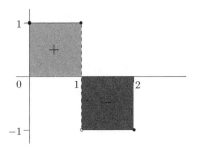

Solution to Exercise 5.4

(A), (C).

Solution to Exercise 5.5

(A), (D). For the falsehood of (B), (C), we note that the inequality $\Phi(P_2) \le \Phi(P_1)$ does *not* imply in general that P_2 is a refinement of P_1. Here is an example. Let $f : [0,1] \to \mathbb{R}$ be defined by

$$f(x) = \begin{cases} 1 & \text{if } x \in [0, 1/2), \\ 0 & \text{if } x \in [1/2, 1]. \end{cases}$$

Let $P_1 := \{0, 1/2, 1\}$ and $P_2 := \{0, 1/3, 2/3, 1\}$. Then $\Phi(P_2) = 1/3 < 1/2 = \Phi(P_1)$. We have

$$\overline{S}(f,P_1) = 1\cdot\frac{1}{2} + 0\cdot\frac{1}{2} = \frac{1}{2}, \text{ and}$$

$$\overline{S}(f,P_2) = 1\cdot\frac{1}{3} + 1\cdot\frac{1}{3} + 0\cdot\frac{1}{3} = \frac{2}{3},$$

and so (B) is false. Also,

$$\underline{S}(f,P_1) = 0\cdot\frac{1}{2} + 0\cdot\frac{1}{2} = 0, \text{ and}$$

$$\underline{S}(f,P_2) = 1\cdot\frac{1}{3} + 0\cdot\frac{1}{3} + 0\cdot\frac{1}{3} = \frac{1}{3},$$

and so (C) is false too.

Solution to Exercise 5.6

Let $\epsilon > 0$. Since $f \in RI[a,c]$, there exists a partition $P_{[a,c]}$ of $[a,c]$ such that

$$\overline{S}(f,P_{[a,c]}) - \underline{S}(f,P_{[a,c]}) < \frac{\epsilon}{2}.$$

Similarly as $f \in RI[c,b]$, there exists a partition $P_{[c,b]}$ of $[c,b]$ such that

$$\overline{S}(f,P_{[c,b]}) - \underline{S}(f,P_{[c,b]}) < \frac{\epsilon}{2}.$$

Let $P_\epsilon := P_{[a,c]} \bigcup P_{[c,b]}$. Then P_ϵ is a partition of $[a,b]$, and clearly

$$\overline{S}(f,P_\epsilon) = \overline{S}(f,P_{[a,c]}) + \overline{S}(f,P_{[c,b]}) \text{ and}$$

$$\underline{S}(f,P_\epsilon) = \underline{S}(f,P_{[a,c]}) + \underline{S}(f,P_{[c,b]}).$$

Thus

$$\overline{S}(f,P_\epsilon) - \underline{S}(f,P_\epsilon) = \overline{S}(f,P_{[a,c]}) - \underline{S}(f,P_{[a,c]}) + \overline{S}(f,P_{[c,b]}) - \underline{S}(f,P_{[c,b]}) < \frac{\epsilon}{2} + \frac{\epsilon}{2} = \epsilon.$$

By the Riemann Condition, it follows that $f \in RI[a,b]$. Also,

$$\underline{S}(f,P_\epsilon) \le \sup_{P \in \mathcal{P}_{[a,b]}} \underline{S}(f,P) = \int_a^b f(x)dx = \inf_{P \in \mathcal{P}_{[a,b]}} \overline{S}(f,P) \le \overline{S}(f,P_\epsilon). \qquad (37)$$

Furthermore,

$$\underline{S}(f,P_{[a,c]}) \le \int_a^c f(x)dx \le \overline{S}(f,P_{[a,c]}) \text{ and}$$

$$\underline{S}(f,P_{[c,b]}) \le \int_c^b f(x)dx \le \overline{S}(f,P_{[c,b]}).$$

Thus

$$\underline{S}(f,P_\epsilon) \le \int_a^c f(x)dx + \int_c^b f(x)dx \le \overline{S}(f,P_\epsilon). \qquad (38)$$

Equations (37) and (38) tell us that the numbers

$$\int_a^b f(x)dx \text{ and } \int_a^c f(x)dx + \int_c^b f(x)dx$$

both lie between the numbers $\underline{S}(f,P_\epsilon)$ and $\overline{S}(f,P_\epsilon)$, and the distance between these latter number is at most ϵ. It thus follows immediately that

$$\left| \int_a^b f(x)dx - \left(\int_a^c f(x)dx + \int_c^b f(x)dx \right) \right| < \epsilon.$$

As $\epsilon > 0$ was arbitrary, we obtain $\int_a^b f(x)dx = \int_a^c f(x)dx + \int_c^b f(x)dx$.

Solution to Exercise 5.7

Let $M > 0$ be such that $|f(x)| \le M$ for all $x \in [a,b]$. Let $\epsilon > 0$. Set $\delta = \frac{\epsilon}{12M}$. (This is obtained by working backward; see the estimates below.) Then as the restrictions of f to $[a,c-\delta]$ and to $[c+\delta,b]$ are continuous, we have $f \in RI[a,c-\delta]$ and $f \in RI[c+\delta,b]$. Thus by the Riemann

Condition, there exist partitions $P_{[a,c-\delta]}$ and $P_{[c+\delta,b]}$ of $[a,c-\delta]$ and $[c+\delta,b]$, respectively, such that

$$\overline{S}(f,P_{[a,c-\delta]}) - \underline{S}(f,P_{[a,c-\delta]}) < \frac{\epsilon}{3}, \quad \text{and } \overline{S}(f,P_{[c+\delta,b]}) - \underline{S}(f,P_{[c+\delta,b]}) < \frac{\epsilon}{3}.$$

Let $P := P_{[a,c-\delta]} \bigcup \{c-\delta, c+\delta\} \bigcup P_{[a,c-\delta]}$. Then P is a partition of $[a,b]$. Also,

$$\overline{S}(f,P) - \underline{S}(f,P) = \overline{S}(f,P_{[a,c-\delta]}) - \underline{S}(f,P_{[a,c-\delta]})$$

$$+ \left(\sup_{x \in [c-\delta,c+\delta]} f(x) - \inf_{x \in [c-\delta,c+\delta]} f(x) \right) \cdot (2\delta)$$

$$+ \overline{S}(f,P_{[c+\delta,b]}) - \underline{S}(f,P_{[c+\delta,b]})$$

$$< \frac{\epsilon}{3} + (2M) \cdot (2\delta) + \frac{\epsilon}{3} < \frac{\epsilon}{3} + 4M \cdot \frac{\epsilon}{12M} + \frac{\epsilon}{3} = \epsilon.$$

By the Riemann Condition, $f \in RI[a,b]$.

Suppose that there are finitely many discontinuities of f, say at the points $c_1, \cdots, c_n$ in (a,b), where $a < c_1 < c_2 < \cdots < c_n < b$. Since $[a, \frac{c_1+c_2}{2}]$ has only one discontinuity (at c_1), it follows from the above that $f \in RI[a, \frac{c_1+c_2}{2}]$. Similarly,

$$f \in RI \left[\frac{c_1+c_2}{2}, \frac{c_2+c_3}{2} \right], \cdots, f \in RI \left[\frac{c_{n-1}+c_n}{2}, b \right].$$

Consequently, $f \in RI[a,b]$ (by Exercise 5.6).

Solution to Exercise 5.8

(1) As $f, g \in RI[a,b]$, also $f - g \in RI[a,b]$ and so the function $|f - g| \in RI[a,b]$. Hence

$$\frac{f + g + |f - g|}{2} \in RI[a,b],$$

that is,

$$x \mapsto \frac{f(x) + g(x) + |f(x) - g(x)|}{2} = \max\{f(x), g(x)\}$$

is in $RI[a,b]$. Thus we conclude that $\max\{f,g\} \in RI[a,b]$.

Since

$$\min\{a,b\} = a + b - \max\{a,b\} \text{ for all } a, b \in \mathbb{R}, \tag{39}$$

it follows that

$$x \mapsto \min\{f(x), g(x)\} = f(x) + g(x) - \max\{f(x), g(x)\} = (f + g - \max\{f,g\})(x)$$

is in $RI[a,b]$. Thus $\min\{f,g\} \in RI[a,b]$.

Finally, we justify the claims in the Hint and (39).

If $a > b$, then $\max\{a,b\} = a$ and $\min\{a,b\} = b = a + b - a = a + b - \max\{a,b\}$.

If $a < b$, then $\max\{a,b\} = b$ and $\min\{a,b\} = a = a + b - b = a + b - \max\{a,b\}$.

If $a = b$, then $\max\{a,b\} = a$ and $\min\{a,b\} = a = 2a - a = a + b - \max\{a,b\}$.

This proves (39).

If $a > b$, then $\max\{a,b\} = a$ and $\dfrac{a + b + |a - b|}{2} = \dfrac{a + b + a - b}{2} = a = \max\{a,b\}$.

If $a < b$, then $\max\{a, b\} = b$ and $\dfrac{a + b + |a - b|}{2} = \dfrac{a + b + b - a}{2} = b = \max\{a, b\}$.

If $a = b$, then $|a - b| = 0$ and $\dfrac{a + b + |a - b|}{2} = \dfrac{a + b}{2} = a = b = \max\{a, b\}$.

(2) Each f_n has only finitely many 'points of discontinuities', at the points $r_1, \cdots, r_n$. It follows that each f_n belongs to $RI[0, 1]$; see Exercise 5.7.

Clearly, if $x = r_m \in [0, 1] \cap \mathbb{Q}$, then $\left(\sup\limits_{n \in \mathbb{N}} f_n \right)(x) = \sup\limits_{n \in \mathbb{N}} f_n(r_m) = 1$.

On the other hand, if $x \in [0, 1] \backslash \mathbb{Q}$, then $\left(\sup\limits_{n \in \mathbb{N}} f_n \right)(x) = \sup\limits_{n \in \mathbb{N}} f_n(x) = \sup\limits_{n \in \mathbb{N}} 0 = 0$.

So $\sup\limits_{n \in \mathbb{N}} f_n$ is the function

$$\mathbb{1}_{[0,1] \cap \mathbb{Q}},$$

which is equal to 1 if its argument belongs to $[0, 1] \cap \mathbb{Q}$, and equal to 0 otherwise. We have seen in Example 5.6 that $\mathbb{1}_{[0,1] \cap \mathbb{Q}}$ is not Riemann integrable.

Solution to Exercise 5.9

Let

$$f(x) = \begin{cases} 1 & \text{if } x \in [0, 1] \cap \mathbb{Q}, \\ -1 & \text{if } x \in [0, 1] \backslash \mathbb{Q} \end{cases} \quad \text{and} \quad g(x) = \begin{cases} -1 & \text{if } x \in [0, 1] \cap \mathbb{Q}, \\ 1 & \text{if } x \in [0, 1] \backslash \mathbb{Q} \end{cases}.$$

Then $f + g \equiv 0, fg \equiv -1, |f| \equiv 1$, and so $f + g, fg, |f| \in RI[0, 1]$.

Let $\mathbb{1}_{[0,1] \cap \mathbb{Q}}$ be the function that is equal to 1 if its argument belongs to $[0, 1] \cap \mathbb{Q}$ and equal to 0 otherwise. Then $f = 2 \cdot \mathbb{1}_{[0,1] \cap \mathbb{Q}} - 1 \notin RI[0, 1]$, since otherwise

$$\mathbb{1}_{[0,1] \cap \mathbb{Q}} = \frac{f + 1}{2} \in RI[0, 1],$$

a contradiction. Similarly $g = -f \notin RI[0, 1]$.

Solution to Exercise 5.10

By the Extreme Value Theorem, for all $x \in [a, b]$,

$$\min_{x \in [a,b]} f(x) =: m \leq f(x) \leq M := \max_{x \in [a,b]} f(x).$$

As ρ is nonnegative, $m\rho(x) \leq \rho(x)f(x) \leq M\rho(x)$ for all $x \in [a, b]$. Hence

$$m \int_a^b \rho(x)dx \leq \int_a^b \rho(x)f(x)dx \leq M \int_a^b \rho(x)dx. \tag{40}$$

If $\displaystyle\int_a^b \rho(x)dx = 0$, then (40) gives $\displaystyle\int_a^b \rho(x)f(x)dx = 0$, and so any $c \in [a, b]$ will do.

If $\displaystyle\int_a^b \rho(x)dx \neq 0$, then by dividing (40) throughout by $\displaystyle\int_a^b \rho(x)dx > 0$, we obtain

$$\min_{x \in [a,b]} f(x) = m \leq \frac{\displaystyle\int_a^b \rho(x)f(x)dx}{\displaystyle\int_a^b \rho(x)dx} \leq M = \max_{x \in [a,b]} f(x).$$

By the Intermediate Value Theorem (applied to the continuous f on the interval with the endpoints taken as the maximiser and minimiser of f on $[a, b]$), it follows that there is a $c \in [a, b]$ such that

$$f(c) = \frac{\int_a^b \rho(x)f(x)dx}{\int_a^b \rho(x)dx},$$

that is, $\int_a^b f(x)\rho(x)dx = f(c) \int_a^b \rho(x)dx$.

That the **continuity of** f cannot be omitted: Take $[a, b] = [0, 1]$, $\rho \equiv 1$, and

$$f(x) = \begin{cases} 0 & \text{if } 0 \leq x \leq \dfrac{1}{2}, \\ 1 & \text{if } \dfrac{1}{2} < x \leq 1. \end{cases}$$

Then $\int_a^b f(x)\rho(x)dx = \int_0^1 f(x)dx = \dfrac{1}{2}$, and $\int_a^b \rho(x)dx = 1$.

But there is no c such that $f(c) = \dfrac{1}{2} = \dfrac{\int_a^b \rho(x)f(x)dx}{\int_a^b \rho(x)dx}$.

That the **nonnegativity of** ρ cannot be omitted: Take $[a, b] = [0, 1]$, $f(x) = x$, and

$$\rho(x) = x - \frac{1}{2} - \epsilon,$$

for $x \in [0, 1]$, where we will specify ϵ later. Then

$$\int_a^b \rho(x)f(x)dx = \int_0^1 x\left(x - \frac{1}{2} - \epsilon\right)dx = \frac{1}{3} - \frac{1}{2}\left(\frac{1}{2} + \epsilon\right) = \frac{1}{12} - \frac{\epsilon}{2},$$

$$\int_a^b \rho(x)dx = \int_0^1 \left(x - \frac{1}{2} - \epsilon\right)dx = -\epsilon.$$

Take $\epsilon = \dfrac{1}{12}$. Then

$$\frac{\int_a^b \rho(x)f(x)dx}{\int_a^b \rho(x)dx} = \frac{\frac{1}{24}}{-\frac{1}{12}} = -\frac{1}{2} < 0,$$

while $f(x) \geq 0$ for all $x \in [0, 1]$.

Solution to Exercise 5.11

(1) We will show by induction on n that the sum of the lengths of the intervals in F_n is $(2/3)^n$. If $n = 0$, then $F_0 = [0, 1]$ has length $1 - 0 = 1 = (2/3)^0$.

Suppose that F_n has length $(2/3)^n$ for some n. Then this F_n has 2^n subintervals, each of length $(1/3)^n$. Since the middle third of each of these intervals is removed in order to get the intervals of F_{n+1}, the sum of the lengths of the resulting intervals is

$$\text{(sum of lengths of the intervals of } F_n) - 2^n \cdot \frac{1}{3} \cdot \left(\frac{1}{3}\right)^n = \left(\frac{2}{3}\right)^n - \frac{1}{3} \cdot \left(\frac{2}{3}\right)^n = \left(\frac{2}{3}\right)^{n+1}.$$

Hence the claim follows for all n by induction.

Since

$$\inf_{P \in \mathcal{P}_{[0,1]}} \overline{S}(1_{F_n}, P) = \overline{S}(1_{F_n}) = \int_0^1 1_{F_n}(x)dx = \left(\frac{2}{3}\right)^n,$$

for $\epsilon > 0$, there exists a partition $P \in \mathcal{P}_{[0,1]}$ such that $\overline{S}(1_{F_n}, P) < \left(\frac{2}{3}\right)^n + \epsilon$. Hence

$$\overline{S}(1_C) = \inf_{P \in \mathcal{P}_{[0,1]}} \overline{S}(1_C, P) \le \inf_{P \in \mathcal{P}_{[0,1]}} \overline{S}(1_{F_n}, P) = \overline{S}(1_{F_n}, P) < \left(\frac{2}{3}\right)^n + \epsilon.$$

Since $\epsilon > 0$ and $n \in \mathbb{N}$ were arbitrary, we obtain $\overline{S}(1_C) \le 0$.

(2) As $1_C \ge 0$, it is clear that $\underline{S}(1_C, P) \ge 0$ for all partitions P of $[0,1]$, and so

$$\underline{S}(1_C) = \sup_{P \in \mathcal{P}_{[0,1]}} \underline{S}(1_C, P) \ge 0.$$

(3) From the Parts (1) and (2) above, we have $0 \le \underline{S}(1_C) \le \overline{S}(1_C) \le 0$, and so

$$\underline{S}(1_C) = 0 = \overline{S}(1_C).$$

Consequently, $1_C \in RI[0,1]$ and $\int_0^1 1_C(x)dx = 0$.

Solution to Exercise 5.12

The indicator function 1_C of the Cantor set $C \subset [0,1]$ is such that $1_C \ge 0$, and moreover $1_C \in RI[0,1]\backslash C[0,1]$ (see Exercise 5.11) with

$$\int_0^1 1_C(x)dx = 0.$$

However, it is not the case that $1_C \equiv 0$.

Solution to Exercise 5.13

Suppose that such a function δ exists. Take any natural number $n > 1$, and $\varphi : [-1,1] \to \mathbb{R}$ by $\varphi(x) = 1, x \in [-1,1]$. Then $\varphi \in RI[-1/n, 1/n]$ for all n, and

$$1 = \varphi(0) = \int_{-1/n}^{1/n} \delta(x)\varphi(x)dx = \int_{-1/n}^{1/n} \delta(x) \cdot 1dx$$

$$\le \left(\sup_{x \in [-1/n, 1/n]} \delta(x)\right)\frac{2}{n} \le \left(\sup_{x \in [-1,1]} \delta(x)\right)\frac{2}{n} \xrightarrow{n \to \infty} 0,$$

a contradiction.

Solution to Exercise 5.14

Let $F(y) := \int_a^y f(t)dt$, $y \in [a, b]$, and $G := F \circ v$, that is,

$$G(x) = F(v(x)) = \int_a^{v(x)} f(t)dt, \quad x \in [a, b].$$

By the Chain Rule and the Fundamental Theorem of Calculus,

$$\frac{d}{dx}\int_a^{v(x)} f(t)dt = G'(x) = F'(v(x)) \cdot v'(x) = f(v(x)) \cdot v'(x), \quad x \in [c, d].$$

Similarly,

$$\frac{d}{dx}\int_a^{u(x)} f(t)dt = f(u(x)) \cdot u'(x), \quad x \in [c, d].$$

Thus

$$\frac{d}{dx}\int_{u(x)}^{v(x)} f(t)dt = \frac{d}{dx}\left(\int_a^{v(x)} f(t)dt - \int_a^{u(x)} f(t)dt\right) = f(v(x)) \cdot v'(x) - f(u(x)) \cdot u'(x),$$

for $x \in [c, d]$.

Solution to Exercise 5.15

We will use Leibniz's Rule for Integrals. We note that the functions

$$t \mapsto \sin(t^2),\ \sin(\sqrt{|t|})$$

are continuous on $\mathbb{R}$, and $x \mapsto 2x, x^2$ are differentiable. Thus for $x \in \mathbb{R}$,

$$F'(x) = \sin((2x)^2) \cdot 2 = 2\sin(4x^2), \quad \text{and}$$

$$G'(x) = \sin(\sqrt{|x^2|}) \cdot 2x = 2x\sin|x|.$$

Solution to Exercise 5.16

(1) Putting $x = 0$, we obtain

$$\int_0^{0^2} f(t)dt = 0 \neq 1 = e^{-0^2},$$

and so no such f exists.

(2) Since $\dfrac{d}{dt}\dfrac{t^3}{3} = t^2$, we have by the Fundamental Theorem of Calculus that

$$\frac{(f(x))^3}{3} - 0 = e^{-x^2}.$$

So $(f(x))^3 = 3e^{-x^2} \geq 0$ for all $x \geq 0$, that is, $f(x) = \sqrt[3]{3}e^{-x^2/3}, x \geq 0$.

(3) By Leibniz's Rule for Integrals, $f(e^{-x^2}) \cdot e^{-x^2} \cdot (-2x) = 2x$. So for $x > 0$,

$$f(e^{-x^2})e^{-x^2} = -1.$$

As f is continuous, and since $\lim\limits_{x\to\infty} e^{-x^2} = 0$, it follows that

$$f(0) \cdot 0 = -1,$$

a contradiction! So no such f exists.

Solution to Exercise 5.17

Using $\sin(\lambda(x - t)) = (\sin(\lambda x))(\cos(\lambda t)) - (\cos(\lambda x))(\sin(\lambda t))$, we obtain

$$y(x) = \frac{\sin(\lambda x)}{\lambda} \int_0^x f(t) \cos(\lambda t)dt - \frac{\cos(\lambda x)}{\lambda} \int_0^x f(t) \sin(\lambda t)dt.$$

Using the Product Rule and the Fundamental Theorem of Calculus, we obtain

$$y'(x) = (\cos(\lambda x)) \int_0^x f(t) \cos(\lambda t)dt + \frac{\sin(\lambda x)}{\lambda} \cancel{f(x) \cos(\lambda x)}$$

$$+ (\sin(\lambda x)) \int_0^x f(t) \sin(\lambda t)dt - \frac{\cos(\lambda x)}{\lambda} \cancel{f(x) \sin(\lambda x)}$$

$$= (\cos(\lambda x)) \int_0^x f(t) \cos(\lambda t)dt + (\sin(\lambda x)) \int_0^x f(t) \sin(\lambda t)dt.$$

Thus

$$y''(x) = -\lambda(\sin(\lambda x)) \int_0^x f(t) \cos(\lambda t)dt + (\cos(\lambda x))f(x) \cos(\lambda x)$$

$$+ \lambda(\cos(\lambda x)) \int_0^x f(t) \sin(\lambda t)dt + (\sin(\lambda x))f(x) \sin(\lambda x)$$

$$= -\lambda \int_0^x f(t) \sin(\lambda(x - t))dt + f(x)((\cos(\lambda x))^2 + (\sin(\lambda x))^2)$$

$$= -\lambda^2 y(x) + f(x) \cdot 1,$$

and so $y''(x) + \lambda^2 y(x) = f(x)$ for all $x \in \mathbb{R}$. Also from the expressions for $y(x)$ and $y'(x)$ above, we see that $y(0) = 0$ and $y'(0) = (\cos 0) \cdot 0 + (\sin 0) \cdot 0 = 0$.

Solution to Exercise 5.18

Suppose that C is a constant such that $V(q) = Cq$ for all q. Then the work done to charge the capacitor to place a charge Q is

$$\int_0^Q V(q)dq = \int_0^Q Cqdq = C \int_0^Q qdq$$

$$= C \int_0^Q \frac{d}{dq} \frac{q^2}{2} dq = C \frac{q^2}{2} \Big|_0^Q = C \frac{Q^2}{2} = Q \frac{CQ}{2} = Q \frac{V(Q)}{2}.$$

Solution to Exercise 5.19

We have, using the Fundamental Theorem of Calculus, that

$$\frac{2^{n+1} - 1}{n + 1} = \frac{1}{n + 1}(1 + x)^{n+1} \Big|_0^1 = \int_0^1 (1 + x)^n dx = \sum_{k=0}^n \binom{n}{k} \int_0^1 x^k dx = \sum_{k=0}^n \binom{n}{k} \frac{1}{k + 1}.$$

Thus $\sum_{k=1}^n \binom{n}{k} \frac{1}{k + 1} = \frac{2^{n+1} - 2 - n}{n + 1}$.

Solution to Exercise 5.20

We have, using the Fundamental Theorem of Calculus and the Chain Rule, that

$$f'(x) = \frac{(\cos x)(\sin x)}{x} \cdot 2x = \sin(2x).$$

As $1 < \pi/2 < 2$, $f'(x) = 0$ in $(1, 2)$ if and only if $x = \pi/2$. Also, $f''(x) = 2\cos(2x) < 0$ in a neighbourhood of $\pi/2$. Hence f is concave in a neighbourhood of $\pi/2$. So f has a local maximiser at $\pi/2$.

Solution to Exercise 5.21

We will use l'Hôpital's Rule. We have for $x > 0$ that

$$0 \le \int_0^x \frac{t^2}{t^6 + 1} dt \le \int_0^x \frac{t^2}{1} dt = \frac{x^3}{3},$$

and so $\lim\limits_{x \to 0+} \int_0^x \frac{t^2}{t^6 + 1} dt = 0$. Also $\lim\limits_{x \to 0+} x^3 = 0$.

By the Fundamental Theorem of Calculus, $\dfrac{d}{dx} \displaystyle\int_0^x \frac{t^2}{t^6 + 1} dt = \frac{x^2}{x^6 + 1}$. We have

$$\lim_{x \to 0+} \frac{1}{3x^2} \frac{d}{dx} \int_0^x \frac{t^2}{t^6 + 1} dt = \lim_{x \to 0+} \frac{1}{3x^2} \frac{x^2}{x^6 + 1} = \lim_{x \to 0+} \frac{1}{3} \frac{1}{x^6 + 1} = \frac{1}{3}.$$

Hence by l'Hôpital's Rule, $\lim\limits_{x \to 0+} \dfrac{1}{x^3} \displaystyle\int_0^x \frac{t^2}{t^6 + 1} dt = \frac{1}{3}$.

Solution to Exercise 5.22

We have

$$\int_1^2 x \log x \, dx = \log x \cdot \frac{x^2}{2} \Big|_1^2 - \int_1^2 \frac{1}{x} \cdot \frac{x^2}{2} dx = (\log 2) \cdot 2 - \frac{1}{2} \int_1^2 x \, dx$$

$$= 2 \log 2 - \frac{1}{2} \cdot \frac{x^2}{2} \Big|_1^2 = 2 \log 2 - \frac{1}{4}(4 - 1) = 2 \log 2 - \frac{3}{4}.$$

Solution to Exercise 5.23

Let $n \in \mathbb{N}$. Then using Integration by Parts, we obtain

$$I(m, n) := \int_0^1 x^m (1 - x)^n dx$$

$$= (1 - x)^n \frac{x^{m+1}}{m + 1} \Big|_0^1 - \int_0^1 n(1 - x)^{n-1}(-1) \frac{x^{m+1}}{m + 1} dx$$

$$= \frac{n}{m + 1} \int_0^1 x^{m+1} (1 - x)^{n-1} dx = \frac{n}{m + 1} I(m + 1, n - 1).$$

Consequently, for $n \in \mathbb{N}$,

$$I(m,n) = \frac{n}{m+1}I(m+1, n-1) = \frac{n}{m+1} \cdot \frac{n-1}{m+2}I(m+2, n-2) = \cdots$$

$$= \frac{n(n-1)(n-1)\cdots 1}{(m+1)(m+2)\cdots(m+n)}I(m+n, 0)$$

$$= \frac{n!}{(m+1)(m+2)\cdots(m+n)}\int_0^1 x^{m+n}dx$$

$$= \frac{n!}{(m+1)(m+2)\cdots(m+n)} \cdot \frac{1}{m+n+1} = \frac{n!m!}{(m+n+1)!}.$$

Solution to Exercise 5.24

We use Integration by Parts and the Fundamental Theorem of Calculus:

$$\int_a^x (x-u)f(u)du = (x-u)\int_a^u f(t)dt\Big|_a^x - \int_a^x (-1)\int_a^u f(t)dtdu$$

$$= 0 - (x-a)\cdot 0 + \int_a^x \left(\int_a^u f(t)dt\right)du = \int_a^x \left(\int_a^u f(t)dt\right)du.$$

Solution to Exercise 5.25

We use induction on n. If $n = 0$, then

$$\frac{1}{0!}\int_a^b (b-t)^0 f^{(0+1)}(t)dt = \int_a^b f'(t)dt = f(b) - f(a),$$

by the Fundamental Theorem of Calculus.

Suppose that the formula has been established for some n. Let $f : [a,b] \to \mathbb{R}$ be such that $f', \cdots, f^{(n+2)}$ exist and $f^{(n+2)} \in C[a,b]$. Then

$$f(b) - \left(f(a) + f'(a)(b-a) + \cdots + \frac{f^{(n)}(a)}{n!}(b-a)^n\right)$$

$$= \frac{1}{n!}\int_a^b (b-t)^n f^{(n+1)}(t)dt$$

$$= \frac{1}{n!}\left(f^{(n+1)}(t)\frac{(b-t)^{n+1}(-1)}{n+1}\Big|_a^b - \int_a^b f^{(n+2)}(t)\frac{(b-t)^{n+1}(-1)}{n+1}\right)$$

$$= \frac{1}{n!}\left(0 + f^{(n+1)}(a)\frac{(b-a)^{n+1}}{n+1} + \frac{1}{n+1}\int_a^b (b-t)^{n+1}f^{(n+2)}(t)dt\right)$$

$$= \frac{f^{(n+1)}(a)}{(n+1)!}(b-a)^{n+1} + \frac{1}{(n+1)!}\int_a^b (b-t)^{n+1}f^{(n+2)}(t)dt.$$

This completes the proof.

Remark 7. In the version of the Taylor's Formula we had learnt earlier, the remainder term was

$$\frac{f^{(n+1)}(c)}{(n+1)!}(b-a)^{n+1},$$

which is somewhat unsatisfactory, as the earlier result just said that such a $c \in (a, b)$ *exists*, without actually telling anything more about it. On the other hand, the present version of the Taylor's Formula has the remainder

$$\frac{1}{n!} \int_a^b (b - t)^n f^{(n+1)}(t) dt,$$

is explicit, and it can be determined, knowing $f^{(n+1)}$.

Solution to Exercise 5.26

In (1), the integrand is odd, and the interval of integration $[-1, 1]$ is symmetric about the origin, so that

$$\int_{-1}^1 x^3 \sqrt{1 - x^2} dx = 0.$$

For (2), we note that as $x \mapsto x^3 \sqrt{1 - x^2}$ is odd and $x \mapsto \sqrt{1 - x^2}$ is even, we obtain

$$\int_{-1}^1 (x^3 + 9) \sqrt{1 - x^2} dx = \int_{-1}^1 x^3 \sqrt{1 - x^2} dx + 9 \int_{-1}^1 \sqrt{1 - x^2} dx = 0 + 9 \cdot \frac{\pi}{2} = \frac{9\pi}{2}.$$

Solution to Exercise 5.27

We use Integration by Substitution/Change of Variables:

$$u = 1 - 4x^2,$$

$$du = -8x dx,$$

$$x = 0 \implies u = 1,$$

$$x = 1/4 \implies u = 3/4.$$

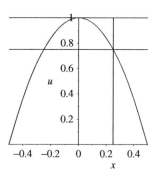

We have

$$\int_0^{\frac{1}{4}} \frac{x}{\sqrt{1 - 4x^2}} dx = \int_1^{\frac{3}{4}} \frac{-1/8}{\sqrt{u}} du = \frac{1}{4} \int_{\frac{3}{4}}^1 \frac{1}{2\sqrt{u}} du$$

$$= \frac{1}{4} \sqrt{u} \Big|_{\frac{3}{4}}^1 = \frac{1}{4} \left(\sqrt{1} - \sqrt{\frac{3}{4}} \right) = \frac{1}{4} \left(1 - \frac{\sqrt{3}}{2} \right).$$

Solution to Exercise 5.28

We use Integration by Substitution/Change of Variables:

$$u = x^{5/4} + 1,$$

$$du = (5/4)x^{1/4}dx,$$

$$x = 0 \implies u = 1,$$

$$x = 16 \implies u = 33.$$

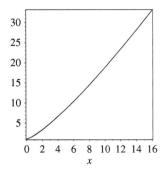

We have

$$\int_0^{16} \sqrt[4]{x} \cdot \sqrt{\sqrt[4]{x^5} + 1}\,dx = \int_1^{33} \frac{4}{5}\sqrt{u}\,du$$

$$= \frac{4}{5}\frac{1}{\frac{1}{2}+1}u^{\frac{1}{2}+1}\Big|_1^{33} = \frac{4}{5}\cdot\frac{2}{3}(33^{3/2} - 1) = \frac{8(33^{3/2} - 1)}{15}.$$

Solution to Exercise 5.29

We have for $x \in \mathbb{R}$ that by Domain Additivity

$$F(x) := \int_x^{x+T} f(t)dt = \int_0^{x+T} f(t)dt - \int_0^x f(t)dt.$$

But

$$\int_0^{x+T} f(t)dt = \int_{-T}^x f(u+T)du \quad \text{(using the substitution } u = t - T)$$

$$= \int_{-T}^x f(u)du.$$

So $F(x) = \int_{-T}^x f(t)dt - \int_0^x f(t)dt = \int_{-T}^0 f(t)dt.$

Thus F is constant, and in particular, $F(a) = F(0)$ for all $a \in \mathbb{R}$, that is,

$$\int_a^{a+T} f(t)dt = \int_0^T f(t)dt$$

for all $a \in \mathbb{R}$.

Solution to Exercise 5.30

We use Integration by Substitution/Change of Variables:

$$u = \cos t,$$
$$du = -\sin t \, dt,$$
$$t = 0 \Rightarrow u = 1,$$
$$t = x \Rightarrow u = \cos x.$$

We have for $0 \leq x < \dfrac{\pi}{2}$ that

$$\int_0^x \tan t \, dt = \int_0^x \frac{\sin t}{\cos t} dt = \int_1^{\cos x} -\frac{1}{u} du$$

$$= -\log u \Big|_1^{\cos x} = -\log(\cos x) - 0 = \log(\sec x).$$

Solution to Exercise 5.31

We use Integration by Substitution/Change of Variables:

$$u = \cos x,$$
$$du = -\sin x \, dx,$$
$$x = 0 \Rightarrow u = 1,$$
$$x = \pi/2 \Rightarrow u = 0.$$

We have

$$\int_0^{\pi/2} (\sin x) \exp(\cos x) \, dx = \int_1^0 -\exp u \, du = \int_0^1 \exp u \, du = \exp u \Big|_0^1 = e - 1,$$

where $e := \exp 1$.

Solution to Exercise 5.32

We use Integration by Substitution/Change of Variables:

$$u = 3x^2,$$
$$du = 6x \, dx,$$
$$x = 0 \Rightarrow u = 0,$$
$$x = 1 \Rightarrow u = 3.$$

We have $\displaystyle \int_0^1 2x \exp(3x^2) \, dx = \int_0^3 \frac{1}{3} \exp u \, du = \frac{1}{3} \exp u \Big|_0^3 = \frac{(\exp 3) - 1}{3}.$

Solution to Exercise 5.33

We have

$$\sum_{k=1}^n \frac{1}{\sqrt{k^2 + n^2}} = \sum_{k=1}^n \frac{1}{\sqrt{(\frac{k}{n})^2 + 1}} \frac{1}{n} = \sum_{k=1}^n f\left(\frac{k}{n}\right) \left(\frac{k}{n} - \frac{k-1}{n}\right),$$

where $f(x) := \dfrac{1}{\sqrt{x^2 + 1}}$, $x \in [0, 1]$. So $\displaystyle\sum_{k=1}^{n} \dfrac{1}{\sqrt{k^2 + n^2}} = S(f, P_n)$, where

$$P_n := \left\{ 0, \frac{1}{n}, \frac{2}{n}, \frac{3}{n}, \cdots, \frac{n}{n} \right\},$$

$$\xi_k = \frac{k+1}{n}, \quad k = 0, 1, 2, 3, \cdots, n-1.$$

As $\displaystyle\lim_{n\to\infty} S(f, P_n) = \int_0^1 f(x)dx$, we obtain $\displaystyle\lim_{n\to\infty} \sum_{k=1}^{n} \frac{1}{\sqrt{k^2 + n^2}} = \int_0^1 \frac{1}{\sqrt{x^2 + 1}} dx.$

Solution to Exercise 5.34

We have

$$\sum_{k=0}^{n-1} \frac{n}{n^2 + k^2} = \sum_{k=0}^{n-1} \frac{1}{n} \cdot \frac{1}{1 + (\frac{k}{n})^2} = S\left(\frac{1}{1+x^2}, P_n\right),$$

where

$$P_n := \left\{ 0, \frac{1}{n}, \frac{2}{n}, \frac{3}{n}, \cdots, \frac{n}{n} \right\},$$

$$\xi_k = \frac{k}{n}, \quad k = 0, 1, 2, 3, \cdots, n-1.$$

Hence

$$\lim_{n\to\infty} \sum_{k=0}^{n-1} \frac{n}{n^2 + k^2} = \lim_{n\to\infty} S\left(\frac{1}{1+x^2}, P_n\right) = \int_0^1 \frac{1}{1+x^2} dx$$

$$= \tan^{-1}x \Big|_0^1 = \frac{\pi}{4} - 0 = \frac{\pi}{4}.$$

Solution to Exercise 5.35

We have

$$\sum_{k=1}^{n} \frac{1}{\sqrt{n^2 + kn}} = \sum_{k=1}^{n} \frac{1}{n\sqrt{1 + \frac{k}{n}}} = S\left(\frac{1}{\sqrt{1+x}}, P_n\right),$$

where

$$P_n := \left\{ 0, \frac{1}{n}, \frac{2}{n}, \frac{3}{n}, \cdots, \frac{n}{n} \right\},$$

$$\xi_k = \frac{k+1}{n}, \quad k = 0, 1, 2, 3, \cdots, n-1.$$

So $\displaystyle\lim_{n\to\infty} \sum_{k=1}^{n} \frac{1}{\sqrt{n^2 + kn}} = \lim_{n\to\infty} S\left(\frac{1}{\sqrt{1+x}}, P_n\right) = \int_0^1 \frac{1}{\sqrt{1+x}} dx = 2\sqrt{1+x} \Big|_0^1 = 2\sqrt{2} - 2.$

Solution to Exercise 5.36

As an example, here we have given code written using MATLAB, but the student might prefer using some other computer package.

```
a=0;                    % right interval endpoint
b=10;                   % left interval endpoint
n=10000;                % number of points in partition
dx=(b-a)/n;             % fineness of the partition
S=0;                    % "initial value" of Riemann sum
for k=0:n-1,            % "for loop" to calculate Riemann sum
    x=a+k*dx;           % current value of x
    y=exp(-x^2);        % value of function at current x
    S=S+y*dx;           % incrementing the Riemann sum by f(x)dx
end                     % end of "for loop"
S                       % display the value of the Riemann sum
```

Running this program in MATLAB gives $\int_0^{10} e^{-x^2} \, dx \approx 0.8867$.

Imagining this to be $\dfrac{\sqrt{\pi}}{2}$, we obtain $\pi \approx (2 \cdot 0.8867)^2 = 3.1451$.

Solution to Exercise 5.37

We have

$$\int_9^y \frac{1}{(x-3)^2} \, dx = -\frac{1}{x-3}\Big|_9^y = -\frac{1}{y-3} + \frac{1}{6} \xrightarrow{y \to \infty} 0 + \frac{1}{6} = \frac{1}{6}.$$

Hence $\int_9^\infty \frac{1}{(x-3)^2} \, dx = \frac{1}{6}$.

Solution to Exercise 5.38

(1) For $n > 1$,

$$\int_0^n \frac{1}{\sqrt{1+x^3}} dx = \int_0^1 \frac{1}{\sqrt{1+x^3}} dx + \int_1^n \frac{1}{\sqrt{1+x^3}} dx$$

$$\leq \int_0^1 \frac{1}{\sqrt{1+x^3}} dx + \int_1^n \frac{1}{\sqrt{x^3}} dx \leq \int_0^1 \frac{1}{\sqrt{1+x^3}} dx + \int_1^\infty \frac{1}{x^{3/2}} dx < \infty.$$

So $\left(\int_0^n \frac{1}{\sqrt{1+x^3}} dx \right)_{n \in N}$ is bounded, and it is clearly increasing. Thus

$$\int_0^\infty \frac{1}{\sqrt{1+x^3}} dx = \lim_{n \to \infty} \int_0^n \frac{1}{\sqrt{1+x^3}} dx$$

exists.

(2) For $n > 1$, we have

$$\int_0^n \frac{x}{1+x^{3/2}}dx = \int_0^1 \frac{x}{1+x^{3/2}}dx + \int_1^n \frac{x}{1+x^{3/2}}dx$$

$$\geq \int_0^1 \frac{x}{1+x^{3/2}}dx + \int_1^n \frac{x}{x^{3/2}+x^{3/2}}dx \geq \int_0^1 \frac{x}{1+x^{3/2}}dx + \frac{1}{2}\int_1^n \frac{1}{\sqrt{x}}dx.$$

Since $\left(\int_1^n \frac{1}{\sqrt{x}}dx\right)_{n \in \mathbb{N}}$ diverges, $\left(\int_0^n \frac{x}{1+x^{3/2}}dx\right)_{n \in \mathbb{N}}$ is unbounded.

Hence $\int_0^\infty \frac{x}{1+x^{3/2}}dx$ does not exist.

Solution to Exercise 5.39

(1) We have

$$\Gamma(1) = \int_{0+}^\infty e^{-t}t^{1-1}dt = \int_{0+}^\infty e^{-t}dt = \int_{0+}^1 e^{-t}dt + \lim_{x \to \infty}\int_1^x e^{-t}dt$$

$$= \frac{e^{-1}-e^0}{-1} + \lim_{x \to \infty}\frac{e^{-x}-e^{-1}}{-1} = -\frac{1}{e} + 1 - 0 + \frac{1}{e} = 1.$$

(2) We have

$$\Gamma(s+1) = \lim_{\epsilon \searrow 0}\int_\epsilon^1 e^{-t}t^{s+1-1}dt + \lim_{x \to \infty}\int_1^x e^{-t}t^{s+1-1}dt$$

$$= \lim_{\epsilon \searrow 0}\int_\epsilon^1 e^{-t}t^s dt + \lim_{x \to \infty}\int_1^x e^{-t}t^s dt$$

Using Integration by Parts, we have

$$\int_\epsilon^1 e^{-t}t^s dt = -t^s e^{-t}\Big|_\epsilon^1 - \int_\epsilon^1 st^{s-1}(-e^{-t})dt = \epsilon^s e^{-\epsilon} - \frac{1}{e} + s\int_\epsilon^1 e^{-t}t^{s-1}dt.$$

As $0 \leq \epsilon^s e^{-\epsilon} \leq \epsilon^s \cdot 1 \xrightarrow{\epsilon \searrow 0} 0$ (as $s > 0$), we obtain

$$\lim_{\epsilon \searrow 0}\int_\epsilon^1 e^{-t}t^s dt = -\frac{1}{e} + s \cdot \lim_{\epsilon \searrow 0}\int_\epsilon^1 e^{-t}t^{s-1}dt. \tag{41}$$

Similarly, using Integration by Parts, we have

$$\int_1^x e^{-t}t^s dt = -t^s e^{-t}\Big|_1^x - \int_1^x st^{s-1}(-e^{-t})dt = \frac{1}{e} - x^s e^{-x} + s\int_1^x e^{-t}t^{s-1}dt.$$

Choose $n \in \mathbb{N}$ such that $n > s$. Then for $x > 0$,

$$e^x = 1 + \frac{x}{1!} + \frac{x^2}{2!} + \cdots \geq \frac{x^n}{n!},$$

and so $e^{-x}x^s x^{n-s} \leq n!$, which gives

$$0 \leq e^{-x}x^s \leq \frac{n!}{x^{n-s}} \xrightarrow{x \nearrow \infty} 0 \quad (\text{as } n > s),$$

and so $\lim_{x \to \infty} x^s e^{-x} = 0$. Thus

$$\lim_{x \to \infty} \int_1^x e^{-t} t^s \, dt = \frac{1}{e} - 0 + s \cdot \lim_{x \to \infty} \int_1^x e^{-t} t^{s-1} \, dt. \tag{42}$$

From (41) and (42), it follows that

$$\Gamma(s+1) = -\frac{1}{e} + s \cdot \lim_{\epsilon \searrow 0} \int_\epsilon^1 e^{-t} t^{s-1} \, dt + \frac{1}{e} + s \cdot \lim_{x \to \infty} \int_1^x e^{-t} t^{s-1} \, dt$$

$$= s \left(\lim_{\epsilon \searrow 0} \int_\epsilon^1 e^{-t} t^{s-1} \, dt + \lim_{x \to \infty} \int_1^x e^{-t} t^{s-1} \, dt \right) = s \left(\int_{0+}^\infty e^{-t} t^{s-1} \, dt \right) = s\Gamma(s).$$

(3) We have $\Gamma(2) = \Gamma(1+1) = 1 \cdot \Gamma(1) = 1 = 1!$, and so the claim holds for $n = 1$. Suppose that for some $k \in \mathbb{N}$, $\Gamma(k+1) = k!$. Then

$$\Gamma((k+1)+1) = (k+1) \cdot \Gamma(k+1) = (k+1) \cdot k! = (k+1)! \,.$$

So the result follows by induction.

Solution to Exercise 5.40

Consider the function f given by

$$f(x) = \begin{cases} n & \text{if } x \in \left[n, \, n + \dfrac{1}{n^3}\right], \, n \geq 2, \\[2mm] 0 & \text{if } x \in \mathbb{R} \setminus \bigcup_{n \geq 2} \left[n, \, n + \dfrac{1}{n^3}\right]. \end{cases}$$

Then $f : [0, \infty) \to [0, \infty)$, and moreover

$$\int_0^\infty f(x)\,dx = \sum_{n \geq 2} n \cdot \frac{1}{n^3} = \sum_{n \geq 2} \frac{1}{n^2} < +\infty.$$

But $f(n) = n$, $n \geq 2$, and so clearly we do not have that $\lim_{x \to \infty} f(x) = 0$.

By the Fundamental Theorem of Calculus, $f(x) - f(0) = \int_0^x f'(x)\,dx$, and so

$$\lim_{x \to \infty} f(x) = f(0) + \int_0^\infty f'(x)\,dx =: L.$$

As $f(x) \geq 0$ for all x, we must have that $L \geq 0$. Suppose that $L > 0$. Then there exists an $R > 0$ such that for all $x > R$, $L - f(x) \leq |f(x) - L| < L/2$, and in particular, $f(x) > L/2$ for all $x > R$. Hence for all $x > R$,

$$\int_0^\infty f(x)\,dx \geq \int_0^x f(x)\,dx \geq \int_R^x f(x)\,dx > (x - R)\frac{L}{2} \xrightarrow{x \to \infty} \infty,$$

which is absurd. Consequently $L = 0$.

Solution to Exercise 5.41

(1) $(1_{[0,1]} * 1_{[0,1]})(t) = \begin{cases} 0 & \text{if } t \le 0, \\ t & \text{if } 0 \le t \le 1, \\ 2-t & \text{if } 1 \le t \le 2, \\ 0 & \text{if } t \ge 2. \end{cases}$

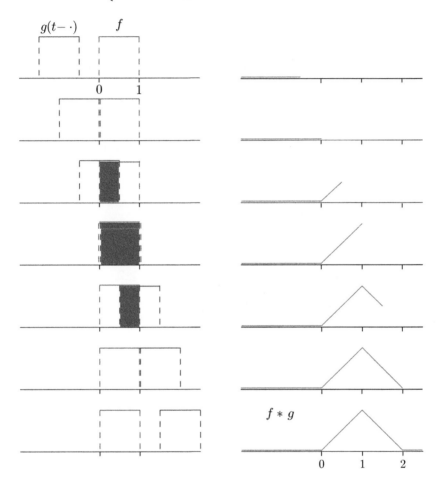

(2) We have for all $t \in \mathbb{R}$ that

$$(g * f)(t) = \lim_{T \to \infty} \int_{-T}^{T} g(\tau)f(t - \tau)d\tau$$

$$= \lim_{T \to \infty} \int_{t+T}^{t-T} g(t - s)f(s)(-1)ds \quad \text{(using the substitution } t - \tau - s)$$

$$= \lim_{T \to \infty} \int_{t-T}^{t+T} f(s)g(t - s)ds = \int_{-\infty}^{\infty} f(\tau)g(t - \tau)d\tau = (f * g)(t).$$

Solution to Exercise 5.42

(1) We have
$$I'(\alpha) = \int_0^\infty \frac{2(\sin(\alpha x)) \cdot \cos(\alpha x) \cdot x}{x^2} dx = \int_0^\infty \frac{\sin(2\alpha x)}{x} dx$$
$$= \int_0^\infty \frac{\sin u}{u} du \quad (\text{using the substitution } u = 2\alpha x)$$
$$= \frac{\pi}{2}.$$

Thus $I(\alpha) = I(\alpha) - 0 = I(\alpha) - I(0) = \frac{\pi}{2}\alpha$, and so $\int_0^\infty \frac{(\sin x)^2}{x^2} dx = I(1) = \frac{\pi}{2} \cdot 1 = \frac{\pi}{2}$.

(2) We have
$$I'(\alpha) = \int_0^\infty e^{-x} \frac{(\cos(\alpha x)) \cdot x}{x} dx = \int_0^\infty e^{-x} \cos(\alpha x) dx$$
$$= \frac{1}{1+\alpha^2} e^{-x}(-\cos(\alpha x) + \alpha \sin(\alpha x))\Big|_0^\infty = \frac{1}{1+\alpha^2}.$$

Thus $I(\alpha) = I(\alpha) - 0 = I(\alpha) - I(0) = \tan^{-1}\alpha$, and so
$$\int_0^\infty e^{-x} \frac{\sin x}{x} dx = I(1) = \tan^{-1} 1 = \frac{\pi}{4}.$$

(3) We have $I'(\alpha) = \int_0^1 \frac{(\log x)e^{\alpha \log x} - 0}{\log x} dx = \int_0^1 x^\alpha dx = \frac{1}{\alpha+1}.$

Thus $I(\alpha) = I(\alpha) - 0 = I(\alpha) - I(0) = \log(\alpha + 1)$, and so $\int_0^1 \frac{x-1}{\log x} dx = I(1) = \log 2.$

(4) We have
$$I'(\alpha) = \int_0^\infty \frac{\frac{1}{1+\alpha^2 x^2} \cdot x - 0}{x} dx = \int_0^\infty \frac{1}{1+\alpha^2 x^2} dx = \frac{1}{\alpha}\tan^{-1}(\alpha x)\Big|_0^\infty = \frac{1}{\alpha} \cdot \frac{\pi}{2}.$$

Thus $I(\alpha) = I(\alpha) - 0 = I(\alpha) - I(1) = \frac{\pi}{2}\log\alpha$, and so
$$\int_0^\infty \frac{\tan^{-1}(\pi x) - \tan^{-1}x}{x} dx = I(\pi) = \frac{\pi}{2}\log\pi.$$

Solution to Exercise 5.43

(1) We have $\int_R^x \frac{GMm}{r^2} dr = -\frac{GMm}{r}\Big|_R^x = -\frac{GMm}{x} + \frac{GMm}{R}$, and so
$$V(R) = \int_R^\infty \frac{GMm}{r^2} dr = \lim_{x \to \infty} \int_R^x \frac{GMm}{r^2} dx$$
$$= \lim_{x \to \infty} \left(-\frac{GMm}{x} + \frac{GMm}{R}\right) = 0 + \frac{GMm}{R} = \frac{GMm}{R}.$$

(2) We must have
$$\text{Kinetic energy} = \frac{1}{2}m(v_e(R))^2 = \frac{GMm}{R} = V(R) = \text{ Potential energy}$$

and so $v_e(R) = \sqrt{\dfrac{2GM}{R}}$.

With the given values, we have that the escape velocity on the surface of the Earth (so that the separation R is just the radius of the Earth),

$$v_e(R) = \sqrt{\frac{2GM}{R}} = \sqrt{\frac{2 \times (6.67384 \times 10^{-11}) \times (5.97219 \times 10^{24})}{6.371 \times 10^6}}$$

$$= \sqrt{1.2512145 \times 10^8} = 1.1185769 \times 10^4 \, \text{m s}^{-1},$$

that is, approximately 11.2 km/s.

(3) Clearly,

$$r_s = \frac{2GM}{c^2},$$

and so with $M = M_\odot = 1.99 \times 10^{30}$ kg, we obtain $r_s = 2.95 \times 10^3$ m, that is about 3 km.

Solution to Exercise 5.44

We have for $y > 0$ that

$$\int_0^y \lambda e^{-\lambda x} dx = \int_0^y -\frac{d}{dx} e^{-\lambda x} dx = -e^{-\lambda x} \Big|_0^y = -e^{-\lambda y} + e^0 = 1 - e^{-\lambda y}.$$

As $e^{\lambda y} = 1 + \lambda y + \cdots \geq \lambda y$, we have

$$0 \leq \lambda e^{-\lambda y} \leq \frac{1}{\lambda y},$$

and so by the Sandwich Theorem, $\lim_{y \to \infty} e^{-\lambda y} = 0$. Consequently,

$$\int_0^\infty e^{-\lambda x} dx = \lim_{y \to \infty} \int_0^y e^{-\lambda x} dx = \lim_{y \to \infty} \frac{1}{\lambda} \int_0^y \lambda e^{-\lambda x} dx = \frac{1}{\lambda} \lim_{y \to \infty} (1 - e^{-\lambda y}) = \frac{1}{\lambda}(1 - 0) = \frac{1}{\lambda}.$$

Solution to Exercise 5.45

We have that e^{-x^2} is a continuous function. Also,

$$e^{x^2} = 1 + \frac{x^2}{1!} + \frac{x^4}{2!} + \cdots \geq 1 + x^2,$$

and so $e^{-x^2} \leq \frac{1}{1+x^2}$. Hence for $y > 1$, we have

$$\int_0^y e^{-x^2} dx = \int_0^1 e^{-x^2} dx + \int_1^y e^{-x^2} dx$$

$$\leq \int_0^1 e^{-x^2} dx + \int_1^y \frac{1}{1 + x^2} dx$$

$$\leq \int_0^1 e^{-x^2} dx + \int_1^y \frac{1}{x^2} dx < \infty$$

since

$$\int_1^\infty \frac{1}{x^2} dx < \infty.$$

Thus $y \mapsto \int_0^y e^{-x^2}\,dx$ is bounded above, and moreover, it is increasing. So

$$\lim_{y \to \infty} \int_0^y e^{-x^2}\,dx$$

exists. Hence the improper integral $\int_0^\infty e^{-x^2}\,dx$ converges.

Solution to Exercise 5.46

We use the substitution $u = \log x$ (so that $du = \frac{1}{x}dx$, and when $x = 2$, $u = \log 2$, while if $x = y$, then $u = \log y$):

$$\int_2^y \frac{1}{x \log x}\,dx = \int_{\log 2}^{\log y} \frac{1}{u}\,du = \log u \Big|_{\log 2}^{\log y} = \log(\log y) - \log(\log 2).$$

As $\log y \xrightarrow{y \to \infty} \infty$, it follows that $\log(\log y) \xrightarrow{y \to \infty} \infty$, and so

$$\int_2^\infty \frac{1}{x \log x}\,dx$$

does not converge. On the other hand,

$$\int_2^y \frac{1}{x(\log x)^2}\,dx = \int_{\log 2}^{\log y} \frac{1}{u^2}\,du = -\frac{1}{u}\Big|_{\log 2}^{\log y} = -\frac{1}{\log y} + \frac{1}{\log 2}.$$

Thus

$$\int_2^\infty \frac{1}{x(\log x)^2}\,dx = \lim_{y \to \infty} \int_2^y \frac{1}{x(\log x)^2}\,dx = \lim_{y \to \infty}\left(-\frac{1}{\log y} + \frac{1}{\log 2}\right) = 0 + \frac{1}{\log 2} = \frac{1}{\log 2}.$$

Solution to Exercise 5.47

(1) We have

$$\log(n + 1) - \log n = \int_1^{n+1} \frac{1}{t}\,dt - \int_1^n \frac{1}{t}\,dt = \int_n^{n+1} \frac{1}{t}\,dt.$$

Clearly

$$\int_n^{n+1} \frac{1}{t}\,dt \le \int_n^{n+1} \frac{1}{n}\,dt = \frac{1}{n} \quad \text{and}$$

$$\int_n^{n+1} \frac{1}{t}\,dt \ge \int_n^{n+1} \frac{1}{n+1}\,dt = \frac{1}{n+1}.$$

(2) We have

$$a_n = 1 + \frac{1}{2} + \frac{1}{3} + \cdots + \frac{1}{n} - \log n$$

$$= 1 + \frac{1}{2} + \frac{1}{3} + \cdots + \frac{1}{n} + (\log 1 - \log 2 + \log 2 - \log 3 + \cdots + \log(n-1) - \log n)$$

$$= \underbrace{(1 + \log 1 - \log 2)}_{\ge 0} + \underbrace{\left(\frac{1}{2} + \log 2 - \log 3\right)}_{\ge 0} + \cdots + \underbrace{\left(\frac{1}{n-1} + \log(n-1) - \log n\right)}_{\ge 0} + \frac{1}{n}$$

$$\ge \frac{1}{n} \ge 0.$$

Also,

$$a_n - a_{n+1} = \left(1 + \frac{1}{2} + \frac{1}{3} + \cdots + \frac{1}{n} - \log n\right) - \left(1 + \frac{1}{2} + \frac{1}{3} + \cdots + \frac{1}{n} + \frac{1}{n+1} - \log(n+1)\right)$$

$$= \log(n+1) - \log n - \frac{1}{n+1} \geq 0.$$

Since $(a_n)_{n \in \mathbb{N}}$ is monotone and bounded, it is convergent with a limit, which we call γ.

Solution to Exercise 5.48

Let $f(x) = \log(1 + x)$, $x \geq 0$. Then $f(0) = \log 1 = 0$, and

$$f'(x) = \frac{1}{1+x}, \quad f'(0) = 1,$$

$$f''(x) = \frac{-1}{(1+x)^2}, \quad f''(0) = -1,$$

$$f'''(x) = \frac{2}{(1+x)^3}.$$

If $x \in (0, \infty)$, then Taylor's Formula gives

$$f(x) = f(0) + f'(0)x + \frac{1}{2!} f''(0)x^2 + \frac{1}{3!} f'''(\theta)x^3,$$

for some $\theta \in (0, x)$. Thus $\log(1 + x) = 0 + x - \frac{x^2}{2} + \frac{1}{3!} \frac{2}{(1+\theta)^3}x^3$, $x \geq 0$.

But clearly $\frac{1}{3!} \frac{2}{(1+\theta)^3}x^3 \geq 0$ for $x \geq 0$ and $\frac{1}{3!} \frac{2}{(1+\theta)^3}x^3 = \frac{x^3/3}{(1+\theta)^3} \leq \frac{x^3}{3}$ for $x \geq 0$.

Hence $x - \frac{x^2}{2} \leq \log(1 + x) \leq x - \frac{x^2}{2} + \frac{x^3}{3}$ for $x \geq 0$.

Solution to Exercise 5.49

We have

$$d(A, B) + d(B, C) = \log\left(\frac{\widehat{AP}}{\widehat{AQ}} \cdot \frac{\widehat{BQ}}{\widehat{BP}}\right) + \log\left(\frac{\widehat{BP}}{\widehat{BQ}} \cdot \frac{\widehat{CQ}}{\widehat{CP}}\right)$$

$$= \log\left(\frac{\widehat{AP}}{\widehat{AQ}} \cdot \frac{\widehat{BQ}}{\widehat{BP}} \cdot \frac{\widehat{BP}}{\widehat{BQ}} \cdot \frac{\widehat{CQ}}{\widehat{CP}}\right) = \log\left(\frac{\widehat{AP}}{\widehat{AQ}} \cdot \frac{\widehat{CQ}}{\widehat{CP}}\right) = d(A, C).$$

Solution to Exercise 5.50

(1) For $y > 0$, using Integration by Parts, we have $\int_y^1 1 \cdot \log x \, dx = -y \log y + y - 1.$

By l'Hôpital's Rule, we have $\lim\limits_{y\to 0} y \log y = \lim\limits_{y\to 0} \dfrac{\log y}{1/y} = \lim\limits_{y\to 0} \dfrac{1/y}{-1/y^2} = \lim\limits_{y\to 0}(-y) = 0.$

Thus $\displaystyle\int_0^1 \log x dx = \lim_{y\to 0}\int_y^1 \log x dx = \lim_{y\to 0}(-y\log y + y - 1) = 0 + 0 - 1 = -1.$

(2) We associate with $\displaystyle\int_0^1 \log x dx$, a 'Riemann type of sum'.

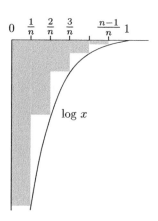

We have

$$-1 = \int_0^1 \log x dx \approx \frac{1}{n}\log\frac{1}{n} + \frac{1}{n}\log\frac{2}{n} + \cdots + \frac{1}{n}\log\frac{n}{n} = \frac{1}{n}\log\frac{1\cdot 2\cdot 3\cdots n}{n^n} = \frac{1}{n}\log\frac{n!}{n^n}.$$

So $\log\dfrac{n!}{n^n} \approx -n$, that is, $\log n! \approx n\log n - n$.

Solution to Exercise 5.51

We have using the Fundamental Theorem of Calculus that

$$\frac{d}{dx}\left(\int_0^x \frac{1}{\sqrt{1+t^2}}dt - \log(x + \sqrt{1+x^2})\right)$$

$$= \frac{1}{\sqrt{1+x^2}} - \frac{1}{x+\sqrt{1+x^2}}\cdot\left(1 + \frac{1}{2\sqrt{1+x^2}}\cdot 2x\right)$$

$$= \frac{1}{\sqrt{1+x^2}} - \frac{1}{x+\sqrt{1+x^2}}\cdot\frac{x+\sqrt{1+x^2}}{\sqrt{1+x^2}} = 0.$$

So for all $x \in \mathbb{R}$,

$$\int_0^x \frac{1}{\sqrt{1+t^2}}dt - \log(x + \sqrt{1+x^2}) = \int_0^0 \frac{1}{\sqrt{1+t^2}}dt - \log(0 + \sqrt{1+0^2}) = 0.$$

Hence $\displaystyle\int_0^x \frac{1}{\sqrt{1+t^2}}dt = \log(x + \sqrt{1+x^2}).$

Also,

$$\frac{d}{dx}\left(\int_0^x \sqrt{1+t^2}dt - \frac{x\sqrt{1+x^2}}{2} - \frac{1}{2}\log(x+\sqrt{1+x^2})\right)$$

$$= \sqrt{1+x^2} - \frac{1\cdot\sqrt{1+x^2}}{2} - \frac{x\cdot x}{2\sqrt{1+x^2}} - \frac{1}{2}\frac{1}{x+\sqrt{1+x^2}}\left(1+\frac{x}{\sqrt{1+x^2}}\right)$$

$$= \sqrt{1+x^2} - \frac{\sqrt{1+x^2}}{2} - \frac{x^2}{2\sqrt{1+x^2}} - \frac{1}{2\sqrt{1+x^2}}$$

$$= \sqrt{1+x^2} - \frac{\sqrt{1+x^2}}{2} - \frac{1+x^2}{2\sqrt{1+x^2}}$$

$$= \sqrt{1+x^2} - \frac{\sqrt{1+x^2}}{2} - \frac{\sqrt{1+x^2}}{2} = 0.$$

Hence for all $x \in \mathbb{R}$,

$$\int_0^x \sqrt{1+t^2}dt - \frac{x\sqrt{1+x^2}}{2} - \frac{1}{2}\log(x+\sqrt{1+x^2})$$

$$= \int_0^0 \sqrt{1+t^2}dt - \frac{0\sqrt{1+0^2}}{2} - \frac{1}{2}\log(0+\sqrt{1+0^2}) = 0.$$

Consequently, $\displaystyle\int_0^x \sqrt{1+t^2}dt = \frac{x\sqrt{1+x^2}}{2} + \frac{1}{2}\log(x+\sqrt{1+x^2})$.

Solution to Exercise 5.52

We use the $\frac{\infty}{\infty}$ form of l'Hôpital's Rule. With

$$f := \log(\log x) \quad \text{and}$$

$$g := \log x,$$

we have

(1) $g, g' > 0$ on $(1, \infty)$,

(2) $\displaystyle\lim_{x\to\infty} g(x) = \lim_{x\to\infty} \log x = \infty$, and

(3) $\displaystyle\lim_{x\to\infty}\frac{f'(x)}{g'(x)} = \lim_{x\to\infty}\frac{\dfrac{1}{\log x}\dfrac{1}{x}}{\dfrac{1}{x}} = \lim_{x\to\infty}\frac{1}{\log x} = 0,$

and so $\displaystyle\lim_{x\to\infty}\frac{f(x)}{g(x)} = \lim_{x\to\infty}\frac{f'(x)}{g'(x)} = 0.$

Solution to Exercise 5.53

(1) We have $b^{\log_b a} = e^{\log_b a \cdot \log b} = e^{\log a} = a.$

(2) We have $(a^b)^c = (e^{b\log a})^c = e^{c\cdot\log(e^{b\log a})} = e^{c\cdot(b\log a)} = e^{(c\cdot b)\cdot\log a} = a^{(b\cdot c)}.$

(3) Suppose that

$$\log_2 3 = \frac{p}{q},$$

where $p, q \in \mathbb{Z}$, and $q \geq 0$. We may assume that $p, q > 0$ (because $3 > 1$, $2 > 1$ imply that $\log 3, \log 2 > 0$, and so $\log_2 3 > 0$ too). Then $2^{p/q} = 3$, that is, $2^p = 3^q$. But as $p \neq 0$, 2^p is even. However, 3^q is odd, a contradiction.

(4) $\sqrt{2}$ is irrational and $2\log_2 3$ is irrational, but $\sqrt{2}^{2\log_2 3} = (\sqrt{2}^2)^{\log_2 3} = 2^{\log_2 3} = 3 \in \mathbb{Q}$.

(5) The flaw is that $\log_{1/2}$ is not an increasing function. Indeed, for $x > 0$,

$$\log_{1/2} x = \frac{\log x}{\log(1/2)},$$

and as $\log$ is strictly increasing and $\log(1/2) < 0$, it follows that $\log_{1/2}$ is strictly decreasing.

(6) See Figure 5.

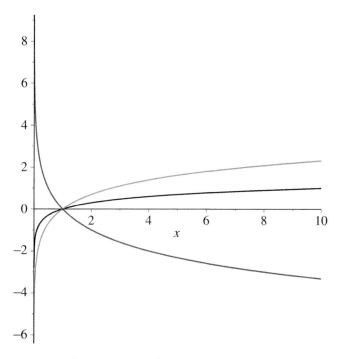

Figure 5. The graphs of $\log$, $\log_{1/2}$, and $\log_{10}$. The decreasing one is the graph of $\log_{1/2}$. Among the increasing ones, the graph of $\log_{10}$ is the one which takes value 1 at $x = 10$.

Solution to Exercise 5.54

For $x > 0$, $\left(\frac{\log x}{x}\right)' = -\frac{1}{x^2} \cdot \log x + \frac{1}{x} \cdot \frac{1}{x} = \frac{1}{x^2}(1 - \log x).$

We have

$$\left(\frac{\log x}{x}\right)' = \frac{1}{x^2}(1 - \log x) \leq 0$$

for $x \in (e, \infty)$ because $\log x > \log e = 1$ (as $\log$ is strictly increasing). As $\pi > 3 > e$, we have that

$$\frac{\log \pi}{\pi} < \frac{\log e}{e},$$

that is, $e \log \pi < \pi \log e$. Hence $\pi^e < e^\pi$, and so e^π is bigger. (A scientific calculator gives $e^\pi \approx 23.1406926328$, while $\pi^e \approx 22.4591577184$.)

Solution to Exercise 5.55

(1) We have

$$\frac{d}{dx}\left(y_0 \exp\left(\int_{x_0}^x a(\xi)d\xi\right)\right) = y_0 \exp\left(\int_{x_0}^x a(\xi)d\xi\right) \frac{d}{dx} \int_{x_0}^x a(\xi)d\xi$$

$$= y_0 \exp\left(\int_{x_0}^x a(\xi)d\xi\right) a(x)$$

$$= a(x) \cdot y_0 \exp\left(\int_{x_0}^x a(\xi)d\xi\right),$$

and so $x \mapsto y_0 \exp\left(\int_{x_0}^x a(\xi)d\xi\right)$ does satisfy $f'(x) = a(x)f(x)$ on the interval I. Moreover,

$$\exp\left(y_0 \int_{x_0}^x a(\xi)d\xi\right)\Big|_{x=x_0} = y_0 \exp\left(\int_{x_0}^{x_0} a(\xi)d\xi\right) = y_0 \cdot \exp 0 = y_0 \cdot 1 = y_0.$$

Next we show uniqueness. To this end, let f be any function such that $f'(x) = a(x)f(x)$ on the interval I and $f(x_0) = y_0$. Then

$$\frac{d}{dx}\left(\exp\left(-\int_{x_0}^x a(\xi)d\xi\right) f(x)\right)$$

$$= \left(\frac{d}{dx} \exp\left(-\int_{x_0}^x a(\xi)d\xi\right)\right) \cdot f(x) + \exp\left(-\int_{x_0}^x a(\xi)d\xi\right) \cdot f'(x)$$

$$= \exp\left(-\int_{x_0}^x a(\xi)d\xi\right) \cdot \frac{d}{dx}\left(-\int_{x_0}^x a(\xi)d\xi\right) \cdot f(x) + \exp\left(-\int_{x_0}^x a(\xi)d\xi\right) \cdot a(x)f(x)$$

$$= \exp\left(-\int_{x_0}^x a(\xi)d\xi\right) \cdot (-a(x)) \cdot f(x) + \exp\left(-\int_{x_0}^x a(\xi)d\xi\right) \cdot a(x)f(x)$$

$$= 0.$$

Thus for all $x \in I$,

$$\exp\left(-\int_{x_0}^x a(\xi)d\xi\right) f(x) = \exp\left(-\int_{x_0}^{x_0} a(\xi)d\xi\right) f(x_0) = \exp(0) \cdot y_0 = 1 \cdot y_0 = y_0,$$

that is, $f(x) = y_0 \exp\left(\int_{x_0}^x a(\xi)d\xi\right)$.

(2) We first check that the given expression for f is indeed a solution. Note that for all $x \in I$, $A'(x) = a(x)$. Thus we have

$$\frac{d}{dx}\left(y_0 e^{A(x)} + e^{A(x)} \int_{x_0}^x b(\xi)e^{-A(\xi)}d\xi\right)$$

$$= y_0 e^{A(x)}a(x) + e^{A(x)}a(x)\int_{x_0}^x b(\xi)e^{-A(\xi)}d\xi + e^{A(x)}b(x)e^{-A(x)}$$

$$= a(x)\left(y_0 e^{A(x)} + e^{A(x)}\int_{x_0}^x b(\xi)e^{-A(\xi)}d\xi\right) + b(x).$$

Moreover, using the fact that $A(x_0) = 0$, we obtain

$$y_0 e^{A(x_0)} + e^{A(x_0)} \int_{x_0}^{x_0} b(\xi)e^{-A(\xi)}d\xi = y_0 \cdot 1 + 1 \cdot 0 = y_0.$$

Next we will establish uniqueness. Suppose that f, g are two solutions. Then $h := f - g$ satisfies

$$h'(x) = f'(x) - g'(x) = a(x)f(x) + b(x) - (a(x)g(x) + b(x)) = a(x)(f(x) - g(x)) = a(x)h(x),$$

for all $x \in I$. Also, $h(x_0) = f(x_0) - g(x_0) = y_0 - y_0 = 0$. So by Part (1), it follows that

$$f(x) - g(x) = h(x) = 0 \cdot \exp\left(\int_{x_0}^x a(\xi)d\xi\right) = 0, \quad x \in I.$$

Hence $f \equiv g$ on I.

(3) We write the given differential equation in the form

$$\frac{y'}{y(y-1)} = \frac{1}{x}.$$

Furthermore,

$$\frac{1}{y(y-1)} = \frac{1}{y-1} - \frac{1}{y}.$$

With

$$P(x) = \log(y(x) - 1) - \log(y(x)),$$

$$Q(x) = \log x,$$

we have that

$$(P - Q)'(x) = \frac{y'(x)}{y(x)-1} - \frac{y'(x)}{y(x)} - \frac{1}{x} = 0,$$

and so $P - Q$ is constant on $(0, \infty)$. Hence

$$\log(y(x) - 1) - \log(y(x)) - \log x = \log\frac{y(x)-1}{xy(x)}$$

must be constant for $x \in (0, \infty)$, and so

$$x \mapsto \frac{y(x)-1}{xy(x)}$$

itself must be a constant, say C. Thus we obtain

$$y(x) = \frac{1}{1 - Cx}, \quad x > 0,$$

where C is a constant.

Solution to Exercise 5.56

Newton's Law of Cooling gives $\Theta'(t) = -k(\Theta(t) - M)$, where $k > 0$ is the constant of proportionality. Then

$$(e^{kt}(\Theta - M))' = e^{kt}(\Theta - M)' + ke^{kt}(\Theta - M) = e^{kt}\underbrace{(\Theta' + k(\Theta - M))}_{=0} = 0.$$

Thus $e^{kt}(\Theta - M)$ is constant, and so $e^{kt}(\Theta(t) - M) = e^{k \cdot 0}(\Theta(0) - M) = 1 \cdot (\Theta_0 - M)$. Rearranging, we obtain $\Theta(t) = e^{-kt}(\Theta_0 - M) + M = e^{-kt}\Theta_0 + M(1 - e^{-kt})$. As $t \to \infty$, $\Theta(t) \to M$, the ambient temperature, as expected.

Solution to Exercise 5.57

(1) We have

$$(e^{ct}A(t))' = e^{ct}A'(t) + ce^{ct}A(t) = e^{ct}\underbrace{(A'(t) + cA(t))}_{=0} = 0.$$

So $e^{ct}A(t)$ is constant, and so $e^{ct}A(t) = e^{c \cdot 0}A(0) = 1 \cdot A_0$. Hence $A(t) = e^{-ct}A_0$ for all t.

(2) (We want in particular that $A(0 + \tau) = A(0)/2$, that is, $A(\tau) = A_0/2$. So

$$e^{-c\tau}A_0 = A_0/2,$$

that is, $e^{-c\tau} = 1/2$. Thus $-c\tau = \log(1/2) = -\log 2$. Thus we must have $\tau = (\log 2)/c$.)
Indeed, with

$$\tau := \frac{\log 2}{c},$$

we have $A(t + \tau) = e^{-c(t+\tau)}A_0 = e^{-ct}e^{-c\tau}A_0 = e^{-ct}\frac{1}{2}A_0 = \frac{1}{2}\underbrace{e^{-ct}A_0}_{=A(t)} = \frac{A(t)}{2}.$

Solution to Exercise 5.58

We have

$$\lim_{m \to \infty} P\left(1 + \frac{r}{m}\right)^{mn} = P\left(\left(1 + \frac{r}{m}\right)^{\frac{m}{r}}\right)^{r \cdot n} = P\left(\lim_{m \to \infty}\left(1 + \frac{r}{m}\right)^{\frac{m}{r}}\right)^{r \cdot n} = Pe^{rn},$$

where to get the last equality, we used the fact that

$$\lim_{h \to 0}(1 + h)^{1/h} = e,$$

which follows from $\lim_{h \to 0}\dfrac{\log(1 + h)}{h} = 1$ and the continuity of exp.

(1) We seek n such that $\dfrac{Pe^{0.06n}}{P} = 2$, that is, $e^{0.06n} = 2$.

Hence $0.06n = \log 2$, that is, $n = \dfrac{\log 2}{0.06} \approx \dfrac{0.6931}{0.06} \approx 11.6$ years.

(2) We seek n such that $\dfrac{P(1 + nr)}{P} = 2$, that is, $1 + 0.06n = 2$, and so $0.06n = 1$.

Thus $n = \dfrac{1}{0.06} \approx 16.7$ years.

Solution to Exercise 5.59

(1) We have $\sinh 0 = \dfrac{e^0 - e^{-0}}{2} = \dfrac{1-1}{2} = 0$, and $\cosh 0 = \dfrac{e^0 + e^{-0}}{2} = \dfrac{1+1}{2} = 1$.

Also, for all $x \in \mathbb{R}$ we have

$$\sinh'x = \frac{d}{dx}\frac{e^x - e^{-x}}{2} = \frac{e^x + e^{-x}}{2} = \cosh x \text{ and}$$

$$\cosh'x = \frac{d}{dx}\frac{e^x + e^{-x}}{2} = \frac{e^x - e^{-x}}{2} = \sinh x.$$

For $x, y \in \mathbb{R}$, we have

$$(\cosh x)(\cosh y) + (\sinh x)(\sinh y)$$

$$= \frac{e^x + e^{-x}}{2} \cdot \frac{e^y + e^{-y}}{2} + \frac{e^x - e^{-x}}{2} \cdot \frac{e^y - e^{-y}}{2}$$

$$= \frac{e^{x+y} + e^{x-y} + e^{-x+y} + e^{-(x+y)}}{4} + \frac{e^{x+y} - e^{x-y} - e^{-x+y} + e^{-(x+y)}}{4}$$

$$= \frac{2e^{x+y} + 2e^{-(x+y)}}{4} = \frac{e^{x+y} + e^{-(x+y)}}{2} = \cosh(x + y).$$

(2) We have $\cosh''x = \cosh \geq 0$ for all x and so $\cosh$ is convex.

Also, $\lim\limits_{x \to \pm\infty} \cosh x = +\infty$. See Figure 6.

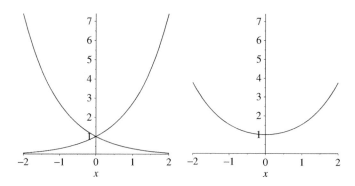

Figure 6. Graphs of e^x, e^{-x} on the left and the graph of $\cosh$ on the right.

We have $\sinh''x = \sinh x > 0$ for $x > 0$, and $\sinh''x < 0$ for $x < 0$. Thus $\sinh$ is convex for $x > 0$, and concave for $x < 0$. Also, $\lim\limits_{x \to \pm\infty} \sinh x = \pm\infty$. See Figure 7.

(3) With $(x, y) = (\cosh t, \sinh t)$, $t \in \mathbb{R}$, we have

$$x^2 - y^2 = (\cosh t)^2 - (\sinh t)^2 = \left(\frac{e^t + e^{-t}}{2}\right)^2 - \left(\frac{e^t - e^{-t}}{2}\right)^2$$

$$= \frac{e^{2t} + 2 + e^{-2t}}{4} - \frac{e^{2t} - 2 + e^{-2t}}{4} = \frac{4}{4} = 1.$$

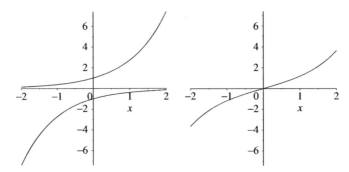

Figure 7.   Graphs of e^x, $-e^{-x}$ on the left, and the graph of cosh on the right.

The curve $t \mapsto (\cosh t, \sinh t)$ is displayed in the following picture. As t increases, the point on the curve moves upward.

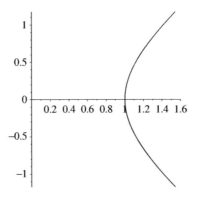

(4) The following picture shows the graph of tanh.

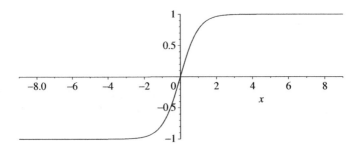

Solution to Exercise 5.60

Let $f(x) = x^x$, $x > 0$. Then

$$h(x) := \log f(x) = \log(\exp(x \log x)) = x \log x.$$

Thus $\lim_{x \to 0+} h(x) = \lim_{x \to 0+} x \log x = \lim_{x \to 0+} \dfrac{\log x}{1/x} = \lim_{x \to 0+} \dfrac{1/x}{-1/x^2} = \lim_{x \to 0+} -x = 0.$

By the continuity of exp, we obtain

$$\lim_{x \to 0+} f(x) = \lim_{x \to 0+} \exp h(x) = \exp\left(\lim_{x \to 0+} h(x)\right) = \exp 0 = 1.$$

Hence also $\lim_{x \to 0+} \dfrac{1}{x^x} = 1.$

So for every $\epsilon > 0$, there exists a $\delta > 0$ such that for all x satisfying $0 < x < \delta$, $\left|\dfrac{1}{x^x} - 1\right| < \epsilon.$

In particular, for all $n \in \mathbb{N}$ such that $n > \dfrac{1}{\delta}$, we have that $x := \dfrac{1}{n}$ satisfies $0 < x < \delta$ and so

$$\left|n^{\frac{1}{n}} - 1\right| = \left|\dfrac{1}{x^x} - 1\right| < \epsilon.$$

Consequently, $\lim_{n \to \infty} n^{1/n} = 1.$

Solution to Exercise 5.61

(1) Define $g(x) := \begin{cases} e^{-x \log x} & \text{for } x > 0, \\ 1 & \text{for } x = 0. \end{cases}$

Then g is continuous by Exercise 5.60.

As exp is continuous, it follows that $\lim_{x \to 0+} g(x) = \lim_{x \to 0+} e^{-x \log x} = e^{-0} = 1 = g(0).$ Thus

$$\lim_{\epsilon \searrow 0} \int_\epsilon^1 \dfrac{1}{x^x} dx = \lim_{\epsilon \searrow 0} \int_\epsilon^1 g(x) dx = \lim_{\epsilon \searrow 0} \left(\int_0^1 g(x) dx - \int_0^\epsilon g(x) dx\right)$$

$$= \int_0^1 g(x) dx - \lim_{\epsilon \searrow 0} \int_0^\epsilon g(x) dx = \int_0^1 g(x) dx,$$

where we have used

$$0 \le \left|\int_0^\epsilon g(x) dx\right| \le \int_0^\epsilon |g(x)| dx \le \left(\max_{x \in [0,1]} |g(x)|\right) \epsilon$$

in order to get the last equality.

(2) We have

$$\int_0^1 \dfrac{1}{x^x} dx = \int_0^1 e^{-x \log x} dx = \int_0^1 \sum_{n=0}^{\infty} \dfrac{(-x \log x)^n}{n!} dx$$

$$= \sum_{n=0}^{\infty} \dfrac{(-1)^n}{n!} \int_0^1 x^n (\log x)^n dx$$

$$= \sum_{n=0}^{\infty} \dfrac{(-1)^n}{n!} \int_\infty^0 e^{-nt} (-t)^n (-e^{-t}) dt,$$

where we used the substitution $x = e^{-t}$ (and so $dx = -e^{-t}dt$, when $x = 0$ we have $t = \infty$, and when $x = 1$ we have $t = 0$) in order to get the last equality. Thus

$$\int_0^1 \frac{1}{x^x}dx = \sum_{n=0}^{\infty} \frac{1}{n!} \int_0^{\infty} e^{-(n+1)t}t^n dt = \sum_{n=0}^{\infty} \frac{1}{n!} \int_0^{\infty} e^{-u} \frac{u^n}{(n+1)^n} \frac{1}{n+1}du,$$

where we used the substitution $u = (n+1)t$ (and so $du = (n+1)dt$, when $t = 0$ we have $u = 0$, and when $t = \infty$ we have $u = \infty$) in order to get the last equality. Hence

$$\int_0^1 \frac{1}{x^x}dx = \sum_{n=0}^{\infty} \frac{1}{n!} \frac{1}{(n+1)^{n+1}} \int_0^{\infty} e^{-u} u^{(n+1)-1} du$$

$$= \sum_{n=0}^{\infty} \frac{1}{n!} \frac{1}{(n+1)^{n+1}} \Gamma(n+1)$$

$$= \sum_{n=0}^{\infty} \frac{1}{n!} \frac{1}{(n+1)^{n+1}} n!$$

$$= \sum_{n=0}^{\infty} \frac{1}{(n+1)^{n+1}} = \sum_{n=1}^{\infty} \frac{1}{n^n}.$$

Solution to Exercise 5.62

With f defined by $f(x) = \log(1 + x)$, $x \in [0, 1]$, we have that

$$S(f, P_n) = \left(\log\left(1 + \frac{1}{n}\right)\right) \frac{1}{n} + \left(\log\left(1 + \frac{2}{n}\right)\right) \frac{1}{n} + \cdots + \left(\log\left(1 + \frac{n}{n}\right)\right) \frac{1}{n}$$

$$= \log\left(\left(1 + \frac{1}{n}\right)\left(1 + \frac{2}{n}\right)\cdots\left(1 + \frac{n}{n}\right)\right)^{\frac{1}{n}}.$$

Since $\lim_{n\to\infty} S(f, P_n) = \int_0^1 f(x)dx$, we obtain

$$\lim_{n\to\infty} \log\left(\left(1 + \frac{1}{n}\right)\left(1 + \frac{2}{n}\right)\cdots\left(1 + \frac{n}{n}\right)\right)^{\frac{1}{n}} = \int_0^1 \log(1 + x)dx$$

$$= (\log(1 + x))x\Big|_0^1 - \int_0^1 \frac{x}{1+x}xdx$$

$$= \log 2 - \int_0^1 \left(1 - \frac{1}{1+x}\right)dx$$

$$= \log 2 - 1 + \log(1 + x)\Big|_0^1 = 2\log 2 - 1$$

$$= \log \frac{4}{e}.$$

Since exp is continuous, it follows that

$$\lim_{n\to\infty} \log\left(\left(1 + \frac{1}{n}\right)\left(1 + \frac{2}{n}\right)\cdots\left(1 + \frac{n}{n}\right)\right)^{\frac{1}{n}} = e^{\log \frac{4}{e}} = \frac{4}{e}.$$

Solution to Exercise 5.63

In decreasing order of growth, we have

$$x^{2x}, \ (2x)^x, \ x^x, \ (\log x)^x, \ e^x, \ 2^x, \ e^{x/2}, \ x^e, \ x^2, \ x^{1/2}, \ (\log x)^2, \ (\log_2 x \ge \log x), \ \log(\log x).$$

Indeed,

(1) $\displaystyle\lim_{x\to\infty} \frac{\log(\log x)}{\log x} = \lim_{x\to\infty} \frac{\frac{1}{\log x}\frac{1}{x}}{\frac{1}{x}} = \lim_{x\to\infty} \frac{1}{\log x} = 0;$

(2) $\displaystyle\lim_{x\to\infty} \frac{\log x}{\log_2 x} = \log 2 < 1;$

(3) $\displaystyle\lim_{x\to\infty} \frac{\log_2 x}{(\log x)^2} = \lim_{x\to\infty} \frac{1}{(\log x)\cdot(\log 2)} = 0;$

(4) $\displaystyle\lim_{x\to\infty} \frac{(\log x)^2}{x^{1/2}} = \lim_{x\to\infty} \frac{\log x}{\sqrt[4]{x}}\cdot\frac{\log x}{\sqrt[4]{x}} = 0\cdot 0 = 0;$

(5) $\displaystyle\lim_{x\to\infty} \frac{\sqrt{x}}{x^2} = \lim_{x\to\infty} \frac{1}{x^{3/2}} = 0;$

(6) $\displaystyle\lim_{x\to\infty} \frac{x^2}{x^e} = \lim_{x\to\infty} \frac{1}{x^{e-2}} = 0;$

(7) $\displaystyle\lim_{x\to\infty} \frac{x^e}{e^{x/2}} = \lim_{x\to\infty} \left(\frac{x^{2e}}{e^x}\right)^{1/2} = \sqrt{0} = 0;$

(8) $\displaystyle\lim_{x\to\infty} \frac{e^{x/2}}{2^x} = \lim_{x\to\infty} \frac{e^{x/2}}{e^{x\log 2}} = \lim_{x\to\infty} e^{x(\frac{1}{2}-\log 2)} = 0 \text{ since } e < 4;$

(9) $\displaystyle\lim_{x\to\infty} \frac{2^x}{e^x} = \lim_{x\to\infty} \frac{e^{x\log 2}}{e^x} = \lim_{x\to\infty} e^{x\log(2/e)} = 0 \text{ since } e > 2;$

(10) $\displaystyle\lim_{x\to\infty} \frac{e^x}{(\log x)^x} = \lim_{x\to\infty} \frac{e^x}{e^{x\log(\log x)}} = \lim_{x\to\infty} e^{x(1-\log(\log x))} = 0;$

(11) $\displaystyle\lim_{x\to\infty} \frac{(\log x)^x}{x^x} = \lim_{x\to\infty} \frac{e^{x\log(\log x)}}{e^{x\log x}} = \lim_{x\to\infty} e^{x\log(\frac{\log x}{x})} = 0;$

(12) $\displaystyle\lim_{x\to\infty} \frac{x^x}{(2x)^x} = \lim_{x\to\infty} \frac{1}{2^x} = 0;$

(13) $\displaystyle\lim_{x\to\infty} \frac{(2x)^x}{x^{2x}} = \lim_{x\to\infty} \frac{2^x e^{x\log x}}{e^{2x\log x}} = \lim_{x\to\infty} e^{x(\log 2-\log x)} = 0.$

Solution to Exercise 5.64

Since $\log_{10} 2 = 0.3010$, it follows that

$$2^{399} = 10^{399\cdot\log_{10} 2} = 10^{120.1109}.$$

But $1 = 10^0 < 10^{0.1109} < 10^1 = 10$, and so 2^{399} is an integer with a decimal representation having 121 digits.

Solution to Exercise 5.65

Using the calculation from Exercise 3.51.(2), we have the partial fraction expansion

$$\frac{x^2 + 3x + 9}{(x+1)(x-2)^2} = \frac{7}{9} \cdot \frac{1}{x+1} + \frac{2}{9} \cdot \frac{1}{x-2} + \frac{19}{3} \cdot \frac{1}{(x-2)^2},$$

and so

$$\int_3^4 \frac{x^2 + 3x + 9}{(x+1)(x-2)^2} dx = \frac{7}{9} \int_3^4 \frac{1}{x+1} dx + \frac{2}{9} \int_3^4 \frac{1}{x-2} dx + \frac{19}{3} \int_3^4 \frac{1}{(x-2)^2} dx$$

$$= \frac{7}{9} \log(x+1) \Big|_3^4 + \frac{2}{9} \log(x-2) \Big|_3^4 + \frac{19}{3} \left(-\frac{1}{x-2} \right) \Big|_3^4$$

$$= \frac{7}{9} \log \frac{5}{4} + \frac{2}{9} \log 2 + \frac{19}{3} \left(-\frac{1}{2} + 1 \right)$$

$$= \frac{7}{9} \log \frac{5}{4} + \frac{2}{9} \log 2 + \frac{19}{6}.$$

Solution to Exercise 5.66

$(x, y) := (3, 2)$ is a solution because

$$3^x - 2^y = 3^3 - 2^2 = 27 - 4 = 23,$$

$$\log_3 x + \log_y 2 = \log_3 3 + \log_2 2 = 1 + 1 = 2.$$

Suppose that (x, y) is another solution where $x \neq 3$. Then we have two cases.

1° $x > 3$. Then $3^x > 3^3 = 27$, and so $2^y < 4$. Thus $y < 2$. But then $\log_y 2 > 1$. Also, $\log_3 x > \log_3 3 = 1$. Hence the second equation cannot be valid, since we have that $\log_3 x + \log_y 2 > 1 + 1 = 2$.

2° $x < 3$. Then $3^x < 3^3 = 27$, and so $2^y > 4$. Thus $y > 2$. But then $\log_y 2 < 1$. Also, $\log_3 x < \log_3 3 = 1$. Hence the second equation cannot be valid, since we have that $\log_3 x + \log_y 2 < 1 + 1 = 2$.

Consequently, $x = 3$. But then $\log_y 2 = 1$, and so $y = 2$. So $(x, y) = (3, 2)$ is the *only* solution.

Solution to Exercise 5.67

From the recurrence relation and the fact that $a_1 = 1$, we see that all the terms a_n are positive. We have

$$\left(1 + \frac{1}{a_1}\right)\left(1 + \frac{1}{a_2}\right)\left(1 + \frac{1}{a_3}\right) \cdots \left(1 + \frac{1}{a_n}\right) = \left(\frac{a_1 + 1}{a_1}\right)\left(\frac{a_2 + 1}{a_2}\right)\left(\frac{a_3 + 1}{a_3}\right) \cdots \left(\frac{a_n + 1}{a_n}\right)$$

$$= \left(\frac{a_2/2}{a_1}\right)\left(\frac{a_3/3}{a_2}\right)\left(\frac{a_4/4}{a_3}\right) \cdots \left(\frac{a_{n+1}/(n+1)}{a_n}\right)$$

$$= \frac{1}{a_1} \cdot \frac{1}{2} \cdot \frac{1}{3} \cdot \frac{1}{4} \cdots \frac{a_{n+1}}{n+1} = \frac{a_{n+1}}{(n+1)!},$$

where we have used the fact that $a_1 = 1$ in order to obtain the last equality. So in order to find the sought for limit, we need to investigate the behaviour of the sequence $(b_n)_{n \in \mathbb{N}}$, where

$$b_n := \frac{a_n}{n!}, \quad n \in \mathbb{N}.$$

Let us try to find if the terms of the sequence $(b_n)_{n\in\mathbb{N}}$ satisfy a recurrence relation. Dividing $a_n = n(1 + a_{n-1})$ on both sides by $n!$ gives

$$\frac{a_n}{n!} = \frac{n(1 + a_{n-1})}{n!} = \frac{1 + a_{n-1}}{(n-1)!} = \frac{1}{(n-1)!} + \frac{a_{n-1}}{(n-1)!},$$

that is, $b_n = \dfrac{a_n}{n!} = \dfrac{1}{(n-1)!} + \dfrac{a_{n-1}}{(n-1)!} = \dfrac{1}{(n-1)!} + b_{n-1}$. Also, $b_1 = \dfrac{a_1}{1!} = \dfrac{1}{1!} = 1$.

Hence it follows that

$$b_n = \left(b_n - b_1\right) + 1 = \left((b_n - b_{n-1}) + (b_{n-1} - b_{n-2}) + \cdots + (b_2 - b_1)\right) + 1$$

$$= \left(\frac{1}{(n-1)!} + \frac{1}{(n-2)!} + \cdots + \frac{1}{1!}\right) + 1 = \sum_{k=0}^{n-1} \frac{1}{k!}.$$

Consequently,

$$\lim_{n\to\infty}\left(1 + \frac{1}{a_1}\right)\left(1 + \frac{1}{a_2}\right)\left(1 + \frac{1}{a_3}\right)\cdots\left(1 + \frac{1}{a_n}\right) = \lim_{n\to\infty}\frac{a_{n+1}}{(n+1)!} = \lim_{n\to\infty} b_{n+1}$$

$$= \lim_{n\to\infty}\sum_{k=0}^{n}\frac{1}{k!} = e.$$

Solution to Exercise 5.68

(1) We use the $\dfrac{0}{0}$ form of l'Hôpital's Rule. With

$$f(x) = 3^{\sin x} - 1,$$
$$g(x) = x,$$

we have

$$\lim_{x\to 0} f(x) = \lim_{x\to 0}(3^{\sin x} - 1) = 3^0 - 1 = 1 - 1 = 0,$$

$$\lim_{x\to 0} g(x) = \lim_{x\to 0} x = 0,$$

and

$$\lim_{x\to 0}\frac{f'(x)}{g'(x)} = \lim_{x\to 0}\frac{e^{(\log 3)(\sin x)}(\log 3)(\cos x)}{1} = e^{(\log 3)\cdot 0}(\log 3)(\cos 0) = \log 3,$$

and so $\lim\limits_{x\to 0}\dfrac{f(x)}{g(x)} = \lim\limits_{x\to 0}\dfrac{f'(x)}{g'(x)} = \log 3$.

Alternately, one could use the Chain rule:

$$\lim_{x\to 0}\frac{3^{\sin x} - 1}{x} = \frac{d}{dx}(3^{\sin x})\Big|_{x=0} = 3^{\sin x}\cdot(\log 3)\cdot\cos x\Big|_{x=0}$$

$$= 3^0\cdot(\log 3)\cdot\cos 0 = 1\cdot(\log 3)\cdot 1 = \log 3.$$

(2) We use the $\dfrac{0}{0}$ form of l'Hôpital's Rule twice. We have

$$\left(\sin x - x + \frac{x^3}{6}\right)\Big|_{x=0} = 0 \text{ and } x^3\Big|_{x=0} = 0,$$

and so

$$\lim_{x \to 0} \frac{\sin x - x + x^3/6}{x^3} = \lim_{x \to 0} \frac{\cos x - 1 + x^2/2}{3x^2}$$

if the latter exists. But

$$\left(\cos x - 1 + x^2/2\right)\Big|_{x=0} = 0 \text{ and } 3x^2\Big|_{x=0} = 0,$$

and so

$$\lim_{x \to 0} \frac{\sin x - x + x^3/6}{x^3} = \lim_{x \to 0} \frac{\cos x - 1 + x^2/2}{3x^2} = \lim_{x \to 0} \frac{-\sin x + x}{6x} = -\frac{1}{6} + \frac{1}{6} = 0.$$

(3) We have $\left(\cos x - 1 + x^2/2\right)\big|_{x=0} = 0$ and $x^4\big|_{x=0} = 0$, and so

$$\lim_{x \to 0} \frac{\cos x - 1 + x^2/2}{x^4} = \lim_{x \to 0} \frac{-\sin x + x}{4x^3},$$

if the latter exists. We have $\left(-\sin x + x\right)\big|_{x=0} = 0$ and $4x^3\big|_{x=0} = 0$, and so

$$\lim_{x \to 0} \frac{-\sin x + x}{4x^3} = \lim_{x \to 0} \frac{-\cos x + 1}{12x^2},$$

if the latter exists. We have $\left(-\cos x + 1\right)\big|_{x=0} = 0$ and $12x^2\big|_{x=0} = 0$, and so

$$\lim_{x \to 0} \frac{-\cos x + 1}{12x^2} = \lim_{x \to 0} \frac{\sin x}{24x} = \frac{1}{24}.$$

Thus $\lim_{x \to 0} \dfrac{\cos x - 1 + x^2/2}{x^4} = \lim_{x \to 0} \dfrac{-\sin x + x}{4x^3} = \lim_{x \to 0} \dfrac{-\cos x + 1}{12x^2} = \lim_{x \to 0} \dfrac{\sin x}{24x} = \dfrac{1}{24}.$

(4) We have $\left(\sin x - x\right)\big|_{x=0} = 0$ and $x^2\big|_{x=0} = 0$, and so

$$\lim_{x \to 0} \frac{\sin x - x}{x^2} = \lim_{x \to 0} \frac{\cos x - 1}{2x},$$

if the latter exists. We have $\left(\cos x - 1\right)\big|_{x=0} = 0$ and $2x\big|_{x=0} = 0$, and so

$$\lim_{x \to 0} \frac{\cos x - 1}{2x} = \lim_{x \to 0} \frac{-\sin x}{2} = 0.$$

Thus

$$\lim_{x \to 0} \left(\frac{1}{x} - \frac{1}{\sin x}\right) = \lim_{x \to 0} \frac{\sin x - x}{x \sin x} = \lim_{x \to 0} \frac{\sin x - x}{x^2} \cdot \frac{x}{\sin x}$$

$$= \lim_{x \to 0} \frac{\sin x - x}{x^2} \lim_{x \to 0} \frac{x}{\sin x} = 0 \cdot 1 = 0.$$

Solution to Exercise 5.69

We note that the integrand f is an odd function, that is,

$$f(x) = -f(-x) \quad \text{for all } x \in [-1/2, 1/2].$$

Indeed,

$$f(-x) = (\cos(-x)) \log \left(\frac{1 - (-x)}{1 + (-x)} \right) = (\cos x) \log \frac{1 + x}{1 - x}$$

$$= (\cos x) \log \left(\left(\frac{1 - x}{1 + x} \right)^{-1} \right) = (\cos x) \left(-\log \frac{1 - x}{1 + x} \right) = -f(x),$$

for all $x \in [-1/2, 1/2]$. Thus by Example 5.24,

$$\int_{-\frac{1}{2}}^{\frac{1}{2}} (\cos x) \cdot \log \left(\frac{1 - x}{1 + x} \right) dx = \int_{-1/2}^{1/2} f(x) dx = 0.$$

Solution to Exercise 5.70

We have for $m, n \in \mathbb{N}$, and $x \in \mathbb{R}$ that

$$\cos((m + n)x) = (\cos(mx))(\cos(nx)) - (\sin(mx))(\sin(nx)),$$

$$\cos((m - n)x) = (\cos(mx))(\cos(nx)) + (\sin(mx))(\sin(nx)),$$

and so $2(\sin(mx))(\sin(nx)) = \cos((m - n)x) - \cos((m + n)x)$.
If $m \neq n$, then

$$\int_{-\pi}^{\pi} (\sin(mx))(\sin(nx)) dx = \frac{1}{2} \int_{-\pi}^{\pi} (\cos((m - n)x) - \cos((m + n)x)) dx$$

$$= \frac{1}{2} \left(\frac{\sin((m - n)x)}{m - n} - \frac{\sin((m + n)x)}{m + n} \right) \Big|_{-\pi}^{\pi} = 0.$$

If $m = n$, then

$$\int_{-\pi}^{\pi} (\sin(mx))(\sin(nx)) dx = \frac{1}{2} \int_{-\pi}^{\pi} \left(1 - \cos((m + n)x) \right) dx$$

$$= \frac{1}{2} \left(2\pi - \frac{\sin((m + n)x)}{m + n} \Big|_{-\pi}^{\pi} \right) = \pi.$$

Similarly

$$\int_{-\pi}^{\pi} (\cos(mx))(\cos(nx)) dx = \frac{1}{2} \int_{-\pi}^{\pi} \left(\cos((m - n)x) + \cos((m + n)x) \right) dx = \begin{cases} 0 & \text{if } m \neq n, \\ \pi & \text{if } m = n. \end{cases}$$

Also, $\int_{-\pi}^{\pi} \underbrace{(\cos(mx))}_{\text{even}} \underbrace{(\sin(nx))}_{\text{odd}} dx = 0.$

Solution to Exercise 5.71

(1) For any integer k and for $x \in \mathbb{R}$, we have

$$\cos(k(x+2\pi)) = \cos(kx + 2\pi k) = \cos(kx) \text{ and}$$
$$\sin(k(x+2\pi)) = \sin(kx + 2\pi k) = \sin(kx).$$

Thus

$$f(x+2\pi) = a_0 + \sum_{k=1}^{n} (a_k \cos(k(x+2\pi)) + b_k \sin(k(x+2\pi)))$$

$$= a_0 + \sum_{k=1}^{n} (a_k \cos(kx) + b_k \sin(kx)) = f(x).$$

(2) We have

$$\frac{1}{2\pi} \int_{-\pi}^{\pi} f(x)dx = \frac{1}{2\pi} \int_{-\pi}^{\pi} \left(a_0 + \sum_{k=1}^{n} (a_k \cos(kx) + b_k \sin(kx)) \right) dx$$

$$= \frac{1}{2\pi} \int_{-\pi}^{\pi} a_0 dx + \frac{1}{2\pi} \sum_{k=1}^{n} \left(a_k \int_{-\pi}^{\pi} \cos(kx)dx + b_k \underbrace{\int_{-\pi}^{\pi} \sin(kx)dx}_{\text{odd}} \right)$$

$$= \frac{1}{2\pi} a_0 \cdot 2\pi + \frac{1}{2\pi} \sum_{k=1}^{n} a_k \left(-\frac{\sin(kx)}{k} \right) \Big|_{-\pi}^{\pi} + 0 = a_0 + 0 = a_0.$$

Using the result from Exercise 5.70, for $\ell = 1, \cdots, n$, we have

$$\frac{1}{\pi} \int_{-\pi}^{\pi} f(x) \cos(\ell x)dx = \frac{1}{\pi} \int_{-\pi}^{\pi} a_0 \cos(\ell x)dx + \sum_{k=1}^{n} a_k \frac{1}{\pi} \int_{-\pi}^{\pi} (\cos(kx))(\cos(\ell x))dx$$

$$+ \sum_{k=1}^{n} b_k \frac{1}{\pi} \int_{-\pi}^{\pi} (\sin(kx))(\cos(\ell x))dx$$

$$= \frac{1}{\pi} a_0 \left(\frac{-\sin(\ell x)}{\ell} \right) \Big|_{-\pi}^{\pi} + a_\ell \frac{1}{\pi} \pi + 0 = a_\ell,$$

and

$$\frac{1}{\pi} \int_{-\pi}^{\pi} f(x) \sin(\ell x)dx = \frac{1}{\pi} \int_{-\pi}^{\pi} a_0 \sin(\ell x)dx + \sum_{k=1}^{n} a_k \frac{1}{\pi} \int_{-\pi}^{\pi} (\cos(kx))(\sin(\ell x))dx$$

$$+ \sum_{k=1}^{n} b_k \frac{1}{\pi} \int_{-\pi}^{\pi} (\sin(kx))(\sin(\ell x))dx$$

$$= 0 + 0 + b_\ell \frac{1}{\pi} \pi = b_\ell.$$

(3) One can use the Maple commands

```
with(plots):
```

$$\texttt{f:=sum}\left(\tfrac{2}{(k \cdot \text{Pi})} \cdot (1-(-1)^k) \cdot \sin(k \cdot x), k = 1..333 \right);$$

```
plot(f,x=-10..10)
```

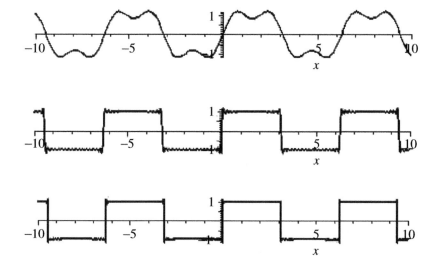

We observe that the graphs seem to look more and more like the graph of the 'square wave' function.

Solution to Exercise 5.72

(1) For $x \in (0, \pi/2]$, we have by the Mean Value Theorem that

$$\frac{\sin x - \sin 0}{x - 0} = \cos c$$

for some $c \in (0, \pi/2)$, and so

$$\frac{\sin x}{x} = \cos c < 1.$$

Thus $\sin x < x$ for $x \in (0, \pi/2]$. For $x > \pi/2$, $x > 1 \geq \sin x$. So $\sin x \neq x$ for all $x \in (0, \infty)$. Hence also $\sin(-x) = -\sin x \neq -x$ for $x \in (0, \infty)$. Thus $\sin x \neq x$ for all $x \in \mathbb{R}\backslash\{0\}$. On the other hand, $\sin 0 = 0$. So 0 is the only fixed point of sin.

(2) Consider $f : \mathbb{R} \to \mathbb{R}$ given by $f(x) = \cos x - x$, $x \in \mathbb{R}$. Then $f(0) = \cos 0 - 0 = 1 > 0$, while $f(\pi/2) = \cos(\pi/2) - \pi/2 = 0 - \pi/2 = -\pi/2 < 0$. Hence by the Intermediate Value Theorem, there exists a $c_* \in (0, \pi/2)$ such that $f(c_*) = 0$, that is, $\cos c_* = c_*$. So there exists a fixed point.

We note that for $x \geq \pi/2$, we have $f(x) = \cos x - x \leq 1 - \pi/2 < 0$. For $x \in (-\pi/2, 0]$, $f(x) = \cos x - x > 0 - x > 0$. For $x \in (-\infty, -\pi/2]$, $f(x) = \cos x - x \geq -1 + \pi/2 > 0$. So there are no fixed points in $(-\infty, 0] \cup [\pi/2, \infty)$. In $(0, \pi/2)$, cos is strictly decreasing, and so cos can have at most one fixed point there (for otherwise if $0 < c_1 < c_2 < \pi/2$ are two fixed points, then we get the contradiction that $\cos c_1 = c_1 < c_2 = \cos c_2$).

(3) Consider the sequence $(x_n)_{n \geq 0}$ defined by $x_0 = $ any real number, and $x_{n+1} = \cos x_n$ for $n > 0$. Then $x_{n+1} - x_n = \cos x_n - \cos x_{n-1}$. We note that $x_1 = \cos x_0 \in [-1, 1]$, and so

$x_2 = \cos x_1 \in [0, 1]$, and $x_{n+1} = \cos x_n \in [0, 1]$ for all $n \geq 2$. Thus there is a $c_n \in (0, 1)$ such that

$$|x_{n+1} - x_n| = |\cos x_n - \cos x_{n-1}| = (\sin c_n)|x_n - x_{n-1}| \leq (\sin 1)|x_n - x_{n-1}| = r|x_n - x_{n-1}|,$$
(43)

where $r := \sin 1 \in (0, 1)$. Using $x_n = x_1 + (x_2 - x_1) + (x_3 - x_2) + \cdots + (x_n - x_{n-1})$ and (43), we can show that $(x_n)_{n \geq 0}$ converges, to say x_*. (See below.) Then

$$f(x_*) = f\left(\lim_{n \to \infty} x_n\right) = \lim_{n \to \infty} f(x_n) = \lim_{n \to \infty} x_{n+1} = x_*.$$

Thus $x_* = c_*$.

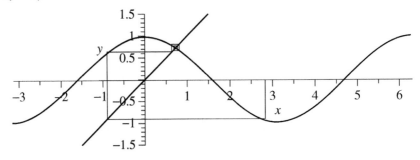

(Proof of the convergence of $(x_n)_{n \geq 0}$: We have

$$|x_{n+k} - x_n| = \left| x_1 + (x_2 - x_1) + \cdots + (x_n - x_{n-1}) + (x_{n+1} - x_n) + \cdots + (x_{n+k} - x_{n+k-1}) \right.$$
$$\left. - \left(x_1 + (x_2 - x_1) + \cdots + (x_n - x_{n-1}) \right) \right|$$

$$= \left| (x_{n+1} - x_n) + \cdots + (x_{n+k} - x_{n+k-1}) \right|$$

$$\leq |x_{n+1} - x_n| + \cdots + |x_{n+k} - x_{n+k-1}|$$

$$\leq r^{n-1}|x_2 - x_1| + \cdots + r^{n+k-1}|x_2 - x_1|$$

$$\leq (r^{n-1} + r^{n-2} + \cdots)|x_2 - x_1|$$

$$= \frac{r^{n-1}}{1 - r}|x_2 - x_1| \xrightarrow{n \to \infty} 0.$$

So the sequence $(x_n)_{n \geq 0}$ is Cauchy, and hence convergent.)

Solution to Exercise 5.73

(1) By Exercise 4.56, we have

$$\cos(x + y) = (\cos x)(\cos y) - (\sin x)(\sin y) \text{ and}$$
$$\sin(x + y) = (\sin x)(\cos y) + (\cos x)(\sin y)$$

for $x, y \in \mathbb{R}$. For $x + y \notin \pi\mathbb{Z} + \pi/2$, dividing the latter equation by $\cos(x + y)$ gives

$$\tan(x + y) = \frac{(\sin x)(\cos y) + (\cos x)(\sin y)}{(\cos x)(\cos y) - (\sin x)(\sin y)}.$$

For $x, y \notin \pi\mathbb{Z} + \pi/2$, we may divide the numerator and the denominator on the right-hand side by $(\cos x)(\cos y)$ to obtain

$$\tan(x+y) = \cfrac{\dfrac{(\sin x)(\cos y)}{(\cos x)(\cos y)} + \dfrac{(\cos x)(\sin y)}{(\cos x)(\cos y)}}{\dfrac{(\cos x)(\cos y)}{(\cos x)(\cos y)} - \dfrac{(\sin x)(\sin y)}{(\cos x)(\cos y)}} = \frac{\tan x + \tan y}{1 - (\tan x)(\tan y)}.$$

(2) Suppose that $\tan 1° \in \mathbb{Q}$. But whenever $\tan n° \in \mathbb{Q}$, for some natural number (<89), it follows from the addition formula

$$\tan (n+1)° = \frac{\tan n° + \tan 1°}{1 - (\tan n°)(\tan 1°)}$$

that also $\tan (n+1)° \in \mathbb{Q}$. Thus $\tan 2°, \tan 3°, \cdots, \tan 60°$ all belong to $\mathbb{Q}$. But from the picture shown below, we have that $\tan 60° = \tan \frac{\pi}{3} = \sqrt{3} \notin \mathbb{Q}$. This contradiction shows that $\tan 1° \notin \mathbb{Q}$.

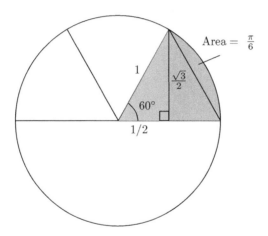

Remark 8.

(1) We remark that $1°$ is the angle $\frac{\pi}{180}$ radians, and so $60°$ is the angle $\pi/3$ radians.

(2) That $\sqrt{3}$ is irrational can be seen as follows. Let $\sqrt{3}$ be rational and suppose that for integers $p, q \neq 0$, $\sqrt{3} = p/q$, where the greatest common divisor of p and q is 1. Then $3q^2 = p^2$ and so $p^2 = 3q^2$. So 3 divides p^2 and so 3 must also divide p. Let $p = 3p'$. Then $9p'^2 = 3q^2$ and so $q^2 = 3p'^2$. So 3 divides q^2 and so 3 must also divide q. But now 3 divides both p and q, contradicting the fact that p and q had no common factor. (Also recall the method given in Exercise 1.2, which gives a different proof of the fact that $\sqrt{3}$ is not rational.)

Solution to Exercise 5.74

Suppose that $x \in \mathbb{Q}$, and let

$$x = \frac{p}{q},$$

where p, q are integers and $q > 0$. Then for $m > q$,

$$m!x = m!\frac{p}{q} = m \cdots (q+1) \cdot q \cdot \frac{p}{q} = \text{an integer.}$$

Thus $2\pi m!x \in 2\pi\mathbb{Z}$, and so $\cos(2\pi m!x) = 1$. Hence we have that for all $n \in \mathbb{N}$, and all $m > q$, $(\cos(2\pi m!x))^n = 1$, so that

$$\lim_{n\to\infty} (\cos(2\pi m!x))^n = \lim_{n\to\infty} 1 = 1.$$

Hence $\lim_{m\to\infty} \lim_{n\to\infty} (\cos(2\pi m!x))^n = \lim_{m\to\infty} 1 = 1.$

Now suppose that $x \notin \mathbb{Q}$. Then for all $m \in \mathbb{N}$, $m!2\pi x \notin \pi\mathbb{Z}$. (Indeed if $m!2\pi x = \pi k$ for some $k \in \mathbb{Z}$, then

$$x = \frac{k}{2m!} \in \mathbb{Q},$$

a contradiction.) So $\cos(m!2\pi x) \neq \pm 1$. Hence $-1 < \cos(m!2\pi x) < 1$ for all m, and so for each m we have that

$$\lim_{n\to\infty} (\cos(m!2\pi x))^n = 0.$$

Consequently $\lim_{m\to\infty} \lim_{n\to\infty} \left(\cos(2\pi m!x)\right)^n = \lim_{m\to\infty} 0 = 0.$

Finally we show the irrationality of e. Fix $m \in \mathbb{N}$. Then

$$m!e = m! \sum_{k\geq 0} \frac{1}{k!} = \text{an integer} + m! \sum_{k>m} \frac{1}{k!}.$$

So $\cos(2\pi m!e) = \cos\left(2\pi \cdot (\text{an integer}) + 2\pi m! \sum_{k>m} \frac{1}{k!}\right) = \cos\left(2\pi m! \sum_{k>m} \frac{1}{k!}\right).$

But

$$m! \left(\frac{1}{(m+1)!} + \frac{1}{(m+2)!} + \cdots + \frac{1}{(m+k)!}\right)$$

$$= \frac{1}{m+1} + \frac{1}{(m+1)(m+2)} + \cdots + \frac{1}{(m+1)\cdots(m+k)}$$

$$\leq \frac{1}{m+1} + \frac{1}{(m+1)^2} + \cdots + \frac{1}{(m+1)^k}$$

$$\leq \frac{1}{m+1} + \frac{1}{(m+1)^2} + \cdots = \frac{1/(m+1)}{1 - 1/(m+1)} = \frac{1}{m}.$$

So it follows that $m! \sum_{k>m} \frac{1}{k!} \leq \frac{1}{m} \xrightarrow{m\to\infty} 0.$

We can choose a m_0 large enough so that for all $m > m_0$, $m! \sum_{k>m} \frac{1}{k!} < \frac{1}{2}.$

Then for $m > m_0$, $(0 <)2\pi m! \sum_{k>m} \frac{1}{k!} < 2\pi \frac{1}{2} = \pi.$

Thus for all $m \geq m_0$, $\cos(2\pi m!e) = \cos\left(2\pi \sum_{k>m} \frac{1}{k!}\right) \in (-1, 1)$, and so for each $m \geq m_0$,

$$\lim_{n\to\infty} (\cos(m!2\pi e))^n = 0.$$

Hence $\lim\limits_{m\to\infty}\lim\limits_{n\to\infty}\left(\cos(2\pi m!e)\right)^n = 0$.

From the first part of the exercise, we can conclude that $e \notin \mathbb{Q}$.

Solution to Exercise 5.75

By the similarity of the two triangles, we have that

$$\frac{d\tan x}{\ell dx} = \frac{\ell}{1},$$

and so

$$\frac{d}{dx}\tan x = \ell^2 = 1 + (\tan x)^2.$$

Solution to Exercise 5.76

We know that

$$\frac{d}{dt}\tan^{-1}t = \frac{1}{1+t^2} \quad \text{for } t \in \mathbb{R}.$$

Thus $\displaystyle\int_0^1 \frac{1}{1+t^2}\,dt = \int_0^1 \frac{d}{dt}\tan^{-1}t\,dt = \tan^{-1}t\Big|_0^1 = \frac{\pi}{4} - 0 = \frac{\pi}{4}$.

Solution to Exercise 5.77

See the following picture.

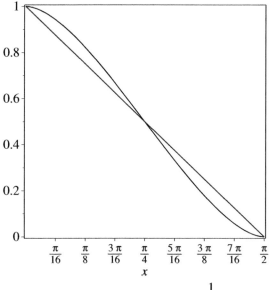

Symmetry in the graph of $\dfrac{1}{1+(\tan x)^{\sqrt{3}}}$.

There is visible symmetry about the point $\left(\frac{\pi}{4}, \frac{1}{2}\right)$. If we cut out the rectangle and cut along the graph of the function, then we get two pieces of paper that overlap perfectly, that is they are congruent.

Let $f(x) := \dfrac{1}{1 + (\tan x)^{\sqrt{3}}}, x \in [0, \pi/2)$. Let $0 < y < \frac{\pi}{4}$. Then using Exercise 5.73(1),

$$f\left(\frac{\pi}{4} + y\right) = \frac{1}{1 + \left(\tan\left(y + \frac{\pi}{4}\right)\right)^{\sqrt{3}}} = \frac{1}{1 + \left(\dfrac{\tan y + 1}{1 - \tan y}\right)^{\sqrt{3}}}$$

$$= \frac{(1 - \tan y)^{\sqrt{3}}}{(1 - \tan y)^{\sqrt{3}} + (1 + \tan y)^{\sqrt{3}}} = 1 - \frac{(1 + \tan y)^{\sqrt{3}}}{(1 + \tan y)^{\sqrt{3}} + (1 - \tan y)^{\sqrt{3}}}$$

$$= 1 - \frac{1}{1 + \left(\dfrac{1 - \tan y}{1 + \tan y}\right)^{\sqrt{3}}} = 1 - \frac{1}{1 + \left(\tan\left(\frac{\pi}{4} - y\right)\right)^{\sqrt{3}}} = 1 - f\left(\frac{\pi}{4} - y\right).$$

This symmetry property implies that

$$\int_0^{\frac{\pi}{2}} \frac{1}{1 + (\tan x)^{\sqrt{3}}}\,dx$$

$$= \frac{1}{2} \cdot \left(\text{Area of the rectangle formed by the four points } (0,0), \left(\frac{\pi}{2}, 0\right), \left(\frac{\pi}{2}, 1\right), (0, 1)\right)$$

$$= \frac{1}{2} \cdot \frac{\pi}{2} \cdot 1 = \frac{\pi}{4}.$$

Solution to Exercise 5.78

By the Differentiable Inverse Theorem, we have that at $y = \sin x \in (-1, 1)$,

$$(\sin^{-1})'(y) = \frac{1}{\sin' x} = \frac{1}{\cos x} = \frac{1}{\sqrt{1 - (\sin x)^2}} = \frac{1}{\sqrt{1 - y^2}}.$$

Thus by the Fundamental Theorem of Calculus,

$$\int_0^y \frac{1}{\sqrt{1 - t^2}}\,dt = \sin^{-1} y - \sin^{-1} 0 = \sin^{-1} y - 0 = \sin^{-1} y.$$

In order to find the other integral, we use Integration by Substitution/Change of Variables:

$$t = \sin u,$$
$$dt = \cos u\, du,$$
$$t = 0 \Rightarrow u = 0,$$
$$t = y \Rightarrow u = \sin^{-1} y.$$

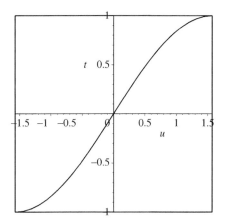

Then

$$\int_0^y \sqrt{1-t^2}dt = \int_0^{\sin^{-1}y} (\cos u) \cdot \cos u\, du = \int_0^{\sin^{-1}y} \frac{\cos(2u)+1}{2}du$$

$$= \left.\frac{\sin(2u)}{4}\right|_0^{\sin^{-1}y} + \frac{\sin^{-1}y}{2}$$

$$= \frac{\sin(2\sin^{-1}y)}{4} + \frac{\sin^{-1}y}{2}$$

$$= \frac{2(\sin(\sin^{-1}y))(\cos(\sin^{-1}y))}{4} + \frac{\sin^{-1}y}{2}$$

$$= \frac{y\sqrt{1-y^2}}{2} + \frac{\sin^{-1}y}{2}.$$

Solution to Exercise 5.79

(1) We have for all real t that $\cos t \leq 1$, and so for $x \geq 0$, we have

$$\int_0^x \cos t\, dt \leq \int_0^x 1 dt,$$

that is, $\sin x - \sin 0 = \sin x - 0 = \sin x \leq x - 0 = x$.

(2) For all $x \geq 0$, we have $\sin x \leq x$. But $\cos$ is decreasing in $[0, \pi/2]$ and so

$$\cos(\sin x) \geq \cos x.$$

Similarly, since $\alpha := \cos x \geq 0$ for $x \in [0, \pi/2]$, we have $\sin\alpha \leq \alpha$, that is,

$$\sin(\cos x) \leq \cos x.$$

For $x \in [0, \pi/2]$, we have that $\cos x \geq 0$, and moreover $\cos x \leq 1 \leq \pi/2$. Thus we also have $\sin(\cos x) \geq 0$ for $x \in [0, \pi/2]$. Hence for $x \in [0, \pi/2]$, we have

$$\cos(\sin x) \geq \cos x = |\cos x| \geq \sin(\cos x) = |\sin(\cos x)|.$$

Now if $x \in [-\pi/2, 0]$, then $\cos(\sin(-x)) \geq |\cos(-x)| \geq |\sin(\cos(-x))|$, and so, using the facts that $\cos$ is even and $\sin$ is odd, we obtain

$$\cos(\sin(-x)) = \cos(-\sin x) = \cos(\sin x) \geq |\cos(-x)| = |\cos x|$$

$$\geq |\sin(\cos(-x))| = |\sin(\cos x)|.$$

Thus $\cos(\sin x) \geq |\cos x| \geq |\sin(\cos x)|$ holds for all $x \in [-\pi/2, \pi/2]$. But we know that each of these functions is periodic with period π, as shown below, and so the inequalities extend to all real x:

$$\cos(\sin(x + \pi)) = \cos((\sin x)(\cos \pi) + (\cos x)(\sin \pi)) = \cos(-\sin x) = \cos(\sin x),$$

$$|\cos(x + \pi)| = |(\cos x)(\cos \pi) - (\sin x)(\sin \pi)| = |-\cos x| = |\cos x|,$$

$$|\sin(\cos(x + \pi))| = |\sin(-\cos x)| = |-\sin(\cos x)| = |\sin(\cos x)|.$$

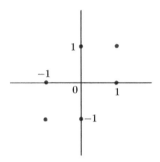

Solution to Exercise 5.80

(1) The point with Cartesian coordinates $(1, 1)$ has polar coordinates $(\sqrt{2}, \frac{\pi}{4})$.

(2) The point with Cartesian coordinates $(1, 0)$ has polar coordinates $(1, 0)$.

(3) The point with Cartesian coordinates $(0, 1)$ has polar coordinates $(1, \frac{\pi}{2})$.

(4) The point with Cartesian coordinates $(-1, 0)$ has polar coordinates $(1, \pi)$.

(5) The point with Cartesian coordinates $(-1, -1)$ has polar coordinates $(\sqrt{2}, -\frac{3\pi}{4})$.

(6) The point with Cartesian coordinates $(0, -1)$ has polar coordinates $(1, -\frac{\pi}{2})$.

Solution to Exercise 5.81

Using $x = r \cos \theta$ and $y = r \sin \theta$, where (r, θ), denote the polar coordinates of the point with Cartesian coordinates (x, y), $y = 3x + 1$ gives

$$r \sin \theta = 3 \cdot r \cos \theta + 1.$$

Using Maple, one can obtain the following picture.

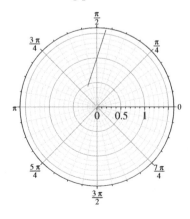

Solution to Exercise 5.82

Multiplying both sides of the equation $r = (\tan\theta)(\sec\theta)$ by r and rearranging, we obtain $(r\cos\theta)^2 = r\sin\theta$, and so $x^2 = y$, where (x, y) are the Cartesian coordinates of the point having polar coordinates (r, θ).

Solution to Exercise 5.83

Since $r = (2 + \cos\theta)^{-1}$, we obtain $2r + r\cos\theta = 1$. Using $x = r\cos\theta$ and $r = \sqrt{x^2 + y^2}$, we have $2\sqrt{x^2 + y^2} + x = 1$ and so $2\sqrt{x^2 + y^2} = 1 - x$. Squaring both sides and rearranging gives $3x^2 + 2x + 4y^2 = 1$, and finally

$$\frac{(x + 1/3)^2}{(2/3)^2} + \frac{y^2}{(1/\sqrt{3})^2} = 1.$$

Using Maple, one can obtain the following picture.

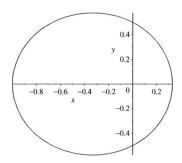

Solution to Exercise 5.84

We have that the area enclosed is

$$4\int_0^a b\sqrt{1 - \frac{x^2}{a^2}}\,dx = 4\int_0^{\frac{\pi}{2}} b\sqrt{1 - (\sin\theta)^2}\, a\cos\theta d\theta \quad \text{(using the substitution } x = a\sin\theta\text{)}$$

$$= 4ab\int_0^{\frac{\pi}{2}} (\cos\theta)^2 d\theta = 4ab\int_0^{\frac{\pi}{2}} \frac{1 + \cos(2\theta)}{2}\,d\theta$$

$$= 4ab\left(\frac{\sin(2\theta)}{4}\Big|_0^{\frac{\pi}{2}} + \frac{1}{2}\frac{\pi}{2}\right) = 4ab\left(0 + \frac{\pi}{4}\right) = \pi ab.$$

When $a = b$, we recover the expression for the area of the disk (πa^2).

Solution to Exercise 5.85

Let (b, c) be the rightmost intersection point, and (a, c) be the leftmost intersection point. We wish to find c such that

$$\int_0^a (c - (2x - 3x^3))dx = \int_a^b (2x - 3x^3 - c)dx,$$

and so $\int_0^b (2x - 3x^3 - c)dx = 0$, that is, $b^2 - \frac{3}{4}b^4 = bc$.

But also as (b, c) is an intersection point, $2b - 3b^3 = c$. Multiplying this equation by b and subtracting $b^2 - \frac{3}{4}b^4 = bc$ gives $b^2 - \frac{9}{4}b^4 = 0$, and so $b = \frac{2}{3}$. Hence

$$c = 2b - 3b^3 = 2 \cdot \frac{2}{3} - 3 \cdot \frac{8}{27} = \frac{4}{9}.$$

To validate the solution, we check that $(2/3, 4/9)$ is indeed the rightmost intersection point of $y = 4/9$ and $y = 2x - 3x^3$: the zeros of $2x - 3x^3 - 4/9 = (2/3 - x)(3x^2 + 2x - 2/3)$ other than $2/3$ are $(-1 \pm \sqrt{3})/2$, which are less than $2/3$.

Solution to Exercise 5.86

(1) By the side-angle-side rule, the two triangles OAB and $OA'B'$ are similar, and thus

$$\frac{\ell(AB)}{\ell(A'B')} = \frac{\ell(OB)}{\ell(OB')} = \frac{d/2}{d'/2} = \frac{d}{d'}.$$

But

$$\frac{\ell(AB)}{\ell(A'B')} \approx \frac{C_d/n}{C_{d'}/n} = \frac{C_d}{C_{d'}},$$

and as $n \to \infty$, we expect the error in the above approximation to tend to 0, so that

$$\frac{C_d}{C_{d'}} = \frac{d}{d'},$$

that is, $\frac{C_d}{d} = \frac{C_{d'}}{d'}$. So the ratio of the circumference of a circle to its diameter is a constant.

(2) The area of the polygon is the area of the shaded parallelogram, and this is approximately the height (which differs from radius $d/2$ by a small amount) times the length of the base (which differs from half the circumference by a tiny amount). As $n \to \infty$, we expect the errors above to go to 0, and so the area of the circle should be

$$\frac{d}{2} \cdot \frac{C_d}{2} = \frac{d}{2} \cdot \frac{\pi \cdot d}{2} = \pi \cdot \left(\frac{d}{2}\right)^2 = \pi \cdot r^2.$$

Solution to Exercise 5.87

We have that the area enclosed is

$$4 \int_0^{\frac{\pi}{4}} \frac{2\cos(2\theta)}{2} d\theta = 4 \int_0^{\frac{\pi}{4}} \cos(2\theta) d\theta = 4 \frac{\sin(2\theta)}{2} \Big|_0^{\frac{\pi}{4}} = 4 \cdot \frac{1}{2} = 2.$$

Solution to Exercise 5.88

The doughnut is the solid of revolution of the planar region bounded by the two curves $x \mapsto R + \sqrt{r^2 - x^2}$ and $x \mapsto R - \sqrt{r^2 - x^2}$. Thus the volume is given by

$$\int_{-r}^r \pi \left((R + \sqrt{r^2 - x^2})^2 - (R - \sqrt{r^2 - x^2})^2 \right) dx$$

$$= \int_{-r}^r \pi \left(r^2 - x^2 + R^2 + 2R\sqrt{r^2 - x^2} - R^2 - (r^2 - x^2) + 2R\sqrt{r^2 - x^2} \right) dx$$

$$= 4\pi R \int_{-r}^r \sqrt{r^2 - x^2} dx.$$

But $\displaystyle\int_{-r}^{r} \sqrt{r^2 - x^2}\,dx = \frac{\pi r^2}{2}$, as the integral is the area of a semicircular disk of radius r.

So the volume of the doughnut is

$$\int_{-r}^{r} \pi\left((R + \sqrt{r^2 - x^2})^2 - (R - \sqrt{r^2 - x^2})^2\right) dx$$

$$= 4\pi R \int_{-r}^{r} \sqrt{r^2 - x^2}\,dx = 4\pi R \cdot \frac{\pi r^2}{2} = 2\pi^2 R r^2.$$

Solution to Exercise 5.89

The volume is given by

$$\int_{-a}^{a} \pi\left(\left(b\sqrt{1 - \frac{x^2}{a^2}}\right)^2 - 0^2\right) dx = \int_{-a}^{a} \pi b^2 \left(1 - \frac{x^2}{a^2}\right) dx = 2\pi b^2 \int_{0}^{a} \left(1 - \frac{x^2}{a^2}\right) dx$$

$$= 2\pi b^2 \left(a - \frac{a^3}{3a^2}\right) = 2\pi b^2 a \frac{2}{3} = \frac{4\pi}{3} b^2 a.$$

If $a = b$, then we recover the expression for the volume of a sphere, $\frac{4}{3}\pi a^3$.

Solution to Exercise 5.90

See the picture below, from which we see that the remaining portion of the ball is a solid of revolution of the planar region bounded by the curves $x \mapsto \sqrt{4 - x^2}$ and $x \mapsto \sqrt{3}$ for $x \in [-1, 1]$.

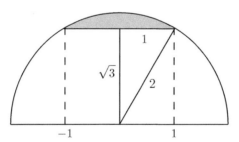

Thus the volume of the remaining portion is

$$\int_{-1}^{1} \pi\left((\sqrt{4 - x^2})^2 - \sqrt{3}^2\right) dx = \int_{-1}^{1} \pi(4 - x^2 - 3)\,dx = \pi \int_{-1}^{1} (1 - x^2)\,dx$$

$$= 2\pi \left(1 - \frac{1}{3}\right) = \frac{4\pi}{3}.$$

Thus the volume cutout is the volume of the sphere minus the volume of the remaining portion, that is,

$$\frac{4}{3}\pi 2^3 - \frac{4\pi}{3} = \frac{28\pi}{3}.$$

Solution to Exercise 5.91

The volume when the height is h is given by

$$V = \int_0^h \pi (y^{1/m})^2 dy.$$

Thus by the Fundamental Theorem of Calculus and by the Chain Rule,

$$\frac{dV}{dt} = \frac{dV}{dh} \cdot \frac{dh}{dt} = \pi (h^{1/m})^2 \cdot \frac{dh}{dt}.$$

But since $\dfrac{dV}{dt}$ is proportional to $\sqrt{h}$ and as we require that

$$\frac{dh}{dt} = \text{constant},$$

(since we want the height to decrease linearly with time), we must have that

$$\frac{2}{m} = \frac{1}{2},$$

that is, $m = 4$.

Solution to Exercise 5.92

The volume is given by

$$V(a) := a^2 + a = \int_0^a \pi (f(x))^2 dx.$$

Thus by the Fundamental Theorem of Calculus,

$$2a + 1 = \pi (f(a))^2.$$

Since f only assumes nonnegative values, it follows from here that

$$f(x) = \sqrt{\frac{2x+1}{\pi}}, \quad x \geq 0.$$

Solution to Exercise 5.93

We have

$$x'(t) = \frac{1}{3}\frac{3}{2}(2t+3)^{1/2}2 = \sqrt{2t+3}, \quad \text{and} \quad y'(t) = \frac{2t}{2} + 1 = t + 1,$$

and so $(x'(t))^2 + (y'(t))^2 = 2t + 3 + t^2 + 2t + 1 = t^2 + 4t + 4 = (t+2)^2$. Hence it follows that the distance travelled is

$$\int_0^3 \sqrt{(x'(t))^2 + (y'(t))^2}dt = \int_0^3 (t+2)dt = \left(\frac{t^2}{2} + 2t\right)\Big|_0^3 = \frac{9}{2} + 6 = \frac{21}{2}.$$

Thus the average speed is $\dfrac{\text{distance travelled}}{\text{time taken}} = \dfrac{21/2}{3} = \dfrac{7}{2}.$

Solution to Exercise 5.94

The curve $t \mapsto \gamma(t) = (x(t), y(t))$, $t \in [-\pi, \pi]$ is given by

$$x(t) = r(t)\cos t = 2(1 + \cos t)\cos t, \text{ and}$$
$$y(t) = r(t)\sin t = 2(1 + \cos t)\sin t.$$

Thus

$$x'(t) = 2(-\sin t)\cos t + 2(1 + \cos t)(-\sin t) = -2\sin t - 4\sin t\cos t$$
$$= -2\sin t - 2\sin(2t),$$
$$y'(t) = 2(-\sin t)\sin t + 2(1 + \cos t)\cos t = 2\cos t + 2((\cos t)^2 - (\sin t)^2)$$
$$= 2\cos t + 2\cos(2t).$$

Hence

$$(x'(t))^2 + (y'(t))^2 = 8\left(1 + (\cos(2t))\cos t + (\sin t)\sin(2t)\right)$$
$$= 8(1 + \cos t) = 8 \cdot 2 \cdot \left(\cos\frac{t}{2}\right)^2.$$

Thus the arc length is $\int_{-\pi}^{\pi}\sqrt{(x'(t))^2 + (y'(t))^2}dt = \int_{-\pi}^{\pi} 4\cos\frac{t}{2}dt = 8\sin\frac{t}{2}\Big|_{-\pi}^{\pi} = 16.$

Solution to Exercise 5.95

Consider the curve $\gamma : [0, 2\pi] \to \mathbb{R}^2$ given by

$$\gamma(t) = (a\cos t, b\sin t), \quad t \in [0, 2\pi].$$

Then $\gamma(t)$ goes round the perimeter of the ellipse once as t increases from 0 to 2π. We have that the arclength of γ is

$$\int_0^{2\pi}\sqrt{(-a\sin t)^2 + (b\cos t)^2}dt = \int_0^{2\pi}\sqrt{b^2 + (a^2 - b^2)(\sin t)^2}dt$$
$$= b\int_0^{2\pi}\sqrt{1 - \left(1 - \frac{a^2}{b^2}\right)(\sin t)^2}dt$$
$$= b\int_0^{2\pi}\sqrt{1 - k^2(\sin t)^2}dt,$$

where $k := \sqrt{1 - \frac{a^2}{b^2}}$.

If $b = a$, then $k = 0$, and we get the circumference $2\pi a$ of a circle with radius a.

Solution to Exercise 5.96

The circular arc BC has the parametrisation $(\cos t, \sin t)$, where $t \in [0, \theta]$. Thus the arc length of the circular arc BC is given by

$$\int_0^{\theta}\sqrt{(-\sin t)^2 + (\cos t)^2}dt = \int_0^{\theta} 1 dt = \theta.$$

Using the fact that $\ell(AC) \le \ell(BC) \le$ circular arc length BC, we obtain

$$\ell(AC) = \sin\theta \le \ell(BC) \le \text{ circular arc length } BC = \theta.$$

Solution to Exercise 5.97

By the Fundamental Theorem of Calculus,

$$(f(x))^2 - (f(1))^2 = \int_1^x \frac{d}{dx}(f(x))^2 dx = \int_1^x 2f(x)f'(x) \le 2\int_1^x f(x)\sqrt{1+(f'(x))^2}dx$$

$$\le 2\int_1^\infty f(x)\sqrt{1+(f'(x))^2}dx = \frac{S}{\pi} < \infty.$$

So for all $x \in [1, \infty)$, we have

$$0 \le f(x) \le \sqrt{f(1)^2 + \frac{S}{\pi}} =: M.$$

Hence we have

$$V = \int_1^\infty \pi(f(x))^2 dx \le \int_1^\infty \pi M \cdot f(x) \cdot 1 dx \le M\pi \int_1^\infty f(x)\sqrt{1+(f'(x))^2}dx = \frac{M}{2} \cdot S < \infty.$$

So V is finite too.

Solution to Exercise 5.98

The surface area is given by

$$\int_0^{2\pi} 2\pi(R + r\sin t)\sqrt{(-r\sin t)^2 + (r\cos t)^2}dt = \int_0^{2\pi} 2\pi(R + r\sin t)rdt$$

$$= 2\pi r\left(R \cdot 2\pi + r(-\cos t)\Big|_0^{2\pi}\right)$$

$$= 2\pi r(R \cdot 2\pi + 0) = 4\pi^2 Rr.$$

Solution to Exercise 5.99

We have that the surface area is

$$A := \int_0^a 2\pi a\left(\cosh\frac{x}{a}\right)\sqrt{1+\left(\sinh\frac{x}{a}\right)^2}dx$$

$$= \int_0^a 2\pi a\left(\cosh\frac{x}{a}\right)\left(\cosh\frac{x}{a}\right)dx = \int_0^a 2\pi a\left(\cosh\frac{x}{a}\right)^2 dx,$$

while the volume is given by $V := \int_0^a \pi a^2\left(\cosh\frac{x}{a}\right)^2 dx$, and so

$$\frac{A}{V} = \frac{\int_0^a 2\pi a\left(\cosh\frac{x}{a}\right)^2 dx}{\int_0^a \pi a^2\left(\cosh\frac{x}{a}\right)^2 dx} = \frac{2}{a}.$$

Solutions to the exercises from Chapter 6

Solution to Exercise 6.1

We have

$$\tan^{-1}\frac{1}{2n^2} = \tan^{-1}\frac{(2n+1)-(2n-1)}{1+(2n+1)(2n-1)}$$

$$= \tan^{-1}\frac{1}{2n-1} - \tan^{-1}\frac{1}{2n+1},$$

and so the partial sums telescope to give

$$\sum_{n=1}^{\infty}\tan^{-1}\frac{1}{2n^2} = \tan^{-1}\frac{1}{1} - \lim_{n\to\infty}\tan^{-1}\frac{1}{2n+1}$$

$$= \frac{\pi}{4} - 0 = \frac{\pi}{4}.$$

Solution to Exercise 6.2

We have

$$\frac{1}{1-x} + \frac{1}{1+x} + \frac{2}{1+x^2} + \frac{4}{1+x^4} + \cdots + \frac{2^n}{1+x^{2^n}}$$

$$= \frac{2}{1-x^2} + \frac{2}{1+x^2} + \frac{4}{1+x^4} + \cdots + \frac{2^n}{1+x^{2^n}}$$

$$= \frac{4}{1-x^4} + \frac{4}{1+x^4} + \cdots + \frac{2^n}{1+x^{2^n}}$$

$$\vdots$$

$$= \frac{2^n}{1-x^{2^n}} + \frac{2^n}{1+x^{2^n}} = \frac{2^{n+1}}{1-x^{2^{n+1}}}.$$

Since $x > 1$, we can write $x = 1 + h$ with $h > 0$, and so for integers $k > 2$, we have

$$x^k = (1+h)^k > \binom{k}{2}h^2 = \frac{k\cdot(k-1)}{2}h^2.$$

Thus $0 < \frac{k}{x^k} < \frac{2}{(k-1)\cdot h^2}$, and so by the Sandwich Theorem, $\lim_{k\to\infty}\frac{k}{x^k} = 0$.

Hence for $x > 1$, we obtain

$$\frac{1}{1-x} + \frac{1}{1+x} + \frac{2}{1+x^2} + \frac{4}{1+x^4} + \cdots + \frac{2^n}{1+x^{2^n}} + \cdots = \lim_{n\to\infty}\frac{2^{n+1}}{1-x^{2^{n+1}}}$$

$$= \lim_{n\to\infty}\frac{\frac{2^{n+1}}{x^{2^{n+1}}}}{\frac{1}{x^{2^{n+1}}}-1}$$

$$= \frac{0}{0-1} = 0,$$

and so $\frac{1}{1+x} + \frac{2}{1+x^2} + \frac{4}{1+x^4} + \cdots + \frac{2^n}{1+x^{2^n}} + \cdots = \frac{1}{x-1}.$

Solution to Exercise 6.3

We have

$$\frac{1}{F_{n-1}F_{n+1}} = \frac{1}{F_{n-1}(F_n + F_{n-1})} = \left(\frac{1}{F_{n-1}} - \frac{1}{F_n + F_{n-1}}\right)\frac{1}{F_n} = \frac{1}{F_{n-1}F_n} - \frac{1}{F_nF_{n+1}}.$$

It is easy to see by induction that $F_n \geq n$ for all $n \in \mathbb{N}$. Indeed, $F_1 = 1 \geq 1$, and if $F_n \geq n$ for some n, then $F_{n+1} = F_{n-1} + F_n \geq 1 + F_n \geq 1 + n$. (Here the first inequality follows from the obvious fact that $n \mapsto F_n$ is increasing, so that $F_1 = 1 \leq F_2 \leq F_3 \leq \cdots \leq F_{n-1}$.) Thus

$$\sum_{k=2}^{n} \frac{1}{F_{k-1}F_{k+1}} = \frac{1}{F_1F_2} - \frac{1}{F_2F_3} + \frac{1}{F_2F_3} - \frac{1}{F_3F_4} + \cdots + \frac{1}{F_{n-1}F_n} - \frac{1}{F_nF_{n+1}}$$

$$= \frac{1}{F_1F_2} - \frac{1}{F_nF_{n+1}} \xrightarrow{n\to\infty} \frac{1}{2}.$$

Solution to Exercise 6.4

We have

$$2\cot x = 2\frac{\cos x}{\sin x} = 2\frac{(\cos(x/2))^2 - (\sin(x/2))^2}{2(\sin(x/2))(\cos(x/2))} = \cot\frac{x}{2} - \tan\frac{x}{2},$$

for $0 < x < \pi/2$, and so $\tan\dfrac{x}{2} = \cot\dfrac{x}{2} - 2\cot x$.

We have

$$\sum_{n=1}^{\infty} \frac{1}{2^n}\tan\frac{\pi/4}{2^n} = \lim_{N\to\infty}\sum_{n=1}^{N}\frac{1}{2^n}\tan\frac{\pi/4}{2^n} = \lim_{N\to\infty}\sum_{n=1}^{N}\frac{1}{2^n}\left(\cot\frac{\pi/4}{2^n} - 2\cot\frac{2\pi/4}{2^n}\right)$$

$$= \lim_{N\to\infty}\sum_{n=1}^{N}\left(\frac{1}{2^n}\cot\frac{\pi/4}{2^n} - \frac{1}{2^{n-1}}\cot\frac{\pi/4}{2^{n-1}}\right)$$

$$= \lim_{N\to\infty}\frac{1}{2^N}\cot\frac{\pi/4}{2^N} - \cot(\pi/4) = -1 + \lim_{N\to\infty}\frac{1}{2^N}\cot\frac{\pi/4}{2^N}$$

$$= -1 + \lim_{N\to\infty}\frac{1}{2^N}\frac{\cos\frac{\pi/4}{2^N}}{\sin\frac{\pi/4}{2^N}}$$

$$= -1 + \lim_{N\to\infty}\cos\frac{\pi/4}{2^N}\cdot\frac{4}{\pi}\cdot\lim_{N\to\infty}\frac{\frac{\pi/4}{2^N}}{\sin\frac{\pi/4}{2^N}}$$

$$= -1 + 1\cdot\frac{4}{\pi}\cdot 1 = \frac{4}{\pi} - 1.$$

Solution to Exercise 6.5

The sequence $(1/n)_{n\in\mathbb{N}}$ is convergent with limit 0. The function $\cos : \mathbb{R} \to \mathbb{R}$ is continuous. Since continuous functions preserve convergence of sequences, it follows that $(\cos(1/n))_{n\in\mathbb{N}}$

is convergent with limit $\cos 0 = 1$. So we can conclude that since

$$\lim_{n\to\infty} \cos \frac{1}{n} = 1 \neq 0,$$

the series $\sum_{n=1}^{\infty} \cos \frac{1}{n}$ does not converge.

Solution to Exercise 6.6

Let $s_n := a_1 + \cdots + a_n$ denote the nth partial sum of the series for $n \in \mathbb{N}$. Since the series converges, $(s_n)_{n\in\mathbb{N}}$ converges to some limit L. So the sequence $(s_{2n} - s_n)_{n\in\mathbb{N}}$ converges to $L - L = 0$. We have

$$s_{2n} - s_n = a_{n+1} + \cdots + a_{2n} \geq \underbrace{a_{2n} + \cdots + a_{2n}}_{n \text{ times}} = n \cdot a_{2n} \geq 0.$$

Thus by the Sandwich Theorem, we obtain $\lim_{n\to\infty} n \cdot a_{2n} = 0$. Hence also

$$\lim_{n\to\infty} 2n a_{2n} = 0. \tag{44}$$

Also, the sequence $(s_{2n+1} - s_{n+1})_{n\in\mathbb{N}}$ converges to $L - L = 0$. We have

$$s_{2n+1} - s_n = a_{n+2} + \cdots + a_{2n+1} \geq \underbrace{a_{2n+1} + \cdots + a_{2n+1}}_{n \text{ times}} = n \cdot a_{2n+1} \geq 0.$$

Thus by the Sandwich Theorem, we obtain $\lim_{n\to\infty} n \cdot a_{2n+1} = 0$. As the sequence $(a_n)_{n\in\mathbb{N}}$ converges to 0, we also have that $(a_{2n+1})_{n\in\mathbb{N}}$ converges to 0. Hence we obtain

$$\lim_{n\to\infty} (2n+1)a_{2n+1} = 2\lim_{n\to\infty} n a_{2n+1} + \lim_{n\to\infty} a_{2n+1} = 2 \cdot 0 + 0 = 0. \tag{45}$$

It follows now from (44) and (45) that $\lim_{n\to\infty} n a_n = 0$.

In order to show that the assumption $a_1 \geq a_2 \geq a_3 \cdots$ cannot be dropped, we consider the *lacunary series* whose n^2th term is $1/n^2$ and all other terms are zero:

$$a_1 = \frac{1}{1^2}, \; a_2 = a_3 = 0, \; a_{2^2} = \frac{1}{2^2}, \; a_{2^2+1} = \cdots = a_{3^2-1} = 0, \; a_{3^2} = \frac{1}{3^2}, \; \cdots.$$

The sequence $(s_n)_{n\in\mathbb{N}}$ of partial sums is clearly increasing since $a_n \geq 0$ for all $n \in \mathbb{N}$. We have

$$s_{n^2} = \frac{1}{1^2} + \frac{1}{2^2} + \frac{1}{3^2} + \cdots + \frac{1}{n^2},$$

which is the nth partial sum of the convergent series

$$\sum_{n=1}^{\infty} \frac{1}{n^2}.$$

It thus follows from the convergence of $(s_{n^2})_{n\in\mathbb{N}}$ that $(s_{n^2})_{n\in\mathbb{N}}$ is bounded, and so $(s_n)_{n\in\mathbb{N}}$ is bounded too (after all, given any $m \in \mathbb{N}$, there is a perfect square N^2 exceeding it, so that $s_m \leq s_{N^2} \leq M$, where M is a bound for the sequence $(s_{n^2})_{n\in\mathbb{N}}$). As the sequence $(s_n)_{n\in\mathbb{N}}$ is monotone and bounded, it is convergent. Note however that the sequence $(na_n)_{n\in\mathbb{N}}$ has as a subsequence

$$(n^2 a_{n^2})_{n\in\mathbb{N}} = (n^2 \frac{1}{n^2})_{n\in\mathbb{N}} = (1)_{n\in\mathbb{N}},$$

and so it follows that $(na_n)_{n\in\mathbb{N}}$ does not converge to 0.

Solution to Exercise 6.7

For $n \in \mathbb{N}$, define $s_n := 1 + 2r + 3r^2 + \cdots + nr^{n-1}$.

Then $rs_n = r + 2r^2 + \cdots + (n-1)r^{n-1} + nr^n$, and so

$$(1-r)s_n = s_n - rs_n = 1 + r + r^2 + \cdots + r^{n-1} - nr^n$$

$$= \frac{(1-r)(1 + r + r^2 + \cdots + r^{n-1})}{1-r} - nr^n = \frac{1-r^n}{1-r} - nr^n.$$

Thus $s_n = \dfrac{1-r^n}{(1-r)^2} - \dfrac{nr^n}{1-r}$. Let $h := \dfrac{1}{|r|} - 1 > 0$. Then for $n \geq 2$,

$$(1+h)^n = 1 + \binom{n}{1}h + \binom{n}{2}h^2 + \cdots + \binom{n}{n}h^n \geq \binom{n}{2}h^2 = \frac{n(n-1)}{2}h^2.$$

So for $n \geq 2$, $0 \leq n|r|^n = \dfrac{n}{(1+h)^n} \leq n\dfrac{2}{n(n-1)h^2} = \dfrac{2}{(n-1)h^2}$.

Hence by the Sandwich Theorem, $\lim\limits_{n \to \infty} |nr^n| = 0$. So $\lim\limits_{n \to \infty} nr^n = 0$ as well, and thus

$$\lim_{n \to \infty} s_n = \lim_{n \to \infty} \left(\frac{1-r^n}{(1-r)^2} - \frac{nr^n}{1-r} \right) = \frac{1-0}{(1-r)^2} - \frac{0}{1-r} = \frac{1}{(1-r)^2}.$$

Solution to Exercise 6.8

(1) We have $0.999 \cdots = \dfrac{9}{10} + \dfrac{9}{10^2} + \dfrac{9}{10^3} + \cdots = \dfrac{\frac{9}{10}}{1 - \frac{1}{10}} = 1 = 1.000 \cdots$.

(2) If $x = N + \dfrac{d_1}{10} + \cdots + \dfrac{d_K}{10^K} = \dfrac{10^K + d_1 10^{K-1} + \cdots + d_{K-1}10 + d_K}{10^K}$, then

$$x = \frac{p}{q},$$

where $p := 10^K + d_1 10^{K-1} + \cdots + d_{K-1}10 + d_K$ is a nonnegative integer and the denominator $q := 10^K \in \mathbb{N}$. So x is a nonnegative rational number.

(3) Let $x = N.d_1 \cdots d_n d_{n+1} \cdots d_{n+m} d_{n+1} \cdots d_{n+m} d_{n+1} \cdots d_{n+m} \cdots$. Then

$$10^n x = Nd_1 \cdots d_n . d_{n+1} \cdots d_{n+m} d_{n+1} \cdots d_{n+m} d_{n+1} \cdots d_{n+m} \cdots, \quad \text{and}$$

$$10^{n+m} x = Nd_1 \cdots d_n d_{n+1} \cdots d_{n+m} . d_{n+1} \cdots d_{n+m} d_{n+1} \cdots d_{n+m} \cdots d_{n+m} \cdots.$$

Note that $m > 0$ since we are given that the decimal expansion is *non*terminating. Upon subtracting one from the other, we obtain

$$10^{n+m} x - 10^n x = Nd_1 \cdots d_n d_{n+1} \cdots d_{n+m} - Nd_1 \cdots d_n =: p \in \mathbb{N}.$$

Thus with $q := 10^n(10^m - 1) \in \mathbb{N}$, we obtain $x = \dfrac{p}{q}$ is a positive rational number.

(4) We have $0.123123123 \cdots = \dfrac{123}{10^3} + \dfrac{123}{10^6} + \dfrac{123}{10^9} + \cdots = \dfrac{\frac{123}{1000}}{1 - \frac{1}{1000}} = \dfrac{123}{999} = \dfrac{41}{333}$.

(5) It is not rational, since the string of digits contains the digits of 10^k, having a 1 followed by a block of zeros of increasing size, making the decimal expansion nonterminating and nonrepeating. As every rational number has either a terminating or nonterminating and repeating decimal expansion, it follows that $0.12345678910111213 \cdots$ is irrational.

Solution to Exercise 6.9

(1) Consider the interval $[1, n]$, and let $\underline{\sigma}_n$ and $\overline{\sigma}_n$ be the step functions defined by

$$\underline{\sigma}_n(x) = f(k+1),$$
$$\overline{\sigma}_n(x) = f(k),$$

for $x \in [k, k+1), k = 1, \cdots, n$. As f is decreasing, for all $x \in [1, n]$ that $\underline{\sigma}_n(x) \leq f(x) \leq \overline{\sigma}_n(x)$, and so

$$\sum_{k=2}^{n} f(k) = \int_1^n \underline{\sigma}_n(x)dx \leq \int_1^n f(x)dx \leq \int_1^n \overline{\sigma}_n(x)dx = \sum_{k=1}^{n} f(k).$$

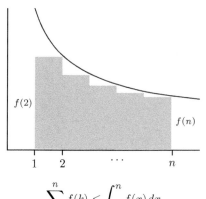

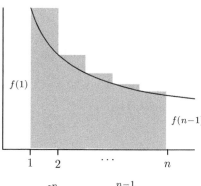

$$\sum_{k=2}^{n} f(k) \leq \int_1^n f(x)\,dx \qquad\qquad \int_1^n f(x)dx \leq \sum_{k=1}^{n-1} f(k)$$

Thus for all $n \in \mathbb{N}$, $\displaystyle\sum_{k=2}^{n} f(k) \leq \int_1^n f(x)dx \leq \sum_{k=1}^{n-1} f(k)$. We have the following cases:

1° $\displaystyle\int_1^\infty f(x)dx$ converges. The first inequality above shows that the partial sums

$$\sum_{k=1}^{n} f(k)$$

are bounded above by $f(1) + \displaystyle\int_1^\infty f(x)dx.$

Also as $f(k) \geq 0$ for all k, the partial sums are increasing. Hence

$$\sum_{n=1}^{\infty} f(n)$$

converges.

2° If $\displaystyle\int_1^\infty f(x)dx$ diverges, then from the fact that for all $n \in \mathbb{N}$

$$\int_1^n f(x)dx \leq \sum_{k=1}^{n-1} f(k),$$

it follows that the partial sums

$$\sum_{k=1}^{n-1} f(k)$$

can't form a bounded sequence, and so $\sum_{n=1}^{\infty} f(n)$ diverges.

(2) Let $f(x) := \dfrac{1}{x \log x}, x \geq 2$.

We see that $f : [2, \infty) \to (0, \infty)$, f is decreasing, and from Exercise 5.46, we know that the improper integral

$$\int_2^\infty \frac{1}{x \log x} dx$$

diverges. Hence by the Integral Test, $\sum_{n=2}^{\infty} \dfrac{1}{n \log n}$ diverges too.

(Note that we start the sum with $n = 2$ to avoid n being 1 when $\log n = 0$.)

On the other hand, with

$$g(x) := \frac{1}{x(\log x)^2}, \quad x \geq 2,$$

we have that $g : [2, \infty) \to (0, \infty)$, g is decreasing, and from Exercise 5.46, we know that the improper integral

$$\int_2^\infty \frac{1}{x(\log x)^2} dx$$

converges. Hence by the Integral Test $\sum_{n=2}^{\infty} \dfrac{1}{n(\log n)^2}$ converges too.

Solution to Exercise 6.10

Since $-1 \leq \sin n \leq 1$ for all $n \in \mathbb{N}$, we have $\left| \dfrac{\sin n}{n^2} \right| \leq \dfrac{1}{n^2}$.

Thus the sequence $(s_n)_{n \in \mathbb{N}}$ of the partial sums of the series $\sum_{n=1}^{\infty} \left| \dfrac{\sin n}{n^2} \right|$ is bounded:

$$s_n \leq \frac{1}{1^2} + \frac{1}{2^2} + \cdots + \frac{1}{n^2} \leq \sum_{n=1}^{\infty} \frac{1}{n^2} < +\infty.$$

Moreover, it is an increasing sequence, and so by the Bolzano–Weierstrass Theorem $(s_n)_{n \in \mathbb{N}}$ is convergent, that is,

$$\sum_{n=1}^{\infty} \left| \frac{\sin n}{n^2} \right| < +\infty.$$

As the series converges absolutely, it is convergent, that is, $\sum_{n=1}^{\infty} \dfrac{\sin n}{n^2}$ converges.

(When we learn the 'Comparison Test' later, we can give a one line justification.)

Solution to Exercise 6.11

We have

(1) $a_n := \dfrac{1}{n^s} \geq 0$ for all $n \in \mathbb{N}$.

(2) As $s > 0$, $(n+1)^s > n^s$ and so $\dfrac{1}{n^s} > \dfrac{1}{(n+1)^s}$ for all $n \in \mathbb{N}$.

(3) $\lim\limits_{n \to \infty} \dfrac{1}{n^s} = 0$.

Thus by the Leibniz Alternating Series Theorem, $\displaystyle\sum_{n=1}^{\infty} \dfrac{(-1)^n}{n^s}$ converges.

Solution to Exercise 6.12

We have

(1) $a_n := \dfrac{\sqrt{n}}{n+1} \geq 0$ for all $n \in \mathbb{N}$.

(2) (We want to know if for all $n \in \mathbb{N}$, $\frac{\sqrt{n}}{n+1} \geq \frac{\sqrt{n+1}}{n+2}$. In order to see this, we square both sides, and rearrange terms to arrive at the equivalent inequality $n^2 + n \geq 1$, which is clearly true.)

We note that for all $n \in \mathbb{N}$, $n^2 + n \geq 1$, and so

$$n(n+2)^2 = n^3 + 4^2 + 4n \geq n^3 + 3n^2 + 3n + 1 = (n+1)^3.$$

Rearranging, we obtain

$$\frac{n}{(n+1)^2} \geq \frac{n+1}{(n+2)^2},$$

and finally, taking square roots, we obtain $\dfrac{\sqrt{n}}{n+1} \geq \dfrac{\sqrt{n+1}}{n+2}$ for all $n \in \mathbb{N}$.

(3) $\lim\limits_{n \to \infty} \dfrac{\sqrt{n}}{n+1} = \lim\limits_{n \to \infty} \dfrac{\frac{1}{\sqrt{n}}}{1 + \frac{1}{n}} = \dfrac{0}{1+0} = 0.$

Thus by the Leibniz Alternating Series Theorem, $\displaystyle\sum_{n=1}^{\infty} (-1)^n \dfrac{\sqrt{n}}{n+1}$ converges.

Solution to Exercise 6.13

We have

(1) We know that $\sin x \geq 0$ for all $x \in [0, \pi]$. Since $0 \leq \frac{1}{n} \leq 1 < \pi$ for all $n \in \mathbb{N}$, it follows that $a_n := \sin \frac{1}{n} \geq 0$ for all $n \in \mathbb{N}$.

(2) The function $x \mapsto \sin x : [0, \frac{\pi}{2}]$ is increasing. Since $0 < \frac{1}{n} < \frac{1}{n+1} < 1 < \frac{\pi}{2}$, it follows that $\sin \frac{1}{n} \leq \sin \frac{1}{n+1}$ for all $n \in \mathbb{N}$.

(3) Finally, the function $x \mapsto \sin x : \mathbb{R} \to \mathbb{R}$ is continuous, and the sequence $\left(\frac{1}{n}\right)_{n \in \mathbb{N}}$ is convergent with limit 0, and so it follows that

$$\lim_{n \to \infty} \sin \frac{1}{n} = \sin \left(\lim_{n \to \infty} \frac{1}{n} \right) = \sin 0 = 0.$$

Thus by the Leibniz Alternating Series Theorem $\displaystyle\sum_{n=1}^{\infty} (-1)^n \sin \frac{1}{n}$ converges.

Solution to Exercise 6.14

(1) Let $(s_n)_{n \in \mathbb{N}}$ be the partial sums of the original series, and $(\sigma_n)_{n \in \mathbb{N}}$ denote the partial sums of the series obtained by inserting parentheses in the original series. Then $(\sigma_n)_{n \in \mathbb{N}}$ is a subsequence of $(s_n)_{n \in \mathbb{N}}$. As $(s_n)_{n \in \mathbb{N}}$ converges to, L, say, so does $(\sigma_n)_{n \in \mathbb{N}}$.

(2) Consider the convergent series $0 + 0 + 0 + \cdots$, with sum 0, which we can think of a series obtained by inserting parentheses in the divergent series $-1 + 1 - 1 + 1 - 1 + 1 \cdots$ giving $(-1 + 1) + (-1 + 1) + (-1 + 1) + \cdots = 0 + 0 + 0 + \cdots$.

(3) If the Harmonic Series converges to $H \in \mathbb{R}$, then

$$H = 1 + \frac{1}{2} + \frac{1}{3} + \frac{1}{4} + \frac{1}{5} + \frac{1}{6} + \frac{1}{7} + \frac{1}{8} + \frac{1}{9} + \frac{1}{10} + \cdots,$$

$$= 1 + \left(\frac{1}{2} + \frac{1}{3} + \frac{1}{4} \right) + \left(\frac{1}{5} + \frac{1}{6} + \frac{1}{7} \right) + \left(\frac{1}{8} + \frac{1}{9} + \frac{1}{10} \right) + \cdots$$

$$> 1 + 3 \cdot \frac{1}{3} + 3 \cdot \frac{1}{6} + 3 \cdot \frac{1}{9} + \cdots = 1 + 1 + \frac{1}{2} + \frac{1}{3} + \cdots = 1 + H,$$

and so $0 > 1$, a contradiction.

Solution to Exercise 6.15

We will use the Ratio Test in each part.

(1) We have that $a_n := \dfrac{n^2}{2^n} \neq 0$ for each $n \in \mathbb{N}$, and

$$\left| \frac{a_{n+1}}{a_n} \right| = \frac{(n+1)^2}{2^{n+1}} \cdot \frac{2^n}{n^2} = \frac{1}{2} \left(1 + \frac{1}{n} \right)^2,$$

and so $\displaystyle\lim_{n \to \infty} \left| \frac{a_{n+1}}{a_n} \right| = \lim_{n \to \infty} \frac{1}{2} \left(1 + \frac{1}{n} \right)^2 = \frac{1}{2}(1 + 0)^2 = \frac{1}{2} < 1.$

Thus the series $\displaystyle\sum_{n=1}^{\infty} \frac{n^2}{2^n}$ converges (absolutely).

(2) We have that $a_n := \dfrac{(n!)^2}{(2n)!} \neq 0$ for each $n \in \mathbb{N}$, and

$$\left| \frac{a_{n+1}}{a_n} \right| = \frac{((n+1)!)^2}{(2n+2)!} \cdot \frac{(2n)!}{(n!)^2} = \frac{(n+1)^2}{(2n+1)(2n+2)} = \frac{n+1}{2(2n+1)} = \frac{1 + \frac{1}{n}}{2(2 + \frac{1}{n})},$$

and so $\lim\limits_{n\to\infty}\left|\dfrac{a_{n+1}}{a_n}\right| = \lim\limits_{n\to\infty}\dfrac{1+\dfrac{1}{n}}{2(2+\dfrac{1}{n})} = \dfrac{1+0}{2(2+0)} = \dfrac{1}{4} < 1.$

Thus the series $\sum\limits_{n=1}^{\infty}\dfrac{(n!)^2}{(2n)!}$ converges (absolutely).

(3) We have that $a_n := \left(\dfrac{4}{5}\right)^n n^5 \neq 0$ for each $n \in \mathbb{N}$, and

$$\left|\dfrac{a_{n+1}}{a_n}\right| = \left(\dfrac{4}{5}\right)^{n+1}(n+1)^5 \cdot \left(\dfrac{5}{4}\right)^n \dfrac{1}{n^5} = \dfrac{4}{5}\left(\dfrac{n+1}{n}\right)^5 = \dfrac{4}{5}\left(1+\dfrac{1}{n}\right)^5,$$

and so $\lim\limits_{n\to\infty}\left|\dfrac{a_{n+1}}{a_n}\right| = \lim\limits_{n\to\infty}\dfrac{4}{5}\left(1+\dfrac{1}{n}\right)^5 = \dfrac{4}{5}(1+0)^5 = \dfrac{4}{5} < 1.$

Thus the series $\sum\limits_{n=1}^{\infty}\left(\dfrac{4}{5}\right)^n n^5$ converges (absolutely).

Solution to Exercise 6.16

We have that

$$\dfrac{n}{n^4 + n^2 + 1} \leq \dfrac{n}{n^4} = \dfrac{1}{n^3}$$

for all $n \in \mathbb{N}$. Since $\sum\limits_{n=1}^{\infty}\dfrac{1}{n^3} < +\infty$, by the Comparison Test,

$$\sum\limits_{n=1}^{\infty}\dfrac{n}{n^4 + n^2 + 1}$$

converges too. We have

$$\sum\limits_{n=1}^{\infty}\dfrac{n}{n^4 + n^2 + 1} = \sum\limits_{n=1}^{\infty}\dfrac{n}{(n^2+1)^2 - n^2}$$

$$= \sum\limits_{n=1}^{\infty}\dfrac{n}{(n^2 + 1 - n)(n^2 + 1 + n)}$$

$$= \dfrac{1}{2}\sum\limits_{n=1}^{\infty}\left(\dfrac{1}{n^2 + 1 - n} - \dfrac{1}{n^2 + 1 + n}\right)$$

$$= \dfrac{1}{2}\sum\limits_{n=1}^{\infty}\left(\dfrac{1}{(n-1)\cdot n + 1} - \dfrac{1}{n\cdot(n+1)+1}\right).$$

But the partial sum of this last series is the 'telescoping sum'

$$\dfrac{1}{0\cdot 1 + 1} - \cancel{\dfrac{1}{1\cdot 2 + 1}} + \cancel{\dfrac{1}{1\cdot 2 + 1}} - \cancel{\dfrac{1}{2\cdot 3 + 1}} + \cdots + \cancel{\dfrac{1}{(n-1)\cdot n + 1}} - \dfrac{1}{n\cdot(n+1)+1}$$

$$= 1 - \dfrac{1}{n\cdot(n+1)+1}.$$

Consequently,

$$\sum_{n=1}^{\infty} \frac{n}{n^4 + n^2 + 1} = \frac{1}{2} \lim_{n \to \infty} \left(1 - \frac{1}{n \cdot (n+1) + 1} \right) = \frac{1}{2}.$$

Solution to Exercise 6.17

As $\sum_{n=1}^{\infty} a_n^{2014}$ converges, $\lim_{n \to \infty} a_n^{2014} = 0$.

But then $\lim_{n \to \infty} |a_n|^{2014} = 0$ as well.

Thus there is an $N \in \mathbb{N}$ such that for all $n > N$, $|a_n|^{2014} < 1$, and so $|a_n| < 1$ too. But now for $n > N$,

$$|a_n^{2015}| = a_n^{2014}|a_n| < a_n^{2014},$$

and so it follows from the Comparison Test that $\sum_{n=1}^{\infty} a_n^{2015}$ converges as well.

Solution to Exercise 6.18

(1) True.

Since the series $\sum_{n=1}^{\infty} |a_n|$ converges, we have necessarily that $\lim_{n \to \infty} |a_n| = 0$.

Thus there is an $N \in \mathbb{N}$ such that for all $n > N$, $|a_n| < 1$. But now for $n > N$,

$$|a_n^2| = |a_n||a_n| < |a_n|,$$

and so it follows from the Comparison Test that $\sum_{n=1}^{\infty} a_n^2$ converges too.

(2) False.

By the Leibniz Alternating Series Theorem, $\sum_{n=1}^{\infty} (-1)^n \frac{1}{\sqrt{n}}$ converges.

But $\sum_{n=1}^{\infty} \left((-1)^n \frac{1}{\sqrt{n}} \right)^2$ is the harmonic series $\sum_{n=1}^{\infty} \frac{1}{n}$, which is divergent.

(3) False.

The harmonic series $\sum_{n=1}^{\infty} \frac{1}{n}$ is divergent, but $\lim_{n \to \infty} \frac{1}{n} = 0$.

(4) True.

The partial sums converge to 0, and so $\sum_{n=1}^{\infty} a_n$ converges to 0, by definition.

(5) False.

The partial sum

$$s_n = \log \frac{2}{1} + \log \frac{3}{2} + \cdots + \log \frac{n+1}{n} = \log \frac{2 \cdot 3 \cdot \cdots \cdot (n+1)}{1 \cdot 2 \cdot \cdots \cdot n} = \log(n+1).$$

Since $(\log(n+1))_{n\in\mathbb{N}}$ diverges, it follows that $\displaystyle\sum_{n=1}^{\infty}\log\frac{n+1}{n}$ diverges.

(6) True.

We know that increasing sequences that are bounded above are convergent. As the sequence of partial sums of the given series is increasing and bounded above, the series is convergent.

(7) True.

Since $\displaystyle\sum_{n=1}^{\infty}a_n$ converges, $\displaystyle\lim_{n\to\infty}a_n=0$.

So $\left(\dfrac{1}{a_n}\right)_{n\in\mathbb{N}}$ is unbounded, and hence can't be convergent to 0. Thus $\displaystyle\sum_{n=1}^{\infty}\frac{1}{a_n}$ diverges.

Solution to Exercise 6.19

Let $\epsilon>0$. Since the series

$$S:=\sum_{n=1}^{\infty}|a_n|+|b_n|$$

is convergent, there exists an $N\in\mathbb{N}$ such that for all $n>N$,

$$\left|\sum_{k=1}^{n}(|a_k|+|b_k|)-S\right|=\sum_{k=n+1}^{\infty}(|a_k|+|b_k|)<\epsilon.$$

As

$$\left|a_k\cos\left(\frac{2\pi k}{T}x\right)+b_k\sin\left(\frac{2\pi k}{T}x\right)\right|$$

$$\leq|a_k|\left|\cos\left(\frac{2\pi k}{T}x\right)\right|+|b_k|\left|\sin\left(\frac{2\pi k}{T}x\right)\right|$$

$$\leq|a_k|\cdot1+|b_k|\cdot1,$$

we have that $\displaystyle\sum_{k=n+1}^{\infty}\left|a_k\cos\left(\frac{2\pi k}{T}x\right)+b_k\sin\left(\frac{2\pi k}{T}x\right)\right|\leq\sum_{k=n+1}^{\infty}(|a_k|\cdot1+|b_k|\cdot1)<\epsilon.$

Thus if $n>N$, then for all $x\in\mathbb{R}$, we have

$$\left|f(x)-\left(a_0+\sum_{k=1}^{n}\left(a_k\cos\left(\frac{2\pi k}{T}x\right)+b_k\sin\left(\frac{2\pi k}{T}x\right)\right)\right)\right|$$

$$=\left|\sum_{k=n+1}^{\infty}a_k\cos\left(\frac{2\pi k}{T}x\right)+b_k\sin\left(\frac{2\pi k}{T}x\right)\right|$$

$$\leq\sum_{k=n+1}^{\infty}\left|a_k\cos\left(\frac{2\pi k}{T}x\right)+b_k\sin\left(\frac{2\pi k}{T}x\right)\right|<\epsilon.$$

This shows that the Fourier series converges uniformly to f.

The plots are displayed on page 319. From the plots, we do see that the partial sum functions look increasingly like the square wave, but we also notice the persistent overshoot at the points $-1,0,1,2,3$, suggesting nonuniform convergence.

Solution to Exercise 6.20

Clearly if $\sum_{n=1}^{\infty} a_n$ converges, then so does $\sum_{n=1}^{\infty} a_{n+1}$.

Thus $\sum_{n=1}^{\infty} \dfrac{a_n + a_{n+1}}{2}$ converges too. By the Arithmetic Mean-Geometric Mean inequality

$$\frac{a+b}{2} \geq \sqrt{ab}$$

for $a, b \geq 0$ (this is just a rearrangement of the trivial observation that $(\sqrt{a} - \sqrt{b})^2 \geq 0$), it follows that for all $n \in \mathbb{N}$,

$$\sqrt{a_n a_{n+1}} \leq \frac{a_n + a_{n+1}}{2},$$

and so the result follows from the Comparison Test.

Solution to Exercise 6.21

('If' part) Suppose that $\sum_{n=1}^{\infty} \dfrac{a_n}{1 + a_n}$ converges. Then

$$\lim_{n \to \infty} \frac{a_n}{1 + a_n} = 1 - \lim_{n \to \infty} \frac{1}{1 + a_n} = 0,$$

and so $\lim_{n \to \infty} \dfrac{1}{1 + a_n} = 1$. Thus there exists an $N \in \mathbb{N}$ such that for all $n > N$, $\dfrac{1}{1 + a_n} > \dfrac{1}{2}$.

But this implies that for $n > N$, $\dfrac{a_n}{1 + a_n} > \dfrac{a_n}{2}$.

By the Comparison Test, it follows that $\sum_{n=1}^{\infty} \dfrac{a_n}{2}$ converges, and so $\sum_{n=1}^{\infty} a_n$ converges as well.

('Only if' part) Suppose that $\sum_{n=1}^{\infty} a_n$ converges. Since the a_ns are all positive, we have

$$\frac{a_n}{1 + a_n} \leq a_n \quad (n \in \mathbb{N}),$$

and so by the Comparison Test, it follows that $\sum_{n=1}^{\infty} \dfrac{a_n}{1 + a_n}$ converges.

Solution to Exercise 6.22

If the series $\sum_{n=1}^{\infty} |a_n|$ converges, we have $\lim_{n \to \infty} |a_n| = 0$. Thus there is an $N \in \mathbb{N}$ such that for all $n > N$, $|a_n| < 1$. But now for $n > N$, $|a_n^2| = |a_n||a_n| < |a_n|$, and so by the Comparison Test

$$\sum_{n=1}^{\infty} a_n^2$$

converges too. So every element of ℓ^1 belongs to ℓ^2.

The sequence $(1/n)_{n \in \mathbb{N}}$ belongs to ℓ^2 since

$$\sum_{n=1}^{\infty} \left(\frac{1}{n}\right)^2 = \sum_{n=1}^{\infty} \frac{1}{n^2} < +\infty,$$

but it does not belong to ℓ^1 (since the harmonic series diverges). So $\ell^1 \subsetneq \ell^2$.

Solution to Exercise 6.23

We have

$$s_n := 1 + \frac{1}{2} + \frac{1}{3} + \cdots + \frac{1}{n} < \underbrace{1 + 1 + 1 + \cdots + 1}_{n \text{ times}} = n,$$

and so $\dfrac{1}{n} < \dfrac{1}{s_n}$ for all $n \in \mathbb{N}$. Hence by the Comparison Test, the series $\displaystyle\sum_{n=1}^{\infty} \frac{1}{s_n}$ diverges.

Solution to Exercise 6.24

We have that

$$\sqrt[n]{\frac{1}{n^n}} = \frac{1}{n} \leq \frac{1}{2} = r < 1$$

for all $n \geq 2$. So it follows from the Root Test that the series $\displaystyle\sum_{n=1}^{\infty} \frac{1}{n^n}$ converges.

Solution to Exercise 6.25

We know that the Fibonacci sequence is strictly increasing, since each term is clearly seen to be positive, and the recurrence gives

$$F_{n+1} = F_n + F_{n-1} > F_n + 0 = F_n.$$

From here, we obtain $F_{n+1} = F_n + F_{n-1} > F_{n-1} + F_{n-1} = 2F_{n-1}$.

Thus $\dfrac{1/F_{n+1}}{1/F_{n-1}} = \dfrac{F_{n-1}}{F_{n+1}} < \dfrac{1}{2} =: r < 1$.

So by the Ratio Test, both $\dfrac{1}{F_0} + \dfrac{1}{F_2} + \dfrac{1}{F_4} + \cdots$ and $\dfrac{1}{F_1} + \dfrac{1}{F_3} + \dfrac{1}{F_5} + \cdots$ converge. So if

$$s_n := \frac{1}{F_0} + \frac{1}{F_1} + \frac{1}{F_2} + \cdots + \frac{1}{F_n}, \quad n \in \mathbb{N},$$

then both $(s_{2n})_{n \in \mathbb{N}}$ and $(s_{2n+1})_{n \in \mathbb{N}}$ converge to the same limit, namely

$$\sum_{n=0}^{\infty} \frac{1}{F_{2n}} + \sum_{n=0}^{\infty} \frac{1}{F_{2n+1}},$$

and so $(s_n)_{n \in \mathbb{N}}$ converges, that is, $\displaystyle\sum_{n=0}^{\infty} \frac{1}{F_n}$ converges.

Solution to Exercise 6.26

We have

$$\sqrt{1+n^2} - n = \frac{1+n^2-n^2}{\sqrt{1+n^2}+n} = \frac{1}{\sqrt{1+n^2}+n} \geq \frac{1}{\sqrt{n^2+n^2}+n} = \frac{1}{\sqrt{2}+1} \cdot \frac{1}{n} \geq 0.$$

As the Harmonic Series $\sum_{n=1}^{\infty} \frac{1}{n}$ diverges, it follows by the Comparison Test that also

$$\sum_{n=1}^{\infty} (\sqrt{1+n^2} - n)$$

diverges.

Solution to Exercise 6.27

As k^2 is an integer, we have

$$\left| \sin\left(\pi\sqrt{k^4+1}\right) \right| = \left| \sin\left(\pi k^2\sqrt{1+1/k^4}\right) \right| = \left| \sin\left(\pi k^2\sqrt{1+1/k^4} - \pi k^2\right) \right|$$

$$= \left| \sin\left(\pi k^2\left(\sqrt{1+1/k^4} - 1\right)\right) \right| = \left| \sin\left(\frac{\pi}{k^2}\left(\frac{\sqrt{1+1/k^4} - 1}{1/k^4}\right)\right) \right|.$$

But by the Mean Value Theorem, we have for $x > 0$ that there exists a $c_x \in (0,1)$ such that

$$\frac{\sqrt{1+x} - 1}{x} = \frac{1}{2\sqrt{1+c_x}} < \frac{1}{2}.$$

Consequently, $\dfrac{\sqrt{1+1/k^4} - 1}{1/k^4} < \dfrac{1}{2}.$

Also, for each $x \neq 0$, we have that there exists a c between 0 and x such that

$$\left| \frac{\sin x - \sin 0}{x - 0} \right| = |\cos c| \leq 1,$$

so that $|\sin x| \leq |x|$ for all $x \in \mathbb{R}$. Hence

$$\left| \sin\left(\pi\sqrt{k^4+1}\right) \right| = \left| \sin\left(\frac{\pi}{k^2}\left(\frac{\sqrt{1+1/k^4} - 1}{1/k^4}\right)\right) \right| \leq \left| \frac{\pi}{k^2}\left(\frac{\sqrt{1+1/k^4} - 1}{1/k^4}\right) \right| \leq \frac{\pi}{k^2} \cdot \frac{1}{2}.$$

As $\sum_{k=1}^{\infty} \frac{1}{k^2}$ converges, by the Comparison Test, $\sum_{k=1}^{\infty} |\sin(\pi\sqrt{k^4+1})|$ converges as well.

Hence $\sum_{k=1}^{\infty} \sin(\pi\sqrt{k^4+1})$ converges absolutely.

Solution to Exercise 6.28

(1) We have $L := \lim_{n\to\infty} \left|\frac{c_{n+1}}{c_n}\right| = \lim_{n\to\infty} \left|\frac{1}{1}\right| = 1$.

So the radius of convergence is indeed 1, by Theorem 6.15.

When $x = \pm1$, $x^n = \pm1$, thus the sequence $(x^n)_{n\in\mathbb{N}}$ is not convergent with limit 0. Thus series is not convergent at these points.

(2) We have $L := \lim_{n\to\infty} \left|\frac{c_{n+1}}{c_n}\right| = \lim_{n\to\infty} \left|\frac{n^2}{(n+1)^2}\right| = 1$.

So the radius of convergence is indeed 1, by Theorem 6.15.

When $x = \pm1$, we have that

$$\left|\frac{x^n}{n^2}\right| = \frac{1}{n^2},$$

and so it follows from the Comparison Test that the series is convergent at these points.

(3) We have $L := \lim_{n\to\infty} \left|\frac{c_{n+1}}{c_n}\right| = \lim_{n\to\infty} \left|\frac{n}{n+1}\right| = 1$.

So the radius of convergence is indeed 1, by Theorem 6.15.

When $x = +1$, we have that the series $\sum_{n=1}^{\infty}\frac{x^n}{n} = \sum_{n=1}^{\infty}\frac{1}{n}$ diverges.

When $x = -1$, $\sum_{n=1}^{\infty}\frac{x^n}{n} = \sum_{n=1}^{\infty}\frac{(-1)^n}{n}$ converges by the Leibniz Alternating Series Test.

(4) We have $L := \lim_{n\to\infty} \left|\frac{c_{n+1}}{c_n}\right| = \lim_{n\to\infty} \left|\frac{n}{n+1}\right| = 1$.

So the radius of convergence is indeed 1, by Theorem 6.15.

When $x = -1$, $\sum_{n=1}^{\infty}(-1)^n\frac{x^n}{n} = \sum_{n=1}^{\infty}\frac{1}{n}$ diverges.

When $x = 1$, $\sum_{n=1}^{\infty}(-1)^n\frac{x^n}{n} = \sum_{n=1}^{\infty}\frac{(-1)^n}{n}$ converges by the Leibniz Alternating Series Test.

Solution to Exercise 6.29

We have

$$f'(x) = \sum_{n=1}^{\infty} 2n\frac{x^{2n-1}}{(2n)!} = \sum_{n=1}^{\infty}\frac{x^{2n-1}}{(2n-1)!} = \sum_{m=0}^{\infty}\frac{x^{2(m+1)-1}}{(2(m+1)-1)!} = \sum_{m=0}^{\infty}\frac{x^{2m+1}}{(2m+1)!} = g(x).$$

Also, $g'(x) = \sum_{n=0}^{\infty}(2n+1)\frac{x^{2n}}{(2n+1)!} = \sum_{n=0}^{\infty}\frac{x^{2n}}{(2n)!} = f(x)$.

Let $h : \mathbb{R} \to \mathbb{R}$ be defined by $h(x) = (f(x))^2 - (g(x))^2$, $x \in \mathbb{R}$. Then

$$h'(x) = 2f(x)f'(x) - 2g(x)g'(x) = 2f(x)g(x) - 2g(x)f(x) = 0, \quad x \in \mathbb{R},$$

and so h is constant. Consequently for all $x \in \mathbb{R}$,

$$h(x) = (f(x))^2 - (g(x))^2 = h(0) = (f(0))^2 - (g(0))^2 = 1^2 - 0^2 = 1.$$

Solution to Exercise 6.30

Consider the series $e^x = \displaystyle\sum_{n=0}^{\infty} \frac{x^n}{n!}, \quad x \in \mathbb{R}.$

Multiplying by x, we obtain $xe^x = \displaystyle\sum_{n=0}^{\infty} \frac{x^{n+1}}{n!}, \quad x \in \mathbb{R}.$

Differentiating with respect to x gives $e^x + xe^x = (1+x)e^x = \displaystyle\sum_{n=0}^{\infty} \frac{(n+1)x^n}{n!}.$

Multiplying again by x, we get $(x + x^2)e^x = \displaystyle\sum_{n=0}^{\infty} \frac{(n+1)x^{n+1}}{n!}$, and differentiating this gives

$$(1 + 3x + x^2)e^x = \sum_{n=0}^{\infty} \frac{(n+1)^2 x^n}{n!}.$$

Setting $x = 1$, we find $\displaystyle\sum_{n=0}^{\infty} \frac{(n+1)^2}{n!} = 5e.$

Solution to Exercise 6.31

We have

$$c_1 + 2c_2 x + 3c_3 x^2 + \cdots = 2x(c_0 + c_1 x + c_2 x^2 + \cdots) = 2c_0 x + 2c_1 x^2 + 2c_2 x^3 + 2c_3 x^4 + \cdots.$$

But using the relation between the coefficients of a power series and the derivatives of the function that the power series defines, we see that we can equate the coefficients of powers of x. Hence

$$c_1 = 0, \quad 2c_2 = 2c_0, \quad 3c_3 = 2c_1, \quad 4c_4 = 2c_2, \quad \cdots.$$

From here it follows that

$$c_1 = c_3 = c_5 = \cdots = 0,$$

$$c_2 = c_0, \quad c_4 = \frac{c_2}{2} = \frac{c_0}{2!}, \quad c_6 = \frac{c_4}{3} = \frac{c_0}{3!}, \cdots.$$

Hence $f(x) = c_0 \left(1 + \dfrac{x^2}{1!} + \dfrac{x^4}{2!} + \dfrac{x^6}{3!} + \cdots \right) = c_0 e^{x^2}$, where $c_0 \in \mathbb{R}$ is arbitrary.

Solution to Exercise 6.32

(1) For $x \approx 0$, $-1/x^2$ is negative with a large absolute value, and so $e^{-1/x^2} \approx 0$. For $x \to \pm\infty$, $-1/x^2 \to 0$, and so $e^{-1/x^2} \to 1$.

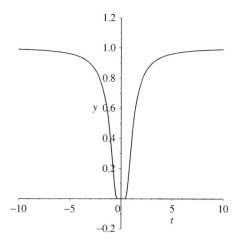

(2) Let $m > 0$. Choose $n \in \mathbb{N}$ such that $n > m/2$. For $x \neq 0$,

$$e^{1/x^2} = \sum_{k=0}^{\infty} \frac{1}{k!} \frac{1}{x^{2k}} \geq \frac{1}{n!} \frac{1}{x^{2n}},$$

and so $e^{-1/x^2} \leq n! x^{2n}$. So $\left| \dfrac{e^{-1/x^2}}{x^m} \right| \leq \dfrac{n! |x|^{2n}}{|x|^m} = n! |x|^{2n-m} \xrightarrow{x \to 0} 0.$

(3) We prove this by induction on n. For $x \neq 0$,

$$f'(x) = e^{-1/x^2} \frac{2}{x^3} = e^{-1/x^2} P_1 \left(\frac{1}{x} \right),$$

where $P_1(t) := 2t^3$. Suppose that the claim holds for all natural numbers $\leq k$. We now show this for $k + 1$. We know

$$f^{(k)}(x) = e^{-1/x^2} P_k \left(\frac{1}{x} \right),$$

where P_k is a polynomial. Hence

$$f^{(k+1)}(x) = e^{-1/x^2} \frac{2}{x^3} P_k \left(\frac{1}{x} \right) + e^{-1/x^2} P_k' \left(\frac{1}{x} \right) \left(-\frac{1}{x^2} \right)$$

$$= e^{-1/x^2} \left(\frac{2}{x^3} P_k \left(\frac{1}{x} \right) + P_k' \left(\frac{1}{x} \right) \left(-\frac{1}{x^2} \right) \right)$$

$$= e^{-1/x^2} P_{k+1} \left(\frac{1}{x} \right),$$

where $P_{k+1}(t) := 2t^3 P_k(t) + P_k'(t)(-t^2)$, which is clearly a polynomial.

(4) We have $f'(0) = \lim\limits_{x \to 0} \frac{e^{-1/x^2} - 0}{x} = 0$. If $f^{(n)}(0) = 0$ for some n, then

$$f^{(n+1)}(0) = \lim_{x \to 0} \frac{f^{(n)}(x) - f^{(n)}(0)}{x}$$

$$= \lim_{x \to 0} \frac{e^{-1/x^2} P_n \left(\frac{1}{x}\right) - 0}{x}$$

$$= \lim_{x \to 0} \frac{e^{-1/x^2}}{x} \left(\frac{c_d}{x^d} + \cdots + \frac{c_1}{x} + c_0\right)$$

(where P_n is given by $P_n(t) = c_d t^d + \cdots + c_1 t + c_0$)

$$= \lim_{x \to 0} \left(\frac{e^{-1/x^2}}{x^{d+1}} c_d + \cdots + \frac{e^{-1/x^2}}{x^2} c_1 + \frac{e^{-1/x^2}}{x} c_0\right) = 0.$$

Bibliography

[A]   T.M. Apostol. *Calculus. Volume I. One-Variable Calculus, with an Introduction to Linear Algebra.* Second Edition. John Wiley & Sons, Inc., New York, 1991.

[A2] T.M. Apostol. *Mathematical Analysis: A Modern Approach to Advanced Calculus.* Addison-Wesley Publishing Company, Reading, MA, 1957.

[B]   V. Bryant. *Yet Another Introduction to Analysis.* Cambridge University Press, Cambridge, 1990.

[B2] B. Bukhovtsev, V. Krivchenkov, G. Myakishev, and V. Shalnov. *Problems in Elementary Physics.* Mir Publishers, Moscow, 1978.

[G]   S. Ghorpade and B. Limaye. *A Course in Calculus and Real Analysis.* Undergraduate Texts in Mathematics. Springer, New York, 2006.

[J]   K. Joshi. *Introduction to General Topology.* A Halsted Press Book. John Wiley and Sons, New York, 1983.

[K]   S. Kim and K. Kwon. Smooth (C^∞) but nowhere analytic functions. *American Mathematical Monthly*, **107**, 264–266, 3, 2000.

[K2] L. Klosinski, G. Alexanderson, L. Larson. The fifty-fourth William Lowell Putnam Mathematical Competition. *American Mathematical Monthly*, **101**, 725–734, 4, 1994.

[L]   L. Larson. *Problem-Solving Through Problems.* Problem Books in Mathematics. Springer-Verlag, New York, 1983.

[N]   V. Naik. *Calculus* teaching resources at the University of Chicago. Available at http://math.uchicago.edu/ vipul/teaching.html.

[N2] D. Newman. *A Problem Seminar.* Problem Books in Mathematics. Springer-Verlag, New York, Berlin, 1982.

[N3] D. Newman. An interesting limit. Elementary problems and solutions E924 [1950, 416]. *American Mathematical Monthly*, **58**, 190–191, 3, 1951.

[N4] D. Newman and T. Parsons. On monotone subsequences. *American Mathematical Monthly*, **95**, 44–45, 1, 1988.

[N5] I. Niven. A simple proof that π is irrational. *Bulletin of the American Mathematical Society*, **53**, 509, 6, 1947.

The How and Why of One Variable Calculus, First Edition. Amol Sasane.
© 2015 John Wiley & Sons, Ltd. Published 2015 by John Wiley & Sons, Ltd.

[P]   C. Pugh. *Real Mathematical Analysis*. Undergraduate Texts in Mathematics. Springer, New York, 2002.

[R]   W. Rudin. *Principles of Mathematical Analysis*. Third edition. McGraw-Hill, Singapore, 1976.

[R2]   T.-L. Rădulescu, V. Rădulescu and T. Andreescu. *Problems in Real Analysis. Advanced Calculus on the Real Axis*. Springer, New York, 2009.

[S]   G. Simmons. *Introduction to Topology and Modern Analysis*. McGraw-Hill, New York, 1963.

[SW]   J. Snow and K. Weller. *Exploratory Examples for Real Analysis*. The Mathematical Association of America, Washington, DC, 2003.

[S2]   M. Spivak. *Calculus*. Third edition. Cambridge University Press, Cambridge, 2006.

Index

The How and Why of One Variable Calculus, First Edition. Amol Sasane.
© 2015 John Wiley & Sons, Ltd. Published 2015 by John Wiley & Sons, Ltd.

Printed and bound by CPI Group (UK) Ltd, Croydon, CR0 4YY

09/10/2024

14571431-0001